AF245498

Mechanoreceptors

Development, Structure, and Function

Mechanoreceptors
Development, Structure, and Function

Edited by

Pavel Hník
Tomáš Soukup
Richard Vejsada and
Jiřina Zelená

Institute of Physiology
Czechoslovak Academy of Sciences
Prague, Czechoslovakia

Plenum Press • New York and London

Library of Congress Cataloging in Publication Data

International Symposium on Mechanoreceptors (1987: Prague, Czechoslovakia)
  Mechanoreceptors: development, structure, and function.

  "Proceedings of the International Symposium on Mechanoreceptors, held August
11–15, 1987, Prague, Czechoslovakia" —
  Bibliography: p.
  Includes index.
  1. Mechanoreceptors — Congresses. I. Hník, Pavel. II. Title.
QP369.I56 1987   599′.01852   88-5796
ISBN 0-306-42832-6

Proceedings of the International Symposium on Mechanoreceptors,
held August 11–15, 1987, Prague, Czechoslovakia

© 1988 Plenum Press, New York
A Division of Plenum Publishing Corporation
233 Spring Street, New York, N.Y. 10013

Printed in the United States of America

F O R E W O R D

The occasion of the second World Congress of Neuroscience
provided an opportunity to combine a celebration of the 200th
Anniversary of Jan Evangelista Purkyne (Purkinje) with an International Symposium on Mechanoreceptors, hosted by the Institute of Physiology of the Czechoslovak Academy of Sciences.

The Symposium held on 11th - 15th August, 1987 brought
together an international group of experts from 17 countries
and provided an exhilarating opportunity for mutual interaction
and discussion on the development, structure, regeneration and
function of muscle and joint receptors, cutaneous and other
mechanoreceptors. By concentrating on mechanoreceptors from
different sites and tissues in the body the symposium was able
to explore both the diversity and similarity of the underlying
biological mechanisms.

This volume embodies the individual contributions but does
not fully convey the flavour of the discussions nor of course,
the excitement of meeting in an original building, the Caroli-
num, of Charles University founded in 1348 by the Emperor
Charles IV. As befits a Symposium, in its original classical
Greek meaning of a drinking party, there were opportunities for
relaxed informal meetings to carry forward discussions of topics
raised during the formal sessions. The venue of the Symposium,
Prague, an ancient and fascinating city, now undergoing specta-
cular renovation of its remarkable architectural heritage of
medieval and baroque buildings in the old city, added a unique
flavour to the meeting. It reminded us all of the continuity of
scholastic endeavour and of the international fraternity of
scientists; of whom Purkyně was an elder Czech master and the
late Ernest Gutmann whose own interests initiated the sustained
investigation of neuromuscular physiology that characterised
the host departments, is a contemporary Czech exemplar.

Our warm thanks and congratulations go to all those involved in initiating and realising the Symposium. This volume
stands as a permanent reminder of their efforts and will provide a springboard for further advances in knowledge of a particular kind of sensory receptor.

                                        Ainsley Iggo
                                        University of Edinburgh

                                        October, 1987

P R E F A C E

This volume contains the Proceedings of the Symposium on "Mechanoreceptors - Development, Structure and Function" which was organized by the Institute of Physiology, Czechoslovak Academy of Sciences in collaboration with the Faculty of Medicine, Charles University, as a Satellite meeting of the Second World Congress of Neuroscience in Budapest. The symposium was held in Prague on August 11-15, 1987 with more than 80 participants from all over the world.

The aim of this meeting was to bring together experienced and "budding" scientists studying various aspects of mechanoreceptor development, reinnervation, regeneration, grafting, structure and function. It was an interdisciplinary conference, with the participation of morphologists and physiologists working in the field of muscle, tendon, joint and cutaneous mechanoreceptors. The meeting gave the participants an opportunity of exchanging ideas and discussing various aspects regarding mechanoreceptor structure and function.

The papers contained in this volume were not refereed and hence only the authors are responsible for their scientific contents. The editors have tried to conform with the Plenum Press requirements and have had the manuscripts retyped with appropriate language corrections, where necessary. The volume is divided into eight sections and reflects the topics of individual sessions and the sequence of papers presented. The contributions of the poster session, which are presented as shorter articles of 2-4 pages, have been inserted into the related sections of the book according to their contents.

Part I. deals with the respective role of the periphery and sensory axon terminals in morphogenetic cell interactions during mechanoreceptor development. Part II. describes the immunocytochemical, structural and functional development of muscle spindles in relation to the process of their innervation. The fate of mechanoreceptors after their destruction by freezing and after various types of grafting or impairment is considered in Part III. Part IV. contains the results of reinnervation of mechanoreceptors after nerve injury, with special attention being paid to sensory reinnervation following nerve crush early in development, which results in a loss of sensory neurons and of most of the immature receptors. Theoretically important problems of intrinsic mechanisms of mechanoreceptor function are dealt with in Part V. Part VI. on the structure and function of mature muscle receptors includes electrophysiological, electron microscopical, immunocytochemical, stereological and quantitative studies of muscle re-

ceptors. The videoprograms presented by the late prof. I.A. Boyd
belonged to the highlights of the symposium; unfortunately, the
fascinating experience of watching living muscle spindle function cannot be transmitted by the written text. Part VII. on
the functional morphology of sensory receptors comprises contributions concerning immuno- and histochemistry, electron
microscopy and the effect of aging, and topographical studies
of the type and distribution of mechanoreceptors in distinct
regions of the body. The last section, Part VIII., presents
the results on the functional significance of the feed-back
from skeletal muscles, skin and joint afferents.

This meeting was dedicated to the 200th Anniversary of
the birth of a famous Czech physiologist, histologist and psy-
chophysiologist, Jan Evangelista Purkyně (known as J.E. Purkinje
in world literature), whose anniversary was included in the
calendar of UNESCO for 1987. The organizers also remembered
the 10th anniversary of the premature death of Ernest Gutmann,
who founded the Department of the Physiology and Pathophysio-
logy of the Neuromuscular System at the Physiological Institute
of the Czechoslovak Academy of Sciences in Prague. He was a
world known champion of neurotrophic relations in neuromuscular
physiology and admired by his many pupils and friends.

Our thanks, as organizers, are due to Mrs. J. Pištorová
for her meticulous help with the typing and to the staff of
Plenum Press for their help in preparing this publication.

                              P. Hník   T. Soukup
                              R. Vejsada   J. Zelená

CONTENTS

PART I

MORPHOGENETIC INTERACTIONS
IN RECEPTOR DEVELOPMENT

MORPHOGENETIC INTERACTIONS IN THE

DEVELOPMENT OF AVIAN CUTANEOUS SENSORY RECEPTORS

Raymond Saxod

Developmental Neurobiology, University of Grenoble

France

INTRODUCTION

The skin of birds presents an extremely complex network of nerves, containing mainly somatic and visceral sensory fibers. These fibers terminate by different types of cutaneous end-organs which have been thoroughly described: Herbst, Grandry, Merkel and Ruffini corpuscles, and free nerve terminals (for review, see Chouchkov, 1978; Saxod, 1978).

Epidermis contains only free nerve endings, and all the other sensory end organs are situated in the dermis, and are mainly located around the feather follicles, in the dermis of foot pads and in the beak skin where they are abundant, especially in aquatic species.

One of the fundamental problems regarding sensory receptors is the relationship between the sensory nerves and their end-organs, that is the questions of specification of receptor formation and of factors involved in their development and maintenance.

Birds offer an excellent material for an embryonic experimental approach of these problems, mainly because they possess different kinds of cutaneous receptors, with a regional and species-specific distribution, and easy access to the embryo. Therefore we performed several types of experiments in chick, quail and duck embryos to analyze the morphogenetic sequence of cutaneous sensory receptors and the developmental origin of their components, and to try to find out the factors involved in their specification.

In our experiments, only three types of end organs were taken into account: Herbst, Grandry and Merkel corpuscles (fig. 1). All of them are formed by the association of 1- a nerve ending (peripheral process of a spinal ganglionic neuron) accompanied by more or less specialized cells (so-called sensory or accessory cells), this central part functioning as the mechano-transducting unit, 2- enveloping cells forming lamellar structures with an outer capsule consisting of joined flat cells.

The technique used was embryonic transplantation of pieces of beak-skin or of frontal buds, analysis of the grafts for the presence of corpuscles and comparison of types of receptors found with those which form during normal development, that is duck beak type Herbst corpuscle and

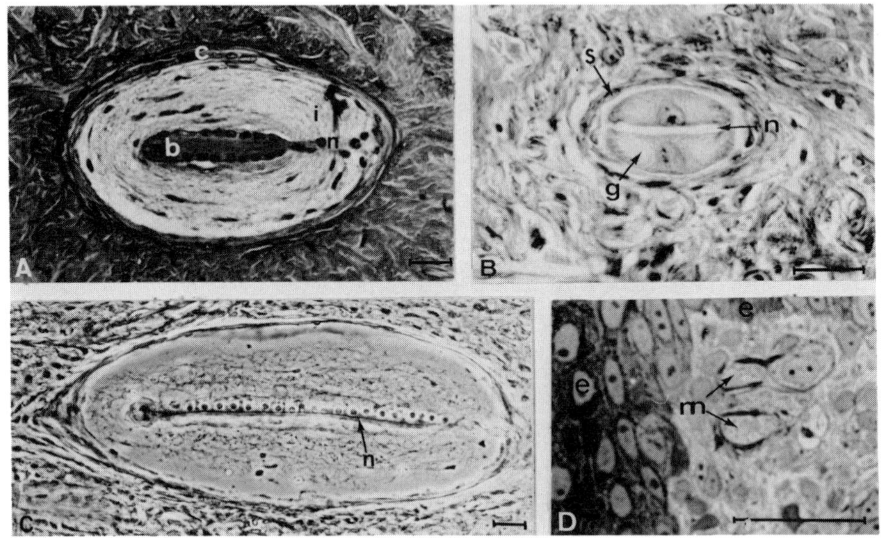

Fig. 1. A and B: adult duck beak (A) duck beak type Herbst corpuscle;
(B) Grandry corpuscle. C and D: adult chick beak (C) chick
beak type Herbst corpuscle; (D) Merkel corpuscles. Longitudinal sections; b, inner bulb; c, outer capsule; e, epidermis;
g, Grandry cell; i, inner space; n, nerve ending; m, Merkel
cell; s, satellite cell. Bars = 20 µm.

Grandry corpuscle for the duck beak, and chick beak type Herbst corpuscle
and Merkel corpuscle for the chick and the quail beak.

NORMAL DEVELOPMENT

Before undertaking the experimental study, stages of normal development of Herbst, Grandry and Merkel corpuscles have been traced at the
ultrastructural level (Saxod, 1970a, 1970b, 1980). From the morphological
data, it appears that corpuscles develop during the last third of embryonic development, that constituents are assembled according to an inside
out sequence, and that the nerve ending is present from the beginning of
corpuscle formation, as confirmed by Idé (1978), Malinovsky and Pac (1985)
and Pac and Malinovsky (1985).

ROLE OF NERVE FIBERS

The first group of experiments (fig. 2) was performed to modify the
normal relationship between the beak skin territory and the normal source
of innervation (trigeminal nerve).

Development of grafts without any innervation was obtained by chorio-
allantoic grafting. Whatever the stage of explantation (table 1), the non-
innervated grafts studied were devoid of corpuscles, and if corpuscles
were already developing at the time of grafting, they regressed and disappeared. The conclusions are that the nerve ending is necessary for the
onset of development of corpuscles, their further histogenesis and maintenance of their organisation.

Innervation of frontal buds by different sources of cutaneous nerves
was performed by orthotopic grafting (on frontal bud), or by heterotopic

4

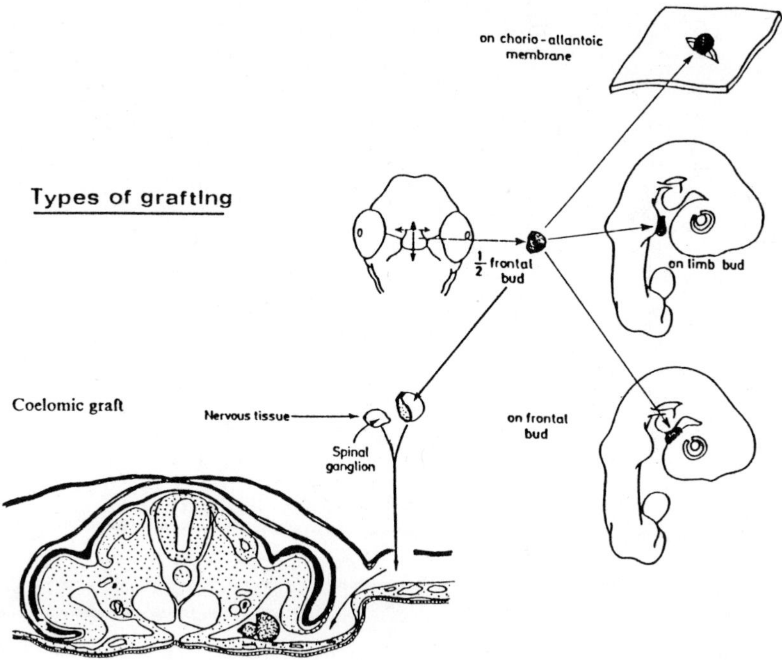

Fig. 2. Embryonic transplantations of hemifrontal buds. Coelomic associations were put into the embryo via the extra-embryonic coelom.

grafting (on limb bud) and coelomic association with a spinal ganglion (innervation by spinal nerves) either in the same species (homoplastic experiment) or between two species (xenoplastic innervation).

Histological analysis of the half-beak obtained from these experiments show that formation of corpuscles is obtained in all kinds of grafting. Thus a heterotopic source of innervation and even a xenoplastic one, between the chick and the duck, allows the histogenesis of cutaneous receptors. To test the role of the different categories of nerve fibers (sensory, motor, somatic, sympathetic), frontal buds were associated, in coelomic graft, with different restricted sources of innervation. The results (table 1) show that no corpuscles develop if only sympathetic or motor fibers are provided. Therefore, only the somato-sensory nerve endings are able to ensure the development of cutaneous end-organs, and this development does not require any central connection (conclusion from coelomic grafts).

Moreover, as no corpuscles were formed in presence of mouse or lizard spinal ganglion, such competence is restricted to the somato-sensory nerve fibers from the same class of animals as that of the dermal mesenchyme.

ROLE OF CUTANEOUS TISSUES

To answer the questions regarding the determination of the specificity of corpuscle formation, that is the respective roles of nerve endings and cutaneous tissues, the xenoplastic grafts were analyzed by determining what kind of corpuscle was formed. In all cases (table 1), the results show that corpuscles which develop in the explants are in

Table 1: Embryonic transplantation and recombination experiments

Donor of[a] frontal bud	Source of innervation (TYPES OF GRAFTING)		Chick type Herbst	Chick type Merkel	Duck type Herbst	Duck type Grandry
chorioallantoic C 8-14	D 8-18	0	0	0	0	0
on frontal bud C 5-13	D 4.5	trigeminal nerve	+	+	0	0
D 5-14	C 3.5		0	0	+	+
on limb bud C 5-9	D 4.5	brachial plexus	+	+	0	0
D 3-12	C 3.5		0	0	+	+
coelomic C 5-6	D 6	spinal ganglion	+	+	0	0
D 6	C 6		0	0	+	+
C or D 6-7	C or D 6-7	neural tube, sympathetic ganglion	0	0	0	0
D 6	Mouse 12-15, Lizard 22[b]	spinal ganglion				

a,C: chick embryo, D: duck embryo, numbers refer to the age of embryos in number of days
of incubation or gestation.
b, embryos of Lacerta muralis Laur (eggs incubated at 26°C).

conformity with the origin of the frontal bud which has been grafted, namely duck type Herbst corpuscle and Grandry corpuscle in a duck graft, and
chick type Herbst corpuscle and Merkel corpuscle in a chick (or quail) graft.
Moreover, the number and the spatial distribution of receptors formed
correspond to the origin of the skin. The same results were obtained when
a pure frontal bud mesenchyme was grafted. Therefore, the corpuscle type
and distribution are determined by the regional and specific quality of
the innervated dermal mesenchyme, and this determination occurs very early
in development.

To modify the relative age of nerve endings and beak skin, heterochronic experiments were performed by transplanting frontal buds and pieces
of beak skin from embryos of various ages to hosts at a fixed age,
and the developmental stages of corpuscles of the grafts were then compared
to the normal ones. The results show that the stage at which development of corpuscles begins depends on the absolute age of cutaneous tissue,
but is not related to the total duration of its innervation.

EMBRYONIC ORIGIN

In order to interpret from a morphogenetic point of view the results
of xenoplastic and heterochronic experiments, it was necessary to know
the developmental origin of the cells constituting the corpuscles. For the
Herbst corpuscle, it has been possible to use the quail nuclear marker
(Saxod, 1973) and to obtain chimeric corpuscles by coelomic grafting of
quail and duck or chick associations (fig. 3). The results are very clear
and consistent: the inner bulb cells, situated along the nerve ending, are
always of the type of the spinal ganglion providing the nerve fibers,
whereas inner space cells and capsular cells conform to the specific origin of the frontal bud.

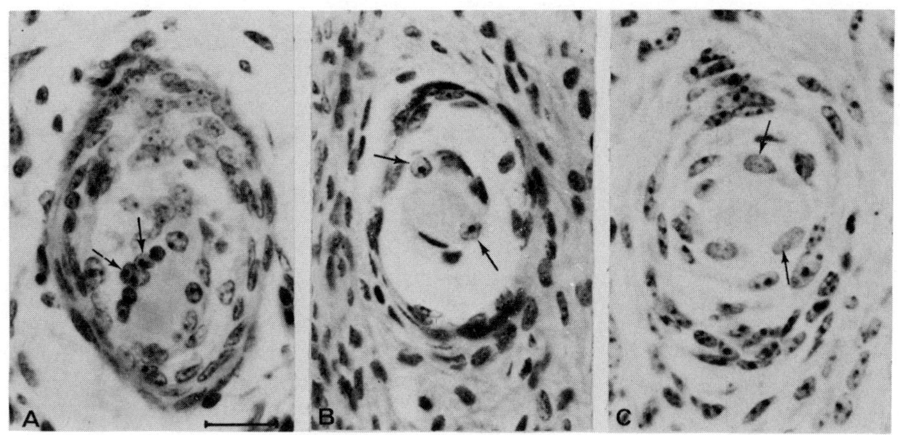

Fig. 3. Chimeric Herbst corpuscles obtained in coelomic graft. A and
B: association of a quail embryo sensory ganglion and chick
frontal bud, inner bulb cells display the characteristic quail-
type nuclear granules (arrows); (A) oblique section; (B) transverse section. C: reverse association (chick sensory ganglion
and quail frontal bud), inner bulb cells are of the chick type
(arrows). Feulgen staining, same magnification, bar = 15 μm.

Thus, Herbst corpuscle is formed by cells of two distinct developmental origin: 1- inner bulb cells originate from cells accompanying the
nerve during its outgrowth, and on cytological and histochemical arguments (Saxod, 1978), they can be interpreted as specialized and modified
Schwann cells, 2- other cells are provided by the surrounding predermal
mesenchyme.

MORPHOGENETIC INTERACTIONS

Taken all together, ultrastructural analysis of Herbst corpuscle
development and experimental data reveal the existence of two phases of
development of this end organ, an initial "phase of determination", during
which the kind of association between the nerve terminal and the surrounding cells is determined, followed by a "phase of histogenesis" corresponding to the building up of the structure.

Finally, for the Herbst corpuscle, the interactions and collaborations between somato-sensory nerve fibers, which are indispensable, and
dermal mesenchyme, which is responsible for the corpuscle specificity,
can be envisaged according to the following scheme. Early in development,
as skin innervation proceeds, somato-sensory nerves "stimulate" the dermal
mesenchyme, which, in return and later on, at a certain stage of its development, exerts a "specific morphogenetic influence" on the somatosensory nerve endings and companion cells. Then, according to the species
and regional origin of the dermal mesenchyme, these companion cells orga-
nize themselves to form a specific inner bulb, with a determined number
of cells and a particular organization of lamellae. Afterwards, the capsule and inner space are progressively formed around the inner core by
interactions between the inner bulb and the surrounding dermal cells which
divide to envelop, with a specific spatial organization, the whole structure.

Concerning other corpuscles, their embryonic origin is not yet clearly established, but similar morphogenetic sequences can be envisaged.

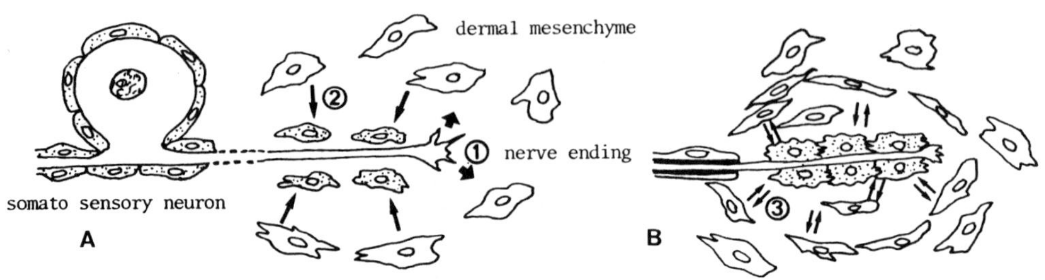

Fig. 4. Morphogenetic sequence of Herbst corpuscle development:
A, the somato-sensory nerve ending stimulates the dermal mesenchyme (1), which, in return, exerts (2) a specific morphogenetic influence on companion cells; B, other interactions
(3) take place during the phase of histogenesis.

        In conclusion, it is clear that development of corpuscles requires
interactions and collaborations between the two embryonic partners of the
association and that morphogenetic messages are elaborated, exchanged and
interpreted from the early determination stage to the completion of histogenesis. But for the moment, we have absolutely no idea about the bioche-
mical nature of these messages.

REFERENCES

Chouchkov, Ch., 1978, Cutaneous receptors, Adv. Anat. Embryol. Cell Biol.,
        54: 1-62.
Idé, C., 1978, A cytologic study of Grandry corpuscle development in
        chicken toe skin, J. Comp. Neurol., 179: 301-324.
Malinovsky, L., and Pac, L., 1985, Ultrastructural development of the
        Herbst corpuscle in the skin of the beak of the domestic duck,
        Folia Morphologica, 33, 150-155.
Pac, L., and Malinovsky, L., 1985, Development of the Grandry corpuscles
        in the skin of the beak of the domestic duck, Folia Morphologica,
        33: 379-384.
Saxod, R., 1970a, Etude au microscope électronique de l'histogenèse du
        corpuscule sensoriel cutané de Herbst chez le canard, J. Ultrast.
        Res., 33: 463-492.
Saxod, R., 1970b, Etude au microscope électronique de l'histogenèse du
        corpuscule sensoriel cutané de Grandry chez le canard, J. Ultrast.
        Res., 32: 477-496.
Saxod, R. 1973, Developmental origin of the Herbst cutaneous sensory
        corpuscle. Experimental analysis using cellular markers, Develop.
        Biol., 32: 167-178.
Saxod, R., 1978, Development of cutaneous sensory receptors in birds. In:
        Handbook of Sensory Physiology, M. Jacobson, ed., 9: 338-417,
        Springer, Berlin.
Saxod, R., 1980, Development of Merkel corpuscles in the chicken beak.
        The problem of their origin and identity, Biol. Cell., 37: 61-66.

CRITICAL PERIODS IN THE DEVELOPMENT OF MECHANORECEPTORS

J. Zelená and I. Jirmanová

Institute of Physiology
Czechoslovak Academy of Sciences
Vídeňská 1083, Prague, Czechoslovakia

INTRODUCTION

Mammalian encapsulated mechanoreceptors depend on sensory innervation for their development and maintenance (Zelená, 1964, 1976). When their nerves are crushed or transected during an early stage of development, not only the nerve terminals, but also the non-neuronal cells of the receptors rapidly degenerate and disintegrate. The early developmental stage is termed the critical period of development, as the loss of sensory innervation leads to the destruction of mechanoreceptors, and the neogenesis of a new set of receptors is seriously hampered or impossible (Zelená and Hník, 1963). During maturation, the auxiliary structure of sensory receptors becomes gradually resistant to denervation (Werner, 1973; Schiaffino and Pierobon Bormioli, 1976) until fully differentiated mechanoreceptors of adult animals survive after axotomy and can be reinnervated again when their nerves regenerate (Zelená and Hník, 1963; Schröder, 1974).

The question has arisen whether a critical period also exists in the development of those mechanoreceptors supplied with a lamellar core that is derived from Schwann cells which, as a rule, do not perish after axonal degeneration and are available during nerve regeneration, when they either guide or accompany the regenerating axons. To answer this question, we have studied rat Pacinian and digital corpuscles after nerve crush performed at different stages of their development and maturation, and investigated the denervated tissue at the beginning of and at different time intervals after nerve regeneration. This article presents a brief review of the experiments concerning Pacinian corpuscles and gives a preview of the preliminary results obtained on digital lamellar corpuscles.

PACINIAN CORPUSCLES

<u>Structure</u>. Pacinian corpuscles are rapidly adapting mechanoreceptors which consist, in the rat and other laboratory animals, of a cylindrical axon terminal with 2 rows of lateral processes

and 1 to 3 bulbous enlargements in the ultraterminal region; the
terminal is surrounded by a lamellar inner core bisected by 2
symmetrical radial clefts; the core is enclosed by a capsule
composed, in the rat, of approximately 30 concentric lamellar
layers of capsular cells divided by interspaces (Munger, 1971;
Hunt, 1974; Ilyinsky, 1975; Otelin et al., 1976). We have studied
rat Pacinian corpuscles in the crural region where 40-50 corpuscles are attached to the terminal branches of the interosseous
nerve beneath the interosseous membrane and at the fibula (Milburn, 1973), similarly as in the cat (Hunt, 1974) and mouse (Ide;
Nava, this volume).

Development. In rat fetuses, the crural Pacinian corpuscles
begin to develop 1-2 days before birth (Zelená, 1978). Their
development is similar to that of avian Herbst corpuscles as
described by Saxod (Saxod, 1978; this volume). At the initial
stage of development on the 20th day of gestation, a rat corpuscle consists of an oblong axon terminal with lateral processes
that contain dense core vesicles; the terminal is flanked by
Schwann cells which differentiate into lamellar inner core cells
under the morphogenetic influence of the terminal. Concomitantly,
the developing inner core cells affect the differentiation of
the sensory ending and later contribute to determining the final
symmetrical localization of lateral processes in two rows along
the cylindrical terminal. In newborn rats, the full number of
corpuscles is already present in the crural region, but their
structure is immature. They have only a few loose lamellae of
the inner core around the terminal, and only a couple of capsular
layers around the core. In 5-day-old rats, the corpuscles already possess a small lamellar inner core which increases in
size and exhibits 2 straight radial clefts in the 2nd postnatal
week (Zelená, 1978). A full complement of about 30 capsular
layers has differentiated 3-4 weeks after birth.

Denervation. When the sciatic nerve is crushed or transected
in the rat during the first postnatal days, when crural Pacinian
corpuscles are still immature, their further development is ar-
rested and they rapidly degenerate and disintegrate. After nerve
section in 1-day-old rats, the axon terminal degenerates within
12 hours and the debris is taken up by lamellar cells; the inner
core lamellae break down and retract during the first 2 days
after denervation, the cells begin to degenerate both in the
core and in the capsule, and macrophages at the circumference
of degenerating corpuscles engulf the cell debris (Zelená, 1980).
The corpuscles shrink and disintegrate, and their number decreases to zero values 5 days after nerve section. There is a
fundamental difference in the behavior of the inner core cells
and the Schwann cells in the nerve branches: while the lamellar
cells degenerate after denervation, Schwann cells proliferate
and form Büngner bands to support axonal regeneration (Zelená,
1980). The corpuscles denervated in 5-day-old rats degenerate
at a slower rate. The lamellar core is destroyed (Zelená et al.,
1978), but some of the capsules apparently persist to become reinnervated. After denervation performed in 10-day-old rats, Pacinian corpuscles cease growing (Zelená et al., 1978), but both
their inner core and the capsule survive during the transient
time interval preceding reinnervation. In adult rats (Zelená,
1982) and cats (Zelená, 1984a), crural Pacinian corpuscles remain
unchanged for several months after denervation.

Regeneration and reinnervation. A quantitative study of re-

generation and reinnervation of rat Pacinian corpuscles after
nerve crush shows that the outcome of nerve regeneration depends
upon the age at which the nerves are crushed: the earlier the
nerve crush, the worse the result. After crushing the nerve in
newborn rats, primary sensory neurons are seriously impaired by
retrograde axonal reaction and more than 50 % of them die (Aldskogius et al., 1985). The surviving axotomized neurons regenerate and sprout, but axonal regeneration is defective. Despite
the presumed sprouting, the number of myelinated axons of the
regenerated interosseous nerve is reduced by 60 % 6 months after
nerve crush at birth, and axonal diameters are by one half smaller than those of the control nerve (Zelená, 1981). The regenerating axons induce neoformation of Pacinian corpuscles, but the
number and size of regenerated corpuscles are drastically reduced
to 20 % of the control parameters, and the ultrastructure of corpuscles is considerably altered. Since several axons converge
to each site of corpuscle neoformation and each of the axon terminals induces formation of its inner core from the adjacent
Schwann cells, the regenerated corpuscles are composed of multiple terminals and multiple inner cores, all enclosed in a thin
capsule that has a reduced number of capsular lamellae (Zelená,
1981). This altered ultrastructure of corpuscles is not remodeled
later on, but persists for life. After nerve crush in 5-day-old
rats, the regenerating axons reinnervate the preserved capsules
in which they form again multiple axon terminals with multiple
inner cores, as after crushing the nerve in newborn rats; the
mean number of regenerated corpuscles attains 40 % of the control
number, and their size is still greatly reduced. The critical
period terminates 7 days after birth. From this time onwards,
over 90 % of Pacinian corpuscles survive in the crural region
after the sciatic nerve crush and recover after subsequent reinnervation (Fig. 1). However, the ultrastructure of Pacinian
corpuscles reinnervated after nerve crush performed in 7 to 15-
day-old rats still differs from reinnervated corpuscles of adult
rats, as each of the multiple axon terminals induces the differentiation of several layers of new lamellae, and the newly
formed miniature cores are incorporated into the original structure. In 20-day-old rats, reinnervation of Pacinian corpuscles
after nerve crush proceeds as in adult animals (Zelená, 1984b).

In mature corpuscles of adult rats, the original axon terminal degenerates soon after nerve crush, but otherwise the corpuscles do not even undergo atrophy during the transient denervation period which lasts approximately 3 weeks (Zelená, 1982,
1984b). During reinnervation, the corpuscles become initially
hyperinnervated by several axonal sprouts. However, only 1-3
axons become myelinated in each corpuscle and reinnervate it
permanently, whereas other redundant axonal sprouts retract.
Although 62 % of reinnervated corpuscles become supplied by a
single axon, only 10 % are monoterminal as before the nerve
crush. In all other instances, regenerated axons form multiple
terminals which become accomodated between the lamellae of the
original core. New lamellae are only formed around axonal
branches in the core or around ectopic axon terminals, e.g. at
the end of the capsular channel. A remodeling of a multiterminal
into a monoterminal innervation pattern, as has been suggested
to occur in cat mesenterial corpuscles several months after reinnervation (Chouchkov, 1978), does not take place in crural
Pacinian corpuscles of the rat or cat; the latter receive either
monoterminal or multiterminal innervation during the first 2
months of nerve regeneration, and retain it during further life
(Zelená, 1984b).

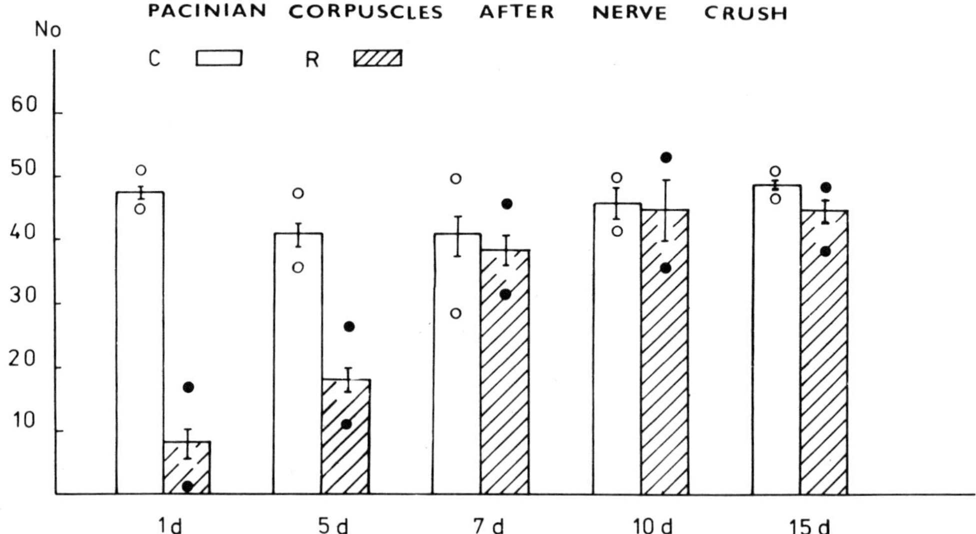

Fig. 1. The number of crural Pacinian corpuscles in control and
reinnervated hind limbs of rats after early denervation.
The pairs of columns represent mean numbers of crural
Pacinian corpuscles $\pm$ standard error (vertical bars) in
control (white) and reinnervated (hatched) legs 2 or
more months after crushing the right sciatic nerve in
1 to 15-day-old rats. Abscissa: the age in days (d) at
which the rats were operated; ordinate: the number of
corpuscles. Individual columns represent mean counts of
crural corpuscles removed from 5 to 8 animals and stained
for cholinesterase <u>in toto</u>. Circles show the range of
values in each group. Note the decrease in the number of
corpuscles following reinnervation in rats operated at
1 and 5 days of age; there is no significant difference
in the number of control and reinnervated corpuscles
after nerve crushing at the age of one week or more after
birth.

DIGITAL CORPUSCLES

<u>Development: The effect of denervation and reinnervation.</u>
Digital corpuscles of the rat toe pads are analogous to Meissner
corpuscles in primates and man, which subserve the perception
of moving touch (Dellon, 1981). In the rat, their structure and
development resemble those of the mouse toe pads described by
Ide (1976, 1977). The corpuscles are situated in the dermal papillae beneath the epidermis; the supplying myelinated axon
gives rise to axon terminals with lateral processes surrounded
by approximately 10 lamellae; the corpuscle is enclosed by 1 or
2 capsular layers which are often incomplete. The digital corpuscles begin to differentiate postnatally. Tiny axon terminals
filled mainly with vesicles can be found in the dermal papillae
already 2 days after birth, but the first lamellae appear at the

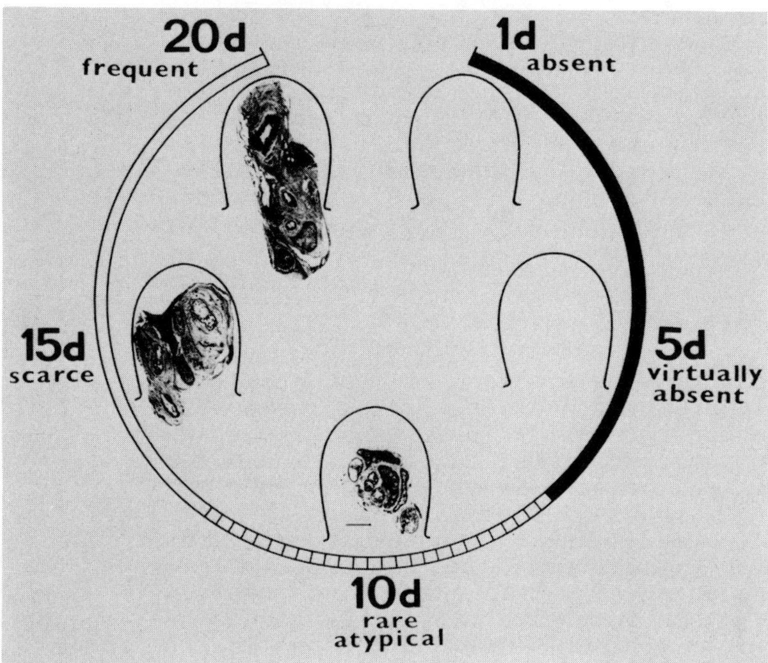

Fig. 2. The dependence of rat digital corpuscles upon sensory
innervation during the critical period of development.
The corpuscles are absent in toe pads following reinnervation after sciatic nerve crush in 1 to 5-day-old
rats (black segment of the dial); occasional atypical
corpuscles appear after nerve crush at 10 days of age
(hatched segment); the structure, size and incidence of
reinnervated corpuscles become gradually normal when
the sciatic nerve is crushed at 15 to 20 days after
birth (white segment) and later. Sections of representative corpuscles photographed at the same magnification indicate changes in their structure and size after
operations at 10 to 20 days of age. Bar: 5 μm.

terminals 8-12 days after birth, and the corpuscles become
structurally mature in 20-day-old rats. The development of digital corpuscles is prevented by crushing the sciatic nerve in
young rats up to 5 days of age (Fig. 2). This irreversible elimination of corpuscles is apparently due both to a great loss
of primary sensory neurons following axotomy and to the changes
in the periphery, mainly the retraction and disintegration of
developing Schwann cell pathways in the papillae. The few sensory axons that regenerate and reinnervate the foot after neonatal nerve crush grow into the deep dermis but do not find
their way to the papillae which remain permanently devoid of
axons and Schwann cells, as has been verified by the examination

of serially sectioned toe pads removed one year after nerve
crush. Occasional regenerated corpuscles were first observed
in rats after nerve crush performed 10 days after birth; these
newly formed corpuscles are, however, atypical, as their terminals are surrounded by only a couple of lamellae, and the lamellar coat remains thus reduced even a year after the operation.
More corpuscles of an almost normal ultrastructure occur in the
toe pads reinnervated after nerve crush performed in 15 and 20-
day-old rats (Fig. 2). When the sciatic nerve is crushed in
1-month-old rats, the corpuscles survive denervation and fully
recover after reinnervation, as has been previously found to be
the case in adult mice (Ide, 1982).

CONCLUSIONS

     Injury of sensory axons at the early stage of the critical
period before or immediately after the onset of differentiation
of corpuscular receptors results in the arrest of their development and their irreversible elimination due both to the loss and
impairment of primary sensory neurons and to alterations of the
periphery. Crushing of sensory axons at the later stage of the
critical period, when the receptors are already formed but still
structurally immature, leads to complete or partial destruction
of the receptor structure and to the neoformation of atypical
receptors upon induction by regenerating axons. The extent of
the critical period is related to the time course of normal receptor development.

     The critical period for the development of the somatosensory
periphery has its counterpart in the critical period for the
cytoarchitectonic differentiation of somatotopic brain maps along
the sensory pathways and in the sensory cortex (Van der Loos and
Woolsey, 1973). The differentiation of brain maps proceeds from
the periphery towards the cortex, but the causal and temporal
relation of the development of peripheral receptors to the differentiation in the central nervous system is still not clear
(Brenowitz et al., 1980; Nurse and Farraway, this volume). The
key link between the periphery and the nerve centers is apparently the primary sensory neuron. Peripheral axons of primary sensory neurons grow to the predilected areas where they induce the
differentiation of nerve-dependent encapsulated receptors or contact their target cells, while central processes of the neurons
reflect, by their fasciculation and oriented growth, the spatial
distribution of peripheral axons and thus transmit the image of
the periphery to the brain. The lesions of peripheral axons
during the critical period of development impair the structural
differentiation both in the periphery and in the brain centers.

REFERENCES

Aldskogius, H., Arvidsson, J., and Grant, G., 1985, The reaction
        of primary sensory neurons to peripheral nerve injury
        with particular emphasis on transganglionic changes,
        Brain Res. Rev., 10:27-46.
Brenowitz, G.L., Tweedle, C.D., and Johnson, J.I., 1980, The de-
        velopment of receptors in the glabrous forepaw skin of
        pouch young opossums, Neurosci., 5:1303-1310.
Chouchkov, Ch., 1978, Cutaneous Receptors, Adv.Anat.Embryol.Cell
        Biol., 54, Springer, Berlin.

Dellon, A.L., 1981, "Evaluation of Sensibility and Re-education of Sensation in the Hand", Williams and Wilkins, Baltimore.

Hunt, C.C., 1974, The Pacinian corpuscle, in: "The Peripheral Nervous System", J.I. Hubbard, ed., Plenum Press, London, pp. 374-404.

Ide, C., 1976, The fine structure of the digital corpuscle of the mouse toe pad, with special reference to nerve fibers, Am. J. Anat., 147:329-356.

Ide, C., 1977, Development of Meissner corpuscle of mouse toe pad, Anat. Rec., 188:49-67.

Ide, C., 1982, Regeneration of mouse digital corpuscles, Am. J. Anat., 163:59-72.

Ilyinsky, O.B., 1975, "Fiziologia Sensornykh Sistem", Nauka, Leningrad.

Milburn, A., 1973, The development of muscle spindles in the rat, PhD Thesis, Durham.

Munger, B.L., 1971, Patterns of organization of peripheral sensory receptors, in: "Handbook of Sensory Physiology", Vol. 1, W.R. Lowenstein, ed., Springer-Verlag, New York, pp. 523-556.

Otelin, A.A., Mashansky, V.F., and Mirkin, A.S., 1976, "Teltse Vater-Pacini", Nauka, Leningrad.

Saxod, R., 1978, Development of cutaneous sensory receptors in birds, in: Development of Sensory Systems. Handbook of Sensory Physiology, Vol. IX, M. Jacobson, ed., Springer-Verlag, Berlin, Heidelberg, New York, pp. 337-417.

Schiaffino, S., and Pierobon Bormioli, S., 1976, Morphogenesis of rat muscle spindles after nerve lesion during early postnatal development, J. Neurocytol., 5:319-336.

Schröder, J.M., 1974, The fine structure of de- and reinnervated muscle spindles. I. The increase, atrophy and hypertrophy of intrafusal muscle fibres, Acta neuropathol. (Berl.), 30:109-128.

Van der Loos, H., and Woolsey, T.A., 1973, Somatosensory cortex: structural alterations following early injury to sense organs, Science, 179:395-398.

Werner, J.K., 1973, Duration of normal innervation required for complete differentiation of muscle spindles in newborn rats, Exp. Neurol., 41:214-217.

Zelená, J., 1964, Development, degeneration and regeneration of receptor organs, Prog. Brain Res., 13:175-213.

Zelená, J., 1976, The role of sensory innervation in the development of mechanoreceptors, Prog. Brain Res., 43:59-64.

Zelená, J., 1978, The development of Pacinian corpuscles, J. Neurocytol., 7:71-91.

Zelená, J., 1980, Rapid degeneration of developing rat Pacinian corpuscles after denervation, Brain Res., 187:97-111.

Zelená, J., 1981, Multiple innervation of rat Pacinian corpuscles regenerated after neonatal axotomy, Neurosci., 6:1675-1686.

Zelená, J., 1982, Survival of Pacinian corpuscles after denervation in adult rats, Cell Tis. Res., 224:673-683.

Zelená, J., 1984a, The effect of long-term denervation on the ultrastructure of Pacinian corpuscles in the cat, Cell Tis. Res., 238:387-394.

Zelená, J., 1984b, Multiple axon terminals in reinnervated Pacinian corpuscles of adult rat, J. Neurocytol., 13:665-684.

Zelená, J., and Hník, P., 1963, Effect of innervation on the

increase after day 45, even though the size of the papilla and the total number of taste buds continued to increase. This suggested that the development of new taste buds awaited the expansion of the gustatory epithelium (Fig. 1; Hosley and Oakley, 1987). The IXth nerve induced the development of a mean of 91 foliate taste buds. Regardless of the presence or absence of the IXth nerve, the chorda tympani induced a mean of 35 foliate taste buds during development. We concluded that the chorda tympani and the IXth nerves induce two spatially intermingled, but developmentally independent, pools of taste buds whose sum equals the total number of foliate taste buds present in normal animals.
The virtual absence of mature taste buds and the paucity of immature taste buds on the tongue of newborn rats suggest that taste buds do not assist in mediating suckling behavior in altricial rodents. Thus, we believe, on the basis of our anatomical observations and other investigators' behavioral assessments of neonatal rat's responses to smell and taste, that olfactory, rather than gustatory, cues play the dominant role among chemosensory stimuli driving suckling behavior.

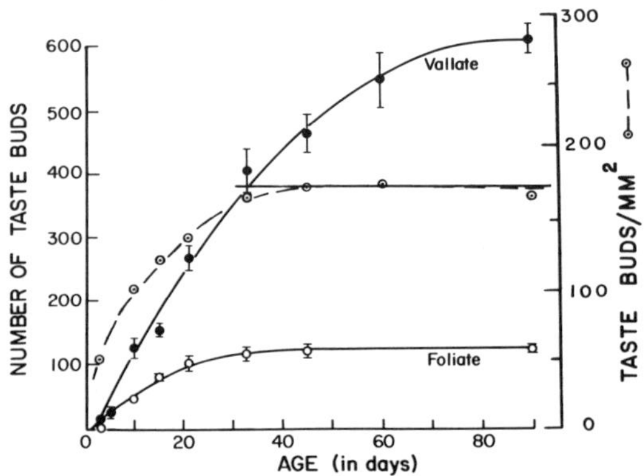

Figure 1. For both the vallate and foliate papillae the number of mature taste buds is a function of postnatal age. The data points are means±1sem with 6-22 animals per data point. The dashed line shows that the density of vallate taste buds in the gustatory epithelium reaches a ceiling of 178 buds/mm² at about day 45.

To quantitate developmental changes in immature vallate and foliate taste buds we made camera lucida tracings of all taste bud profiles in each histological section of the tongue. With a graphics tablet we traced a total of 15,000 profiles whose areas were entered into a microcomputer to allow determination of the volume and position of each mature and immature taste bud.

On the average an immature taste bud appeared in 1-2 sections and a mature taste bud in 3-4 sections. Immature taste buds decreased from 18% to only 2% of the mature taste bud population from day 15 to day 90. Throughout this period, mature and immature taste buds displayed a wide range of volumes. We argue from this and other data that there is no set-point volume at which taste pore formation is triggered. Moreover, even after a pore has formed, mature taste buds continue to enlarge.

From the number of immature taste buds present on a given day and the rate of rise of the population curve of mature taste buds, we calculated that 10.5±0.9 days (mean±1sd) would be required for that many buds to mature.

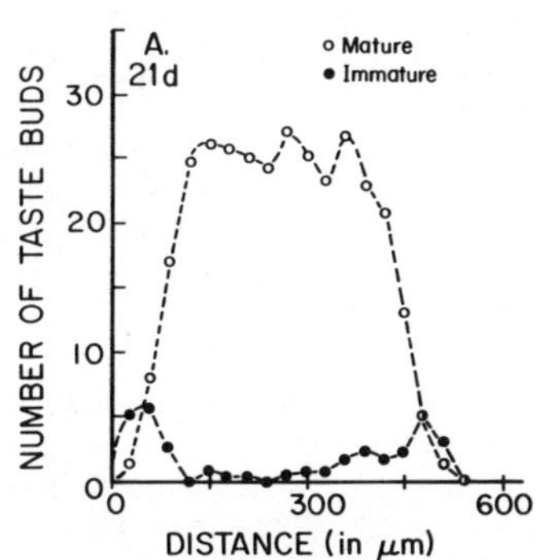

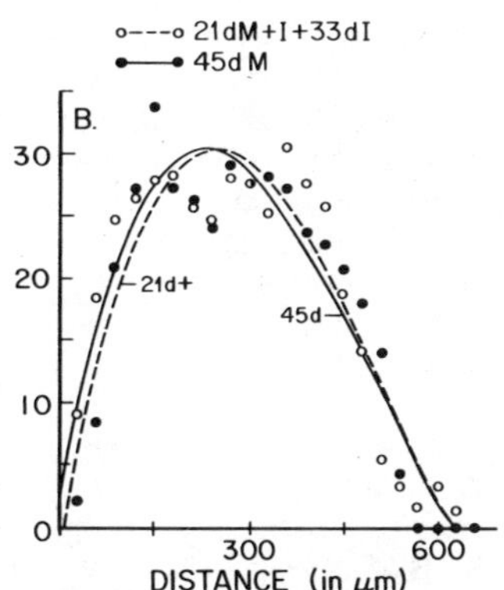

Figure 2. (A) The spatial distribution of mature taste buds (open circles) and immature taste buds (filled circles) plotted in 30 μ increments along the rostro-caudal axis of the vallate papilla. The dashed lines are second order polynomial interpolations between points that are means of three 21 day old rats. (B) The spatial distribution of mature vallate taste buds at day 45 (solid curve and solid circles) is closely approximated (99.86% of the area under the curve) by the sum of three spatial distributions: mature and immature taste buds at day 21 and immature taste buds at day 33 (dashed curve labeled 21d+, open circles). The two spatial distributions of taste buds are represented by third order polynomial curves fitted by the least squares procedure where each data point is the mean of three animals. The correlation coefficients are +.96 for the solid curve and +.95 for the dashed curve.

From days 21-45 immature buds were found predominantly in the rostro-caudal extremes of the vallate trench where the trench continues to lengthen and there are the fewest mature taste buds. At ages 15, 21, 33 and 45 days we plotted the rostro-caudal spatial distributions of the immature and mature taste buds (example in Fig. 2A). As predicted from a 10.5 day maturation time, the sum of the spatial distributions of both mature and immature taste buds at a given age equaled the spatial distribution of mature taste buds 10.5 days later. Moreover, this was true when two cohorts of immature taste buds were considered as in Figure 2B. We found that the foliate taste buds also matured in about 10 d. From the early rapid increase in the population of mature taste buds and the small number of immature taste buds present 1-3 days after birth, we estimate that in the first postnatal week vallate and foliate taste buds matured rapidly, perhaps within 2-3 days.

Apparently new vallate taste buds do not form by division of mature taste buds. We found no buds with more than one pore. And since most immature taste buds were clustered together in the rostro-caudal extremes of the trench, immature taste buds were generally not near a mature taste bud from which they might have arisen by division. Finally, a constant 10.5 day maturation time after day 15 suggested a single program for development, which we believe to be a continuation of the *de novo* formation of taste buds.

19

<u>The Neural Induction of Taste Buds and a Sensitive Period</u>

The century-old conjecture that taste axons induce taste buds in development has been a difficult proposition to prove because early removal of all axons would lead to a lack of taste buds in adults merely from the lack of neurotrophic support. No vallate taste buds remain at day 90 after complete denervation at day 3 or day 75; there is never a time when vallate taste buds can survive complete denervation. We capitalized upon the bilateral innervation of the vallate papilla to try to evaluate the neural induction of taste buds. If one IXth nerve is removed from adults, 496 of 610 taste buds continue to be maintained by the remaining IXth nerve. If one IXth nerve is removed at birth, only 230 taste buds develop. Assuming additivity, with two IXth nerves present from birth, adult rats should have only 460 taste buds, rather than normal 610 taste buds. This leaves a shortfall of 150 (610-460=150 buds) that can only be explained by a synergistic interaction between the two IXth nerves. Since only 30 immature and no mature vallate taste buds are present at birth, the absence of the 150 taste buds when only one IXth nerve is present during development must be due to the failure of a developmental synergism, rather than the failure of a neutrophic synergism in the maintenance of existing taste buds (Hosley et al., 1987a). Probably all vallate and foliate taste buds innervated by the IXth nerve are neurally induced in development.

Removal of one IXth nerve reveals a sensitive period from 0 to 10d postpartum when the development of vallate taste buds is most profoundly impaired. For example, at day 90 there were only 234 taste buds after one IXth nerve had been removed at birth but 430 taste buds after one IXth nerve had been removed at day 20. Some taste buds develop after day 20. They probably have late sensitive periods (Hosley et al., 1987b).

<u>Early Denervation Alters the Gustatory Epithelium</u>

In two groups of 3 day old rats the right IXth nerve was crushed. In group I the left IXth nerve was also removed. Subsequently, on day 75 in half of the animals in each group the regenerated right IXth nerve was transected and the chorda tympani sutured to the IXth nerve's distal stump to passively guide chorda tympani axons to the vallate papilla. With <u>only</u> the cross-regenerated chorda tympani innervating the vallate, there were four times as many taste buds present in the vallate papilla which had had one normal IXth nerve throughout development (Group II; 214$\pm$22 taste buds, mean$\pm$1sem; n=7) compared with the vallate papilla that lacked all innervation during the period (day 3 to ca. day 13) when the crushed right IXth nerve was absent (Group I, 51$\pm$10 taste buds, n=8, p<.001) or compared with control half of Group I in which nerve crossing was not done (48$\pm$12, n=10, p<.001).

A schematic diagram shows a minimal sequence of steps in the development, regeneration, and turnover of taste receptor cells (Fig 3). Progenitor cells are believed to

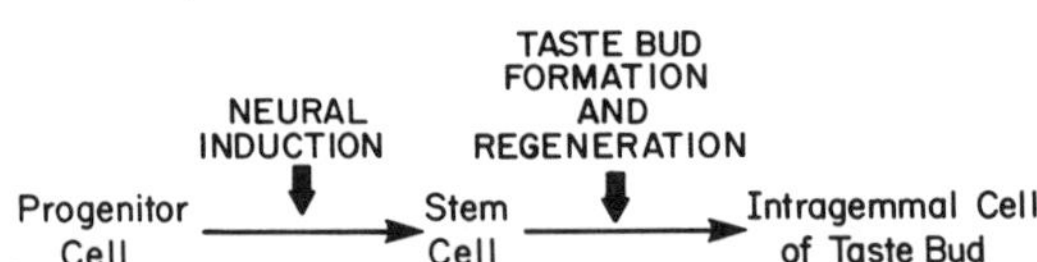

Figure 3. A model for the role of taste axons in mammalian taste bud development. Stem cells are induced from progenitor cells by taste axons. Taste bud formation, taste cell turnover, and regeneration of denervated taste buds is mediated by the differentiated daughters of stem cells.

be present in developing gustatory epithelium. The stem cell (basal cells for taste buds) is the mother cell that gives rise to daughters that both renew taste bud cells during normal cell turnover, and restore taste buds following regeneration of interrupted taste axons. We do not know what fraction of the stem cell population cycles to mediate normal receptor cell turnover. Preliminary evidence suggests that taste axons act to shut down the production of keratin in daughters of taste stem cells.

SUMMARY AND CONCLUSIONS

The following points summarize several observations and conclusions from the studies which have been briefly outlined above.

1. Our data reveal a prolonged postnatal maturational period of several weeks during which foliate and vallate taste buds mature.

2. The newborn rat's tongue lacks mature taste buds (Fig. 1). It is olfaction, not taste, that chemically guides suckling.

3. After day 15, immature vallate taste buds require 10.5 ± 0.9 days to mature. These taste buds arise *de novo* in epithelial growth zones in the rostro-caudal extensions of the vallate trench.

4. Vallate taste buds are neurally induced during a developmental sensitive period that is maximal from 0 to 10 days postpartum. Stem cells are probably postnatally induced from progenitor cells by taste axons present during the sensitive period.

5. Early denervation alters the vallate taste epithelium: it will no longer support proper numbers of taste buds, even if it is reinnervated by a normal taste nerve (chorda tympani). Hence, progenitor cells which fail to receive innervation during their sensitive period probably die or irreversibly adopt alternative developmental fates.

6. We propose as a minimal sequence in taste bud development, that progenitor cells are neurally induced to stem cells that give rise to daughter cells, which differentiate into the cells of the taste bud. Since the gustatory epithelium is a renewing epithelium, daughters of stem cells replace aged taste receptor cells during normal cell turnover. In adults the dividing stem cells are responsible for reconstituting taste buds that regenerate following reinnervation.

7. Taste bud regeneration does not recapitulate development. The IXth nerve will maintain more taste buds in the adult than it will induce during development. In addition the *de novo* development of a taste bud is a more prolonged, and apparently more elaborate sequential process than the regeneration of a taste bud from stem cells.

ACKNOWLEDGEMENTS

I would like to acknowledge the numerous contributions of several collaborators, notably M. A. Hosley and S. E. Hughes. The responsibility is entirely mine for any insufficiently qualified remarks in this synopsis of our work. This research program has been supported by the National Institutes of Health and most recently by a Javits Neuroscience Investigator Award from NINCDS.

REFERENCES

1. Hosley, M. A. and Oakley, B., 1987, Postnatal development of the vallate papilla and taste buds in rats, <u>Anat. Rec</u>., 218:216-222.
2. Hosley, M. A., Hughes, S. E., and Oakley, B., 1987a, Neural induction of taste buds, <u>J. Comp. Neurol</u>., 260:224-232.
3. Hosley, M. A., Hughes, S. E., Morton, L. L, and Oakley, B., 1987b, A sensitive period for the neural induction of taste buds, <u>J. Neurosci</u>., 7:2075-2080.
4. Oakley, B., 1985, Trophic competence in mammalian gustation, in: "Taste, Olfaction and the Central Nervous System," D. Pfaff, ed., Rockefeller Univ. Press, New York., pp 92-103.
5. Sloan, H. E., Hughes, S. E., and Oakley, B., 1983, Chronic impairment of axonal transport eliminates taste responses and taste buds, <u>J. Neurosci</u>., 3:117-123.
6. Zelená, J., 1957, The morphogenetic influence of innervation on the ontogenetic development of muscle spindles, <u>J. Embryol. Exp. Morph</u>., 5:283-292.
7. Zelená, J., 1980, Rapid degeneration of developing rat Pacinian corpuscles after denervation. <u>Brain Res</u>., 187:97-111.

DO DEVELOPING MERKEL CELLS IN VIBRISSAL TACTILE RECEPTORS PROVIDE A

TEMPLATE FOR PATTERN FORMATION IN THE TRIGEMINAL SYSTEM?

Colin Nurse and Laura Farraway

Department of Biology
McMaster University
Hamilton, Ontario, Canada, L8S 4K1

INTRODUCTION

The trigeminal pathway in rodents has served as an important model for studying pattern formation in the central nervous system. In this pathway there is a topographic point-to-point relation between the array of mystacial vibrissae on the whisker pad and their central projections in the barrel-field of the somatosensory cortex (Woolsey and Van der Loos, 1970; Welker, 1976; Belford and Killackey, 1980). Among the peripheral receptor sites the aggregates of Merkel cells, which are involved in tactile sensation in the vibrissal follicles, have been proposed as candidates for the master pattern or template for the central maps (Killackey, 1980). Support for this idea would be strengthened if the development of Merkel cells in the vibrissae should occur independently of retrograde influences within the whisker-to-barrel pathway. To test this we have investigated in the rat the effects of neonatal deafferentation of the whisker pad on Merkel cell development in the vibrissae using the quinacrine fluorescence technique (Nurse et al., 1983). Merkel cells are also present in the overlying whisker pad epidermis and these provided an additional population for comparing the effects of denervation. Since the method permits rigorous quantification of Merkel cell populations, this study was expected to address further the controversial issue of the role of sensory nerves on the development and maintenance of Merkel cells in different epithelial locations (Nurse et al., 1984).

Operative Procedure and Merkel Cell Quantification

Unilateral transection of the infraorbital nerve was performed on 1 day-old rats under ice hypothermia. Excision of a piece of the severed nerve trunk ensured there was no nerve regeneration to the whisker pad, at least over the first two weeks of postnatal life. In all cases the contralateral whisker pad served as the innervated control. Merkel cells were quantified by the quinacrine fluorescence technique applied to whole mounts as previously described (Nurse et al., 1983, 1984). Samples from the four largest and most caudal vibrissae and the immediately overlying skin of the whisker pad were examined for Merkel cells in both the operated and control side. To obtain the total cell number in the cylindrical vibrissal follicles, the cell count obtained in one-half of the follicle was doubled since the cell distribution is approximately symmetrical. Due to the increased difficulty in obtaining a clean isolation of the Merkel cell

Table 1. Merkel Cell Number in Innervated and Denervated
          Vibrissae[a]

Age (days)	Innervated	Denervated
6	651 ± 32, n = 14	592 ± 35, n = 9
9	700 ± 26, n = 15	758 ± 16, n = 16
14	770 ± 70, n = 10	774 ± 40, n = 7

[a] All animals received unilateral denervation of the whisker
pad at postnatal day 1. Values are mean $\pm$ S.E.; n = number
of vibrissae sampled from 4 - 6 rat pups.

clusters in older pups reliable quantification of Merkel cell number was
limited to the first two postnatal weeks.

Merkel Cell Development in Innervated and Denervated Vibrissae

At the time of denervation (i.e., postnatal day 1), each of the four
caudal vibrissae contained approximately 485 ± 10 Merkel cells (mean $\pm$
S.E.; n = 51) in the follicular epithelium. Control (innervated) and
denervated vibrissae were examined in operated pups of ages 6, 9 and 14
days. The results of this study are summarized in table 1, where it is
apparent that denervation did not affect the subsequent development of
Merkel cells over the first two postnatal weeks; in fact, there was no significant difference in Merkel cell number between denervated and innervated
vibrissae at each of the ages examined. In normal unoperated animals
vibrissal Merkel cells reach a plateau of about 750 - 800 cells around the
end of the first postnatal week (not shown), as is the case for the operated
animals in table 1. Fluorescence micrographs of Merkel cells in innervated
and denervated vibrissal follicles are shown in Figs. 1A, B for a 6 day-old
pup. Evidence that denervation was successful was obtained by conventional
silver staining and by use of certain styrylpyridinium dyes, which appear
to stain both sensory nerve fibers as well as Merkel cells (unpublished
observations). In addition, electron microscopy confirmed the absence of
nerve profiles in denervated vibrissal samples but their presence in
innervated ones (Figs. 2A, B); however, in both preparations Merkel cells,
with their characteristic dense cored granules were readily identified
(Fig. 2).

Effect of Denervation on Merkel Cells of Whisker Pad Epidermis

In addition to the cylindrical cuff of Merkel cells described above,
each vibrissa or sinus hair in the rat is surrounded by a conical ridge or
collar of Merkel cells as it penetrates the overlying whisker pad epidermis
(see Nurse et al., 1983; Renehan and Munger, 1986). These Merkel cell
collars provided an additional population for assaying the effect of sensory
denervation. In marked contrast to the results described above the majority
of these Merkel cells disappeared within a few days following denervation
(cf. Figs. 1C and D). The results from these experiments are summarized in
table 2. Since at the time of denervation each collar contained approximately 105 ± 3 Merkel cells (mean $\pm$ S.E.; n = 40), there was a subsequent
loss of about 80% of the cells over the next 5 post-operative days in the
denervated samples. This appeared to be the result of a genuine loss of
Merkel cells rather than a failure of the cells to accumulate the dye since
in the electron microscope no Merkel cells were found in the 6 and 9 day-
old denervated samples but were readily found in the innervated ones (not
shown).

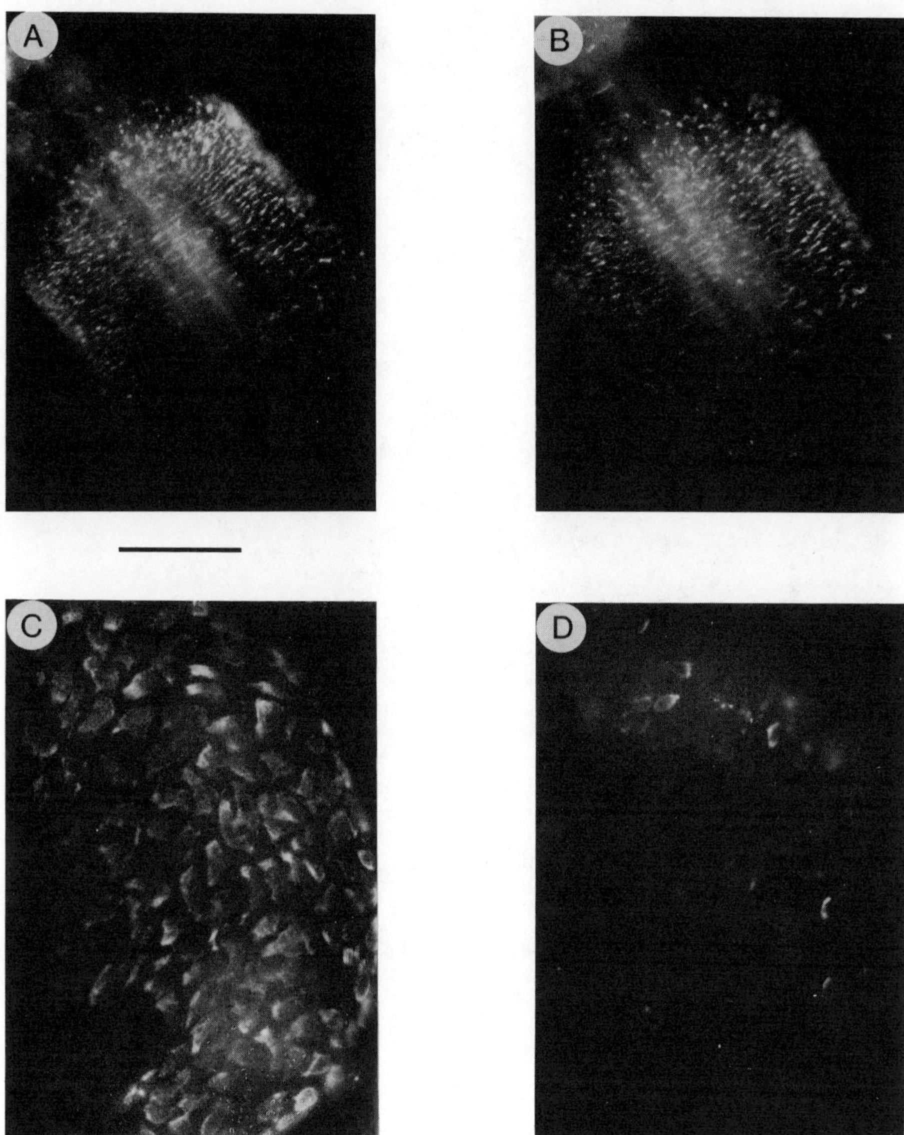

Fig. 1. Quinacrine fluorescent Merkel cells in whole mounts of innervated and denervated regions of the rat whisker pad. Samples obtained from pups (ca. 6 days old) which received a unilateral transection of the infraorbital nerve at postnatal day 1. A. Control (innervated) vibrissal follicle; B. Corresponding denervated vibrissal follicle; C. Control (innervated) Merkel cell collar in whisker pad epidermis; D. Corresponding denervated Merkel cell collar. Calibration bar represents 125 μm (A, B) and 60 μm (C, D).

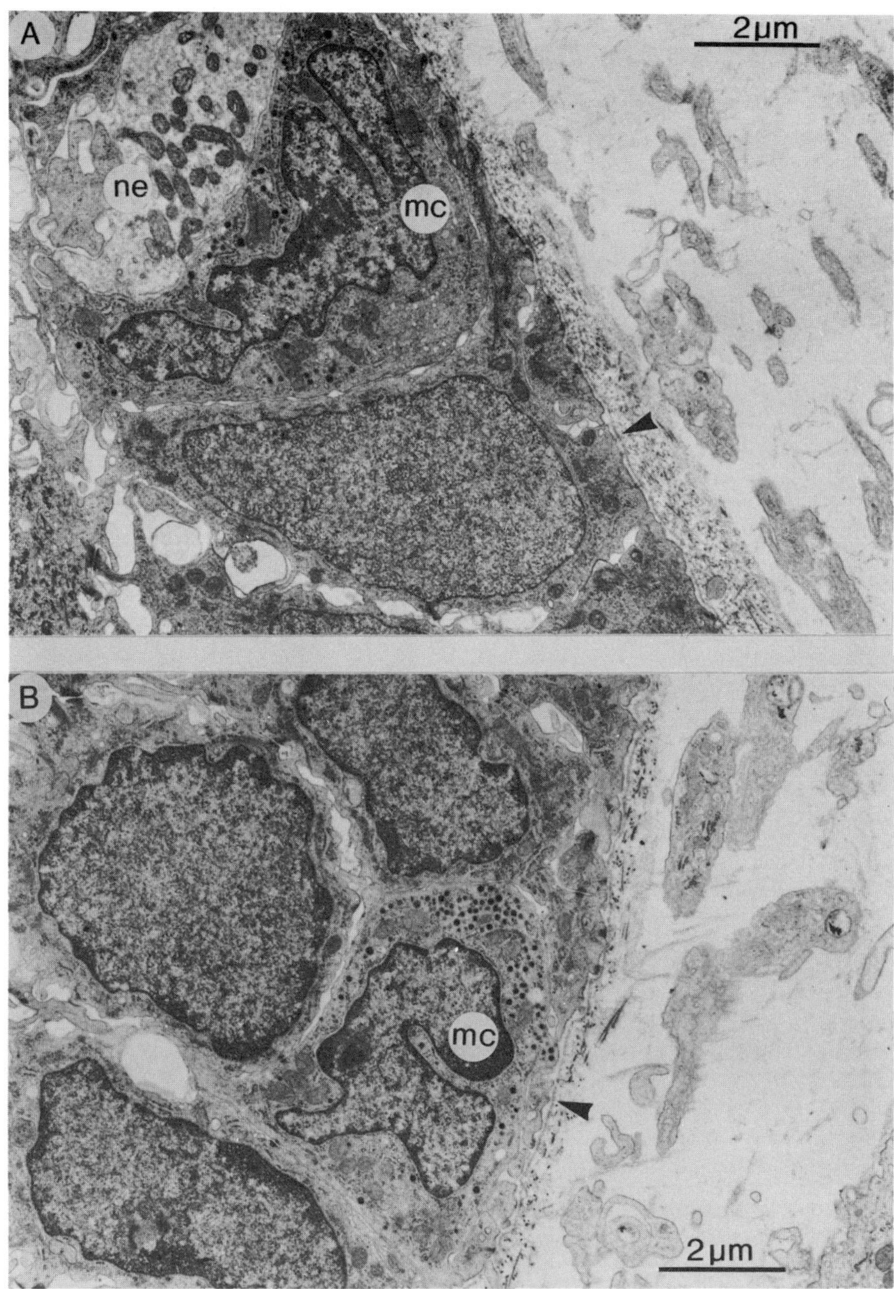

Fig. 2.   Electron micrographs of innervated and denervated rat vibrissae. <u>A</u>. Innervated region from a 14 day-old pup. <u>B</u>. Denervated region from a 9 day-old pup, operated at postnatal day 1. Merkel cells (mc) contain characteristic dense cored granules; nerve ending (ne) contains numerous mitochondria. Arrow heads on the right locate the basal lamina or glassy membrane.

Table 2. Merkel Cell Number in Innervated and Denervated
Collars in the Whisker Pad Epidermis[a]

Age (days)	Innervated	Denervated
4	130 ± 3, n = 19	71 ± 3, n = 19
6	128 ± 5, n = 23	18 ± 3, n = 25
9	131 ± 4, n = 14	14 ± 2, n = 27

[a] Samples obtained from Merkel cell collars that surround the
four caudal vibrissal hairs as they penetrate the overlying
skin. Values are mean $\pm$ S.E.; n = number of collars sampled
from 4 - 6 pups. All animals received unilateral denervation
at postnatal day 1.

DISCUSSION

The main conclusion from this study is that neonatal deafferentation
has drastically different effects on two separate populations of Merkel
cells in the rat whisker pad; those located in the outer root sheath of the
mystacial vibrissae continued to develop in normal numbers whereas the majority of those in the overlying whisker pad epidermis rapidly disappeared.
These results, based on rigorous quantification of the entire Merkel cell
population, indicate that real differences do exist between Merkel cells in
different epithelial locations. Such differences may already be inherent
in their precursors cells, which on the basis of growing evidence appear to
be epithelial in nature (English, 1977; Nurse et al., 1984). Interestingly,
vibrissal follicles appear to develop from a more differentiated epithelium
compared to the surrounding whisker pad epidermis (Van Exan and Hardy,
1980).

Our results support and extend previous studies indicating that Merkel
cells of the vibrissae survive denervation in normal numbers (Hartschuh and
Weihe, 1977: Benkenstein, 1979; Renehan and Munger, 1986). In all these
studies however, the denervation was performed in adult animals, after the
Merkel cells had completed their development and their central connections
established. This is an important distinction since in the present study
Merkel cells were still developing at the time of denervation, which occurred within the critical period during which sensory input from the periphery
is known to be required for the formation of the specialized neuronal as-
semblies in the barrel-field of the somatosensory cortex (Belford and
Killackey, 1980). Thus the vibrissal follicles were able to develop a
normal pattern of Merkel cells though deprived of an early sensory input,
which would be expected to prevent the formation of a proper central map.
This result is consistent with the idea that the array of vibrissal Merkel
cells contains the template or master pattern for specifying the barrel-
field in the rodent somatosensory cortex (Killackey, 1980). The implication
of these arguments is that Merkel cell populations that develop in the
absence of sensory nerves may play an important role in cortical organization. In this respect it is noteworthy that Merkel cells in a different
skin region, the glabrous skin of the hindpaw, also develop postnatally in
the absence of sensory nerves (Mills et al., 1984; see also Diamond, 1982),
and their cortical projections terminate in a specialized region of the
barrel-field (Welker, 1976).

The striking nerve dependence of the Merkel cells in the whisker pad
epidermis agrees with the findings of Palmer (1965) on the snout skin of

27

the oppossum; these Merkel cells behave similarly to those in the touch
domes of hairy skin (English, 1977; Nurse et al., 1984). Whether these
nerve-dependent Merkel cells are less important for the organization of the
somatosensory cortex in general, remains to be determined.

Acknowledgements

We acknowledge the expert technical assistant of Kathy Vollmer with
the electron microscopy. This study was supported by N.S.E.R.C. Canada.

REFERENCES

Belford, G.R., and Killackey, H.P., 1980, The sensitive period in the
     development of the trigeminal system of the neonatal rat. J. Comp.
     Neurol., 193: 335-350.
Benkenstein, M., 1979, Veränderungen der Ultrastruktur der Merkelschen
     Nervenendigungen an Sinushaaren von Ratten nach Denervation. Acta
     Anat. 105: 409-422.
Diamond, J., 1982, Modelling and competition in the nervous system: clues
     from the sensory innervation of skin. Curr. Top. Develop. Biol.,
     17: 147-205.
English, K.B., 1977, The ultrastructure of cutaneous type 1 mechanorecep-
     tors (Haarscheiben) in cats following denervation. J. Comp. Neurol.,
     172: 137-164.
Hartschuh, W., and Weihe, E., 1977, The effect of denervation on Merkel
     cells in cats. Neurosci. Letts. 5: 327-332.
Killackey, H.P., 1980, Pattern formation in the trigeminal system of the
     rat. Trends Neurosci. 3: 303-306.
Mills, L., Nurse, C.A., and Diamond, J., 1984, Regional differences in the
     sensory nerve dependence of Merkel cell development in rat skin. Soc.
     Neurosci. Abstr. 10: 1058.
Nurse, C.A., Mearow, K.M., Holmes, M., Visheau, B., and Diamond, J., 1983,
     Merkel cell distribution in the epidermis as determined by quinacrine
     fluorescence. Cell Tiss. Res. 228: 511-524.
Nurse, C.A., MacIntyre, L., and Diamond, J., 1984, A quantitative study of
     the time course of the reduction in Merkel cell number within
     denervated rat touch domes. Neurosci., 11: 521-533.
Palmer, P. 1965. Ultrastructural alterations of Merkel cells following
     denervation. Anat. Rec., 151: 396-397.
Renehan, W.E., and Munger, B.L., 1986, Degeneration and regeneration of
     peripheral nerve in the rat trigeminal system. II. Response to
     nerve lesions. J. Comp. Neurol., 249: 429-459.
VanExan, R.J., and Hardy, M.H., 1980, A spatial relationship between
     innervation and the early differentiation of vibrissa follicles in
     the embryonic mouse. J. Anat., 131: 643-656.
Welker, C., 1976, Receptive fields of barrels in the somatosensory neo-
     cortex of the rat. J. Comp. Neurol. 166: 173-190.
Woolsey, T.A., and Van der Loos, H., 1970, The structural organization of
     layer IV in the somatosensory region (SI) of mouse cerebral cortex.
     Brain Res. 17: 205-242.

COATED VESICLES IN DEVELOPING MUSCLE SPINDLES

Heather Stephens, Jan Kucera and Jon Walro

Department of Neurology (HS & JK)   Department of Anatomy(JW)
School of Medecine   College of Medecine
Boston University   Northeastern Ohio Univ.
80 East Concord St.   Rootstown, Oh 44272
Boston, Ma 02118

INTRODUCTION

Clathrin-coated vesicles mediate the transport of materials such as growth factors by endocytosis, exocytosis and membrane recycling (Willingham and Pastan, 1985). Coated vesicles are distributed in a variety of tissues including skeletal muscle, where they are thought to partake in both endo- and exocytosis (Benson et al., 1985; Haye et al; 1986). The presence of coated invaginations under the sarcolemma of sensory regions of intrafusal fibers in muscle spindles of 18-day gestational rats led to speculation that the development of intrafusal fibers may be influenced by neurotrophic factors released by sensory terminals and internalized via coated vesicles (Landon, 1972;Zelená & Soukup, 1973). We tested the hypothesis that coated organelles are involved in the exchange of information in the developing muscle spindle both between incoming axons and intrafusal fibers and between the different types of intrafusal fibers. If these two modes of communication do operate, then one would expect that the incidence of coated organelles would be higher 1) in developing versus adult spindles, 2) under sensory rather than motor terminals and 3) under apposed surfaces of intrafusal fibers rather than under the free sarcolemmal surface.

MATERIALS AND METHODS

Histological procedures

Spraque-Dawley rats of three age groups were studied: 0-day, 4-day and young adult. The rats were perfused with 2.5% glutaraldehyde and 1% paraformaldehyde, the soleus muscles excised, post-fixed in osmium tetroxide and embedded in Epon 812 as previously described (Walro and Kucera, 1985).

Identification criteria

Intrafusal fibers. Three intrafusal fiber categories were recognized: nuclear bag 2, bag 1, and chain. The adult fibers were distinguished by commonly accepted morphological and size criteria (Walro and Kucera, 1985). Additional features used for fiber type identification in the developing muscle spindle included: 1) the bag 2 fiber was morphologically more mature than the bag 1 or chain fibers and 2) the bag 1 fiber separates from the bag 2- chain fiber ensemble at the polar regions (Milburn, 1973).

Nerve fibers. The innervation patterns were reconstructed as previously described (Walro and Kucera, 1985). The absence of basal lamina interposed between the axon terminal and the muscle fiber was a feature used to differentiate sensory from motor terminals (Landon, 1972).

Coated vesicles. Because clathrin-coated membranes can assume a variety of shapes such as invaginations, pits, tubules or vesicles (Willingham and Pastan, 1985), all characterized by the presence of electron-dense, bristlelike projections on their cytoplasmic surface, they were grouped into the one 'coated vesicle' category for analysis.

Quantitative analysis

The incidence of coated vesicles under the sarcolemma of three membrane surfaces in two spindle regions was evaluated for each type of intrafusal fiber in three muscle spindles. The membrane surfaces compared were: 1) the post-synaptic or junctional sarcolemma under the nerve terminal endings; 2) the appositional sarcolemma of adjoining intrafusal fibers 3) the free or non-appositional surface of the intrafusal fibers. The spindle regions compared were :1) the sensory and 2) the motor region.

Three muscle spindles from three animals at each of the three age groups were analysed. The sensory zone of each muscle spindle was cut transversely at five different levels separated by intervals of ten microns. At each level, sets of eighteen ultrathin sections, separated by a series of nine 1 micron thick sections, were collected. Three of the eighteen ultrathin sections were photographed and micrographs of each intrafusal fiber were scored for the presence of coated organelles using a Zeiss Makro dissecting microscope at a final magnification of 160,000X. The length of each membrane surface expressed per hundred microns was determined by morphometric analysis. The results were tabulated as number of coated vesicles per hundred microns of sarcolemmal membrane. Thus, the basic sampling procedure included: 3 age groups X 3 muscle spindles per age group X 2 regions per spindle X 5 levels per muscle spindle region X 3 sections per level X 4 intrafusal fibers per muscle spindle section X 3 surfaces per intrafusal fiber. This sampling protocol was modified slightly in the polar regions because the motor terminals are not uniformly distributed along the length of the intrafusal fiber surface. Thus, once motor endings were detected by light microscopy, ten-fourty ultrathin sections were sampled. The protocol was also varied to take into account other circumstances such as fewer intrafusal fibers (0-day spindles) and lack of appositional surfaces (in polar regions of developing and in all regions of adult spindles).

Statistical analysis

The number of coated invaginations per hundred microns of muscle membrane was analysed among groups of intrafusal fibers categorized according to age of animal, spindle region, fiber type and surface. Combinations of the aforementioned factors were compared pair-wise by 2 way analysis of variance. Differences between individual groups were elucidated by a Student's Newman Keuls test for significance at the 95% confidence level (Sokal and Rohlf, 1969).

RESULTS

Distribution of coated vesicles in muscle spindles

Coated vesicles were localized under the sarcolemma of both sensory and motor terminals and in developing as well as adult spindles(Figs. a-f). They were also found under the appositional sarcolemma of adjoining intrafusal fibers in the 0 and 4-day spindles. At both junctional and appositional

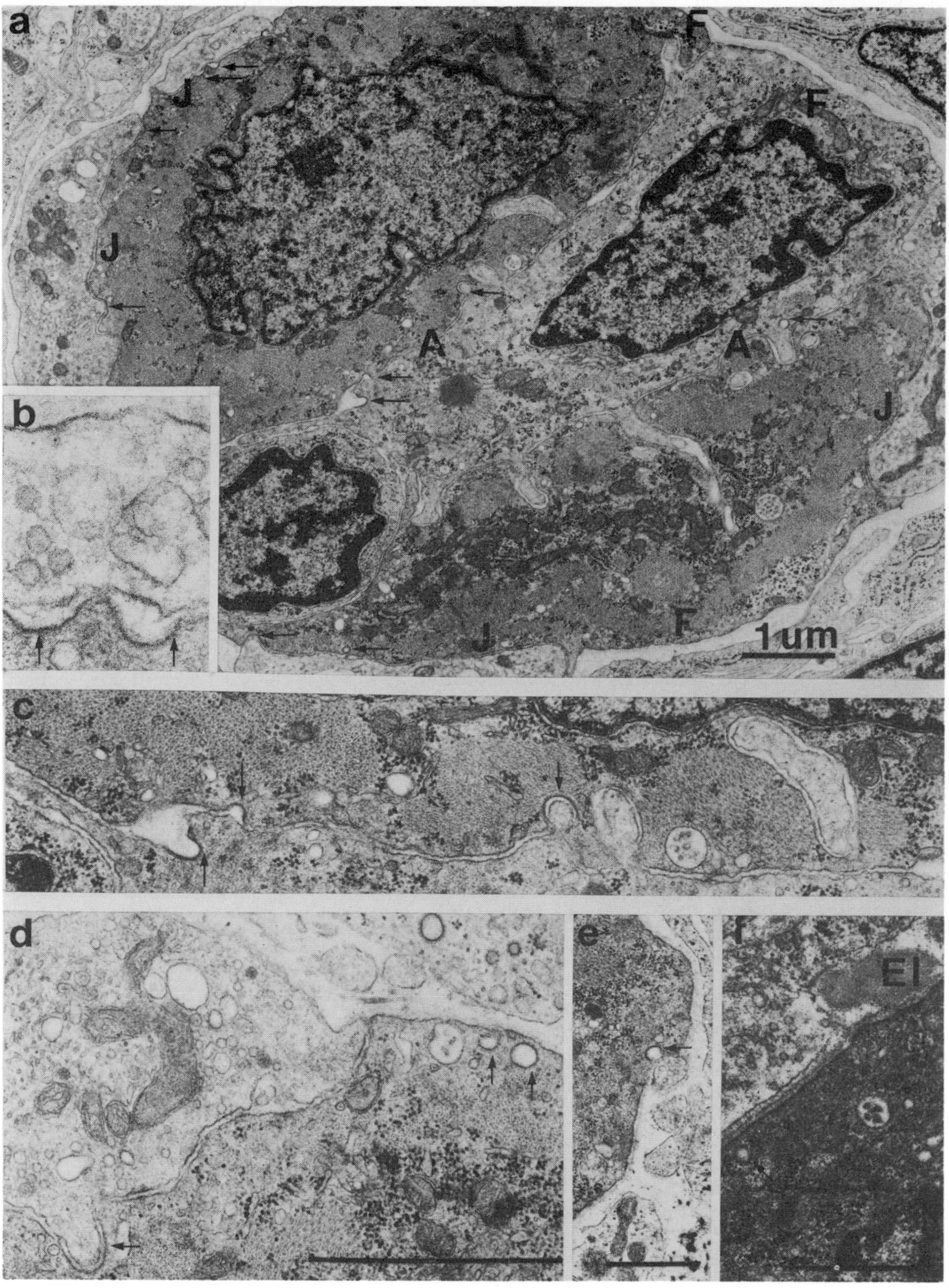

Figure 1

Table 1. Incidence of coated vesicles in the sensory region

Age	Bag 1 fiber		Bag 2 fiber		Chain fiber	
	free	junctional	free	junctional	free	junctional
0-day	5.0[a]	23.0	5.3	21.7	1.8	18.7
4-day	4.7	29.0	6.3	28.3	4.3	14.7
Adult	2.0	3.7	2.7	4.7	2.0	5.3

Table 2. Incidence of coated vesicles in the motor region

Age	Bag 1 fiber		Bag 2 fiber		Chain fiber	
	free	junctional	free	junctional	free	junctional
0-day	2.4[a]		3.1	8.5	1.7	
4-day	3.9	11.0	3.6	14.6	4.0	11.4
Adult	1.5	4.4	2.1	5.7	1.1	4.7

a The incidence of coated vesicles per hundred microns of sarcolemma.
The standard error of the mean was less than 10% for all cases.

sarcolemma the coated vesicles or invaginations were frequently associated
with fingerlike projections (Figs. a-d). At all ages they were rare
occurences under free, non-junctional sarcolemma. The diameter of coated
vesicles under sensory terminals was 102 ± 8.3 nm (n=10).

Coated vesicles under sensory terminals

The incidence of coated vesicles in the spindle sensory region was
significantly higher under junctional than under free sarcolemma (Table 1).
The number of coated vesicles per hundred microns of junctional membrane was
also significantly higher in developing versus adult muscle spindles, but did
not differ between 0- and 4-day spindles (Table I). This relationship was
consistent for each of the three types of intrafusal fiber examined. There
did not appear to be any differences among the fiber types for any of the
surfaces examined. The incidence of coated vesicles under junctional membrane
was highest for the bag fibers in 4-day spindles (a ten-fold difference as
compared to the free, non-junctional surface of adult intrafusal fibers).

Coated vesicles under motor terminals

The incidence of coated vesicles in the spindle motor region was highest
in the developing spindles, as compared to the adult spindles (Table 2).

The coated vesicles were more common under junctional versus free sarcolemma. Their incidence was, however, lower than that found for the sensory terminals in all types of intrafusal fiber at 0- and 4-days but was not significantly different between the two categories of terminals in the adult. Only bag 2 fibers exhibited sufficient motor terminals for quantification in the 0-day spindles; motor terminals were extremely rare on bag 1 fibers and not present at all on the still immature chain fiber.

Coated vesicles under appositional sarcolemma

The incidence of coated vesicles was much higher under appositional than free or non-appositional surfaces (results not shown). In 4-day spindles it reached a level comparable with the junctional sarcolemma.

DISCUSSION

The coated vesicles found in the muscle spindle resemble by several morphological criteria, including appearance and diameter, those described in other tissues. This suggests that the coated vesicles of the muscle spindle may also functionally mimic their counterparts in other tissues by taking part in receptor-mediated endocytosis (Willingham and Pastan, 1985). The coated vesicles found in the muscle spindles were distributed under both sensory and motor terminals. They were also found under non-junctional sites where their incidence was greater under appositional than non-appositional sarcolemma. The differential distribution of coated vesicles at diverse sites in developing and adult spindles indicates that they may be carrying out a variety of strategic functions at site-specific locations.

The greater incidence of coated vesicles in developing versus adult spindles of the rat strongly supports the hypothesis that they mediate neurohumoral transport at critical stages of development (Zelena, 1957; Landon, 1972; Zelená and Soukup, 1973). The age-dependant incidence of coated vesicles is consistent with their mediating the ontogenetic influence of both sensory and motor nerves. However, coated vesicles are still present at low levels under both nerve terminals in adult intrafusal fibers, which suggests that they continue to mediate neurotrophic input in mature muscle.

Coated vesicles are significantly more common under sensory than motor terminals. These data suggest that sensory innervation exerts a greater influence than motor on the developing spindle. Indeed, the influence of sensory innervation on the differentiation of intrafusal fibers is powerful, because spindles will not develop in deafferented muscles (Zelená, 1957). The greater incidence of coated vesicles under sensory as opposed to motor terminals suggests that more trophic factors would be released at equatorial rather than polar regions. This information gradient may enable the developing muscle fiber to differentiate between sensory and motor terminals. The presence of coated vesicles under fingers of the arriving axon may also be a mechanism for nerve-muscle recognition to establish the appropriate neuromuscular specificity (Landmesser, 1980). Chemical signals could be read by both the axon and muscle cell, perhaps coded by means of a concentration gradient, which would enable them to recognize each other and establish a feedback system. This would entail a 2-way communication system; indeed coated vesicles were also observed under the axonal presynaptic membrane.

The incidence of coated organelles was as high at appositional sites between intrafusal fibers as under the sarcolemma at sensory terminals. This is concordant with coated vesicles being structural correlates of information transfer between the developing intrafusal fibers. This would be most important for the muscle spindle during its formation, when the fibers are intimately connected by interlocking pseudopodial, filiform processes (Landon,

1972). The intrafusal fibers separate during development, at this time direct
communication between them is no longer needed. However, apposing surfaces
were occasionally noted between bag 1 and 2 fibers in the adult, with coated
vesicles present under the sarcolemma. Variations on the distribution pattern
of the coated organelles suggest how intercellular communication may be
operating. For instance, the presence of coated organelles which directly
face one another is indicative of information exchange between developing
fibers rather than transfer solely from one fiber to another. In turn, a
series of coated pits underneath the plasma membrane suggests that there may
be a precipitation site or centre of crystallization from which coated
vesicles are pushed or slid along. Coated organelles are more numerous at
points where interdigitating fingers of one intrafusal fiber penetrated its
neighbour than at linear portions of apposing membrane. These invaginations
may be sites of high information exchange between forming fibers.

This study has systematically examined the hypothesis that coated
vesicles may be the mediators of neurohumorally mediated transmission, which
would modulate gene expression. The findings are consistent with a role for
the coated vesicle, probably via receptor-mediated endocytosis, as a
structural correlate for the transport of neurohormones to intrafusal fibers
from sensory terminals and, to a lesser extent, from motor terminals. Such
information channelling may be important for nerve-muscle recognition and
sorting of respective motor and sensory domains. The nature of the substances
that mediate the neurotrophic influence and the pathways which are followed
by the trophic material within the cell after its incorporation remain to be
ascertained. Finally, the results extend the hypothesis by suggesting that
coated vesicles are involved in another form of communication taking place in
the muscle spindle, between the intrafusal fibers, themselves.

REFERENCES

Benson, J. J., Porter-Jordan, K., Buoniconti, P., and Fine, R.E., 1985,
    Biochemical and cytochemical evidence indicates that coated vesicles in
    chick embryo myotubes contain newly synthesized acetylcholinesterase,
    J. Cell Biol., 101:1930-40.
Haye, K.R., Foster, R.F., Goff, J.P., and Kaufman, S.J., 1986, Endocytosis of
    α-2 macroglobulin is developmentally regulated during myogenesis,
    Dev. Biol., 114:470-474.
Landon, D.N., 1972, The fine structure of developing muscle spindles in the
    rat, J. Neurocytol., 1:189-210.
Landmesser, L.,1980, The generation of neuromuscular specificity, Ann. Rev.
    Neurosci., 3:279-302.
Milburn, A., 1973, The early development of muscle spindles in the rat,
    J. Cell Sci., 12:175-195.
Sokal, R.R., and Rohlf, F.J., 1969, in "Biometry," W.H. Freeman,
    San Francisco, pp. 387-402.
Walro, J.M., and Kucera, J., 1985, Motor innervation of intrafusal fibers in
    rat muscle spindles: Incomplete separation of dynamic and static
    systems, Am. J. Anat., 173: 55-68.
Willingham, M.C., and Pastan, I.,1985, Morphologic methods in the study of
    endocytosis in cultured cells. Ch. 10 in "Endocytosis," M.C.
    Willingham and I., Pastan, eds., Plenum Press, New York.
Zelena, J., 1957, The morphogenetic influence of innervation on the
    ontogenetic development of muscle spindles, J. Embryol. Exp.
    Morph., 5: 283-292.
Zelena, J., and Soukup, T., 1973, Development of muscle spindles deprived of
    fusimotor innervation, Z. Zellforsch. 144: 435-452.

THE ONTOGENETIC DEVELOPMENT OF SENSORY CORPUSCLES

Libor Páč and Lubomír Malinovský

Department of Anatomy, Faculty of Medicine
Purkyně University
Komenského nám. 2, Brno, Czechoslovakia

The development of sensory nerve endings correlates with
the different maturity of new-born individuals in nidifugous
and nidicolous species of mammals and birds (Malinovský and
Páč, 1986). This assumption is based upon previous studies of
the Herbst corpuscle development in the duck (Malinovský and
Páč, 1985) and of the development of simple sensory corpuscles
in the cat (Malinovský and Páč, 1982). The study of the development of sensory nerve endings in other species of mammals
and birds could corroborate or refute this assumption.

MATERIAL AND METHODS

The development of Herbst corpuscles was studied in the
beak skin of the blackbird (Turdus merula), the development
of the simple sensory corpuscles was studied in the lip skin
of sheep foetuses (Ovis ammon, f.aries). The material was fixed
with 3 % glutaraldehyde and postfixed with 2 % OsO_4. Ultrathin
sections were examined in a Tesla BS 500 electron microscope.

RESULTS AND DISCUSSION

Sensory corpuscles (both Herbst and simple sensory corpuscles) develop in five different stages (the grouping of
Schwann cells around the dendritic zone - stage 1, the differentiation of the inner core - stage 2, the formation of the
capsule - stage 3, the structural maturation of the sensory
corpuscle - stage 4, the growth of the whole corpuscle - stage
5). The development of Herbst corpuscles in the duck (nidifugous species) and in the blackbird (nidicolous species) is
of the same duration (i.e. 10 days). The Herbst corpuscle development in the duck is completed (stages 1 to 4) before
hatching (Malinovský and Páč, 1985). The development of Herbst
corpuscles in the blackbird is terminated after hatching.
Simple sensory corpuscles in the sheep (nidifugous species)
and in cat (nidicolous species) also develop during the same
time (i.e. three months). The developmental stages 1 to 4
take place prenatally in the sheep. The development of simple

sensory corpuscles in the cat is completed postnatally (Malinovský and Páč, 1982).

     Sensory corpuscles in nidifugous mammals and birds are
structurally mature at birth or hatching (the sheep and the
duck). The development of sensory corpuscles in nidicolous
mammals and birds (the cat and the blackbird) starts postnatally since their neonates are still immature. The assumption
about the dependence between the maturity of new-born (or newly-
hatched) individuals in nidifugous and nidicolous mammals and
birds and the onset of development of sensory nerve endings has
been corroborated.

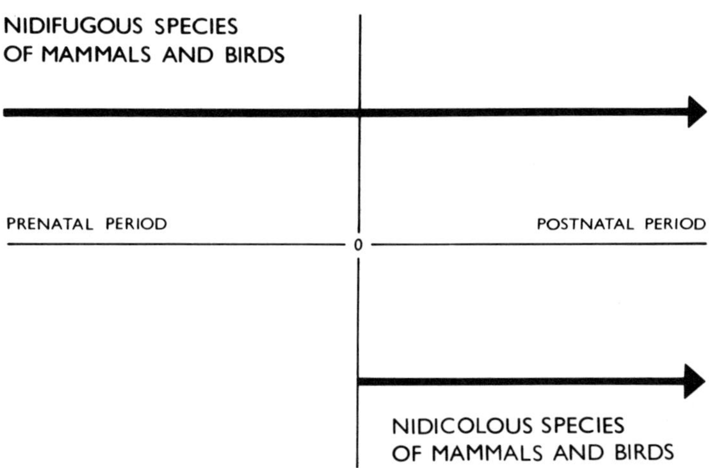

Fig. 1. A schematic representation of the ontogenetic develop-
        ment of sensory nerve formations in nidifugous and
        nidicolous species of mammals and birds.

REFERENCES

Malinovský, L., and Páč, L., 1982, Morphology of sensory cor-
        puscles in mammals, <u>Acta Fac. Med. Univ. Brunensis</u>,
        Brno.
Malinovský, L., and Páč, L., 1985, Ultrastructural development
        of the Herbst corpuscles in the skin of the beak of the
        domestic duck (Anas platyrhynchos, f. domestica). <u>Folia
        morphol.</u>, (Prague) 33:150-155.
Malinovský, L., and Páč, L., 1986, The development of sensory
        corpuscles in vertebrates with different pattern of onto-
        geny. <u>Verh. Anat. Ges.</u>, 80:845-846.

PART II

DEVELOPMENT OF MUSCLE SPINDLES

HUMAN MUSCLE SPINDLE DEVELOPMENT

Thornell L-E[1], Eriksson P-O[1,2], Fischman DA[3], Grove BK[1],
Butler-Browne GS[4] and Virtanen I[5]

(1) Department of Anatomy and (2) Clinical Oral Physiology,
University of Umeå, S-901 87 Umeå, Sweden; (3) Department of
Cell Biology, Cornell University, New York, USA; (4) Depart-
ment de Biologie Moléculaire, Institut Pasteur, 25 Rue du Dr
Roux, F-757 24 Paris, France; and (5) Department of Pathology,
University of Helsinki, Helsinki, Finland

INTRODUCTION

Myosin, a major component of skeletal muscle, is encoded by a multigene
family. Three major isoforms of the heavy chain subunits, one slow twitch
and two fast twitch, have been distinguished in adult skeletal muscles of
mammals (Whalen, 1985). These isoforms are related to specific, physiologically defined fiber types and their expression is clearly nerve-dependent.
In chicken and amphibia, there is also another slow myosin heavy chain
isoform - slow tonic myosin, present in multiply innervated, slow contracting fibers (Pierobon-Bormioli et al., 1980). This isoform is also
expressed in mammalian skeletal muscle, but only in extraocular muscles and
in some intrafusal fibers of muscle spindles (Pierobon-Bormioli et al.,
1980; Rowlerson et al., 1985).

Developing muscles contain at least two myosin heavy chain isoforms, an
embryonic or fetal form and a neonatal form, which are different from the
adult forms (Whalen et al., 1981). Since that discovery, considerable effort
has gone into exploring the developmental regulation of myosin expression
and fiber specialization (Whalen, 1985).

By using antibodies against some myosin isoforms in combination with
immunocytochemical methods, we have been able to show that fiber type
differences in human skeletal muscle can be distinguished at 15-16 weeks of
gestation especially with respect to the neonatal and slow isoforms
(Thornell et al., 1984), whereas in most reports using enzyme histochemical
techniques for demonstration of ATPase activity, fiber type differentiation
could not be demonstrated before 19-20 weeks of gestation (Dubowitz, 1965;
Fenichel, 1966; Colling-Saltin, 1978).

How and when muscle spindles differentiate in human muscles on the
basis of enzyme- or immunohistochemical characteristics has not to our
knowledge been previously determined, although it has been reported, on the
basis of morphological examinations of serially sectioned and silver
impregnated human fetuses, that muscle spindles can be detected at 11 weeks
of gestation (Cuajunco, 1940; Bowden, 1963; Przedpelska-Ober, 1982). We have
therefore examined embryonic and fetal human limb muscles by immunocyto-

chemical analysis using antibodies to a variety of different isomyosins in
combination with standard enzyme histochemical techniques. Furthermore, as
it is well-known that one distinctive feature of intrafusal fibers is the
presence or absence of a myofibrillar M band, we have also examined whether
there are differences in expression of the myofibrillar M-band proteins, MM-
creatine kinase (MM-CK), M-protein and myomesin during muscle spindle
development.

MATERIALS AND METHODS

Specimens

        Specimens were obtained from abortions. Gestational age was estimated
by comparing the time of the last menstrual period in relation to the crown-
rump and crown-head lengths of the fetuses. The investigation has been
examined and approved by the Medical Ethical Committee of the University of
Umeå, Sweden.

        Samples from the limbs were rapidly frozen in propane chilled in liquid
nitrogen. Serial frozen sections were cut in a Zeiss Cryostat.

Immunocytochemical methods

        The antibodies to slow, fast and neonatal myosins used in the present
study have previously been described in detail (Butler-Browne and Whalen,
1984). A monoclonal antibody to slow tonic myosin was obtained by
immunization with chicken ALD myosin (Sawchak et al., 1985). This antibody
distinctly recognizes slow tonic myosin and shows no crossreactivity with
slow twitch myosin, whereas at high concentrations it also binds to
mammalian fast myosin. The production and specificity of monoclonal
antibodies against the two high molecular weight M-band proteins, M-protein
165 kD and myomesin 185kD, have been reported (Grove et al. 1984). A
polyclonal antibody against MM-creatine kinase was purchased from Merck,
Darmstadt, West Germany. The secondary antibodies and PAP reagents were
obtained from Dakopatts, Copenhagen, Denmark. Standard PAP techniques were
used for the immunolabelling and the sections were examined in a Leitz
Dialux microscope or in an Olympus Vanox microscope.

ATPase Activity

        The sections were stained to demonstrate myofibrillar ATPase activity
at pH 9.4 and after acid preincuation at pH 4.6 and 4.3 (for references see
Thornell et al., 1984).

RESULTS

        No differences in staining between myotubes were seen in sections
stained for ATPase until 15-16 week of gestation, whereas a difference in
staining was seen between the primary and secondary generations of myotubes
already in 10 week old fetuses with the former more strongly labelled with
antibodies against slow myosin and the latter with antibodies against
neonatal myosin (Figs. 1,2). In later stages there was a gradual transition
to stronger staining with antibodies against slow myosin.

        In sections stained with anti-slow tonic myosin, a few scattered large
individual myotubes were distinctly stained in fetuses of 10-12 week
gestation, whereas in 14 week old fetuses both scattered primary and
secondary myotubes showed staining (Fig. 3). These myotubes could not
otherwise be differentiated from the rest of the myotubes.

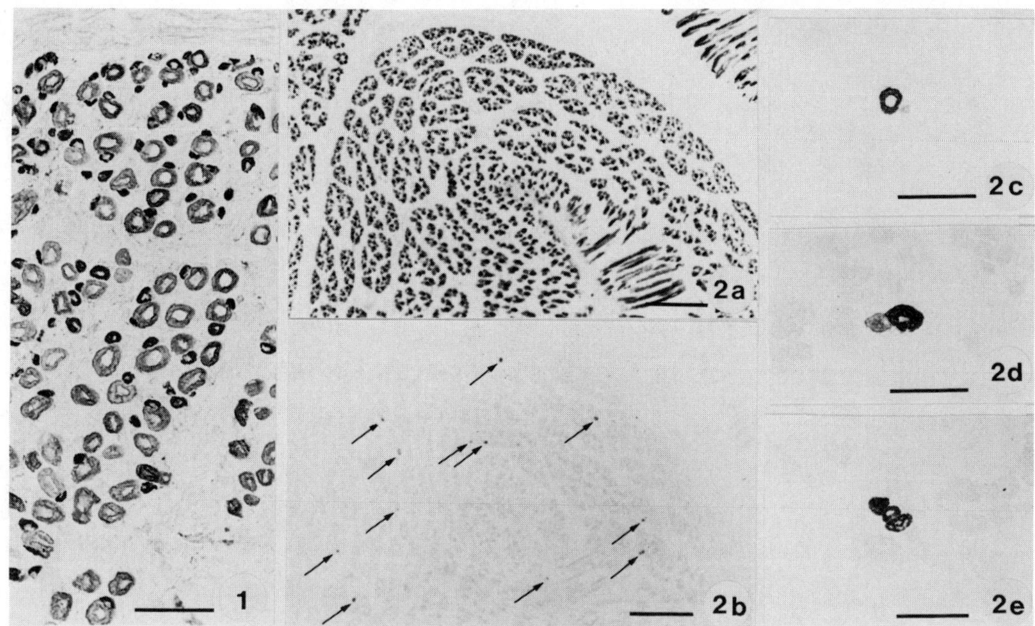

Fig. 1. Cross section of the upper arm of a fetus aged 10 weeks stained with antibodies against neonatal myosin. Note that there is a difference in staining of primary (large size) and secondary (small size) generation fibers; bar 50 μm.

Fig. 2. Serial cross sections of limb muscle of human fetus aged 14 weeks stained with a) anti-slow twitch myosin and b-e) anti-slow tonic myosin. Note that all primary myotubes are stained in a) whereas in b) scattered myotubes are stained (arrows). In c-e) slow tonic myosin labelled myotubes are shown in higher magnification. Note that the secondary generation of myotubes is also labelled, whereas all other surrounding myotubes are completely unstained. No capsules are present; bars a-b) 250 μm, c-e) 50 μm.

Typical muscle spindles were first recognized in 15-16 week old fetuses (Fig. 3). Slow tonic myosin was only present in the intrafusal fibers.

All myotubes in the 10-14 week fetuses contained all three M-band proteins, MM-CK, M-protein and myomesin. In fetuses of 16-18 week gestation (Fig. 3), one or two intrafusal fibers per spindle appeared to lack staining for M-protein.

In newborn infants, the only fibers which were stained with the anti-slow tonic myosin antibodies were the intrafusal fibers of large diameter, whereas the intrafusal fibers of small diameter and the extrafusal fibers were consistently negative. All intrafusal fibers contained MM-CK and myomesin, whereas the staining for M-protein varied (Fig. 4). All extrafusal fibers were, on the other hand, stained for all three M-band proteins (Fig. 4).

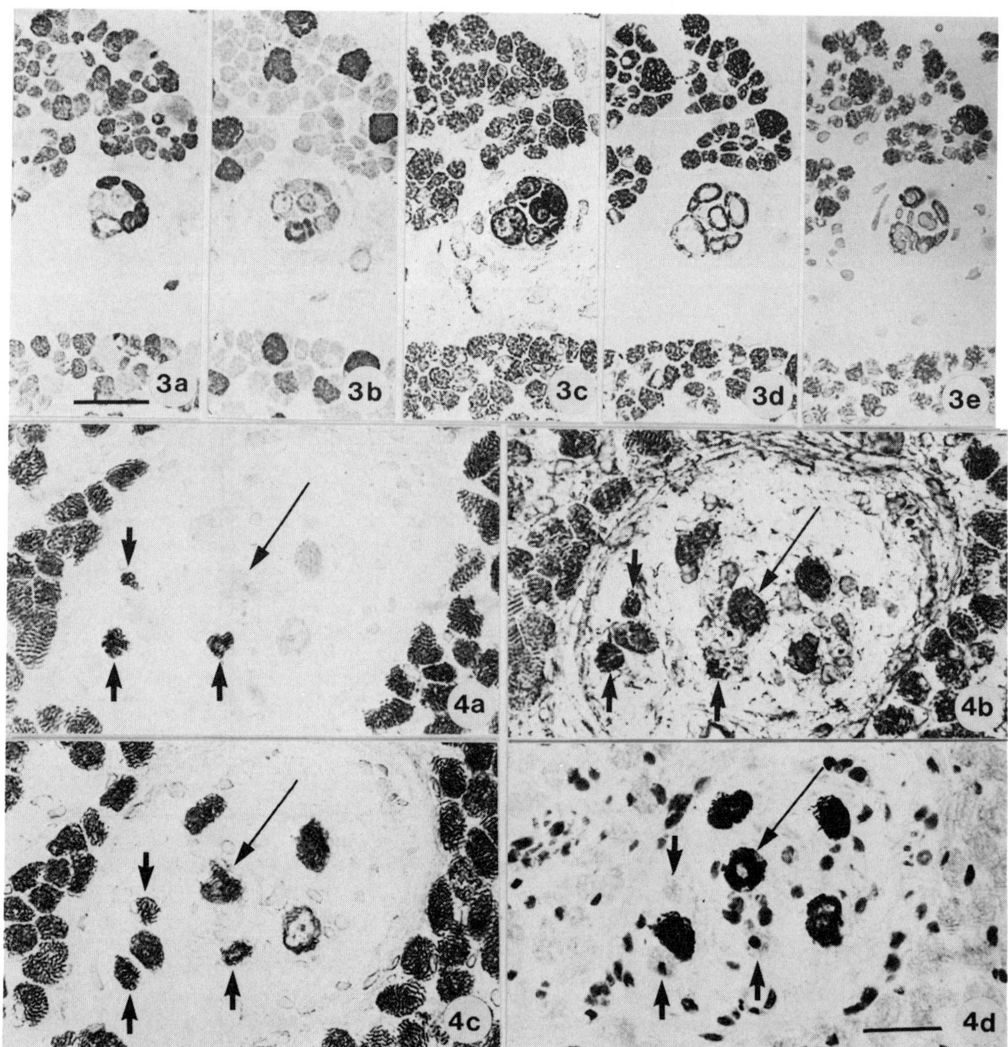

Fig. 3. Serial cross sections of the anterior tibial muscle of a fetus aged
16 weeks stained for a) neonatal myosin, b) slow twitch myosin, c)
MM-CK, d) myomesin, and e) M-protein. A muscle spindle composed of
seven fibers and surrounded by a capsule is seen. Note differences
in staining among the intrafusal fibers as well as the extrafusal
fibers in a and b. The primary generation of extrafusal fibers shows
low anti-neonatal myosin staining but high anti-slow myosin
staining, whereas the secondary generation myotubes show the
reversed pattern. All fibers are stained for MM-CK and myomesin,
whereas two fibers seem to lack staining for M-protein; bar 25 μm.

Fig. 4. Cross sections of muscle spindle in human masseter of a newborn
child stained for a) anti-M-protein, b) anti-MM-CK, c) anti-myomesin
and d) anti-slow tonic myosin and haematoxylin to stain nuclei. Only
nuclear chain fibers (small arrows) are labelled with anti-M-
protein. All fibers are labelled with anti-MM-CK and anti-myomesin.
In contrast, bag fibers are labelled with anti-slow tonic myosin
(one is indicated with a long arrow), but unlabelled by anti-M-
protein whereas the nuclear chain fibers (small arows) and the
extrafusal fibers are unlabelled by anti-slow tonic myosin, but
labelled with anti-M-protein; bar 25 μm.

DISCUSSION

There has been considerable interest in exploring how myogenic cells
develop into muscle fibers. From 1960 until the early 1980s it was widely
accepted that the developing motor neurons determined the future type of
initially undifferentiated myotubes; however, during the past few years this
concept has been challenged. Miller and Stockdale (1986) found, using antibodies against slow and fast myosins, that primary chicken myotubes
exhibited different phenotypes, some contained only fast myosins, some slow
and some both and concluded that three innervation-independent lineages of
myoblasts and myotubes are present during early myogenesis in chicken.
Multiple myoblasts lineages forming primary myotubes have not been
distinguished in mammals, however, Dhoot (1986) and Narusawa et al. (1987)
have shown that nerve-independent expression of the slow myosin heavy chain
isoform occurs in all primary generation myotubes, whereas in secondary
myotubes slow myosin is lacking or is occasionally transiently present.

Thus this is the first study to show myotube heterogeneity within the
primary generation of myotubes in mammals. One set of primary myotubes
appears to be destined to become the primordium of the muscle spindle
fibers, whereas the rest of the myotubes become extrafusal fibers. Within
this latter group heterogeneity with respect to myosin content was seen
between the primary and secondary generation of myotubes as has been
proposed earlier (Thornell et al., 1984; Pons et al., 1986). The primary
myotubes containing slow tonic myosin seem to be destined to become nuclear
bag fibers. Sometimes they had a larger diameter, but otherwise appeared
identical to the other unstained primary myotubes. In favourable longitudinal sections, however, an accumulation of nuclei proved that the stained
fibers were spindle fibers under development. This is in accordance with
previous morphological studies on development of human muscle spindles
(Cuajunco, 1940; Bowden, 1963; Przedpelska-Ober,1982). Further indirect
support for this suggestion is that in older fetuses in which spindles were
unequivocally morphologically distinguished slow tonic myosin was found only
in intrafusal fibers.

In contrast to extrafusal fibers where primary and secondary myotubes
differed in isomyosin composition, the secondary generation myotubes as well
as the primary myotubes of the presumptive muscle spindles contained slow
tonic myosin. Not until later in development was fiber diversification
within the intrafusal fibers seen with respect to isomyosins (fetuses 16-18
weeks gestation). However, diversification with respect to M-band proteins
appeared still later in development. We have obtained similar results for
developing muscle spindles in rat and cat indicating a general scheme for
the development of spindles in mammals.

These observations raise a number of questions. Do the primary myotubes
which express slow tonic myosin arise from a special cell lineage or is
sensory innervation tuning the genes of the primary myotubes? Since the
secondary generation of myotubes requires innervation for formation and is
generally considered to be derived from cells of a separate lineage
(Milburn, 1984), does sensory innervation likewise affect those cells which
are formed close to the myotubes containing slow tonic myosin? What factors
further regulate the diversification of the intrafusal fibers into types
with different isomyosin and M-band compositions?

The present observations add further information on the complexity of
muscle spindle development (recently reviewed by Milburn, 1984; Barker and
Banks, 1986). Experiments are now in progress to answer the above questions.

Supported by the Swedish Medical Research Council (3934 and 6874), the
University of Umeå and the Swedish Dental Society.

REFERENCES

Barker, D., and Banks, R. W., 1986, The muscle spindle, in: "Myology", A. G. Engel and B. Q Barker, eds., Mc Graw-Hill Co, New York, pp. 309-340.
Bowden, R. E. M., 1963, Muscle spindles in the human foetus, Acta Biol. Univ. Szeged IX, 35-59.
Butler-Browne, G. S., and Whalen, R. G., 1984, Myosin isozyme transitions occurring during the post-natal development of the rat soleus muscle. Dev. Biol.,102:324-334.
Colling-Saltin, A.-S., 1978, Enzyme histochemistry on skeletal muscle of the human foetus, J. Neurol. Sci., 39:169-185.
Cuajunco, F., 1940, Development of the neuro-muscular spindle in human fetuses, Carnegie Inst. Wash. Pub. No 173 Contrib. to Embryology, 28:95-128.
Dhoot, G. K., 1986, Selective synthesis and degradation of slow skeletal myosin heavy chains in developing muscle fibers, Muscle & Nerve, 9:155-164.
Dubowitz, V., 1965, Enzyme histochemistry of skeletal muscle, Part 2: Developing human muscle, J. Neurol. Neurosurg. Psychiat., 28:519-524.
Fenichel, G. M., 1966, A histochemical study of developing human skeletal muscle, Neurology (Minneap.), 16:741-745.
Grove, B. K., Kurer, V., Lehner, C., Doetschman, T. C., Perriard, J.-C., and Eppenberger, H. M., 1984, A new 185,000-dalton muscle protein detected by monoclonal antibodies, J. Cell Biol.,98:518-524.
Milburn, A., 1984, Stages in the development of cat muscle spindles, J. Embryol. exp. Morph., 82:177-216.
Miller, J. B., and Stockdale, F. E., 1986, Developmental regulation of the multiple myogenic cell lineages of the avian embryo, J. Cell Biol., 103:2197-2208.
Narusawa, M., Fitzsimons, R. B., Izumo, S., Nadal-Ginard, B., Rubinstein, N. A., and Kelly, A. M., 1987, Slow myosin in developing rat skeletal muscle, J. Cell Biol.,104:447-459.
Pierobon-Bormioli, S., Sartore, S., Vitadello, M., and Schiaffino, S., 1980, "Slow" myosins in vertebrate skeletal muscle. An immuno-fluorescence study, J. Cell Biol., 85:672-681.
Pons, F., Léger, J. O. C., Chevallay, M., Tomé, F. M. S., Fardeau, M., and Léger, J. J., 1986, Immunocytochemical analysis of myosin heavy chains in human fetal skeletal muscles, J. Neurol. Sci., 76:151-163.
Przedpelska-Ober, E. 1982, The development of muscle spindles in human fetuses. Anat. Anz. (Jena), 152:371-382.
Rowlerson, A., Gorza, L., and Schiaffino, S, Immunohistochemical identification of spindle fibre types in mammalian muscle using type-specific antibodies to isoforms of myosin, in: "The Muscle Spindle," I. A. Boyd, and M. H. Gladden, eds., Stockton Press, Glasgow (1985).
Sawchak, J. A., Leung, B., and Shafiq, S. A., 1985, Characterization of a monoclonal antibody to myosin specific for mammalian and human type II muscle fibers, J. Neurol. Sci., 69:247-254.
Thornell, L.-E., Billeter, R., Butler-Browne, G. S., Eriksson, P. O., Ringqvist, M., and Whalen R. G., 1984, Development of fiber types in human fetal muscle, An immunocytochemical study, J. Neurol. Sci., 66:107-115.
Whalen, R. G., 1985, Myosin isoenzymes as molecular markers for muscle physiology, J. exp. Biol., 115:43-53.
Whalen, R. G., Sell, S. M., Butler-Browne, G. S., Schwartz, K., Bouveret, P., and Pinset-Hänström, I., 1981, Three myosin heavy-chain isozymes appear sequentially in rat muscle development, Nature (Lond.), 292:805-809.

EARLY TYPE-DIFFERENTIATION OF INTRAFUSAL FIBERS

Anthea Rowlerson

Institute of Physiology
The University
Glasgow, G12 8QQ, U.K.

INTRODUCTION

The importance of afferent innervation for the development
of intrafusal fibers is well known (Zelená, 1957, 1976; Milburn,
1973, 1984). In contrast, the loss of the motor supply to
spindles has relatively little effect on spindle morphology
and, until recently, motor innervation of spindles was thought
to be established relatively late in development (Skoglund,
1960; Milburn, 1973). However, subsequent examination of muscle
development and spindle function in fetal and neonatal cat
muscle has thrown new light on this subject. Taken together:
(i) the presence of simple motor end-plates on nascent spindle
bag fibers very early in development (Milburn, 1984), (ii) a
higher incidence of functional beta-innervation in neonatal
cat muscle (compared to the adult) (Gregory and Proske, 1985)
and (iii) variable functional gamma-innervation at the same
stage (some spindles lacked a fusimotor supply, in others it
was present but not classifiable into distinct $gamma_s$ and
$gamma_d$ types (Gregory and Proske, 1986); these suggest that
the motor supply of intrafusal fibers is initially non-selective beta-innervation, which is then gradually replaced by
the selective gamma system during the late fetal and early
postnatal stages.

How is this selectivity of intrafusal fiber innervation
achieved? Recent work on the motor innervation of extrafusal
muscle fibers may provide a clue. Differentiation of extrafusal fibers into the main fast and slow types occurs very
early in development, and their motor innervation is initially
non-selective (i.e. single motor units contain both fiber types)
(Jones et al., 1987a). It is suggested that this pre-existing muscle fiber type differentiation helps to guide elimination of mismatched neuromuscular contacts (e.g. between "slow"
muscle fibers and "fast" motoneurons), leading to the adult
state of selective innervation of fast and slow fibers (Jones
et al., 1987b). If a similar mechanism exists for intrafusal
fiber type differentiation and innervation, then the future
spindle fibers must become committed to the intrafusal fiber
pathway early enough in development for them to be recognized

as such by inappropriate alpha- and gamma-motoneurons.

METHODS

As part of a wider study of muscle spindle development, I have looked for evidence of this commitment of nascent 1° myotubes to intrafusal fiber formation in developing muscle from several mammalian species. Fortunately, intrafusal fibers differ from extrafusal fibers in their myosin composition, and this can be detected immunohistochemically by the use of suitably isoform-specific antibodies to myosin (teKronnie et al., 1981; Rowlerson et al., 1985). Characteristically, D-bag_1 and S-bag_2 fibers react strongly and moderately respectively with an antibody specific for tonic fiber myosin (kindly supplied by Prof. S. Schiaffino), whereas chain fibers and extrafusal fibers give no reaction (see Table 1). This antibody, and in some cases histochemical myosin ATPase activity, were used to examine: (i) several muscles, including triceps surae, from fetal and neonatal cats in the age range about 2 weeks before birth to 1 1/2 weeks after birth, (ii) triceps surae in the rat at 17 and 19 days gestation, and after birth, and (iii) triceps surae, peroneus longus and other muscles from human fetuses of about 16 weeks gestation. Muscle samples were combined into composite blocks and snap frozen; sections were cut serially at 10 μm thickness and stained for myosin ATPase activity after alkali preincubation, and with antibodies specific for tonic myosin only (antitonic) (Pierobon-Bormioli et al., 1980) and for slow myosins generally (anti-I) (as used in Jones et al., 1987a,b).

RESULTS

By 1 1/2 weeks after birth, cat muscle spindles resembled adult spindles in the immunoreactivity of their bag fibers and in myosin ATPase activity (compare reactions shown in Figure 1 with summary in Table 1).

Table 1. Some Type Characteristics of Intrafusal and Extrafusal Muscle Fibers in Adult Cat Muscle

| | Intrafusal fibers[a] | | | Extrafusal fibers | |
	D-bag_1	S-bag_2	Chain	Type I	Type II
mATPase after:					
high alkali	−	−	++	−	++
mild alkali	−	+/++	++	−	++
Reaction with					
anti-I	++	++	−	++	−
anti-tonic	++	+	−	−	−

(a): in the sleeve region

At about 5 days before birth the spindles were obviously less mature, but these same type characteristics were already present (see Figure 2). At about 50 days gestation distinct chain fibers were hard to identify, but bag fibers could be distinguished from each other, and from extrafusal fibers, immunohistochemically (Figure 3a). The difference in myosin ATPase activity between the bag fibers was also just detectable.

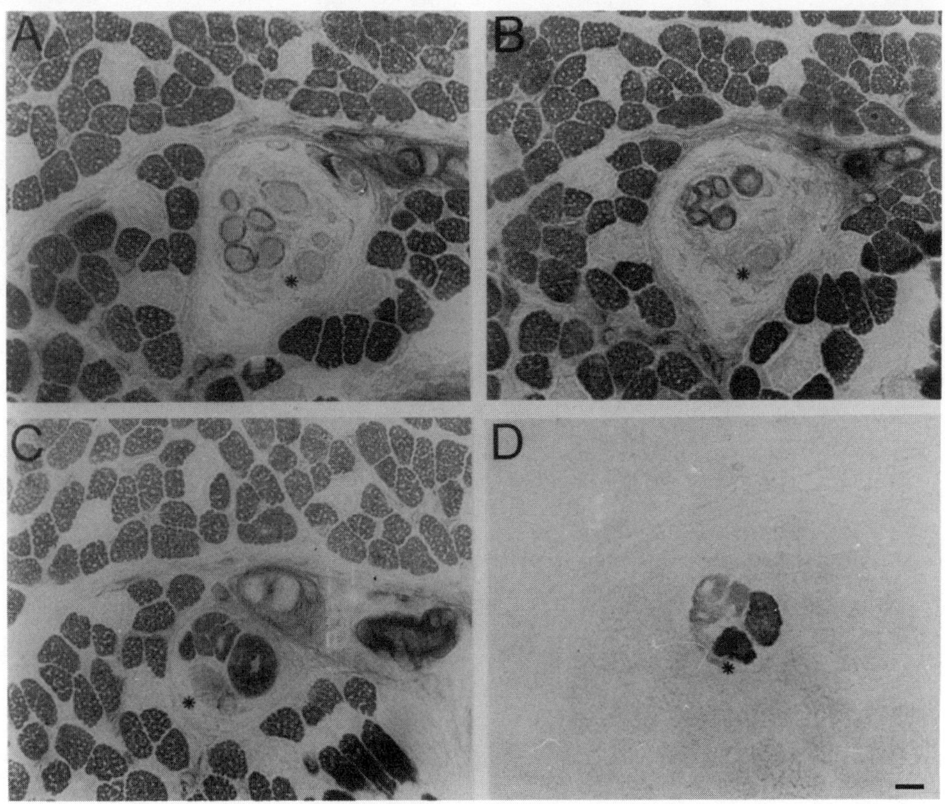

Figure 1. Cat tenuissimus, 1 1/2 weeks. (A) (B) (C) mATPase
activity after high (A) and mild (B), (C) alkali preincubation,(D) immunoperoxidase, with anti-tonic antibody. Asterisk indicates D-bag$_1$ fiber. Scale bar 10 μm.

     At 17 days gestation all hind-limb muscles in the rat ap-
peared to consist of primary myotubes only, and at this stage
even differentiation of extrafusal fibers into those with (S)
and without (F) slow myosin was barely evident. No fibers reacting with the anti-tonic antibody were found in these muscles.
By 19 days gestation, differentiation of extrafusal fibers in
the triceps surae was very marked, and single myotubes (usually
confined to regions of these muscles rich in S fibers) reacted
with the anti-tonic antibody. Occasionally, presumptive "spindles" could be identified by other morphological clues, but with
one exception, even when they appeared to consist of two closely
apposed fibers, only one reacted with the anti-tonic antibody.
Myosin ATPase activity was not able to distinguish between fiber
types at this age.

     The fetal human muscles examined appeared to be at a deve-
lopmental stage very slightly more mature than the 19 gestational
day rat, but clearly less mature than the 50 gestational day cat.
Extrafusal fibers were differentiated into F and S types, and
fibers reacting with the anti-tonic antibody usually occurred
singly (see Figure 3b), though pairs, of which one reacted more
strongly than the other, were occasionally present. The single
anti-tonic positive fibers in the 19 gestational day rats and

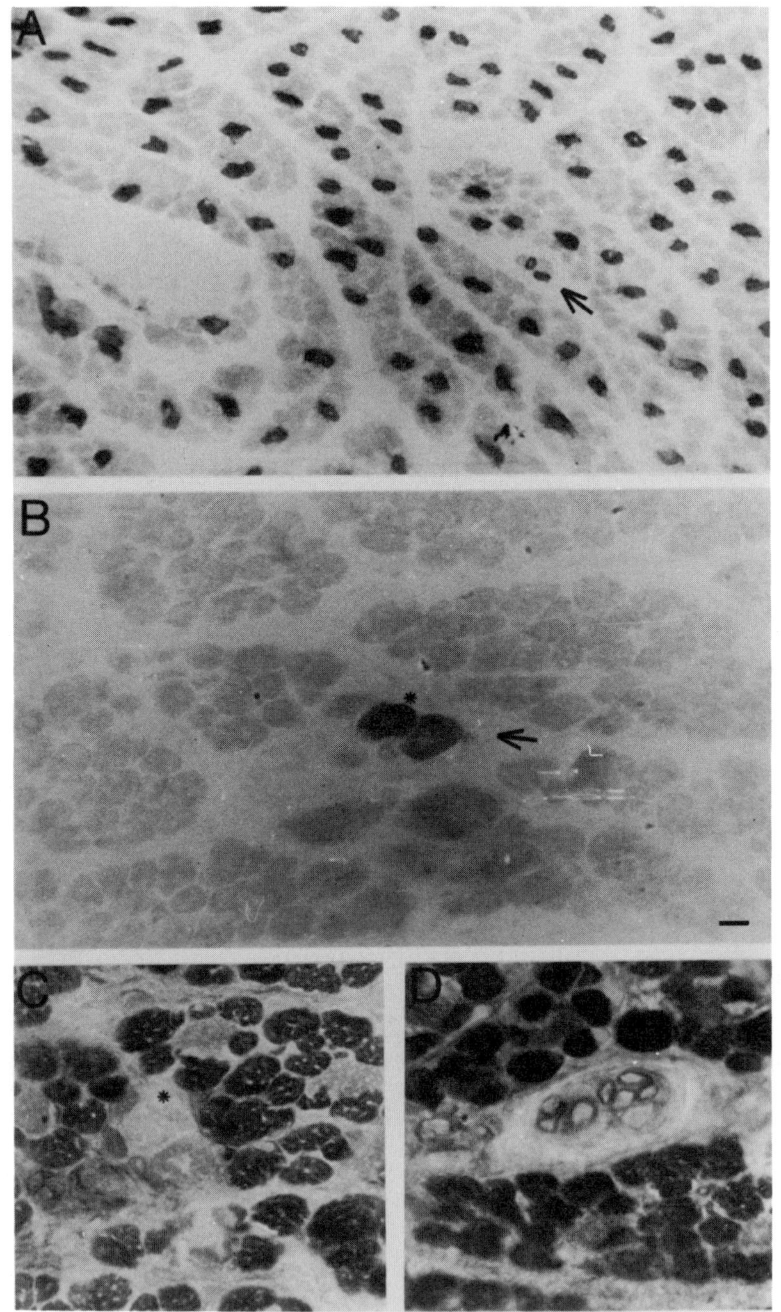

Figure 2. Cat peroneus longus, about 5 days before birth.
(A) Immunoperoxidase with anti-I, (B) Immunoperoxidase
with anti-tonic (C) and (D) mATPase. Asterisk indicates
D-bag$_1$ fiber. Scale bar 10 μm.

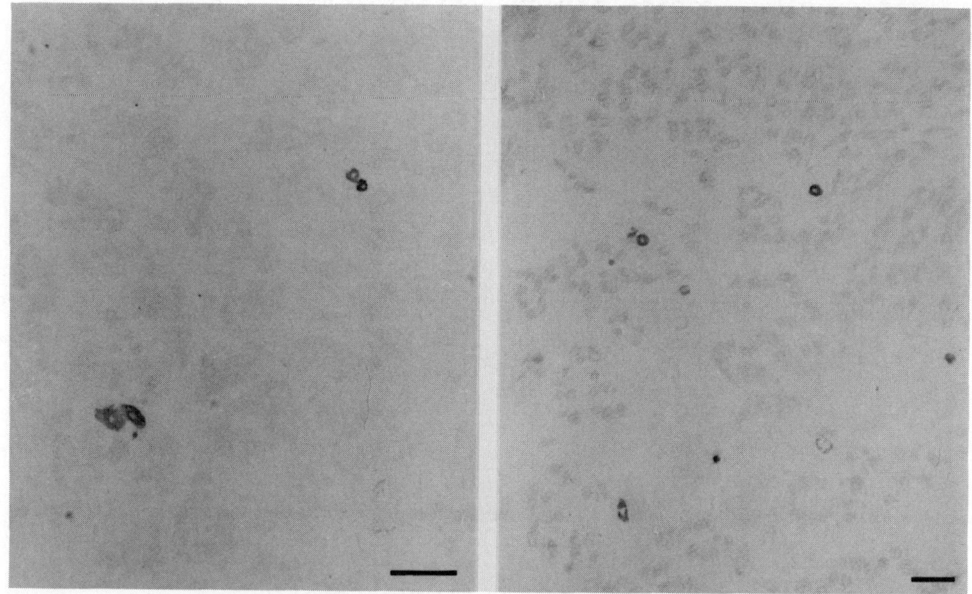

Figure 3. (a) Cat triceps surae, 50 gestational days, immunoperoxidase with anti-tonic, (B) human triceps surae,
about 16 weeks gestation, immunoperoxidase with anti-
tonic. Scale bar 50 μm.

fetal human muscles were presumably the future bag$_2$ fibers,
which are the earliest to develop (Milburn, 1984).^{2}

DISCUSSION

Comparing these results for the developing rat, cat and
human muscle with each other and with the development of morphological characteristics of extrafusal and intrafusal fibers in
the cat muscle described by Milburn (1984), it seems that the
developmental stages reached by the 19 gestational day rats and
16 gestational week human muscle correspond to about 38 and 39
gestational days in the cat. In some cases it was possible to
follow fibers which reacted positively with the anti-tonic antibody through many serial sections, and the positive reaction
was found to be maintained over several hundred microns in
length. Thus in these cases the commitment to intrafusal fiber
formation, as seen here reflected in the characteristic myosin
composition of the future bag fibers, is expressed throughout
the developing intrafusal fiber, not just in the region of contact with the 1° afferent terminal (which is presumably responsible for inducing the commitment in the first place).

Pre-existing differentiation of presumptive intrafusal bag
fibers therefore occurs both sufficiently early, and in the parts
of the fibers likely to carry motor end-plates, for it to be
able to help establish the selective innervation of intrafusal
fibers by influencing the remodelling of neuromuscular contacts
known to occur in developing muscle. The mechanisms used could
include both elimination of mis-matched neuromuscular contacts

(as suggested for extrafusal fibers) and, possibly, active sta-
bilization of contacts with appropriate gamma axons (cf. Arbuthnott et al., 1982).

ACKNOWLEDGEMENTS

I thank Prof. S. Schiaffino (Padua) for a gift of the anti-
tonic antibody, Dr. M.H. Gladden for collecting and freezing
some of the muscle samples examined, and the Wellcome Trust for
financial support. I am also grateful to the Royal Society, the
Physiological Society (Rushton Fund) and the Wellcome Trust for
travel grants enabling me to attend this Symposium on Mechanoreceptors.

REFERENCES

Arbuthnott, E.R., Ballard, K.J., Boyd, I.A., Gladden, M.H., and
     Sutherland, F.I., 1982, The ultrastructure of cat fusi-
     motor endings and their relationship to foci of sarcomere
     convergence in intrafusal fibres, J. Physiol., 331:285-309.
Gregory, J.E., and Proske, U., 1985, Responses of muscle recep-
     tors in the kitten, J. Physiol., 366:27-46.
Gregory, J.E., and Proske, U., 1986, Fusimotor axons in the
     kitten, J. Neurophysiol., 56:1462-1473.
Jones, S.P., Ridge, R.M.A.P., and Rowlerson, A., 1987a, The non-
     selective innervation of muscle fibres and mixed composi-
     tion of motor units in a muscle of neonatal rat, J. Phy-
     siol., 386:377-394.
Jones, S.P., Ridge, R.M.A.P., and Rowlerson, A., 1987b, Rat
     muscle during post-natal development: evidence in favour
     of no interconversion between fast- and slow-twitch fibres,
     J. Physiol., 386:395-406.
teKronnie, G., Donselaar, Y., Soukup, T. and van Raamsdonk, W.,
     1981, Immunohistochemical differences in myosin composi-
     tion among intrafusal muscle fibres, Histochemistry, 73:
     65-74.
Milburn, A., 1973, The early development of muscle spindles in
     the rat, J. Cell Sci., 12:175-195.
Milburn, A., 1984, Stages in the development of cat muscle spin-
     dles, J. Embryol. exp. Morph., 82:177-216.
Pierbon-Bormioli, S., Sartore, S., Vitadello, M., and Schiaffino,
     S., 1980, "Slow" myosins in vertebrate skeletal muscle,
     J. Cell Biol., 85:672-681.
Rowlerson, A., Gorza, L., and Schiaffino, S., 1985, Immunohisto-
     chemical identification of spindle fibre types in mammalian
     muscle using type-specific antibodies to isoforms of myo-
     sin, in: "The Muscle Spindle", I.A. Boyd and M.H. Gladden,
     eds., MacMillan, London.
Skoglund, S., 1960, The activity of muscle receptors in the kit-
     ten, Acta physiol. scand., 50:203-221.
Zelená, J., 1957, Morphogenetic influence of innervation on the
     ontogenetic development of muscle spindles, J. Embryol.
     exp. Morph., 5:283-292.
Zelená, J., 1976, The role of sensory innervation in the deve-
     lopment of mechanoreceptors, Progr. Brain Res., 43:59-64.

INNERVATION OF IMMATURE MUSCLE SPINDLES IN THE RAT

Jan Kucera and Jon M. Walro

Department of Neurology, Boston University, Boston, MA 02118
Department of Anatomy, Northeastern Ohio Universities
College of Medicine, Rootstown, OH 44272 USA

INTRODUCTION

Three aspects of the development of spindle nerve supply in the rat
soleus muscle are addressed. First, the innervation of developing
spindles is described and analyzed in a more systematic manner than was
possible in studies (Landon, 1972; Milburn, 1973) which preceded detailed
analysis of the neural organization of adult spindles (Walro and Kucera,
1985). Second, whether remodelling of neuromuscular synapses occurs in
immature intrafusal fibers similar to that described for developing
extrafusal fibers (Thompson, 1986) is investigated. Third, the temporal
relationship between the maturation of intrafusal fibers and establishment
of fusimotor innervation is examined at different stages of spindle
development.

MATERIALS AND METHODS

Virgin Sprague-Dawley female rats were placed overnight in individual
breeding cages with males. The day of appearance of sperm on vaginal
swabs taken at 0900 was designated as day 0 of gestation. Gestation
lasted 21-23 days in this strain of rats, with 63% of fertilized rats
delivering on the 22nd day of pregnancy. This method of estimating the
time of fertilization and determining the embryonic age of rat fetuses is
similar to that used by Duxson et al. (1986), and is accurate to
approximately 12 hours. Postnatal rats were staged in days after
parturition, with the day of birth designated as day 0.

Fetal or neonatal rats of three litters from timed pregnant rats were
used. Rats anesthetized with sodium pentobarbital (35 mg/kg i.p.) were
perfused with 3% glutaraldehyde and 2% paraformaldehyde in 0.1M cacodylate
buffer via the abdominal aorta. Soleus muscles were excised, postfixed in
1% OsO_4, and embedded in Epon. Transverse sections of the muscles were
cut on an LBK III Ultratome at a thickness of either 1 μm or 60 nm and
stained for examination by light or electron microscopy, respectively.

Spindles were cut transversely into serial ultrathin sections,
beginning in the extracapsular polar region and proceeding through the
equator into the other pole. Eight sections, spanning approximately 0.5
μm of tissue, were placed on each grid. Sequential electron micrographs

of the cross-sectioned spindles were taken at intervals of 0.5 or 1.0 μm at magnifications ranging from 1,000 to 10,000x.

Bag$_2$ fibers were discriminated from bag$_1$ fibers according to three criteria (Milburn, 1973; Kucera et al., 1978): i) bag$_2$ fibers display more nuclear profiles at the equator; ii) bag$_2$ fibers consistently have a greater cross-sectional area in the juxtaequatorial region; iii) bag$_2$ fibers are nearly always longer in the soleus muscle. Nuclear chain fibers were shorter and thinner than either bag$_2$ or bag$_1$ fibers. Afferents and efferents were distinguished based on the form of their endings. Motor, but not sensory, endings on intrafusal fibers contain basal lamina interposed between the axon terminals and muscle fibers. Primary afferents terminate in the equatorial region of the intrafusal fibers, whereas secondary afferents terminate in the juxtaequatorial regions of fibers (Landon, 1972; Milburn, 1973). The intracapsular region where the motor and sensory axons terminate in adult spindles (Walro and Kucera, 1985) was examined by skip-serial electron microscopy in its entirety for each of the spindles. The examination was continued into the extracapsular region, as long as the intrafusal fibers devoid of capsular envelopes could be confidently distinguished from extrafusal fibers. The extreme extracapsular region was surveyed for motor innervation by light microscopy of serial transverse sections, and no motor endings were encountered in this region of developing spindles. Indeed, no motor endings are present in the extreme extracapsular region of adult rat intrafusal fibers (Walro and Kucera, 1985).

RESULTS

At least three soleus muscle spindles were cut into serial transverse sections for each of the following three age groups: 20 days of gestation (20e), the day of birth (0dp), and four days postpartum (4dp).

Intrafusal fibers

Spindles in each age group contained both relatively mature myofibers with well-organized myofibrils and immature multinucleated myotubes with sparce myofibrils. The number of myofibers ranged from one (bag$_2$) in 20e spindles to three (bag$_2$, bag$_1$, chain) in 4dp spindles (Table 1). Only one myotube was present in a spindle regardless of age. The myotube was apposed to the bag$_2$ fiber, and extensions of its sarcoplasm interdigitated with the bag$_2$ fiber. The immature myotubes represented a nascent bag$_1$ fiber in 20e spindles, and a nascent chain fiber in 0dp and 4dp spindles.

Sensory innervation

Every spindle was innervated by one afferent identified as the primary (Fig. 1). About two-thirds of spindle poles also received one or two secondary afferents. The proportion of spindle poles innervated by secondary afferents was comparable among spindles of the three age groups and mature spindles (Walro and Kucera, 1985). The bag$_2$ fiber received more sensory endings than either bag$_1$ or chain fibers. Individual axon terminals often formed cross-terminals between two or three intrafusal fibers. The cross-terminals involved all combinations of the three types of intrafusal fibers, including the immature chain fibers, and were encountered throughout the extent of the sensory synaptic region. Chain fibers received all of their primary sensory innervation through cross-terminals from adjacent bag$_2$ or bag$_1$ fibers. Sites of sensory endings were occupied by single profiles of terminal afferent axons. The number of afferents or density of sensory endings in a spindle correlated with neither the number nor maturity of the intrafusal fibers.

Fiber	Age		
	20e	Odp	4dp
bag$_1$	mt/0	mf/0-3	mf/1-14
bag$_2$	mf/0-1	mf/1-5	mf/2-6
1st chain	absent	mt/0	mf/0-1
2nd chain	absent	absent	mt/0

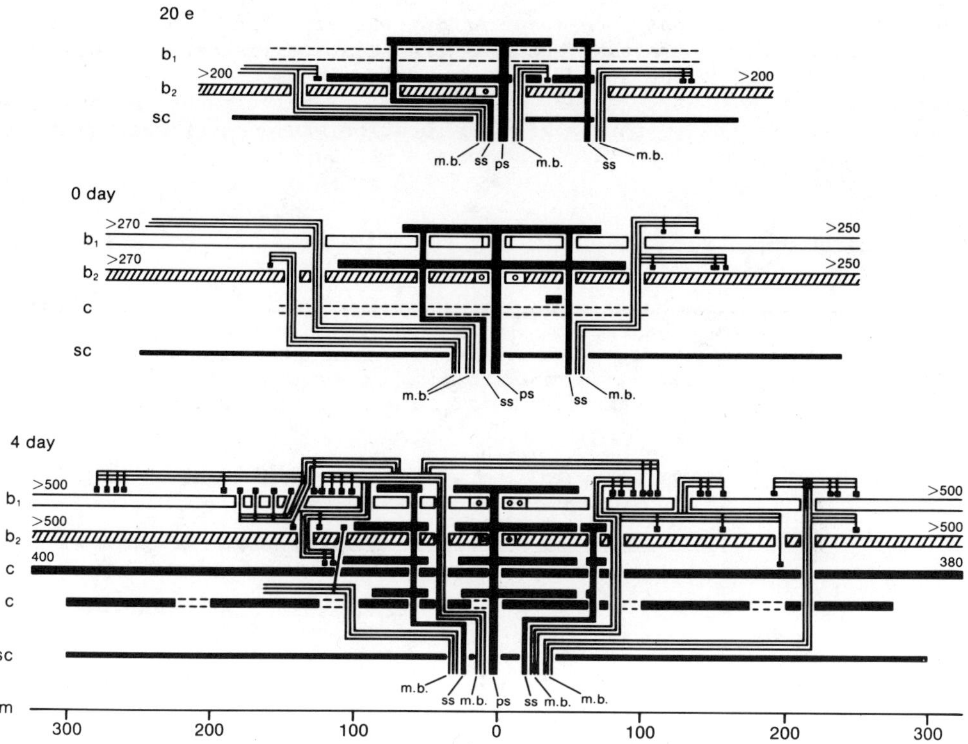

Fig. 1. Schematic representation of the nerve
supplies to representative 20e, Odp and 4dp spindles
as reconstructed from skip-serial transverse ultrathin
sections. The spindles contain nuclear bag$_1$ (empty),
bag$_2$ (hatched), and chain (solid) myofibers or immature
myotubes (dashed lines). The equatorial region of bag
fibers is denoted by circles. Primary afferents (ps),
secondary afferents (ss) and bundles of motor axons
(mb) innervate the intrafusal fibers. The breadth of
sensory domains and location of motor endings are shown.
The extent of the capsule (sc) is indicated by a solid
line. Scale shows distance relative to the equator
(zero).

<u>Motor innervation</u>

The density of motor innervation on intrafusal fibers increased with age. One 20e spindle was devoid of motor innervation in both poles, whereas all Odp or 4dp spindles received motor innervation. Nerve bundles composed of 5-20 unmyelinated motor axons penetrated the midportion of the capsule together with afferents and coursed toward the polar regions. The number of motor bundles per spindle pole ranged from 0-2 in 20e spindles to 1-4 in 4dp spindles. The bundles and/or their branches ultimately terminated on one or more intrafusal fibers, although some branches composed of only a few motor axons ended in the extreme encapsulated polar spindle region without forming a synapse. Most of the motor bundles coinnervated two or three types of intrafusal fibers, although an occasional bundle gave rise to motor endings confined to one fiber. The extent of coinnervation of different types of intrafusal fibers by individual motor axons that composed the motor bundles could not be determined. Motor endings were found only on the relatively mature intrafusal myofibers. The immature myotubes invariably lacked motor innervation. Both bag_1 and bag_2 fibers carried multiple motor endings on both poles, whereas only one motor ending was ever found in each pole of a chain fiber. The intrafusal endings were relatively short (5-10μm), and were located within 100-200 μm of the equator irrespective of the presence or absence of secondary sensory innervation. Rarely, motor endings were observed within the sensory region, a location where motor endings never occur in adult spindles (Walro and Kucera, 1985).

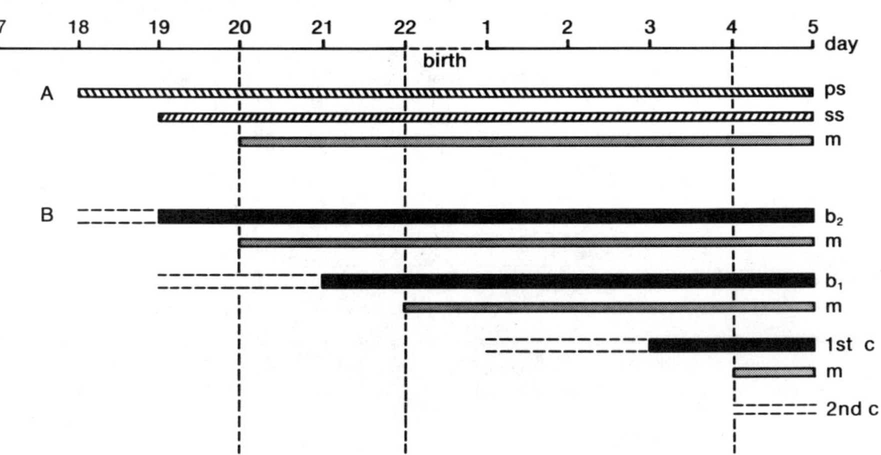

Fig. 2. A. Chronology of the arrival of primary (ps), secondary (ss) and motor (m) innervation to spindles of the developing rat soleus muscle, based on the work of Zelena (1957), Milburn (1973) and the present study.

B. Innervation of nuclear bag_1 (b_1), bag_2 (b_2) and chain (c) fibers by motor axons relative to the chronology of formation of the intrafusal fibers. Solid bars represent the myofiber stage and dashed lines the myotube stage. Presence of motor innervation is denoted by cross-hatched bars (m). The vertical dashed lines indicate times at which spindles were examined.

The intrafusal motor endings typically contained several profiles of terminal axons. The occurrence of multiple axon terminals at single neuromuscular junctions suggested that individual motor endings were innervated polyaxonally. The polyaxonal innervation of single motor endings was most obvious for those endings which were the sole site of termination of entire motor bundles on intrafusal fibers.

DISCUSSION

The first neuromuscular contacts in immature rat spindles (Fig. 2) are observed on the 18th day of gestation and involve primary afferents (Zelená, 1957). The present study suggests that the secondary afferents contact spindles between the 18th and 20th day of gestation because the proportion of spindle poles innervated by secondary afferents did not increase substantially after the 20th day of gestation. Motor axons contact spindles on the 20th day of gestation or shortly afterwards because some 20e spindles lacked motor endings whereas all 0dp spindles received motor innervation. The number and relative density of motor endings on the three types of intrafusal fiber in 4dp rats were comparable to those observed in adult spindles (Walro and Kucera, 1985). Thus, the process of differentiation of intrafusal fibers must be advanced enough in 4dp spindles to include determinants for the approximate number of endings on each fiber type. However, motor endings of 4dp spindles were located much closer to the equator and in a more narrow zone of intrafusal fibers than in adult spindles. Motor endings may be displaced further from the equator as a consequence of the intercalation of new plasma membrane with its overlying basal lamina as intrafusal fibers grow in length.

Bundles of motor axons innervate perinatal spindles, whereas individual motor axons innervate adult spindles (Walro and Kucera, 1985). Moreover, single motor endings were innervated by multiple axons in immature spindles as evidenced by termination of entire motor axon bundles within single intrafusal endings. Hence retraction of supernumerary motor axons from spindles and intrafusal motor endings must accompany maturation of the fusimotor system. This retraction of axons might be associated with a degree of reorganization of motor connections because the distribution of motor bundles among the immature intrafusal fibers was less specific than reported for adult spindles (Walro and Kucera, 1985). Elimination of axons from intrafusal motor synapses might represent a mechanism whereby individual axons adjust the number of spindles they innervate. In contrast, single afferents innervated spindles, and single axon terminals occupied sites of individual sensory endings in all three age groups. Rearrangement of sensory endings in the course of spindle development is limited to the elimination of most cross-terminals between intrafusal fibers.

Intrafusal fibers acquire motor innervation in the same sequence as they form – bag_2, bag_1, and chain. At each stage of development only the more mature myofibers possessed motor endings. These fibers had well-organized, densely packed myofibrils. Fibers in the myotube stage never carried motor endings, irrespective of whether motor axons were abundant within the intracapsular space. Hence differentiating intrafusal fibers must attain a certain stage of maturity characterized by the presence of relatively well-organized and densely packed myofibrils before they are innervated by motor axons.

The primary myotubes of the extrafusal muscle fibers begin their assembly prior to the arrival of motor axons, and can develop in the absence of motor innervation (Harris, 1981). In contrast, the secondary extrafusal myotubes receive motor innervation by a lateral transfer of

axon terminals from the primary myotubes at the earliest stages of
their assembly (Duxson et al., 1986), and cannot develop in the absence of
motor axons and muscle activity (Harris, 1981). Intrafusal fibers
resemble primary rather than secondary extrafusal myotubes in their
ability to assemble in the absence of motor innervation. Spindles of all
age groups contained nascent intrafusal fibers with myofilaments but no
motor endings. Moreover, cross-terminals of motor axons reflective of
lateral transfer of axon terminals from primary myotubes (bag_2) to
secondary myotubes (bag_1, chain) were not encountered. Intrafusal fibers
of spindles also differentiate in deefferented muscles (Zelená and Soukup,
1973); hence formation of spindles is independent of motor innervation.

The observation that all intrafusal fibers, including those derived
from secondary myotubes, received motor innervation relatively late in
their assembly casts doubt on the speculations (Milburn, 1973, 1984) that
motor axons play an important role alongside the sensory innervation in
the early assembly and differentiation of intrafusal fibers. Whether this
independence from motor innervation reflects a special inherent myogenic
potential of intrafusal myotubes, or whether it stems from the innervation
of spindles by sensory axons remains to be determined.

REFERENCES

Duxson, M.J., Ross, J.J., and Harris, A.J., 1986, Transfer of differen-
    tiated synaptic terminals from primary myotubes to new-formed muscle
    cells during embryonic development in the rat, _Neurosci. Lett.,_
    71:147-152.
Kucera, J., Dorovini-Zis, K., and Engel, W.K., 1978, Histochemistry of rat
    intrafusal muscle fibers and their motor innervation, _J. Histochem._
    Cytochem., 26:971-988.
Landon, D.N., 1972, The fine structure of equatorial regions of developing
    muscle spindles in the rat, _J. Neurocytol._, 1:189-210.
Milburn, A., 1973, The early development of muscle spindles in the rat, _J._
    Cell Sci., 12:175-195.
Milburn, A., 1984, Stages in the development of cat muscle spindles,
    J. Embryol. exp. Morph., 82:177-216.
Thompson, W.J., 1986, Changes in the innervation of mammalian skeletal
    muscle fibers during postnatal development, _Trends in Neurosci._,
    9:25-28.
Walro, J.M., Kucera, J., 1985, Motor innervation of intrafusal fibers
    in rat muscle spindles: Incomplete separation of dynamic and static
    systems, _Am. J. Anat._, 173:55-68.
Zelená, J., 1957, The morphogenetic influence of innervation on the
    ontogenetic development of muscle spindles, _J. Embryol. exp._
    Morphol., 5:283-292.
Zelená, J., and Soukup, T., 1973, Development of muscle spindles
    deprived of fusimotor innervation, _Z. Zellforsch. Mikrosk. Anat._,
    144:435-452.

THE MOTOR INNERVATION OF NEWBORN KITTEN MUSCLE SPINDLES

M.H. Gladden[+] and A. Milburn[*]

[+]Institute of Physiology, University of Glasgow, Glasgow, UK
[*]Department of Zoology, University of Durham, Science
  Laboratories, South Road, Durham, UK

INTRODUCTION

In neonatal kittens an appreciable number of soleus spindle afferents
did not respond when the ventral root filaments were stimulated at strengths
sufficient to recruit fusimotor axons (Gregory and Proske, 1986). This may
imply that the fusimotor innervation is not yet functionally developed, as
Gregory and Proske suggest. However, Milburn (1984) did find some fusimotor
endings in peroneal muscles of newborn kittens which had some of the postsynaptic features of fusimotor endings in adult cats (see Boyd and Gladden,
1985, Part 1 review). We are presently engaged in a more systematic review
of the development of fusimotor innervation, using the tenuissimus muscle.
By birth in these muscles also post-junctional elaboration was already well
advanced at some fusimotor endings, while post-junctional folding was
absent or rudimentary at extrafusal motor endings.

In newborn kittens extrafusal motor end plates are still polyneuronally
innervated (Bagust, Lewis and Westerman, 1973). In contrast we found that
polyneuronal innervation could be ruled out at least for a proportion of
intrafusal motor end plates.

METHODS

Tenuissimus muscles from three newborn kitten littermates were
processed for light and electron microscopy, and for immunohistochemistry.
Muscles for light microscopy were stained with silver (Gladden, 1970), and
14 muscles teased from them. One silver-stained spindle was embedded in
resin and serially sectioned for light microscopy to trace the axons and
intrafusal fibres. The muscles for electron microscopy were stained
conventionally (see Milburn, 1984). A pole of one spindle was transversely
sectioned serially, ultrathin sections being taken at 5 µm intervals.

There was no difficulty in distinguishing nuclear chain (nc) from
nuclear bag fibres in transverse sections because of their distinctly
smaller diameter. In each spindle the static bag_2 (Sb_2) fibre was closely
associated with the nc fibres in the equatorial region, while the dynamic
bag_1 (Db_1) fibre lay apart from this grouping (Fig. 1a) as noted by Barker
and Milburn (1981). The prominent ring of elastic fibres which in adult
cat spindles identifies the Sb_2 fibres in transverse sections of the polar

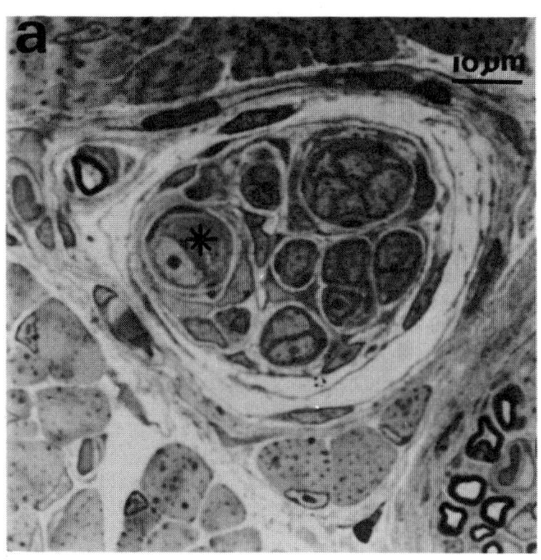

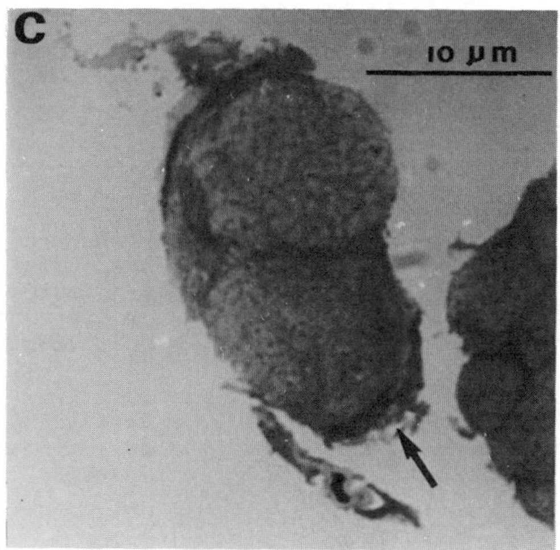

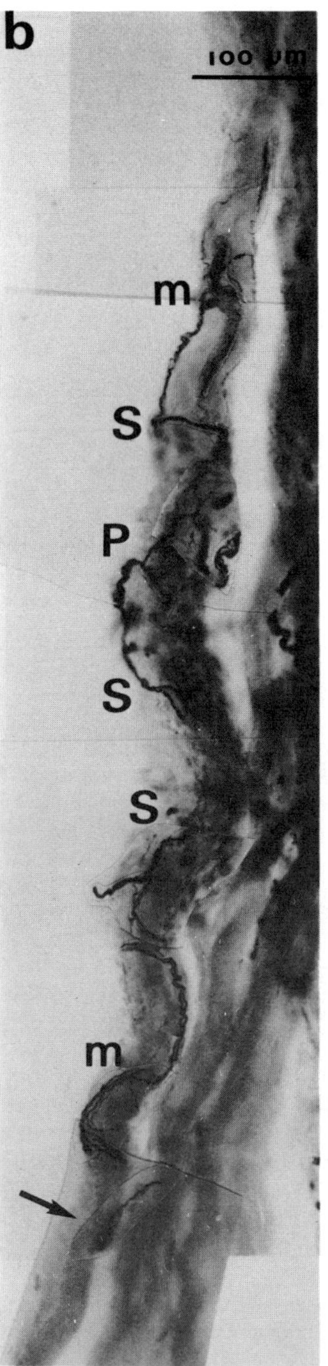

Fig. 1   (a) T.S. of equatorial region of 1 day old kitten tenuissimus
muscle spindle; Db_1 fibre(*) lies apart from Sb_2 and nc fibres.
(b) Teased silver-stained spindle with a primary (P) and three
secondary (S) sensory endings. Fusimotor axons and endings (m).
Arrow indicates motor ending shown in (c).
(c) T.S. of two nuclear bag fibres, motor ending present on lower
with axon terminal branches shown by arrow.

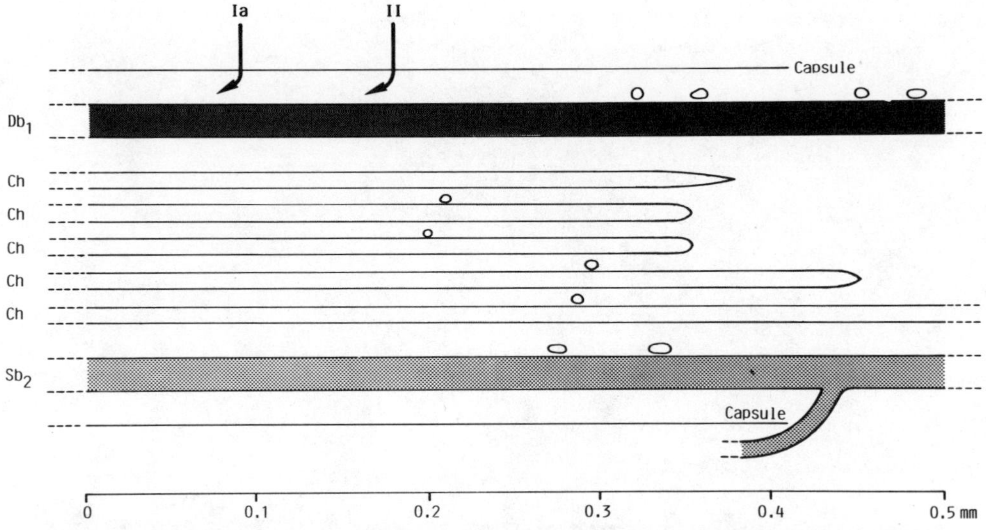

Fig. 2   Reconstruction of spindle pole from serial transverse sections.
Circles and ovals indicate the position of endings, but not their
nature.

regions (Gladden, 1976) was not yet sufficiently developed in these young
animals.

The immunohistochemical methods and results have been described elsewhere (Rowlerson, 1987a&b).

RESULTS

The motor innervation of all poles of the silver-stained spindles
could not be completely analysed because of the complexity of innervation,
overlying nerve trunks, overstaining and occasional dissection damage. Nevertheless, it was clear that all poles did have a motor innervation. One
to five motor axons supplied each pole, each axon having between one and
four endings. The sensory innervation consisted of a primary, and zero to
three secondary sensory endings. Autonomic axons, which were also present,
were of smaller diameter and did not terminate in distinct endings, unlike
the fusimotor axons.

An example of a fusimotor ending for which polyneuronal innervation
could be unequivocally excluded is indicated by an arrow in Fig. 1b. A
cross section of this terminal is shown in Fig. 1c. The large black dot
below one bag fibre is the axon, while the finer diameter lines more closely
applied to the fibre (arrows in Fig. 1c) are the terminal branches of the
axon.

Reconstruction of the spindle pole sectioned for electron microscopy
(Fig. 2) showed that the Db_1 fibre had four endings, more than would be
expected in an adult cat tenuissimus spindle. The number of motor endings
on the Sb_2 and nc fibres was within the normal adult range. It has not
been possible to trace the unmyelinated axons supplying these endings
through successive sections with confidence, and so determine whether any
polyneuronal innervation exists in this spindle.

The axon terminals lay superficially in both fusimotor endings on the
Sb_2 fibre. Post-junctional folding was absent or rudimentary as in adult
cat spindles. Three of the endings on the Db_1 fibre were superficial,

59

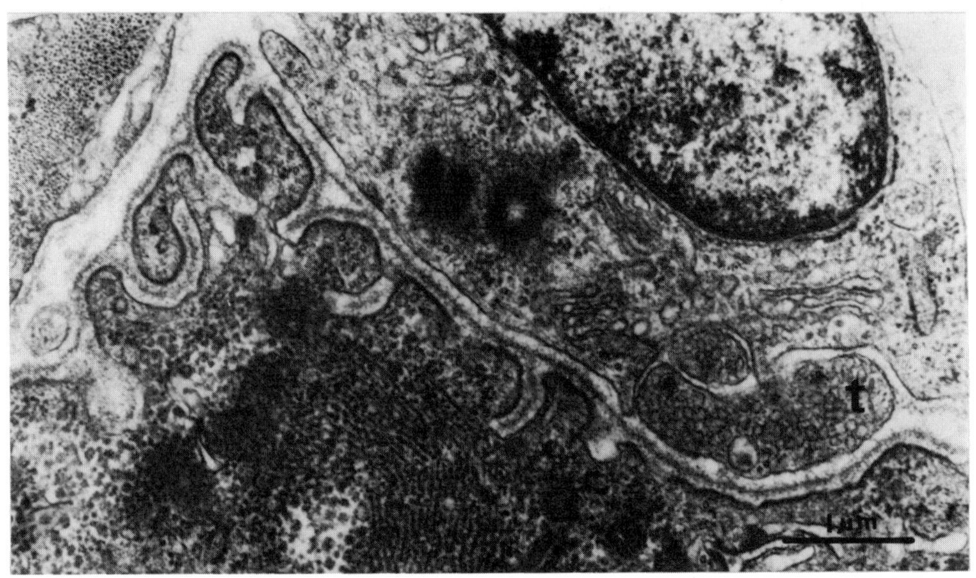

Fig. 3 Motor ending on nc fibre. t: axon terminal.

whilst one was markedly indented. Two of the endings on nc fibres were
similar to those on the Sb_2 fibres. The two other endings on the nc fibres
had post-junctional surfaces with deep, narrow folds, even in areas not
directly beneath the axon terminals (Fig. 3). These resembled the "Mc"
endings thought by Arbuthnott et al., (1981) to comprise a distinct group
of fusimotor endings on nc fibres. None of the extrafusal motor endings
encountered in the tenuissimus muscle had the extensive subjunctional
folding characteristic of adult cat muscle.

DISCUSSION

Clearly the development of the intrafusal motor innervation in newborn
kittens is more advanced than has generally been assumed. The failure of
spindle afferents to respond to fusimotor stimulation at this young age
(Gregory and Proske, 1986) may be due to functional immaturity of the
afferents, rather than the efferents. However structural maturity cannot
necessarily be equated with functional maturity, and possibly the intrafusal muscle fibres do not contract when fusimotor axons are stimulated.

The support of the Medical Research Council for this work is gratefully
acknowledged.

REFERENCES

Arbuthnott, E. R., Ballard, K. J., Boyd, I. A., Gladden, M. H., and
    Sutherland, F. I., 1982, The ultrastructure of fusimotor nerve endings
    in the cat and their relationship to the foci of local contraction in
    intrafusal muscle fibres, J. Physiol., 331:285-309.
Bagust, J., Lewis, D. M., and Westerman, R. A., 1973, Polyneuronal inner-
    vation of kitten skeletal muscle, J. Physiol., 229:241.
Barker, D., and Milburn, A., 1982, Development of cat muscle spindles,
    J. Physiol., 325:85P.
Boyd, I. A., and Gladden, M. H., 1985, Part 1 review, in: "The Muscle
    Spindle", I. A. Boyd and M. H. Gladden, eds., Macmillan, London.

Gladden, M. H., 1970, A modified pyridine-silver stain for teased preparations of motor and sensory nerve endings in skeletal muscle, Stain Technol., 45:161.

Gladden, M. H., 1976, Structural features relative to the function of intrafusal muscle fibres in the cat, Prog. in Brain Res., 44:51.

Gregory, J. E., and Proske, U., 1986, Fusimotor axons in the kitten, J. Neurophysiol., 56:1462.

Milburn, A., 1984, Stages in the development of cat muscle spindles, J. Embryol. exp. Morphol., 82:177.

Rowlerson, A., 1987a, Type-differentiation of intrafusal fibres during muscle development in the cat, J. Physiol., in press.

Rowlerson, A., 1987b, Early type-differentiation of intrafusal fibres, this vol.

DEVELOPMENTAL ASPECTS OF MUSCLE STRETCH RECEPTOR FUNCTION
IN THE RAT AND THE CAT

R. Vejsada[1], P. Hník[1], L. Jami[2], R. Payne[3]
and D. Zytnicki[2]

[1]Institute of Physiology, Czechoslovak Academy
of Sciences, Prague, Czechoslovakia
[2]Laboratoire de Neurophysiologie, Collège de
France, Paris, France
[3]Department of Physiology, Marischal College
University of Aberdeen, Aberdeen, Scotland

INTRODUCTION

It was already Skoglund in 1960 who pointed out that the
functional properties of muscle spindles of young kittens differ in several aspects from those of adult cats. One of the most
conspicuous features in new-born kitten spindles was their phasic activity upon a ramp-and-hold stretch of the parent muscle.
According to Skoglund (1960 b), this fast-adapting response prevailed in gastrocnemius muscle spindles until 6 - 10 days after
birth.

In recent years, interest in the functional maturation of
muscle stretch receptors was revived and spindle function has
been studied in more detail in rat pups (Vejsada et al., 1985a)
and kittens (Gregory and Proske, 1985; Jami et al., 1986a,b;
Gregory and Proske, 1986, 1987; Proske and Gregory, this volume).
We report here on the main characteristics of the passive spindle response to stretch in these two species, combining the
results obtained in the Prague laboratory (Vejsada et al.,
1985a) and at Collège de France (Jami et al., 1986a,b).

METHODS AND RESULTS

Development of function of muscle receptors in rats

The experiments were performed on 72 rat pups, one to 19
days old, anesthetized with urethane (2.2 mg/g i.p.). Most of
the hind limb muscles were denervated except for the gastrocnemius and soleus. The rat was placed in a warmed pool of paraffin oil and the rectal temperature and heart rate were monitored
throughout the experiment. Ramp-and-hold stretches lasting 5 s
were applied to the triceps surae muscle through the Achilles
tendon, using a manually operated puller. Afferent activity was
recorded in dorsal root filaments. Only a few units in the

youngest rats showed a typical spindle pause in firing upon
muscle twitch and most stretch receptors thus remained uniden-
tified. However, numerous examples of clearly pausing receptors
in 5-day-old and older rats permitted us to assume that spindle
endings prevailed over tendon organ endings in our random samples
in rats of all the age groups studied.

The prime objective of these experiments was to ascertain
at which age the muscle stretch receptors become capable of a
maintained, slowly adapting response to stretch. We therefore
classified the receptors as being either slowly adapting (SA),
if they discharged till the end of the hold phase of at least
one of the stretches applied, or rapidly adapting (RA), if firing ceased before the end of each stretch. Some receptors gave
only a very phasic "on" response and were pooled into a separate group.

Using these criteria, we could show that even in 1-day-old
rats, twelve units out of the 23 receptors found (52 %) exhibited a SA response and 10 units (44 %) adapted rapidly to all
the stretches, with one unit of the "on" type of response. The
percentage of SA receptors increased markedly with age and by
the 10th day of life, almost 90 % of all units behaved as slowly
adapting receptors and only 10 % were still relatively rapidly
adapting. Not only the ratio between SA and RA stretch receptors
changed in favor of the former but also the threshold to stretch
decreased during the early postnatal period. The unit in Fig. 1A
is from a 3-day-old rat and required a 1 mm stretch (which was
almost 10 % of the resting length of the triceps surae muscle)
to maintain its firing. However, only a 0.5 mm stretch (less
than 3 % of muscle length) was sufficient to elicit a SA response in a 10-day-old animal (Fig. 1B). Nevertheless, the fall
of the firing frequency during the maintained phase of stretch
was relatively large in rats of all ages. The dynamic sensitivity, on the other hand, considerably increased with age, as
can be judged from the fact that there was a five-fold increase
of the dynamic index for stretches of the same amplitude between
the first and 18th postnatal day. The discharge frequencies at
the dynamic peak in rats 18 days old were comparable with those
in adult rats (cf. Hník and Lessler, 1973).

Development of muscle receptor function in kittens

The experiments were carried out on 23 kittens, five to
18 days old, anesthetized with Midatrene (1.2 ml/kg i.m., plus
additional i.v. doses during the experiment). Artificial ven-
tilation was introduced after intubation of the trachea, when
necessary. The heart rate and rectal temperature were monitored
and we were meticulous in maintaining the temperature in the
narrow optimum range of 37.5 - 38.2 °C by means of an electrically heated pad and by enwrapping the animal's body and hind
limbs in cotton wool. Discharges from muscle stretch receptors
were recorded in filaments of the L7 dorsal root. Spindle units
were identified by a pause in firing during muscle twitch and/or
by their dynamic sensitivity to stretch. In some experiments,
the hind limb was denervated as completely as possible leaving
the nerves to the whole triceps or the medial gastrocnemius
only, and the muscle, immersed in paraffin oil, was stretched
with its tendon attached to an electromagnetic puller.

In 11 kittens, including the youngest, we adopted a mo-

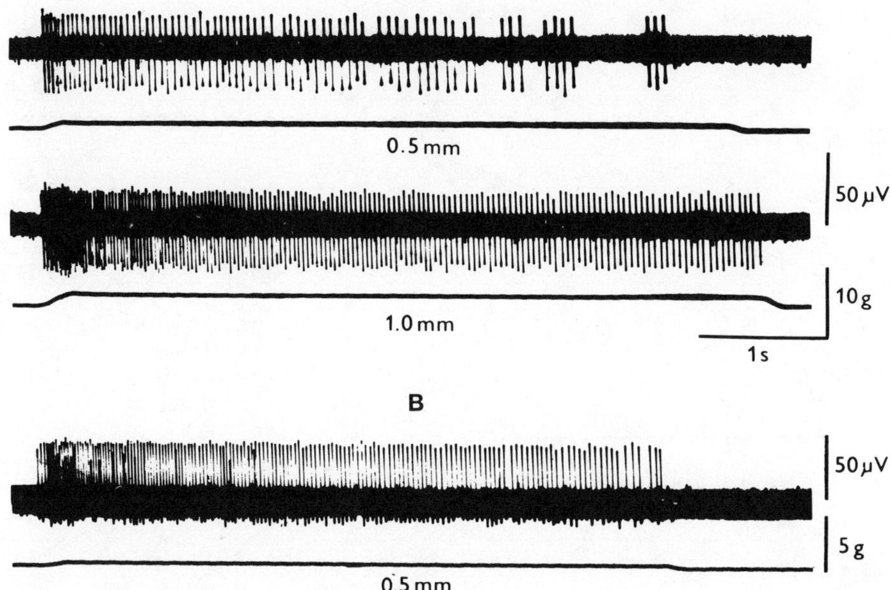

Fig. 1. Activity of muscle stretch receptors in young rats. A: a unit responding to a 0.5 mm and 1.0 mm stretch of the triceps surae muscle in a 3-day-old rat pup. B: a spindle unit in a 10-day-old rat. Time calibration is common for A and B. For further details see text.

dified procedure in order to ensure the maximum viability of the preparation. Virtually no surgery was done on the hind limb at the beginning of the experiment, except for cutting the femoral and obturator nerves. The limb was then fixed so as to make dorsiflexion in the ankle possible. Cotton wool wraps and an electrical heating plaquette were used to keep the limb warm (35 - 37 °C). Several receptor units were recorded that fired when the triceps surae muscle was stretched by ankle dorsiflexion of pre-set amplitudes. The hind limb was then gradually denervated and only stretch receptor units originating from the gastrocnemius or soleus muscle were included in the final score. At the end of the experiment, we usually verified the muscle spindle origin of the units employing the contraction test, and measured the actual stretch of the muscle resulting from ankle dorsiflexion of known magnitudes.

With this method we succeeded in recording receptors that responded by maintained firing to stretch in kittens 5 to 7 days old. These SA units had a low threshold, provided that the parent muscle had been left intact, and could respond to stretches as small as 0.2 - 0.5 mm. In general, however, the majority of stretch receptor units exhibited relatively rapid adaptation to maintained stretch until the second postnatal week.

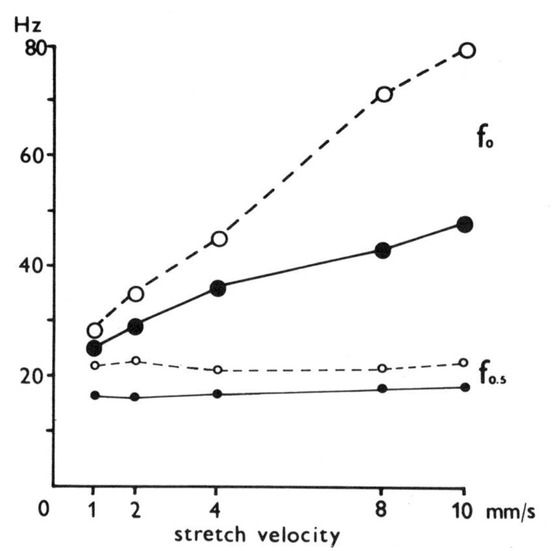

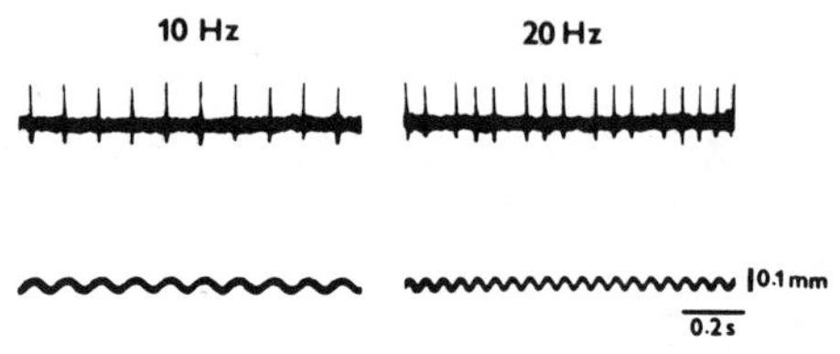

Fig. 3. Encoding of a 0.1 mm sinusoidal muscle stretch by a spindle ending in a 14-day-old kitten. The discharge was driven 1:1 at 10 Hz vibration but failures occurred at 20 Hz. The receptor gave a maintained response to a ramp-and-hold stretch and had conduction velocity 24 m/s.

Fig. 2. An example demonstrating the postnatal increase of the dynamic sensitivity of kitten muscle spindles. Open circles and broken line: 18-day-old kitten, full circles and solid line: 14-day-old kitten. Values of f_0 (large circles) represent the instantaneous firing frequency (in Hz, on the ordinate) at the end of the ramp phase of stretch. Values of $f_{0.5}$ (small circles) correspond to the maintained firing frequencies 0.5 second after completion of the ramp. Stretch velocity was varied from 1 mm/s to 10 mm/s (the abscissa). In both kittens, the triceps surae muscle was stretched by 2 % of its resting length. Note that the velocity-dependent dynamic response is enhanced in the older kitten, whereas the static sensitivity is similar in both cases.

It may be seen from Fig. 2 that also the dynamic sensitivity of kitten spindles increases with age (see also Proske and Gregory, this volume). When we applied sinusoidal stretch (Fig. 3), we found that the ability of spindles to monitor higher frequencies of vibration improved, while the threshold for stretch decreased in older kittens. Yet, 2 - 3 weeks after birth, the threshold for a phase-locked driving was still much higher (100 - 300 µm) and the maximum vibration frequency monitored by the receptor was considerably lower (50 - 150 Hz) when compared with the vibration sensitivity of spindles in adult cats.

DISCUSSION

Comparison of morphological vs. functional development of muscle spindles in the cat and the rat

In Fig. 4, we have attempted to compare the main events of the structural development of muscle spindles in the calf muscles of the cat and the rat. It is evident from the figure that

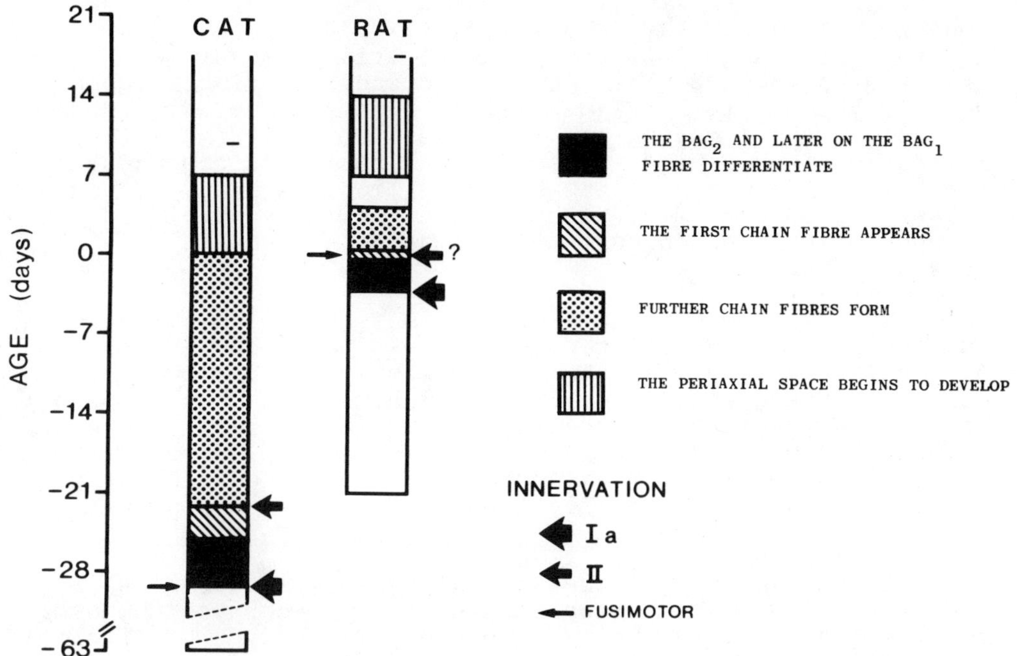

Fig. 4. A schematic comparison of the morphological development
of muscle spindles in the cat and the rat. Four stages
of spindle differentiation have been chosen and depicted
(see description on the right). The first appearance of
primary (Ia) and secondary (II) sensory innervation and
of fusimotor innervation is indicated by arrows. Age
(in seven-day intervals) is plotted on the vertical
axis. Days denoted by minus sign correspond to gestation
period, day 0 is the day of birth. Compiled from data
of Zelená, 1957; Landon, 1972; Milburn, 1973, 1984;
Barker and Milburn, 1984; Kucera and Walro, this volume.

the differentiation of spindles in the kitten already commences
approximately in the middle of gestation and that the full
adult complement of intrafusal fibers, as well as the sensory
and motor innervation of spindles appear before birth. As far
as their function is concerned, we confirmed Skoglund's (1960b)
original finding (as did, in this respect, Gregory and Proske,
1985, and Butler, this volume) that, for the most part, muscle
spindles in neonatal kittens respond phasically to stretch.
However, receptor units capable of maintained firing, including
those rarely occurring in the youngest animals, apparently have
a much lower threshold to stretch than seemed from Skoglund's
experiments. Since the dynamic responsiveness was also found
to be limited, we conclude that monitoring of muscle length and
its changes by kitten muscle spindles is relatively restricted
until the second to third postnatal week.

In the rat, on the other hand, the first gastrocnemius
muscle spindles start to differentiate only three days prenatally, on the average (see Fig. 4). At this age, the nascent spindles consist of one intrafusal myotube innervated by a primary

sensory axon. Yet, Kudo and Yamada (1985) could evoke a phasic
afferent discharge by stretching the triceps surae muscle in
rat fetuses 18 days old. The other stages of spindle morphological development are also attained later in the rat than in
the cat, so that during the first week after birth, spindle
maturation lags behind in the former species as compared with
the latter. In contrast to this assertion, however, a much
larger percentage of stretch receptor units capable of a tonic
response are to be found in rats than in kittens at an early
postnatal period.

It thus seems that the criteria of morphological maturity,
as they are schematically depicted in Fig. 4, are not adequate
for making a direct correlation possible. Other determinants of
spindle passive sensitivity to stretch, such as the membrane
properties involved in receptor potential generation, the degree
of maturity of sensory nerve terminals, or the mechanical properties and the mode of attachment of intrafusal fibers, are
likely to play a role in functional maturation of muscle spindles in these two species.

It has been postulated that kitten muscle spindles
(Skoglund, 1960b) and cutaneous and joint receptors (Ekholm,
1967) exhibited only phasic responses if supplied by afferents
conducting at less than 18 - 20 m/s. We found that, in kittens
of the same age, muscle stretch receptor units with a more maintained response often had a higher conduction velocity than the
phasic ones. In the new-born rat, however, most muscle afferents
are not yet myelinated and conduct at 1.4 m/s, on the average
(Vejsada et al., 1985b), and SA cutaneous afferents have conduction velocities less than 1 m/s (Fitzgerald, 1987). Yet, even
such immaturity does not apparently prevent the nerve fibers
from conveying trains of impulses. Furthermore, Ferrington and
Rowe (1980) reported that SA tactile afferents in new-born kittens had conduction velocities 4 to 8 m/s only. It is hence
evident that low conduction velocity of the sensory nerve does
not, by itself, restrict the receptor to an exclusively phasic
response.

It should be pointed out that the absence of the static
component of the response to stretch is not, of course, the only
criterion of the immaturity of spindle function. But the dynamic sensitivity of kitten spindles has also been shown to mature
relatively late after birth (see also Gregory and Proske, 1987;
Proske and Gregory, this volume). In our experiments in kittens
we even encountered examples suggesting that the dynamic and
static sensitivity of individual muscle spindles may develop
more or less independently of each other.

Physiological implications of the maturation of muscle receptors

The relatively early functional maturation of muscle
stretch receptors in the rat precedes the development of posture
and normal ambulation. The tonic stretch reflex and the basic
postural and locomotor reflexes do not appear until the 2nd to
3rd postnatal week in the rat hind limb muscles (Bursian, 1973;
Altman and Sudarshan, 1975). However, a phasic, monosynaptically
transmitted stretch reflex response can be evoked as early as
on embryonic day 19 (Kudo and Yamada, 1985) and locomotor activity can be induced by administration of biologically active

substances (L-DOPA, aspartate) in neonatal rats prior to the
normal onset of locomotion (Navarrete and Vrbová, 1985; Kudo
and Yamada, 1987). These findings indicate that it is the central part of the reflex arc which matures later and thus delays
(possibly due to synaptic immaturity) the onset of reflex mechanisms. This seems to be an analogous situation to that described by Fitzgerald (1987) concerning the postnatal development of the flexion reflex evoked by nociceptive stimulation
in rats.

In the cat, on the other hand, the gradual improvement of
the functional properties of muscle stretch receptors seems to
coincide with the development of normal posture and locomotion.
Reflexes subserving postural and locomotor functions first appear during the second and third postnatal week (Skoglund,
1960a). According to Sechzer et al. (1984), the hind limb postural reflex mediated by muscle proprioceptors does not attain
its adult-like quality until 35 - 40 days after birth. Again,
as in the rat, a monosynaptically transmitted ventral root discharge in response to stimulation of muscle afferents can be
evoked even before birth (Naka, 1964).

CONCLUSIONS

It may thus be concluded that it is not possible, as yet,
to correlate directly the development of muscle spindle function
during the early postnatal period in kittens and rats with the
literary data about the morphological maturation of these receptors in the two species.

Either the response characteristics employed here (i.e.
designation of the receptors as slowly or rapidly adapting on
the basis of their passive response to stretch) are not an adequate criterion for assessing the degree of functional maturity,
or finer morphological details of spindle development should be
sought.

REFERENCES

Altman, J., and Sudarshan, K., 1975, Postnatal development of
        locomotion in the laboratory rat, Anim. Behav., 23:896-
        -920.
Barker, D., and Milburn, A., 1984, Development and regeneration
        of mammalian muscle spindle, Sci. Progr. (Oxford), 69:
        45-64.
Bursian, A.V., 1973, The development of stretch reflex in grow-
        ing rats (in Russian, with English summary), Zh. evol.
        biokhim. fiziol., 9:600-605.
Butler, R., Functional maturation of muscle spindles in the
        tenuissimus muscles of kittens (this volume).
Ekholm, J., 1967, Postnatal changes in cutaneous reflexes and
        in the discharge pattern of cutaneous and articular
        sense organs, Acta physiol. scand., Suppl. 297:1-130.
Ferrington, D.G., and Rowe, M.J., 1980, Functional capacities
        of tactile afferent fibres in neonatal kittens, J. Phy-
        siol. (London), 307:335-353.
Fitzgerald, M., 1987, Cutaneous primary afferent properties in
        the hind limb of the neonatal rat, J. Physiol. (London),
        383:79-92.

Gregory, J.E., and Proske, U., 1985, Responses of muscle receptors in the kitten, J. Physiol. (London), 366:27-45.
Gregory, J.E., and Proske, U., 1986, Fusimotor axons in the kitten, J. Neurophysiol., 56:1462-1486.
Gregory, J.E., and Proske, U., 1987, Responses of muscle receptors in the kitten to succinyl choline, Exp. Brain Res., 66:167-174.
Hník, P., and Lessler, M.J., 1973, Changes in muscle spindle activity of the chronically de-efferented gastrocnemius of the rat, Pflügers Arch., 341:155-170.
Jami, L., Vejsada, R., and Zytnicki, D., 1986a, Muscle stretch receptors in kittens, Neurosci. Lett., Suppl. 26:S563.
Jami, L., Vejsada, R., and Zytnicki, D., 1986b, Récepteurs musculaires répondant à l´étirment chez le chaton, J. Physiol. (Paris), 81:8A.
Kucera, J., and Walro, J.M., Innervation of immature muscle spindles in the rat (this volume).
Kudo, N., and Yamada, T., 1985, Development of the monosynaptic stretch reflex in the rat: an in vitro study, J. Physiol. (London), 369:127-144.
Kudo, N., and Yamada, T., 1987, N-Methyl-D,L-aspartate-induced locomotor activity in a spinal cord-hindlimb muscles preparation of the newborn rat studied in vitro, Neurosci. Lett., 75:43-48.
Landon, D.N., 1972, The fine structure of the equatorial regions of developing muscle spindles in the rat, J. Neurocytol., 1:189-210.
Milburn, A., 1973, The early development of muscle spindles in the rat, J. Cell Sci., 12:175-195.
Milburn, A., 1984, Stages in the development of cat muscle spindles, J. Embryol. exp. Morphol., 82:177-216.
Naka, K.-I., 1964, Electrophysiology of the fetal spinal cord. I. Action potentials of the motoneuron, J. gen. Physiol., 47:1003-1022.
Navarrete, R., and Vrbová, G., 1985, Precocious activation of locomotor activity by L-DOPA in the neonatal rat, Neurosci. Lett., Suppl. 22:S596.
Proske, U., and Gregory, J.E., The dynamic sensitivity of muscle spindles in the kitten (this volume).
Sechzer, J.A., Folstein, S.E., Geiger, E.H., Mervis, R.F., and Meehan, S.M., 1984, Development and maturation of postural reflexes in normal kittens, Exp. Neurol., 86:493-505.
Skoglund, S., 1960a, On the postnatal development of postural mechanisms as revealed by electromyography and myography in decerebrate kitten, Acta physiol. scand., 49:299-317.
Skoglund, S., 1960b, The activity of muscle receptors in the kitten, Acta physiol. scand., 50:203-221.
Vejsada, R., Hník, P., Payne, R., Ujec, E., and Paleček, J., 1986a, The postnatal functional development of muscle stretch receptors in the rat, Somatosensory Res., 2: 205-222.
Vejsada, R., Paleček, J., Hník, P., and Soukup, T., 1986b, Postnatal development of conduction velocity and fibre size in the rat tibial nerve, Int. J. Devl. Neurosci., 3: 583-595.
Zelená, J., 1957, Morphogenetic influence of innervation on the development of muscle spindles, J. Embryol. exp. Morphol., 5:283-292.

FUNCTIONAL MATURATION OF MUSCLE SPINDLES IN THE TENUISSIMUS

MUSCLES OF KITTENS

Richard Butler

Department of Anatomy
McMaster University
Hamilton, Ontario, Canada

FUNCTIONAL DEVELOPMENT OF RECEPTORS

Almost thirty years ago, Skoglund (1960) published the results of a
series of experiments in which he was able to assess the acquisition of
physiological function in maturing gastrocnemius muscle spindles of
kittens from birth to 45 days of age. He found that all stretch
receptors responded phasically in newborn and young kittens. It was not
until 6-10 days after birth that muscle spindles began to develop tonic
discharges in response to a stretch and hold stimulus. At 17 days of
age, the effect of gamma motoneurons on spindle discharges was
demonstrable and Skoglund presented evidence for an earlier alpha
innervation of the spindles.

The late fetal (55+ days gestation) and early neonatal stages of
development of intrafusal muscle fibers in the tenuissimus muscle of
kittens can be visualized as two anatomically and histochemically
distinct groups (Butler, 1980). One group consists of a single nuclear
bag fiber with an ATPase profile characteristic of slow twitch muscle.
The second group consists of a second nuclear bag fiber in close
association with the nuclear chain fibers and each member of this group
exhibits ATPase activity characteristic of fast twitch muscle. Complete
dehiscence of these groups within maturing spindles required 4-5 weeks.
Milburn (1984) studied the development of spindles in the peroneal
muscles of fetal kittens (34+ days gestation) and proposes a pattern of
assembly which is not consistent with the groupings that I proposed for
tenuissimus spindles.

A study of the development of physiological function in maturing
tenuissimus muscle spindles was carried out in kittens of post-natal
ages 1-45 days and at 4 months. Cats were bred in the University's
animal quarters to ensure a regular supply of kittens. Muscle spindle
afferent responses were recorded in isolated dorsal root filaments in
response to linear stretch of the muscle during selective fusimotor
stimulation. At the 6-8 day stage, it was usual to find mostly phasic
components in the afferent response to stretch. Beyond this, increasing
levels of tonic activity can be recorded with increasing age. Between
19-22 days of age fusimotor effects due to alpha motoneuron stimulation
can be recorded and at the same time gamma motoneuron effects are
present, although they are often not well developed. The afferent
discharge is irregular in these latter cases. It is around this time in

development that fast twitch and slow twitch muscle fibers are
differentiating their contraction characteristics (Buller et al., 1960)
and their non-ATPase histochemical profiles (Nystrom, 1968b). Beyond 25
days, and certainly by 35 days of age, gamma static and gamma dynamic
responses comparable to adults are present and these findings are
consistent with the motor behaviour of kittens of corresponding age.
The interesting point is that the maturation of spindle function
precedes the maturation of the central components of the motor system.

FUNCTIONAL DEVELOPMENT OF THE DESCENDING MOTOR SYSTEM IN THE CAT

     Bruce and Tatton (1980a,b) studied the development and maturation
of the descending motor system in kittens. They report that motor
cortical output to motoneurons innervating forelimb muscles first appear
at 37-45 days and that motor cortical responses to mechanoreceptors do
not appear until 55-65 days of age, i.e. after completion of the output
linkages. They also showed that motor cortical outputs to face,
forelimb and proximal hindlimb muscles appeared in a synchronous rather
than sequential manner across the area 4 homunculus. The same authors
(Bruce and Tatton, 1981) also showed that the major driving systems to
the spinal cord (cortical, rubral, reticular, and vestibular) are
anatomically present by 20 days of age but that at least one, the
corticospinal tract, is not effective at this stage in activating spinal
circuits which influence alphas. Muscle spindles mature functionally at
a rate whereby adult-like responses can be elicited (25 days) before any
descending excitation pathway is functional (37-45 days) and long before
any long-loop reflexes can be demonstrated by cortical recording (55-65
days). The appearance of static and dynamic spindle responses (25-35
days) correlates in time with the appearance of a bimodal distribution
of axon fiber diameters to muscles (10-35 days; Nystrom, 1968a) and the
appearance of a bimodal distribution of motoneuron cell body sizes
(19-24 days; Tatton et al., 1984).

REFERENCES

Bruce, I.C., and Tatton, W.G., 1980a, Sequential output-input
     maturation of kitten motor cortex, Exp. Brain Res., 39: 411-419.
Bruce, I.C., and Tatton, W.G., 1980b, Synchronous development of motor
     cortical output to different muscles in the kitten. Exp. Brain
     Res., 40: 349-353.
Bruce, I.C., and Tatton, W.G., 1981, Descending projections to the
     cervical spinal cord in the developing kitten, Neuroscience
     Letters, 25: 227-231.
Buller, A.J., Eccles, J.C., and Eccles, R.M., 1960, Differentiation of
     fast and slow muscles in the cat hind limb, J. Physiol., 150:399-416.
Butler, R., 1980, The organization of muscle spindles in the
     tenuissimus muscle of the cat during late development, Develop.
     Biol., 77: 191-212.
Milburn, A., 1984, Stages in the development of cat muscle spindles,
     J. Embryol. exp. Morph., 82: 177-216.
Nystrom, B., 1968a, Fibre diameter increase in nerves to "slow-red" and
     "fast white" cat muscles during postnatal development, Acta
     Neurol. Scandinav., 44: 265-294.
Nystrom, B., 1968b, Histochemistry of developing cat muscles, Acta
     Neurol. Scandinav., 44: 405-439.
Skoglund, S., 1960, The activity of muscle receptors in the kitten,
     Acta physiol. scand., 50: 203-221.
Tatton, W.G., Hay, M., and Bruce, I.C., 1984, Postnatal growth of
     medial gastrocnemius motoneurons in the cat, Neurosci. Abst., 10:913.

THE DYNAMIC SENSITIVITY OF MUSCLE

SPINDLES IN THE KITTEN

U. Proske and J.E. Gregory

Department of Physiology, Monash University

Clayton, Victoria, Australia

This is a report on further studies of the dynamic responsiveness of kitten muscle spindles, previously found to be low. We show that the threshold to vibration is higher than in the adult and the responsive frequency range is smaller. The response to large stretches shows an apparent saturation, not seen in the adult, which may have its origin in the mechanical properties of the immature receptor.

We have been studying muscle receptors in kittens for several years, building on the pioneering observations of Skoglund (1960). We now know that muscle spindles in the new-born already have a functional fusimotor innervation and that within the first week of life fusimotor fibres can be identified as having either static or dynamic actions (Gregory and Proske, 1985, 1986). At birth many fusimotor axons are unmyelinated and the process of myelination takes several weeks to complete. There is a higher than normal incidence of skeleto-fusimotor or beta innervation suggesting that during development some motor terminals on intrafusal fibres will be eliminated, as is known to occur for extrafusal fibres (Bagust, et al.,1972).

The topic we would like to discuss in more detail here is the dynamic sensitivity of spindles. In the adult spindle it is the primary ending, with its terminals on the equatorial regions of nuclear bag and nuclear chain fibres, which has a high dynamic sensitivity. Secondary endings, lying in a more polar location on nuclear chain fibres, are predominantly length detectors and are rather insensitive to dynamic stimuli. This difference between the two kinds of endings is recognisable right from birth, both in the conduction velocity of the afferent fibres and the pattern of responses during muscle stretch (Gregory and Proske, 1987). However, while the response of the secondary ending in the kitten typically looks just like a scaled down version of the adult response, that of the primary ending differs in the shape of its dynamic component. Here it should be mentioned that the overall rates of firing of all kitten muscle receptors are much lower than in the adult, presumably because of the immaturity of the innervating axons.

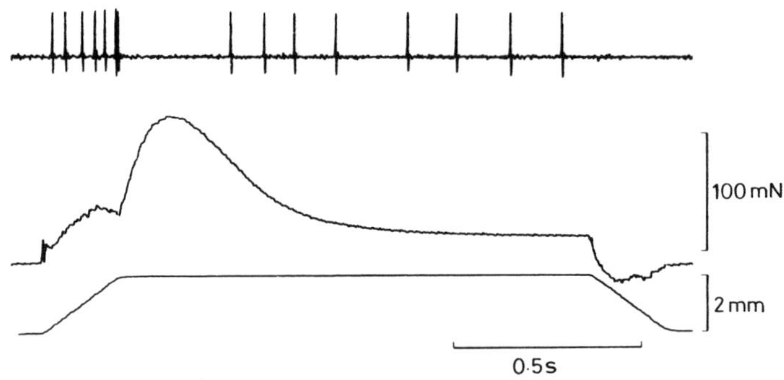

Fig. 1. Discharge (upper trace) of a soleus muscle spindle primary
ending in a 1 day old kitten during a ramp-and-hold stretch
(bottom trace) and muscle twitch elicited by stimulation of
the muscle nerve at the end of the ramp. Tension shown in
middle trace. Note that discharge is initiated by stretch
and silenced during twitch.

During our study of receptors in the new-born we encountered a
problem which Skoglund had already faced twenty-five years earlier. In
very young animals, muscle spindles have no resting discharge.
Furthermore, at muscle lengths up to the optimum for a whole muscle twitch
the response to stretch consists of only a few impulses during the length
change and there is no maintained response during the hold phase of
stretch. A hold response appears only at longer lengths. As a result it
is often quite difficult to determine what kind of receptor one is dealing
with. However, once a maintained response is obtained it is possible to
test whether or not the discharge can be unloaded by a muscle twitch and
thereby distinguish spindles from tendon organs (Fig.1).

Skoglund solved the problem of receptor identification by giving the
animal succinyl choline. At low concentrations this drug selectively
induces a resting discharge in primary endings of muscle spindles. Once a
receptor had acquired a resting discharge Skoglund was able to apply the
twitch test successfully. It is now known that succinyl choline exerts
its action by inducing a contracture in nuclear bag intrafusal fibres
(Gladden, 1976). One important consequence of this is that spindles not
only develop a resting discharge but they show a large increase in dynamic
response during muscle stretch (Rack and Westbury, 1966). When we tried
using succinyl choline we made the unexpected observation that in the
new-born, primary endings did not show a significant increase in dynamic
response (Gregory and Proske, 1987). Not until two or more weeks later
did a clear-cut increase begin to emerge. Here then was an aspect of
receptor responsiveness in the kitten which differed fundamentally from
the adult and an understanding of which might provide new insight into the
mechanism of generation of all dynamic responses. This conclusion forms
the basis of the series of experiments presented here.

Tne experiments were carried out on kittens aged between 1 and 14
days (80-300 g wt) and one adult cat (2.8 kg wt). Anaesthesia was

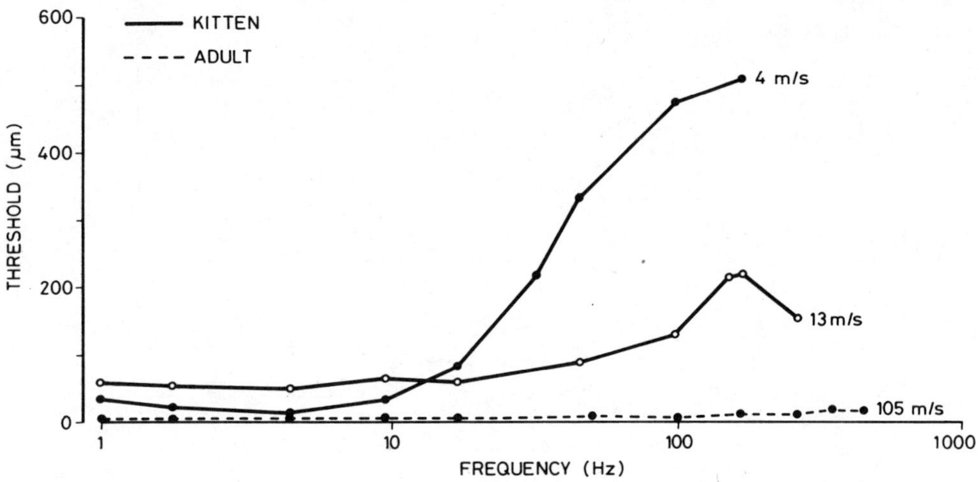

Fig. 2. Relation between threshold amplitude for a 1:1 response and
frequency of vibration plotted for a soleus spindle primary
ending (open circles) and a secondary ending (filled
circles) in a four day old kitten, and for one adult ending
(dashed line).   Conduction velocity of afferent shown at
the right of each curve. Muscle at maximum physiological
length (L$_{max}$).

induced with a ketamine-xylazine mixture injected intramuscularly, and
continued with intravenous pentobarbitone sodium, (40 mg/kg). The soleus
muscle and its nerve supply were dissected free of surrounding tissue and
the tendon of insertion tied to a muscle stretcher. Nerves to all other
muscles in the hind limb were cut.   Responses were recorded from
functionally single afferents isolated in dissected filaments of L$_7$ and
S$_1$ dorsal roots.   Primary endings of soleus spindles were identified by
the conduction velocity of their axons, the presence of a dynamic response
during a ramp-and-hold stretch and an unloading response during a muscle
twitch (Fig.1).

A simple and convenient way to measure the dynamic sensitivity of a
spindle is to use longitudinal vibration applied to the tendon (Brown et
al.,1967).   If the kitten spindle had an abnormally low dynamic
sensitivity, this should show up in the response to vibration. It was
found that the optimum conditions for a vibration response were obtained
only after the muscle had been stretched to its maximum body length
(L$_{max}$).   For the majority of the 27 receptors we tested with vibration
the response range was 1-200 Hz, the upper limit measurable being
restricted by the maximum excursion of the muscle stretcher (600 μm at
200 Hz).   Thresholds for a 1:1 response were plotted against vibration
frequency and this is shown in Fig.2. Here responses of three spindles
have been shown, that of a primary ending from a four day old kitten with
afferent conducting at 13 m/s, a secondary ending from the same animal
conducting at 4 m/s and an adult primary afferent conducting at 105 m/s.
Clearly the adult receptor responded to a much lower amplitude of
vibration and over a wider range of frequencies than any of the kitten
receptors.   The difference is particuarly striking at the higher

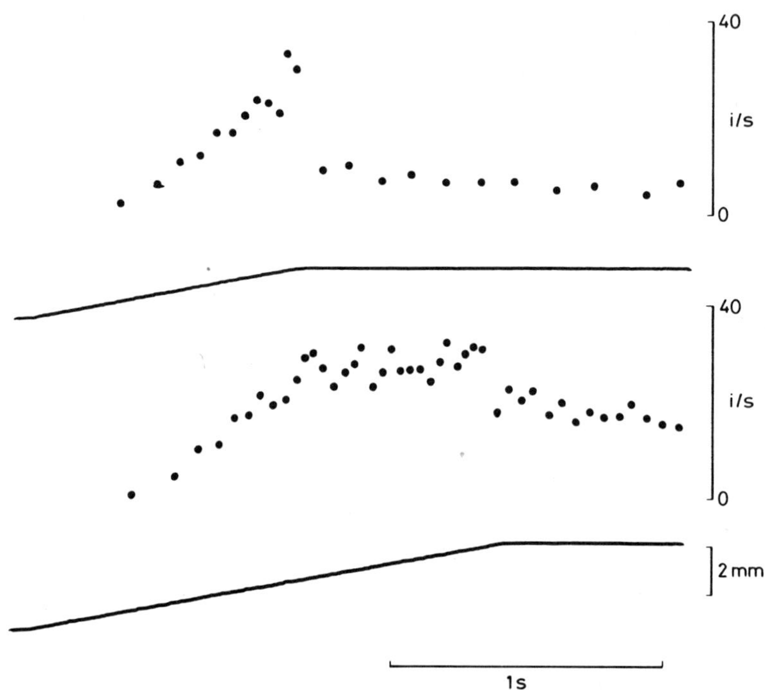

Fig. 3. Response of a soleus spindle primary ending in a 2 day old kitten to ramp-and-hold stretches of different amplitude (2.0 mm and 3.5 mm) but same rate (2 mm/s). Response shown as instantaneous frequency in upper trace in each panel, muscle length in lower. Initial muscle length L_{max} -2.5 mm.

frequencies. Throughout this part of the range the kitten primary ending had a significantly lower vibration threshold than the secondary ending, indicating that a difference in dynamic sensitivity of the two ending types was already established at this early stage.

The conclusion from this experiment is that kitten spindles do have some dynamic sensitivity although it is lower than in the adult. Whether this is due to properties of the transducing elements or to the axon itself remains uncertain. Certainly, rather long axonal refractory periods have been shown in the kitten (Skoglund, 1960). We conclude that, based on their vibration sensitivity, kitten spindles would be expected to show at least some increase in dynamic response in the presence of succinyl choline. The fact that they do not leads us to seek other explanations.

An incidental observation made during these experiments was that when the muscle was held at very long lengths, afferent firing during a stretch increased over only part of the length change and near the peak of stretch it reached a plateau, in a manner somewhat reminiscent of stretch responses in the presence of succinyl choline. It seemed possible that by stretching the muscle out to long lengths, or by infusing succinyl choline to produce an intrafusal contracture, the sensory region of the spindle was being pulled by the test stretch into a length range where dynamic

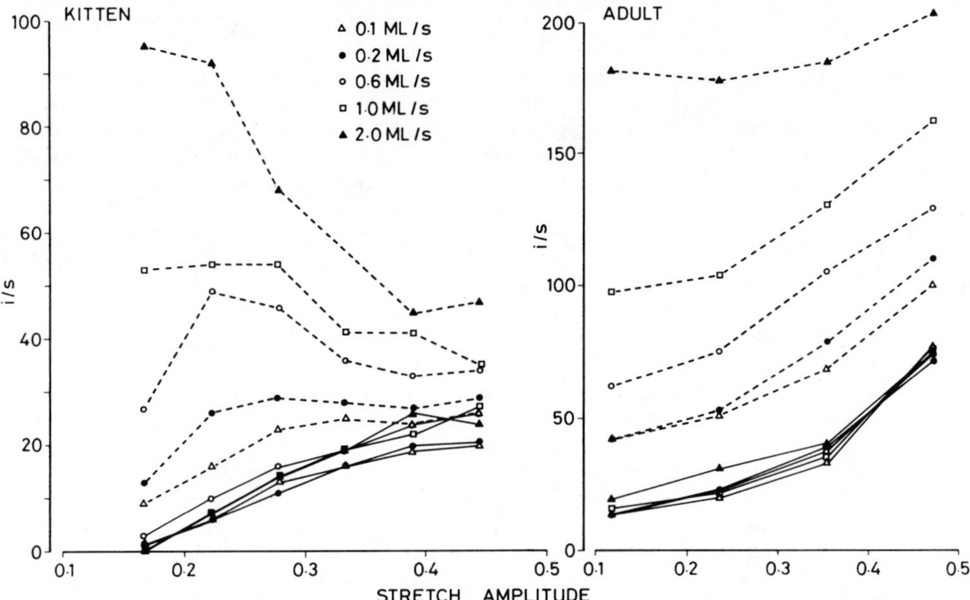

Fig. 4. For one soleus spindle primary ending in a 2 day old kitten and one adult ending, discharge rates during ramp-and-hold stretch at various rates versus stretch amplitude. Broken lines plot firing rate at end of ramp stretch at rates indicated, expressed as muscle fibre lengths/s. Amplitudes shown as fraction of fibre length, measured to be 9 mm for kitten and 34 mm for adult. Solid lines plot firing rate during hold phase of stretch, 0.5 s after end of ramp. Initial muscle length $L_{max}-2.5$ mm for kitten and $L_{max}-10$ mm for adult.

responses were saturating. A simple experiment showed that it was not a saturation in the encoder (axon).

An example of an apparently saturating response is shown in Fig.3. A spindle primary ending of a two day old animal responded to a stretch of 2 mm at 2 mm/s with an increase in firing throughout the length change. The response to a 3.5 mm stretch at the same rate, however, began to show signs of limiting after about 2 mm of length change. Notice that for the two responses peak firing during the length change is about the same while the maintained rate at the new length is significantly higher with the larger stretch.

This experiment was repeated systematically for a number of spindles and over a range of stretch rates. On each occasion as soon as stretch amplitude reached a little over 2 mm, firing rate began to limit. Another way of showing this was to plot the rate of firing at the end of a length change against stretch amplitude (Fig.4). Stretch rates and amplitudes have been normalised by expressing them in terms of muscle fibre lengths.

This has permitted a direct comparison with comparable stretches applied to an adult spindle. Here several points can be made. For both kitten and adult, firing rate measured at the end of stretch is higher when using faster stretches. This for the kitten emphasises the point that any saturation measured at lower stretch rates is not likely to reside in the encoder. Secondly, in adult spindles there is a monotonic increase in afferent discharge with stretch amplitude over the whole range studied. The slope of this relation, which represents the length sensitivity of the spindle, is little different whether measured under static (solid lines) or dynamic (dashed lines) conditions. For the kitten, the dynamic response is clearly limiting, reaching a plateau of about 30 impulses per second at lower stretch rates and actually falling from its initial higher value when using faster stretches. Although not clearly shown here we found that in the kitten the dynamically determined length sensitivity was much higher than that obtained from the static frequency-extension curve and at higher stretch rates it was much higher than for adult spindles.

To summarise our findings, we have observed an apparent saturation of the dynamic response of muscle spindles in the kitten when using large amplitude stretches. The saturation is not due solely to a limiting capability in rate of firing by the afferent axon. We have considered a possible mechanical model which may account for our findings. The dynamic response of the adult spindle is thought to arise from a sensory ending lying on an essentially elastic part of the intrafusal fibre, the nuclear bag, in series with a visco-elastic region, the striated intrafusal poles. If, perhaps because of a lack of elastic fibres around the immature intrafusal fibres, the sensory ending in the kitten is in series with a region which exhibits viscosity but little elasticity, responses like those described here can be generated. Firing rate would not increase with length beyond a certain point simply because the viscosity would begin to yield. Clearly, however, this kind of model goes only part way towards explaining our findings. It does not explain, for example, the fall in firing rate observed in the kitten with stretches at higher rates. But muscle cannot be modelled simply as a visco-elasticity. The presence of cross-bridges within sarcomeres is likely to give intrafusal fibres a stiction-like property and it is this which may help to account in the kitten for the high dynamic length sensitivity.

REFERENCES

Bagust, J., Lewis, D.M., and Westerman, R.A., 1973, Polyneuronal innervation of kitten skeletal muscle, J.Physiol., 229:241.
Brown, M.C., Engberg, I., and Matthews, P.B.C., 1967, The relative sensitivity of muscle receptors of the cat, J.Physiol., 192:773.
Gladden, M.H., 1976, Structural features relative to the function of intrafusal muscle fibres in the cat, Prog. Brain Res., 44:51.
Gregory, J.E., and Proske, U., 1985, Responses of muscle receptors in the kitten, J.Physiol., 366:27.
Gregory, J.E., and Proske, U., 1986, Fusimotor axons in the kitten, J. Neurophysiol., 56:1462.
Gregory, J.E., and Proske, U., 1987, Responses of muscle receptors in the kitten to succinyl choline, Exp. Brain Res., 66:167.
Rack, P.M.H., and Westbury, D.R., 1966, The effects of suxamethonium and acetylcholine on the behaviour of cat muscle spindles during dynamic stretching, and fusimotor stimulation, J.Physiol., 186:698.
Skoglund, S., 1960, The activity of muscle receptors in the kitten, Acta Physiol. Scand., 50:203.

POSTNATAL DEVELOPMENT AND AGING OF MUSCLE SPINDLES IN THE MOUSE

MASSETER MUSCLE AND EFFECTS OF A FINE-GRAINED DIET ON THEM

Norihiko Maeda, Koichi Osawa[x], Tamuro Masuda[x],
Tsuyoshi Kawasaki[xx] and Masayoshi Kumegawa

Departments of Oral Anatomy, [x]Oral Diagnosis and
[xx]Oral Surgery, Josai Dental University
Sakado, Japan

INTRODUCTION

Neuromuscular spindles of skeletal muscles of vertebrates
are encapsulated stretch receptors with structurally specialized
intrafusal muscle fibers, innervated with sensory and fusimotor
nerve endings (Barker and Cope, 1962; Boyd, 1962; Matthews,
1964; Barker et al., 1972). There are many detailed studies on
the development of muscle spindles in several animal species
and man (Sutton, 1915; Tello, 1917, 1922; Cuajunco 1927, 1940;
Hewer, 1935; Zelená, 1957, 1959, 1976; Mavrinskaia, 1960; Bowden, 1963; Schiaffino and Pierobon Bormioli, 1976). Those earlier
works showed that the most important event during the development of muscle spindles was the arrival of the sensory nerve
fibers into the muscle, and the event occurred in the early stage
of gestation. Muscle spindle formation is initiated by the primary afferent nerve contact with a bundle of developing myotubes. In some species, muscle spindle formation is completed
during the intrauterine period, but in rats the onset of muscle
spindle differentiation is observed to occur at a late stage of
gestation (Zelená, 1957, 1959; Marchand and Eldred, 1969; Landon, 1972; Milburn, 1973); and the formation of rat muscle spindles is completed early postnatally. On the other hand, the
postnatal development of rat masseter muscle, one of masticatory
muscles, continues for a comparatively long time period after
weaning (Maeda et al., 1981). We therefore investigated the
postnatal development of the masseter muscle and muscle spindles
of mice (Maeda et al., 1985), whose life cycle is similar to
that of rats. Moreover, we examined the effect of easily chewable diet on morphological and functional changes of muscle
spindles (Maeda et al., 1987; Maeda et al., in press).

MATERIALS AND METHODS

ICR strain mice were used throughout the present experiments.
Pregnant females were kept in separate cages at constant temperature (25 $^{\circ}$C) under alternating 12-hour periods of light and

dark. Newborn mice were kept with their mothers. Litters were reduced to 10 pups on day 3 or 5 after birth. Control mice were killed at various times from 5th to 425th days after birth. Masseter muscles from 5 animals at each time period were treated by the silver method by Barker and Ip (1963), and specimens were stored and cleared in glycerin. Thereafter, 2 or 3 whole specimens of masseter muscle were selected at random. Muscle spindles were dissected under a microtelescope and mounted in glycerin. Other specimens were frozen in liquid nitrogen and were cut into transverse serial sections of 12 μm thickness in a cryostat at -20 ^{o}C, and rapidly air-dried sections were treated for SDH by the method of Nachlas et al. (1957). Half of sections were stained with hematoxylin. Both types of treated sections were mounted in glycerin.

For analysis of effects of easily chewable diet on the masseter muscles, control mice were fed a solid diet (CA1; Clea, Japan Inc., Tokyo, Japan), and experimental animals were given a fine-grained CA1 powder (20 μm particle diameter) from days 20 and 365 after birth. Groups of animals were killed under anesthesia on the 40th, 60th, 90th, 120th, 150th, 180th and 500th day after birth. At each time period, one of the masseter muscles from each animal was treated by the silver method of Barker and Ip (1963) and dissected, and another of them was cut into serial sections of 12 μm thickness in a cryostat at -20 ^{o}C. And sections were also treated for SDH and hematoxylin, and mounted in glycerin.

All types of sections were observed under an Olympus BHC light microscope. The diameters of intrafusal and extrafusal muscle fibers, and primary endings were measured with a micrometer (OSM; Olympus, Japan). From each of three animals, measurements were made on 30-40 cross-sections of intrafusal muscle fibers, on 100 extrafusal muscle fibers with each type of SDH activity and on 124-156 primary endings. Moreover, muscle spindles were classified into 4 types according to the designation described by Maeda et al. (1985). Data concerning diameters were analyzed by the t-test and other data were analyzed by the χ^2.

RESULTS

The postnatal changes in the histochemically determined SDH activity and that of the development of Ia fiber endings showed time courses similar to each other. Before the weaning period, there were a few muscle spindles with complete annulospiral endings of Ia fibers, and the differences of the SDH activity were not observed among intrafusal muscle fibers. With growth, the primary endings of Ia fibers in mice began to develop spiral endings in many muscle spindles, and intrafusal muscle fibers began to have different SDH activities. After weaning, 3 types of intrafusal muscle fibers were detectable on the basis of SDH activity. And by the 40th postnatal day, almost all muscle spindles had completely coiled endings, while the formation of annulospiral endings still continued in some muscle spindles. By day 50 after birth, all muscle spindles had completely coiled endings, and continued to maintain complete coiled endings till the 120th postnatal day. However, after the 150th postnatal day, the ratio of muscle spindles with complete annulospiral endings decreased in the masseter muscle. In aged mice (500-day-old), only 34 % of muscle spindles had complete annulospiral endings in the masseter muscle.

In mice fed a fine-grained diet, decrease of diameters of
intrafusal muscle fibers and both red and white muscle
fibers were induced, and the ratio of muscle spindles with complete annulospiral endings and the diameter of primary endings
decreased significantly. SDH activity in mice fed the diet also
decreased in extrafusal and intrafusal muscle fibers.

DISCUSSION

In a series of studies concerning muscle spindle development, we found that the formation of muscle spindles continued after the weaning period in the mouse masseter muscle
(Maeda et al., 1985), although in almost all of earlier works
it is stated that spindle formation is completed during the
gestation period or in the early period after birth (Sutton,
1915; Cuajunco, 1927; Hewer, 1935; Cuajunco, 1940; Kalugina,
1956; Zelená, 1957, 1959, Mavrinskaia, 1960; Bowden, 1963; Marchand and Eldred, 1969). Additionally, parallel development
between the formation of annulospiral endings of the Ia fiber
and postnatal changes of SDH activity in intrafusal muscle fibers was found (Maeda et al., 1985). These facts indicate that
the postnatal development of muscle spindles depends on muscle
tension of masticatory muscle activated after weaning. In fact,
fine-grained diet suppressed the development of muscle spindles,
and induced precocious degeneration of muscle spindles in the
masseter muscle of developing and adult animals (Maeda et al.,
in press). Moreover, the atrophy of intrafusal muscle fibers
seems to be caused secondarily, but not directly, by the fine-
grained diet, because a significant decrease in the diameter
of intrafusal muscle fibers was observed 20 days after the reduction in the diameter of extrafusal muscle fibers (Maeda et
al., 1987). That is to say, the reduction in the diameter of
intrafusal muscle fibers could result from disuse atrophy of
extrafusal muscle fibers caused by the fine-grained diet. On
the other hand, in our study on mice, both red and white muscle
fibers were significantly affected by the intake of the fine-
grained diet (Maeda et al., 1987). Certainly, mice fed a fine-
grained diet would require a lower level of force for biting
and mastication, because such mice take in their food in much
the same way as they would lap milk from a dish. Accordingly,
red muscle fibers adapted for postural support and chewing may
undergo selective disuse atrophy in mice fed a fine-grained diet
as described for those fibers in the female edentulous monkey
by Maxwell et al. (1980). On the other hand, in contrast to primates, mice rapidly bite off solid food with their sharp upper
and lower incisors, then masticate the pieces of solid food with
their molars. Therefore, the white muscle fibers adapted for
forceful contraction also would show disuse atrophy due to the
lack of the rapid and forceful strokes by the incisors. Previously, since suppression of mastication inhibits the development
of the facial part of the skull in developing rats (Maeda et al.,
1986), the capacity of mastication may drop acceleratingly in
mice; and this drop in the capacity would result in more severe
atrophy of the masticatory muscles and their mechanoreceptor
organs, muscle spindles. If so, these phenomena must be a part
of a vicious circle. Moreover, degeneration of afferent nerve
fibers in many muscle spindles may lead to a disorder of the
mastication mechanism because of decreased sensory inputs from
the masseter muscle, resulting in the loss of normal tension in
the masseter muscle.

        From our present results, people in advanced countries,
who eat easily chewable food every day should be advised that
such practice, if taken to the extreme, may lead to precocious
aging of the masticatory system.

ACKNOWLEDGEMENTS

        This work was supported by a Grant-in-Aid for Special Re-
search of Mechanobiological Research on the Masticatory System
(No. 61134026) from the Ministry of Education, Science and Culture, Japan.

REFERENCES

Banker, B.O., Przybylski, R.J., Van der Meulen, N.P., and Vic-
        tor, M., 1972, "Research in Muscle Development and the
        Muscle Spindle", International Congress Series No. 240,
        Excerpta Medica, Amsterdam.
Barker, D., and Cope, M., (eds.), 1962, "Symposium on muscle
        receptors", Hong Kong University Press, Hong Kong.
Barker, D., and Ip, M.C., 1963, A silver method for demonstrat-
        ing the innervation of mammalian muscle in teased prepa-
        rations, J. Physiol., 169:73P.
Boyd, I.A., 1962, The structure and innervation of nuclear bag
        fiber system and the nuclear chain fiber system in mam-
        malian muscle spindles, Phil. Trans. Roy. Soc. B., 245:81.
Bowden, R.E.M., 1963, Muscle spindles in the human foetus, Acta
        Biol., 9:35.
Cuajunco, F., 1927, Embryology of the neuromuscular spindle,
        Contr. Embryol, 19:45.
Cuajunco, F., 1940, Development of the neuromuscular spindle in
        the human fetuses, Contr. Embryol., 28:97.
Hewer, E.E., 1935, The development of nerve endings in the human
        fetuses, J. Anat., 69:369.
Kalugina, M.A., 1956, On the question of the development of pro-
        prioceptors in the striated muscle of mammals. Arkh. Anat.
        Gistol. Embriol., 33:59.
Landon, D.N., 1972, The fine structure of the equatorial regions
        of developing muscle spindles in the rat, J. Neurocytol.,
        1:189.
Maeda, N., Amano, H., Machino, M., Kaneko, T., and Kumegawa, M.,
        1986, L-Thyroxine, cortisol, and diet affect the postnatal
        development of the facial part of the skull in developing
        rats, Anat. Anz., 161:99.
Maeda, N., Hanai, H., and Kumegawa, M., 1981, Postnatal develop-
        ment of masticatory organs in rats. I. Consecutive changes
        in histochemical properties and diameter of muscle fibers
        of the masseter superficialis, Anat. Anz., 149:319.
Maeda, N., Kawasaki, T., Osawa, K., Yamamoto, Y., Sumida, H.,
        Masuda, T., and Kumegawa, M., 1987, Effects of long-term
        intake of a fine-grained diet on the mouse masseter mus-
        cle, Acta Anat., 128:326.
Maeda, N., Kawasaki, T., Osawa, K., Sumida, H., Yamamoto, Y.,
        Masuda, T., Suwa, T., Watanabe, Y., and Kumegawa, M.,
        Effects of a fine-grained diet on annulospiral endings
        of Ia fibers in muscle spindles of mice, Anat. Anz., (in
        press).

Maeda, N., Osawa, K., Masuda, T., Hakeda, Y., and Kumegawa, M.,
     1985, Postnatal development of the annulospiral endings
     of Ia fibers in muscle spindles of mice, Acta Anat.,
     124:42.
Marchand, E.R., and Eldred, E., 1969, Postnatal increase of in-
     trafusal fibers in the rat muscle spindle, Exp. Neurol.,
     25:655.
Matthews, P.B.C., 1964, Muscle spindles and their motor control,
     Physiol. Rev., 44:219.
Mavrinskaia, L.F., 1960, On the relationship between the deve-
     lopment of the nerve endings of the skeletal muscles and
     the appearance of the movement activity in the human
     foetus, Arkh. Anat. Gistol. Embriol., 38:61.
Maxwell, L.C., McNamara, J.A. Jr., Carson, D.S., and Faulkner,
     J.A., 1980, Histochemistry of fibers of masseter and tem-
     poralis muscles of edentulous monkeys Macaca mulatta,
     Arch. Oral Biol., 25:87.
Milburn, A., 1973, The early development of muscle spindles in
     the rat, J. Cell Sci., 12:175.
Nachlas, M.M., Tsou, K.C., De Souza, E., Cheng, C.S., and Selig-
     man, A.M., 1957, Cytochemical demonstration of succinic
     dehydrogenase by the use of a new p-nitrophenyl substi-
     tuted ditetrazole. J. Histochem. Cytochem., 5:420.
Schiaffino, S., and Pierobon Bormioli, S., 1976, Morphogenesis
     of rat muscle spindles after nerve lesion during early
     postnatal development, J. Neurocytol., 5:319.
Sutton, A.C., 1915, On the development of the neuromuscular
     spindle in the extrinsic eye muscles of the pig, Am. J.
     Anat.. 18:117.
Tello, J.F., 1917, Genesis de las terminaciones nerviosas moto-
     rices y sensitivas, Trab. Lab. Invest. Biol. Univ. Madr.,
     15:101.
Tello, J.F., 1922, Die Entstehung der motorischen und sensiblen
     Nervendigungen, Z. Ges. Anat., 64:348.
Zelená, J., 1957, The morphogenetic influence of innervation on
     the ontogenetic development of muscle spindles, J. Em-
     bryol. Exp. Morph., 5:283.
Zelená, J., 1959, Effect of innervation on the development of
     skeletal muscle. Babákova Sbírka, Avicenum, Praha, 12:1.
Zelená, J., 1976, The role of sensory innervation in the deve-
     lopment of mechanoreceptors. Prog. Brain Res., 43:59.

POSTNATAL DEVELOPMENT OF MUSCLE SPINDLES AND EXTRAFUSAL MUSCLE
FIBERS IN THE MOUSE TEMPORAL MUSCLE AND DIETARY EFFECT

Koichi Osawa, Norihiko Maeda[x], Tsuyoshi Kawasaki[xx],
Tamuro Masuda, Yoshiro Yamamoto[xx] and Masayoshi
Kumegawa[x]

Departments of Oral Diagnosis, [x]Oral Anatomy and
[xx]Oral Surgery, Josai Dental University
Sakado, Japan

INTRODUCTION

In 1985, Maeda et al. found that the formation of annulospiral endings in the masseter muscle of developing mice continued after weaning and was completed by the 50th postnatal day.
Moreover, Maeda et al. (1987) reported that a long-term intake
of a fine-grained diet resulted in degeneration of annulospiral
endings of muscle spindles in the masseter muscle of mice.

Since the temporal muscle plays an important role in mastication in combination with the masseter muscle, we examined the
postnatal development of extrafusal muscle fibers and muscle
spindles in the mouse temporal muscle, and the effects of a fine-
grained diet on the postnatal development of muscle spindles
were assessed (Maeda et al., in press).

MATERIALS AND METHODS

ICR-strain mice were used in the present study. Newborn
mice were kept with their mothers. Litters were reduced to 10
pups on day 5 after birth. On day 20 after birth, infant mice
were removed from their mothers. Control mice were fed a solid
diet (CA 1, Clea, Japan, Inc., Tokyo, Japan), and experimental
animals were given a fine-grained CA 1 powder (20 μm particle
diameter) from the 20th postnatal day. Control animals were killed at various times under anesthesia. Animals fed the fine-
grained diet were killed under anesthesia on the 180th day after
birth. Temporal muscles from 3 animals taken at each time period
were removed, and treated by the silver method of Barker and Ip
(1963). The muscle spindles were dissected under a stereomicro-
scope. All specimens were observed under a light microscope and
muscle spindles were classified into 4 types according to the
designation of Maeda et al. (1985).

Two temporal muscles of other different animals were rapidly
frozen in liquid nitrogen. The specimens were cut into transverse
serial sections of 12 μm thickness in a cryostat at -20°C and

air-dried. The sections were stained with succinic dehydrogenase (SDH) activity, as described by Nachlas et al. (1957).
Several sections were stained with hematoxylin for the measurement. The diameters of extrafusal and intrafusal muscle fibers,
and primary endings were measured with a micrometer.

RESULTS AND DISCUSSION

The postnatal changes in the histochemical property of SDH
activity and those of the development of Ia fiber endings showed
time courses similar to each other. Before the weaning period,
there were a few muscle spindles with complete annulospiral endings, and the differences of the SDH activity were not clearly
detectable among intrafusal muscle fibers. With growth of animals, the primary endings of Ia fibers began to develop spiral
endings in many muscle spindles, and individual intrafusal muscle
fibers began to have different SDH activities. And by the 40th
postnatal day, almost all of muscle spindles had completely
coiled annulospiral endings. Differences in SDH activity between
intrafusal muscle fibers became detectable after the weaning
period.

In mice fed a fine-grained diet, decreased diameters of
primary endings and the number of muscle spindles with complete
annulospiral endings were observed (Maeda et al., in press).

These results are closely similar to those in the masseter
muscle of mice, described by Maeda et al. (1985, 1987), and suggest that normal mastication is very important for maintaining
the homeostatic condition of muscle spindles in the temporal
muscle.

ACKNOWLEDGEMENT

This work was supported by a Grant-in-Aid for Special
Research of Mechanobiological Research on the Masticatory System
(No. 61134026) from the Ministry of Education, Science and Cul-
ture, Japan.

REFERENCES

Barker, D., and Ip M.C., 1963, A silver method for demonstrating the innervation of mammalian muscle in teased preparations, J. Physiol., 169:73P.
Maeda, N., Kawasaki, T., Osawa, K., Yamamoto, Y., Sumida, H.,
Masuda, T., and Kumegawa, M., 1987, Effect of long-term
intake of a fine-grained diet on the mouse masseter muscle, Acta Anat., 128:326.
Maeda, N., Osawa, K., Masuda, T., Hakeda, Y., and Kumegawa, M.,
1985, Postnatal development of the annulospiral endings
of Ia fibers in muscle spindles of mice, Acta Anat., 142:
42.
Maeda, N., Sato, M., Osawa, K., Kawasaki, T., Masuda, T., and
Kumegawa, M., Postnatal development of the mouse temporal
muscle and effects of a fine-grained diet on the muscle
spindles, Anat. Anz., (in press).
Nachlas, M.M., Tsou, K.C., De Sauza, E., Cheng, C.S., and

Seligman, A.M., 1957, Cytochemical demonstration of succinic dehydrogenase by the use of a new p-nitrophenyl substituted ditetrazole, _J. Histochem. Cytochem._, 5:420.

AGE RELATED CHANGES OF MUSCLE SPINDLES OF RAT SOLEUS MUSCLE

Volkmar Hartung and Gerhard Asmussen

Carl-Ludwig Institute of Physiology
Karl-Marx University
Leipzig, GDR

Muscular strength and endurance as well as motor skill decrease in old age. This is accompanied by a loss of motor units, a decrease in the number of extrafusal muscle fibers (eff) and by processes of denervation and reinnervation (for ref. see Vandervoort et al., 1986). However, there are no systematic investigations of the influence of life span on the structure and function of mechanoreceptors involved in motor control, such as muscle spindles (msp) and their intra-fusal muscle fibers (iff). We therefore investigated the influence of age on the morphology of msp of a postural muscle of the rat using conventional histological techniques.

MATERIALS AND METHODS

Fourteen Wistar rats of both sexes aged 14, 21, 43, 55, 70, 129, 368, 714, 915, 1170 and 1458 days were used for the experiments. The lower extremities of urethane anesthetized animals were perfused via the aorta with Tyrode solution and afterwards with a 4 % solution of paraformaldehyde. The soleus muscles of both legs were removed, postfixed for 3-6 days, dehydrated, embedded in paraffin and completely cross-sectioned (10 μm). The serial cross-sections of the muscles were stained with modified trichrome staining and examined step-wise by light microscopy.

RESULTS AND DISCUSSION

The number of msp in rat soleus muscles is independent of age and remarkably constant. There are 15-20 msp per muscle. This result is in contrast to that of Kalugina (1956) obtained by a very restricted random sample describing a considerable increase of the number of msp in rat sternocleidomastoid muscle with age. About 7 % of the total number of 462 msp investigated were exceptional in shape (twin and tandem spindles or msp with more than two polar regions). Their appearance is not related to age, sex or body size.

With the exception of the regions close to the tendons
which are free of msp, the distribution of the msp in soleus
muscles are relatively uniform and independent of age. The
change of the length arrangement of the msp during aging as
well as the divergence of the equatorial regions of tandem
spindles can only be explained by assuming growth of the eff
and iff along their whole length and not only restricted to
the myotendinous junctions.

The length of the msp-capsule, the total length of the
msp given by the length of nuclear-bag fibers (NB) which have
an extracapsular part, and the length of the equatorial region
all increase with age. The total length of soleus msp measures
about 2.5 mm at the age of 2-3 weeks, about 3-5 mm at the age
of 2 months and about 4.0 mm at the age of one year. The proportion of the msp-capsule to the total length of msp increases
with age. It amounts to 35-40 % at the age of 2-3 weeks, to
about 50 % in adult animals and up to 70 % in senile rats.
There is a relatively constant proportion of 10:1 between the
length of the muscle and the length of the msp-capsule during
the life span.

The number of eff in rat soleus muscle is 2700 $\pm$ 300 and
it is relatively constant during the greater part of life.
However, it decreases clearly in muscles of senile animals.
The number of iff amounts to 74 $\pm$ 8 and there is no significant
decrease in old age. The msp usually contain 2 NB and 2 nu-
clear-chain fibers (NC). The relation of the two types of iff
and also the number of msp with a nontypical iff-combination
are not changed with age even in muscles of senile rats.

The average diameter of eff increases during life from
13 μm (14th day of life) to 43 μm (one year) and it stays
fairly constant afterwards. The distribution of the diameter
of eff is not only shifted to higher diameters, but it en-
larges with increasing age; the latter is especially pronounced
in muscles of senile animals. In contrast iff show only a very
small increase in diameter during life and the basis of the
distribution of the calibre of NB and NC is practically unchanged. More frequently, the eff exhibit degenerative changes
with increasing age (fiber splitting, hyaline degeneration,
vacuoles); such degenerative changes are not so prominent, or
are absent altogether in iff.

The occurrence of denervation and reinnervation as well
as atrophy and hypertrophy induced by degenerative processes
in motor neurons during old age are considered to be the cause
of the changes in eff. Such processes obviously concern the
iff to a smaller extent. Our results support the hypothesis
that once they are established, iff and msp are relatively in-
variable structures.

REFERENCES

Kalugina, M.A., 1956, A contribution to the development of
     proprioceptors in striated muscles of mammals (in Rus-
     sian), Arkh. Anat. Gist. Embr., 33:59.
Vandervoort, A.A., Hayes, K.C., and Belanger, A.V., 1986,
     Strength and endurance of skeletal muscle in the elderly,
     Physiotherapy Canada, 38:167.

PART III

REGENERATION AND GRAFTING

SIGNIFICANCE OF THE EXTRACELLULAR MATRIX FOR THE REGENERATION OF SENSORY
CORPUSCLES, WITH SPECIAL REFERENCE TO PACINIAN CORPUSCLES

Chizuka Ide

Department of Anatomy, Iwate Medical University
School of Medicine
Morioka 020, Japan

INTRODUCTION

The Pacinian corpuscle consists of an axon terminal and specialized
non-neuronal cells of an inner and an outer core (Pease and Quilliam, 1957;
Nishi et al.,1969; Munger, 1971; Halata, 1977; Chouchkov, 1978; Malinovsky
and Pac, 1982). The inner core is formed by densely packed thin cytoplasmic
processes (lamellae) of lamellar cells which surround the axon terminal in a
bilaterally symmetrical fashion. These inner core lamellar cells are
thought to be derived from Schwann cells. There are thin spaces between
neighboring lamellae, in which an extracellular matrix including basal
lamina-like material and fine collagen fibrils is present. The outer core
is composed of loosely piled layers of thin cells concentrically encircling
the inner core. There are definite basal laminae on these outer core cells
and thin processes. The outer core cells are extensions of the perineural
epithelial cells of the innervating nerve (Shanta and Bourne, 1968). The
axon terminal is characterized by an accumulation of mitochondria beneath
the axolemma and by small cytoplasmic processes (axonal spines) emanating
from the axolemma (Munger et al.,1987; Ide et al.,1987a). It has been
demonstrated that basal laminae can serve as conduits for regenerating axons
in the peripheral nerve (Ide et al.,1983; Osawa et al.,1986). As described
above there are extracelluar matrices containing basal lamina-like material
in the inner core of Pacinian corpuscles. These matrices are considered to
have specific properties for maintaining the cellular organization of the
corpuscle. Although basal laminae of Schwann cells may only serve as
pathways for the elongation of regenerating axons, the extracellular
matrix including basal lamina-like material in the inner core of Pacinian
corpuscles might act, in addition to providing pathways, as the site of
differentiation of axon terminals and lamellar cells. My concern is whether
such extracellular matrix of the Pacinian corpuscle inner core has in fact
properties for cell differentiation, i.e. the ability to regenerate axons and
Schwann cells so as to differentiate into axon terminals and lamellar cells,
respectively.
In my previous study on the regeneration of murine Meissner corpuscles
(Ide, 1986) the basal laminae and associated extracellular material of
lamellar cells acted as the matrix for regenerating axons and Schwann cells
to differentiate into axon terminals and lamellar cells, respectively.

MATERIALS AND METHODS

Pacinian corpuscles are aligned in the periosteum along the long axis
of the bone at the distal end of the fibula in mice. Skin was cut
open near the ankle joint, and Pacinian corpuscles were exposed under the
dissecting microscope. Pacinian corpuscles were frozen in situ by a light
touch with a forceps precooled with liquid nitrogen. Such cryo-treatment
was repeated 3-5 times to kill cellular constituents of the corpuscle. The
regeneration of axons and lamellar cells was observed up to 37 days after
the treatment. Animals were killed 2, 3, 5, 7, 10, 25 and 37 days after
cryo-treatment by intra-cardiac perfusion with 2.0% paraformaldehyde and
2.5% glutaraldehyde in 0.1 M cacodylate buffer (pH 7.2) and processed for
the conventional electron microscopic observations.

RESULTS

All the cellular constituents including the inner and outer core cells
were destroyed and disintegrated into cell debris within 2-3 days after
cryo-treatment. The extracellular matrix including basal lamina-like
material of the inner core apparently remained intact as it had been before
the treatment. The outer core collapsed losing its original expanded
structure, but the basal laminae of the outer core cells apparently remained
undamaged.
The initial feature of nerve growth as well as Schwann cell migration
into the vacant corpuscles was identified within one week after the
treatment. A regenerating axon entered the corpuscle to grow through the
center of the inner core (Fig. 1). Such growing axons were usually not
accompanied by Schwann cells. Though some basal lamina-like material of the
old corpuscle was attached to the surface of the regenerating axons, no
definite basal lamina tube was formed to guide the axons. Later all the
regenerating axons within the corpuscle came to be covered by Schwann cells.
At the same time cells were found within the inner core matrix, whose cell
bodies were separated from the axons. These cells were thought to be derived
from regenerating Schwann cells. It is noteworthy that these cells extended
their cytoplasmic processes around the axon, basically resembling
the lamellar cells of the inner core (Fig. 2). In the subsequent stages of
regeneration these peculiar cells further extended many cytoplasmic processes
around the axon in a concentric fashion. These thin cellular processes
appeared to be focally attached to the basal lamina-like material of the
inner core matrix. However, it was not determined whether or not these
cell processes extended through the spaces which had been occupied by the
original lamellae. It is evident that these cell processes are developing
lamellae and the cell bodies are lamellar cells. Although the cellular
organization was not as elaborate as in the original normal corpuscles,
the regenerated axon terminals and lamellar cells exhibited the cytology
basically similar to that of normal corpuscles (Fig. 3). Usually 2-3 axon
profiles were seen in the inner core of the regenerated corpuscles (Zelena,
1984).
The outer core cells regenerated themselves through the spaces of the
original outer core processes. This mode of regeneration is the same as that
of regenerating axons which extend through the Schwann cell basal lamina
tubes, indicating that the basal laminae of the outer core cells provide
pathways for the extension of regenerating outer core cells. The collapsed
outer core had recovered by 30 days after cryo-treatment to the original
large sphere around the inner core.

DISCUSSION

It is remarkable that highly differentiated sensory corpuscles such as

Pacinian or Meissner corpuscles (Fig. 4) can regenerate in the acellular
matrix of the corresponding corpuscles (Ide, 1986, 1987). It is evident
that the extracellular matrix of the inner core of the Pacinian corpuscle
has the property to regenerate axons and Schwann cells so as to
differentiate into axon terminals and lamellar cells, respectively. This
means that there is some specific substance in the inner core matrix
responsible for the axon and Schwann cell differentiation. The effect of
the extracellular matrix in cell differentiation is more conspicuous in
Pacinian corpuscles than in murine Meissner corpuscles. Basal laminae have
been though to serve as scaffolds for regenerating cells (Vracko, 1972;
Vracko and Benditt, 1972) and the laminin localization in basal laminae has
been histochemically analyzed (Tohyama and Ide, 1984).

It has been reported that the basal laminae of motor endplates have
the property to differentiate the regenerating axons into axon terminals
which have the characteristics of a presynaptic terminal (Sanes et al.
1978). This is in agreement with the present study regarding the role of
the extracellular matrix in the differentiation of sensory axon terminals.
Particular properties of the inner core matrix which are different from the
ordinary connective tissue matrix have been suggested in several studies:
one by Ide and Saito (1980) suggests that the inner core matrix contains an
intense non-specific ChE activity. This enzyme is produced by lamellar cells
and considered to be secreted into the surrounding matrix. Another study
suggests that the axon terminals can frequently sprout in the inner core
even in normal human Pacinian corpuscles (Fig. 5, Ide et al.,1987b).
The matrices of other sensory corpuscles such as Ruffini and Herbst (Saxod,
1978) may have similar properties.

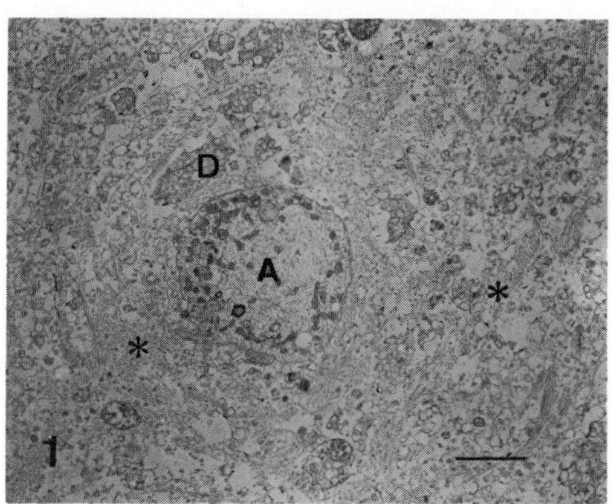

Fig. 1.   Regeneration of a Pacinian
corpuscle in the extracellular
matrix of an old corpuscle. Six
days after cryo-treatment. A
regenerating axon (A) extends
through the center of the inner
core. The inner core contains
degraded cell debris (D) and the
extracellular matrix (*) of an
old corpuscle.

Bar = 2 μm

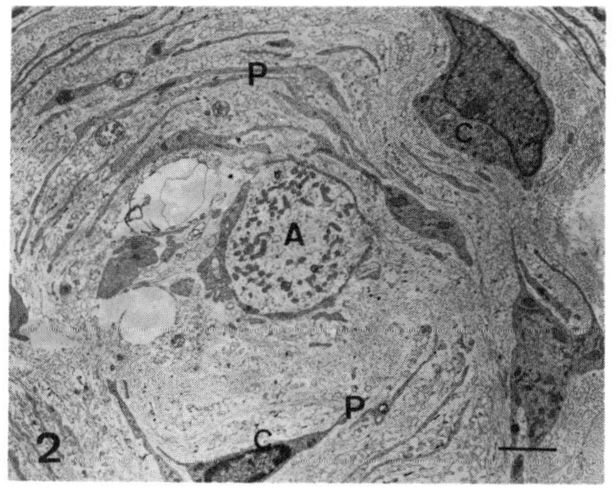

Fig. 2.   Seven days after cryo-
treatment. Cells (C) located
apart from the axon (A) extend
their cytoplasmic processes (P)
through the matrix of the inner
core around the axon terminal,
suggesting the beginning of the
inner core formation.

Bar = 2 μm

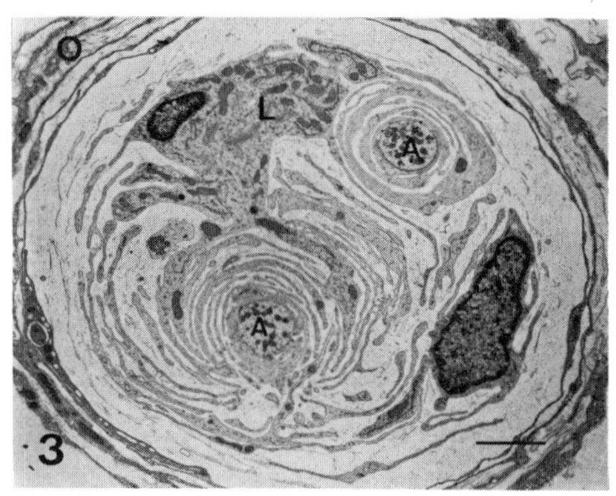

Fig. 3.   Thirty seven days after cryo-treatment. The regenerated inner core has basically the same cytology of axon (A) and lamellar cells (L) as in a normal corpuscle. O, outer core.
Bar = 2 μm

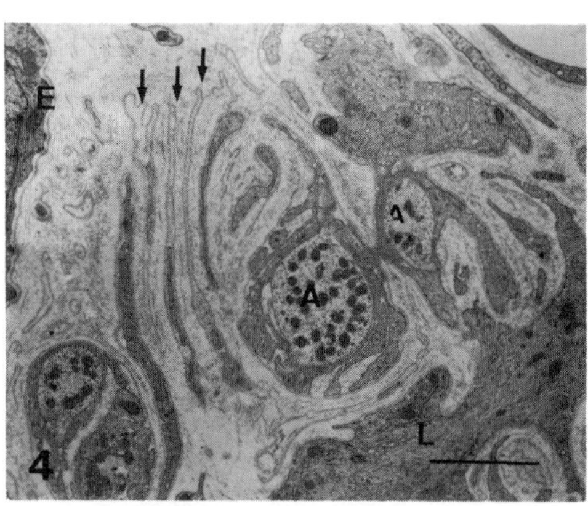

Fig. 4.   Regeneration of a murine Meissner corpuscle. Eight days after cryo-treatment. The developing lamellar cells (L) extend cytoplasmic processes through the basal lamina tubes to surround axons (A). Some basal lamina tubes (arrows) of an old corpuscle remain empty. E, epidermis.
Bar = 2 μm

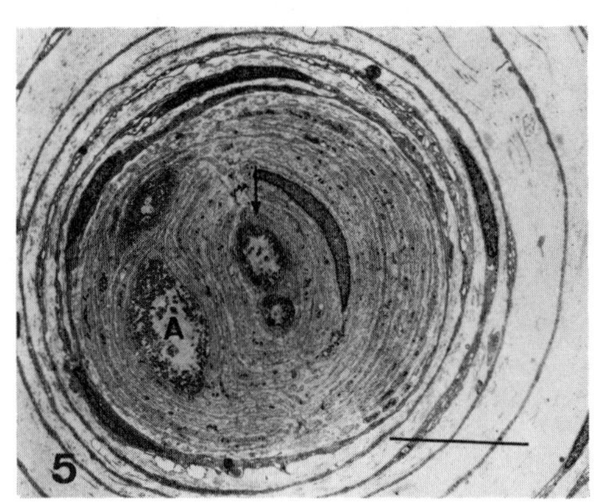

Fig. 5.   Normal human Pacinian corpuscle from the digital skin of an adult. The inner core contains four profiles of axon terminals (A), three of which were identified as branches of the main axon (arrow) by serial sectioning.
Bar = 10 μm

REFERENCES

Chouchkov, Ch., 1978, Cutaneous Receptors, in:A. Brodal, W. Hild, J. van
     Limborgh, R. Ortmann, T.H. Schiebler, G. Tondury, and E. Wolf (Eds),
     Adv. Anat. Embryol. Cell Biol. 54, Springer, Berlin, pp. 41-53.
Halata, Z., 1977, The ultrastructure of the sensory nerve endings in the

articular capsule of the knee joint of the domestic cat (Ruffini corpuscles and Pacinian corpuscles), J. Anat., 124: 717-729.

Ide, C., 1983, Nerve regeneration and Schwann cell basal lamina: observations of the long-term regeneration, Arch. Histol. Jap., 46: 243-257.

Ide, C., 1986, Basal laminae and Meissner corpuscle regeneration, Brain Res., 384: 311-322.

Ide, C., 1987, Role of extracellular matrix on the regeneration of a Pacinian corpuscle, Brain Res., 413; 155-169.

Ide, C., and Saito, T., 1980, Electron microscopic histochemistry of cholinesterase activity of Vater-Pacini corpuscles, Acta Histochem. Cytochem., 13: 298-305.

Ide, C., Tohyama, K., Yokota, R., Nitatori, T.,and Onodera, S., 1983, Schwann cell basal lamina and nerve regeneration, Brain Res., 288: 61-75.

Ide, C., Yoshida, Y., Hayashi, S., Takashio, M.,and Munger, B.L., 1987a, A re-evaluation of the cytology of cat Pacinian corpuscles. II. The extreme tip. Cell Tiss. Res.,(in press).

Ide, C., Nitatori, T., and Munger, B.L., 1987b, The cytology of human Pacinian corpuscles: evidence for sprouting of the central axon. Arch. Histol. Jap.,(in press).

Malinovsky, L., and Pac, L., 1982, "Morphology of Sensory Corpuscle in Mammals." J.E. Purkyne University Brno, Medical Faculty Brno.

Munger, B.L., 1971, Pattern of organization of peripheral sensory receptors, in: "Handbook of Sensory Physiology," W.R. Loewenstein (ed), Vol. 1 Springer, New York, pp. 523-556.

Munger, B.L. Yoshida, Y., Hayashi, S., Osawa, T., and Ide, C., 1987, A re-evaluation of the cytology of Pacinian corpuscles. I. The inner core and clefts, Cell Tiss. Res., (in press).

Nishi, K., Oura, C., and Pallie, W., 1969, Fine structure of Pacinian corpuscles in the mesentery of the cat. J. Cell Biol., 43: 539-553.

Osawa, T., Ide, C., and Tohyama, K., 1986, Nerve regeneration through the allogenic nerve grafts in mice, Arch. Histol. Jap., 49: 68-81.

Pease, D.C., and Quilliam, T.A., 1957, Electron microscopy of the Pacinian corpuscle. J. Biophy. Biochem. Cytol., 3: 331-347.

Sanes, J.R.,Marshall, L.M., and McMahan, U.I., 1978, Reinnervation of muscle fiber basal lamina after removal of myofibers. Differentiation of regenerating axons at original synaptic sites. J. Cell Biol., 78: 176-198.

Saxod, R., 1978, Development of cutaneous sensory receptors in birds. in: M. Jacobson (ed), "Handbook of Sensory Physiology. IX. Development of Sensory System," Springer, Berlin, pp. 337-418.

Shanta, T.R., and Bourne, B.H., 1968, The perineural epithelium. in: G.H. Bourne (ed), "The Structure and Function of Nervous Tissue," Vol. 1, Academic Press, New York, 379-459.

Tohyama, K., and Ide, C., 1984, The localization of laminin and fibronectin on Schwann cell basal lamina, Arch. Histol. Jap., 47: 519-522.

Vracko, R., 1972, Significance of basal lamina for regeneration of injured lung, Virchow Arch. (Pathol. Anat.), 355: 264-274.

Vracko, R., and Benditt, E.P., 1972, Basal lamina: The scaffold for orderly cell replacement, J. Cell Biol., 55: 406-419.

Zelena, J., 1984, Multiple axon terminals in reinnervated Pacinian corpuscles of adult rat, J. Neurocytol., 13: 665-684.

REINNERVATION OF GRAFTED PACINIAN CORPUSCLES BY DORSAL ROOT
AND DORSAL COLUMN AXONS

I. Jirmanová and J. Zelená

Institute of Physiology
Czechoslovak Academy of Sciences
Vídeňská 1083, Prague, Czechoslovakia

INTRODUCTION

Transplantation of various parts of the developing central
nervous system (CNS) into the brain has proved to be a valuable
tool for studying neuronal differentiation, connectivity and
plasticity (Björklund and Stenevi, 1985; Raisman et al., 1987).
In our experiments, we have used nonneuronal grafts from the
periphery for testing the plasticity of dorsal root ganglion
neurons. These pseudounipolar neurons are polarized: only a
single process, crus commune, originates from the neuron, but
it divides not far from the perikaryon into a peripheral and
a central process. Both processes have the ultrastructure of
axons, but differ in their polarity and form fundamentally different terminals. All central processes, i.e. dorsal root axons,
terminate in the spinal cord in transmitter-releasing endings
which contain a large number of synaptic vesicles, whereas peripheral axon terminals are designed to receive various kinds
of stimuli either indirectly, through specialized receptor
cells, or directly as transducing sensory endings which may be
either free or connected with different types of corpuscular
receptors; most of the peripheral sensory terminals are filled
with mitochondria and contain vesicles only at their margins
or extensions. We have attempted to answer the question whether
the dorsal root ganglion neurons would respond to a change of
the target tissue and alter the ultrastructure of their "central" endings, if their processes were connected to peripheral
sensory corpuscles instead of to the spinal cord. To explore
this possibility, we connected a graft of peripheral mechanoreceptors - Pacinian corpuscles - with either dorsal root axons
or dorsal column axons and studied the process of axonal regeneration, the reinnervation of the grafted corpuscles and the
ultrastructure of the regenerated endings.

REINNERVATION OF THE GRAFT BY DORSAL ROOT AXONS: TRANSFORMATION
OF DORSAL ROOT TERMINALS

In this experiment (Zelená and Jirmanová, 1986, 1987), a
group of crural Pacinian corpuscles was dissected out in young

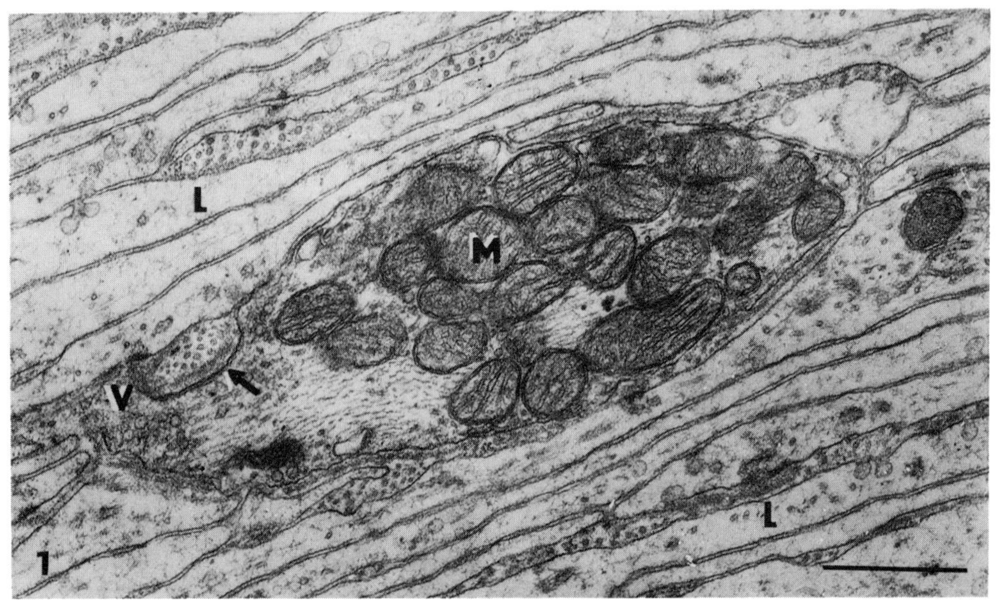

Fig. 1. A terminal of a dorsal root axon in a Pacinian corpuscle
        grafted 2 months previously has the ultrastructure of
        a peripheral sensory ending; it contains mainly mito-
        chondria (M), has a lateral process filled with clear
        vesicles (V) and a membrane specialization: a subaxo-
        lemmal lamina (arrow) at the base of the lateral pro-
        cess. L - the inner core lamellae. Bar: 0.5 μm.

rats with a piece of the interosseous nerve, transferred onto
the surface of the spinal cord, and cross-connected with an exposed and transected L_4 or L_5 dorsal root; the nerve-root anasto-
mosis and the attached corpuscles were inserted under the dura
far from the dural opening to exclude the danger of contamina-
tion by stray peripheral axons. The autografts were examined
1-6 months after the operation following intracardial perfusion
with a fixative composed of 1 % glutaraldehyde and 1 % paraformaldehyde in a phosphate buffer.

     From 2 months after grafting onwards, the transplanted
nerve and its branches were almost completely filled with myelinated and unmyelinated dorsal root axons which elongated up
to the corpuscles and reinnervated most of them by 1-5 myelinated axons; some of the corpuscles were also reinnervated by a
few unmyelinated axons. The axons grew through the capsular
channel, shed their myelin sheath at the end of it, then entered
the inner core, branched and formed multiple axon terminals in
the interlamellar clefts in a similar pattern as do the regenerating peripheral axons of the interosseous nerve (Zelená, 1984).
Almost all the regenerated terminals resembled peripheral sensory endings. They were mainly filled with accumulated mitochondria; many terminals formed thin lateral projections (Fig. 1)
as do various types of lamellated mechanosensitive peripheral
endings. They contained only a few clear vesicles and an occasional small dense core vesicle in marginal areas of lateral
projections. In a few instances, transformed dorsal root ter-

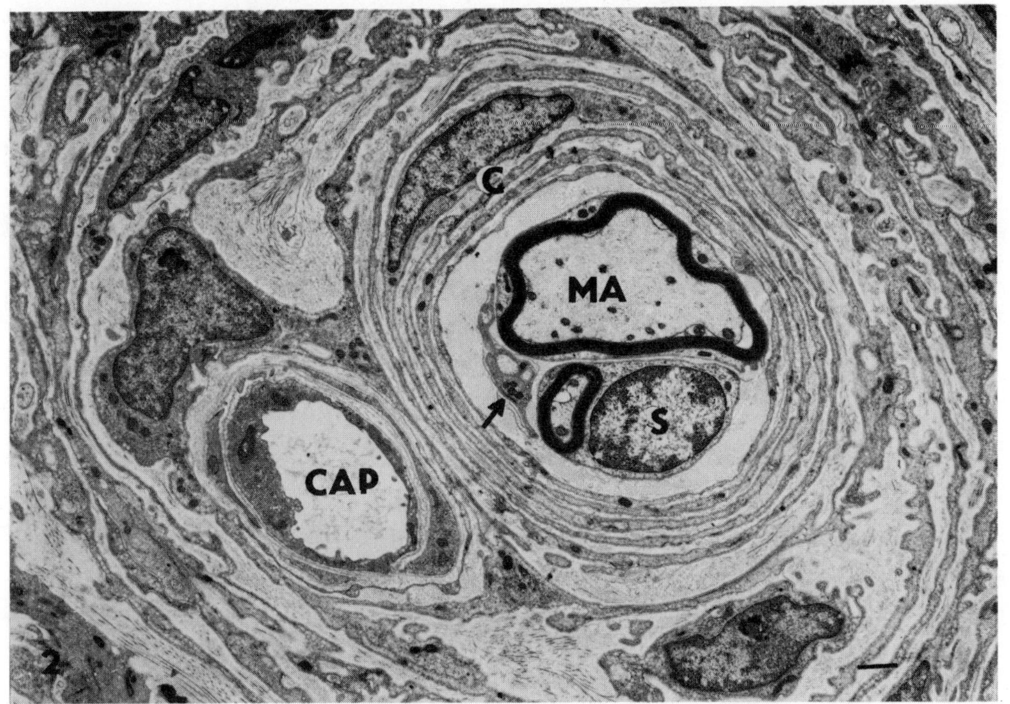

Fig. 2. Nerve entry to a grafted Pacinian corpuscle contains,
        3 months after the operation, two myelinated (MA) and
        three unmyelinated (arrow) dorsal column axons.
        S - Schwann cell, CAP - capillary, C - capsule.
        Bar: 1 μm.

minals even developed membrane specializations (Fig. 1) characteristic for normal (Spencer and Schaumburg, 1973) and regenerated (Zelená, 1981, 1984) peripheral Pacinian endings, and
thought to be significant for the process of transduction. Several corpuscles were also reinnervated, along with the large
terminals of the peripheral type, with small endings loaded
with dense core vesicles 80-120 nm in diameter. None of the
cross-innervated corpuscles contained terminals of the type of
synaptic boutons or glomeruli filled predominantly with clear
synaptic vesicles.

The results have shown that primary sensory neurons can
transform, in the grafted Pacinian corpuscles, their regenerated
dorsal root terminals into endings with the ultrastructure of
peripheral sensory terminals (Zelená and Jirmanová, 1987).

REINNERVATION OF GRAFTED PACINIAN CORPUSCLES BY DORSAL COLUMN
AXONS

The rich reinnervation of the grafts with dorsal root
axons was not surprising, since dorsal roots are similar in
structure to peripheral nerves and regenerate vigorously when
severed in their extraspinal course. The question has arisen
whether the grafted corpuscles would also become reinnervated,
if their nerves were inserted into dorsal columns which pre-

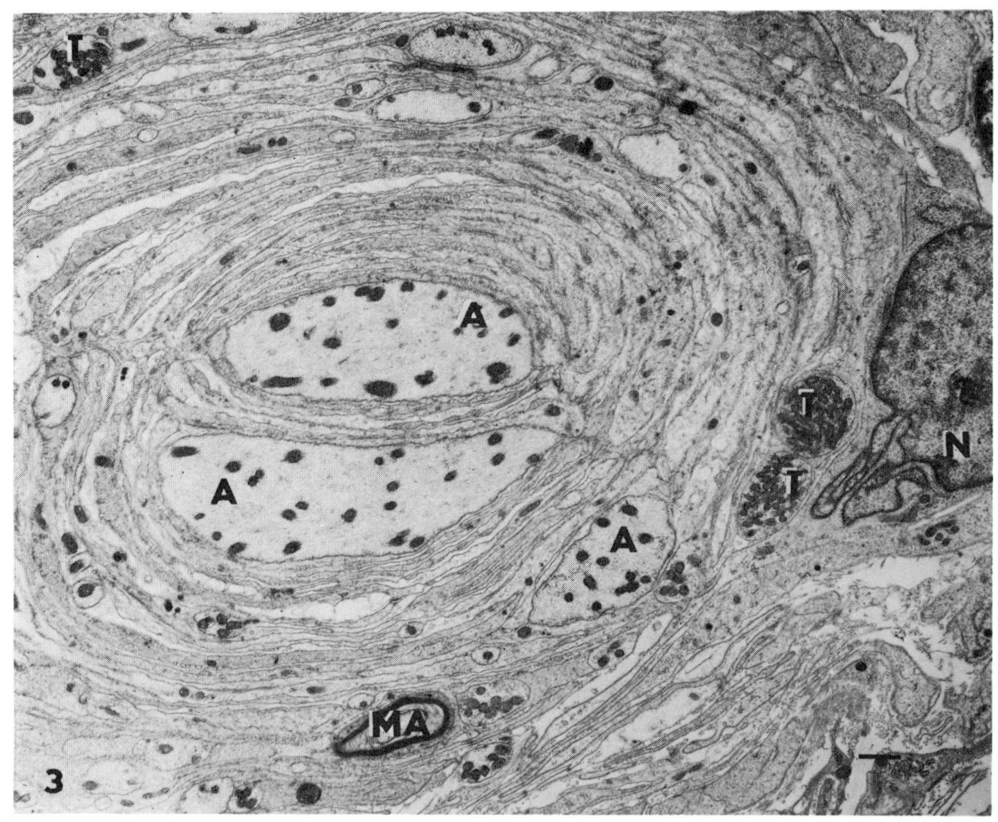

Fig. 3. Transverse section of a Pacinian inner core, which is
        reinnervated, 3 months after grafting, by many branches
        of dorsal column axons (A), and contains several termi-
        nals (T) rich in mitochondria; MA - a myelinated axon,
        N - nucleus of an inner core cell. Bar: 1 μm.

dominantly contain myelinated and unmyelinated central axons of
the dorsal root ganglion neurons. As other CNS axons, dorsal
column axons do not regenerate spontaneously after injury, but
they do regenerate through a transplanted peripheral nerve,
since Schwann cells enhance and support axonal regeneration in
the CNS (Aguayo, 1985). After autotransplantation of the graft
and insertion of the interosseous nerve into the spinal cord
at the level of L_1, the injured dorsal column axons grew through
the grafted nerve into the Pacinian corpuscles attached to it.
Three months after the operation, the grafted nerve and its
branches were profusely reinnervated by myelinated and unmyelinated axons, and most Pacinian corpuscles became supplied each
by one or more myelinated axons (Fig. 2) which formed many
axonal branches and terminals within the inner core. In some
corpuscles, multiple terminals filled with mitochondria prevailed; other corpuscles contained, at the midlevel, mainly
large axonal profiles and only occasional terminals filled with
mitochondria (Fig. 3). Lateral projections or membrane specializations were not found in any of the Pacinian corpuscles examined. Thus the dorsal column axons exhibited a less advanced
transformation of their terminals in the grafted corpuscles
than did the dorsal root axons.

CONCLUDING REMARKS

     The results of our transplantation experiments show that
dorsal root and dorsal column axons are capable of reinnervating
mechanoreceptors grafted onto the spinal cord. Most of the
grafted Pacinian corpuscles survive the transient ischemia due
to rapid revascularization, and accept alien axons that reach
them through the branches of the transplanted interosseous
nerve. Although both dorsal root and dorsal column axons normally form synapses only with the spinal cord neurons, they recognize grafted Pacinian corpuscles as their new target organ.
Apparently, the growth cones of the regenerating central processes of sensory neurons interact with the peripheral cells of
the corpuscles and respond to the signal received from the periphery by arresting elongation and by forming new terminals
which, in most instances, closely resemble peripheral sensory
endings. We do not know how this transformation of the ultrastructure of the "central" endings is achieved, nor do we know
whether the function of the terminals also becomes altered. As
regards the mechanism of the transformation, it is conceivable
that the information molecules taken up by the terminals in
the graft travel by retrograde transport to the perikarya which
acknowledge the changed target of their central axons by altering the composition of the anterograde transport, as if it were
actually directed to the sensory periphery. Instead of the components of synaptic vesicles transported to normal central synapses, the perikaryon apparently supplies the newly formed
terminals with the material necessary for the build-up and maintenance of peripheral sensory endings.

     The structural transformation of the terminals is more
pronounced after reinnervation of the corpuscles by dorsal roots
than by dorsal column axons. Dorsal columns comprise a heterogeneous population of axons of both primary and secondary sensory neurons which may account for the slight difference in the
results obtained in the two sets of experiments. Our data suggest that only the large-type primary sensory neurons which
also innervate sensory receptors in the periphery can transform
their terminals in the graft. The tiny endings filled with
large dense core vesicles, which are probably formed by the
small-type peptidergic neurons, do not change their morphology
after grafting.

     It remains to be established by further studies, whether
the structural transformation of the dorsal root and dorsal
column terminals is also accompanied by their change into mechanosensitive endings. Anomalous neurons would be created by
grafting, if it was established that the terminals also changed
their function: these neurons would be deprived of their polarity, since afferent signals would arrive at their perikarya
along both their peripheral and central processes.

REFERENCES

Aguayo, A.J., 1985, Axonal regeneration from injured neurons in
        the adult mammalian central nervous system, in: "Synaptic
        Plasticity", C.W. Cotman, ed., The Guilford Press, New
        York, London, pp. 457-484.
Björklund, A., and Stenevi, U., 1985, Neural Grafting in the
        Mammalian CNS, Fernström Foundation Series Vol. 5,

Elsevier, Amsterdam, New York, Oxford.
Raisman, G., Morris, R.J., and Zhou, C.-F., 1987, Specificity in
    the reinnervation of adult hippocampus by embryonic
    hippocampal transplants, in: "Neural Regeneration", F.J.
    Seil, E. Herbert, and B.M. Carlson, eds., Progress in
    Brain Research Vol. 71, Elsevier, Amsterdam, pp. 325-333.
Spencer, P.S., and Schaumburg, H.H., 1973, An ultrastructural
    study of the inner core of Pacinian corpuscle, J. Neuro-
    cytol., 2:217-235.
Zelená, J., 1981, Multiple innervation of rat Pacinian cor-
    puscles regenerated after neonatal axotomy, Neuroscience,
    6:1675-1686.
Zelená, J., 1984, Multiple axon terminals in reinnervated Paci-
    nian corpuscles of adult rat , J. Neurocytol., 13:665-684.
Zelená, J., and Jirmanová, I., 1986, Transplantation of Pacinian
    corpuscles into the central nervous system, Verh. Anat.
    Ges., 80:849-850.
Zelená, J., and Jirmanová, I., 1987, Grafts of Pacinian cor-
    puscles reinnervated by dorsal root axons, Brain Res.,
    in press.

EVIDENCE FOR TRANSDIFFERENTIATION OF ALPHA MOTONEURON TERMINALS DURING
REINNERVATION OF MUSCLE SPINDLES

Richard Butler

Department of Anatomy
McMaster University
Hamilton, Ontario, Canada

INTRODUCTION

It is quite clear that during synaptogenesis, the transformation of
growing axons into specialized nerve terminals is largely regulated by
their synaptic targets. This is especially true for vertebrate
neuromuscular junctions where evidence points to local control of
peripheral axon terminal differentiation by the target's basal lamina.
Such dynamic influence by the target apparently exists in mature adult
tissue as well. For example, the results earlier experiments on cat
muscle indicate that sometimes single <u>regenerating</u> alpha motoneurons can
reinnervate two different types of intrafusal muscle fibers
simultaneously (slow twitch and fast twitch) with focal plates and 'en
grappe' terminals occurring on different branches of the same
motoneuron, simultaneously. This led to the converse series of
experiments (using rats) in which motoneurons were left <u>intact</u> while
their targets were made to degenerate and subsequently allowed to
regenerate around the axons. By transplanting a very different target
(slow tonic extraocular eye muscle) into the normal regenerating leg
muscle target, the intact neurons and their terminals have the option to
innervate this very different target as well and my observations
indicate that they do. The results also suggest that some existing
plate terminals are modified to multiple 'en grappe' terminals on the
transplanted target, thus providing evidence for the continuing dynamic
influence of targets on adult neuromuscular junctions. However, the
stability of the new junctions over time is unknown.

REGENERATING ALPHA MOTONEURONS AND INTACT INTRAFUSAL MUSCLE TARGETS

In a series of experiments on the reinnervation of cat muscle
spindles, the most interesting results involved a series of single
regenerating alpha motoneurons (for methods see Brown and Butler, 1976).
These single alphas had beta branches producing static actions in at
least one spindle and, surprisingly, dynamic actions in a different
spindle, simultaneously. The implication is that each of these motor-
neurons functionally innervated 2 different types of muscle fiber, i.e.,
fast twitch (bag_2 and chains) and slow twitch (bag_1), simultaneously.
Normally, the nuclear bag_1 intrafusal muscle fiber is innervated by

dynamic gamma motoneurons with focal plate endings, whereas, the bag_2 and nuclear chain intrafusal muscle fibers possess static gamma motoneurons with trail (i.e. 'en grappe') endings. Further implications are that either each of these alpha motoneurons produced two different types of endings simultaneously, with focal plate endings on slow twitch muscle fibers and trail endings on fast twitch, or that one of the two types of intrafusal muscle fiber can accept two different types of motor endings.

Serial reconstructions of the spindles showed only focal plates on presumptive bag_1 fibers. Only 'modified' trails were found on presumptive bag_2 and chain fibers. These were not as extensive as gamma trails and contained fewer synaptic terminals. The conclusions drawn were: firstly, these single regenerating alpha motoneurons functionally innervated two different types of intrafusal muscle fiber, simultaneously; secondly, different branches of these single alphas had two morphologically different types of ending, simultaneously (Butler, 1980).

I found 14 certain and 3 probable regenerating alphas in different cats which exhibited this phenomenon. Bennett et al. (1973) showed that regenerating motoneurons in avian muscle can modify terminal morphology (from focal plates to their multiple innervation) when forced to innervate a different type of muscle. These studies suggest that terminal morphology can be regulated by muscle fiber type in mature muscle. The common factor here (and with the classic Eccles' papers of the 1960's, e.g. Buller et al., 1960) is that the behaviour of regenerating neurons was being assessed. During the regrowth and redifferentiation of terminals in adult animals, the same phenomena likely exist as during developmental growth.

Up to ten years ago, it was widely accepted that motoneurons could determine the characteristics of the muscle they innervated. However, newer evidence is causing this view to be re-evaluated (Sanes, 1987). Rees (1978) reviewed evidence regarding interneuronal synaptogenesis and concluded that the transformation of growing axons into specialized nerve terminals is clearly regulated by their synaptic targets. This is especially true for vertebrate neuromuscular junctions where the evidence points to local control of peripheral motor axon terminal differentiation by certain postsynaptic membrane specializations, especially the basal lamina (Sanes et al., 1978; McMahan et al., 1980; Glicksman and Sanes, 1983). All of these studies deal with developing or regenerating axons to intact targets or regenerating axons to regenerating targets.

The influence of synaptic targets on intact neurons is less well understood. When necrosis is experimentally produced in the endplate region of muscle fibers, sprouting from the axon terminal can occur (Huang and Keynes, 1983). There is a growing body of evidence that the phenomenon of sprouting from intact terminals is qualitatively not the same as axonal regeneration and that the former should not be classified as a regeneration. Even without damage to the postsynaptic elements, evidence for a dynamic nerve-target relationship exists. Wernig et al., (1984) report an increase in the number of nerve branches of unmyelinated nerve terminals with increasing age in normal adult motor endplates of the mouse. They conclude that there is continual nerve sprouting in intact neuromuscular junctions and the sprouts form synaptic contacts. Furthermore, possible signs of nerve retraction indicate that these synaptic contacts are undergoing continual remodelling.

INTACT ALPHA MOTONEURONS AND REGENERATING MUSCLE TARGETS

   An obstacle in studying the influence of synaptic targets on intact
neurons has been the production of a preparation. In the past, a common
way that experimenters have used to induce new neuromuscular
relationships (in addition to interrupting the nerve supply and allowing
subsequent regeneration) is the technique of muscle minces where an
animal's muscle is minced by repeated crushing with forceps. Following
this, the muscle degenerates and then regenerates (e.g., Studitsky,
1959; Carlson, 1972). Unfortunately, the technique also minces the
nerve terminals within the muscle and the regenerating muscle becomes
innervated by regenerating neurons.

   There is a method available which will induce only muscle to
(rapidly) degenerate and then regenerate. It is now well established
that treatment of a muscle by the local anaesthetic drug bupivacaine
will induce degenerative changes throughout the whole muscle, followed
by regeneration such that at three weeks following treatment, muscle
fiber calibre and muscle cytoarchitecture are restored to a nearly
normal state (e.g. Benoit and Belt, 1970; Hall-Craggs, 1974; and
Carlson, 1976). Two possible objections to this technique have already
been addressed by other authors. Firstly, apart from a period of 10 or
so hours following treatment with bupivacaine when transmission is
blocked by the anaesthetic action of the drug, there is little evidence
that the nerve supply to the muscle is either interrupted or adversely
affected by bupivacaine (Sokoll, Sonesson and Thesleff, 1968; Jirmanova
1975; Hall-Craggs, 1980). Milburn (1976) confirms the lack of
bupivacaine effects on fusimotor terminals but reports that rat muscle
spindle sensory nerve terminals (Ia and group II afferents) can be
reduced in number and abnormal in structure. Secondly, histochemical
characterization of the fiber types of regenerated muscles also suggests
that the regeneration pattern is little disturbed by the degeneration of
the original muscle fibers (Libelius, Sonesson, Stamenovic and Thesleff,
1970; Hall-Craggs and Seyan, 1975) and I have recently confirmed this
(Butler, 1986).

   The fate of motoneuron terminals during bupivacaine treatment is
crucial for the interpretation of experiments. The literature reports
few signs of damage to motor nerve terminal organelles. While end
plates (both extrafusal (Jirmanova, 1975) and intrafusal (Milburn,
1976) appear to withdraw from necrotic muscle fibers, evidence for such
withdrawal has not been observed for intrafusal trail terminals
(Milburn, 1976). Purves (1975) also reports that intact neuron
terminals will 'lift off' degenerating targets.

   The innervation of regenerating muscle cells seems to occur quite
soon after bupivacaine treatment. Jirmanova and Thesleff (1972) report
intracellularly recorded miniature end plate potentials near nerve twigs
only 3 days after treatment. A great deal of work has been reported on
selective synaptic connections, especially motor innervation in
vertebrates, and the reader is referred to any of a number of excellent
reviews such as that by Purves and Lichtman (1985).

   I have developed a preparation which presents intact alpha
motoneurons (bearing focal plate endings) with a choice of two
distinctly different types of muscle to innervate – fast twitch and slow
tonic (Butler, 1985a,b & 1986). An important difference between the two
groups is that slow fibers possess a multiple innervation via
motoneurons with 'en grappe' terminal arborizations while twitch fibers
are singly innervated by motoneurons with focal plate terminals.

Both fiber types exist in mammals with the twitch group almost omnipresent. However a few muscles, notably extraocular eye muscles, do contain slow fibers and they exhibit the innervation and physiological characteristics of slow muscle (Mayr, 1971).

The basic idea is: (1) to take an extraocular muscle (medial rectus, MR) from a rat and transplant it into a fast twitch muscle in its own leg; (extensor digitorum longus, EDL); (2) treat the transplant and host with bupivacaine; and (3) study the resulting regenerated muscle at selected times post treatment for evidence that the transplanted slow muscle fibers have regenerated and have been successfully innervated. Also sought is evidence that their motoneurons (which originally had plate endings) now exhibit some modification towards 'en grappe' endings, thus indicating that the target had a substantial effect in remodelling terminal morphology. The results indicated that both the host muscle <u>and</u> the transplant degenerated and subsequently regenerated as expected and the transplant was readily identified microscopically, lying within the host, in both transverse and longitudinal sections, at all stages. Within the first week, the existing muscle fibers reduce in calibre with fragmented and irregular outlines and become separated from each other. Nuclei surrounded by cytoplasmic rings can be seen. By 7 days, fiber diameters are increasing and greater organization among myofibers is present. Vascularization in the transplant region is evident. At 14 days, the cells are larger and organization is at a higher level. Some endplates on EDL are seen. By 21 days, fiber calibers have returned almost to normal in both host and transplant regions and some degree of hypervascularization exists. ATPase histochemical profiles indicate that slow tonic muscle fibers have regenerated in the region where the transplant is found. At this stage, nerves can be seen entering the bundle of regenerated MR fibers. A few course over the surface of the myofibers and rather than terminating in focal end plates, as they do on regenerated EDL, they can be seen to terminate in a series of fine ramifications on individual regenerated MR fibers. This could be interpreted as evidence for: (1) the conversion of intact alpha motoneuron plate terminals to "engrappe" terminals by the presence of a different target; or (2) by the sprouting from existing alpha branches and the sprouts innervating the transplanted muscle; (3) the innervation of the MR fibers by static gamma motoneurons.

It is unlikely that these endings represent sprouted sensory neurons innervating the extraocular palisade (Ruskell, 1978) since the former occur towards the middle of the fibers rather than at the old myotendinous junctions. I feel that it is also unlikely that these endings are due to static gamma fibers which have 'en grappe' terminals although this point has not yet been firmly established. Milburn (1976) found these gamma trails to be resistant to withdrawal from old basal lamina.

COMMENT

The distinction between sprouting from intact neuromuscular junctions and regeneration has been made. The real question is whether these two events are differentially regulated. During regeneration, the growing axon and its branches reinnervate targets free of the influence of any direct target-derived factor, and directed by means arising from an existing innervation site for that neuron. These regenerating axons can innervate two different types of muscle with two morphologically different terminals but the results could indicate that the mechanism is either local responses within each branch or that the alpha motoneuron

is at least bideterminate and can recognize at least two different types
of target simultaneously, and respond appropriately.

The case with intact neurons is quite different. Whether those
branches which innervate the regenerated transplant muscle are converted
existing plate endings or sprouts, the fact that their terminal
morphology differs from the existing endings suggests a very local
response.

REFERENCES

Bennett, M.R., Pettigrew, A.G., and Taylor, R.S., 1973, The formation of
    synapses in reinnervated and cross-reinnervated adult avian muscle,
    J. Physiol., 230: 331-357.
Benoit, P.W., and Belt, W.D., 1970, Destruction and regeneration of
    skeletal muscle after treatment with a local anaesthetic,
    bupivacaine (Marcaine), J. Anat., 107: 547-556.
Brown, M.C., and Butler, R.G., 1976, Regeneration of afferent and
    efferent fibres to muscle spindles after nerve injury in adult
    cats, J. Physiol., 260: 253-266.
Buller, A.J., Eccles, J.C., and Eccles, R.M., 1960, Interactions
    between motoneurons and muscles in respect of the characteristic
    speeds of their responses, J. Physiol., 150: 417-439.
Butler, R., 1980, Can a regenerating alpha motoneuron innervate two
    different types of intrafusal muscle fibre simultaneously?
    Neurosci. Abst., 6:92.
Butler, R., 1985a, Evidence for the continuing role of targets in
    controlling motoneuron terminals in adult mammals, Neurosci. Abst.,
    11: 1102.
Butler, R., 1985b, Plasticity of regenerating and intact alpha
    motoneuron terminals, Proc. Can. Fed. Biol. Sci.,28: 143.
Butler, R., 1986, Motoneuron-target interactions during regeneration,
    Abst. for "The Current Status of Peripheral Nerve Regeneration", a
    satellite symposium to XXX IUPS Congress, Edmonton, July 1986.
Carlson, B.M., 1972, "The Regeneration of Minced Muscles", Karger,
    Basel, Switzerland.
Carlson, B.M., 1976, A quantitative study of muscle fiber survival and
    regeneration in normal, predenervated, and Marcaine-treated free
    muscle grafts in the rat, Exp. Neurol., 52: 421-432.
Glicksman, M.A., and Sanes, J.R., 1983, Differentiation of motor nerve
    terminals formed in the absence of muscle fibres, J. Neurocytol.,
    12: 661-671.
Hall-Craggs, E.C.B., 1974, Rapid degeneration and regeneration of a
    whole skeletal muscle following treatment with bupivacaine
    (Marcain), Exp. Neurol.,43: 349-358.
Hall-Craggs, E.C.B., 1980, Early ultrastructural changes in skeletal
    muscle exposed to the local anaesthetic bupivacaine (Marcaine),
    Br. J. exp. Path., 61: 139-149.
Hall-Craggs, E.C.B., and Seyan, H.S., 1975, Histochemical changes in
    innervated and denervated skeletal muscle fibers following
    treatment with bupivacaine (Marcain), Exp. Neurol.,46: 345-354.
Huang, C.L.-H., and Keynes, R.J., 1983, Terminal sprouting of mouse
    motor nerves when the post-synaptic membrane degenerates, Brain
    Res., 274: 225-229.
Jirmanova, I., 1975, Ultrastructure of motor endplates during
    pharmacologically induced degeneration and subsequent regeneration
    of skeletal muscle, J. Neurocytol., 4: 141-155.
Jirmanova, I., and Thesleff, S., 1972, Ultrastructural study of
    experimental muscle degeneration and regeneration in the adult rat,
    Z. Zellforsch., 131: 77-97.

Libelius, R., Sonneson, B., Stamenovic, B.A., and Thesleff, S.,1970,
     Denervation-like changes in skeletal muscle after treatment with a
     local anaesthetic (Marcaine), J. Anat., 106: 297-309.
Mayr, R., 1971, Structure and distribution of fibre types in the
     external eye muscles of the rat, Tissue & Cell, 3: 433-462.
McMahan, U.J., Edgington, D.R., and Kuffler, D.P., 1980, Factors that
     influence regeneration of the neuromuscular junction, J. exp.
     Biol., 89: 31-42.
Milburn, A., 1976, The effect of the local anaesthetic bupivacaine on
     the muscle spindle of the rat, J. Neurocytol., 5: 425-446.
Purves, D., 1975, Functional and structural changes in mammalian
     neurones following interruption of their axons, J. Physiol., 252:
     429-463.
Purves, D., and Lichtman, J.W., 1985, Principles of Neural Development,
     Sinauer Associates Inc., Sunderland, Mass.
Rees, R.P., 1978, The morphology of interneuronal synaptogenesis - a
     review, Fed. Proc., 37: 2000-2004.
Ruskell, G.L., 1978, The fine structure of innervated myotendinous
     cylinders in extraocular muscles of rhesus monkeys, J.
     Neurocytol., 7: 693-708.
Sanes, J.R., 1987, Cell lineage and the origin of muscle fiber types,
     Trends in Neurosciences, 10: 219-221.
Sanes, J.R., Marshall, L.M., and McMahan, U.J., 1978, Reinnervation of
     muscle fiber basal lamina after removal of myofibers, J. Cell
     Biol., 78: 176-198.
Sokoll, M.D., Sonesson, B., and Thesleff, S., 1968, Denervation changes
     produced in an innervated skeletal muscle by long-continued
     treatment with a local anaesthetic, Eur. J. Pharmacol., 4: 179-187.
Studitsky, A.N., 1959, Experimental Surgery of Muscles (In Russian),
     Izdatel, Akad. Nauk. S.S.S.R., Moscow.
Wernig, A., Carmody, J.J., Anzil, A.P., Hansert, E., Marciniak, M. and
     Zucker, H., 1984, Persistence of nerve sprouting with features of
     synaptic remodelling in soleus muscles of adult mice, Neuroscience,
     11: 241-253.

REGENERATION OF MUSCLE SPINDLES IN GRAFTED EXTENSOR DIGITORUM
LONGUS MUSCLE OF THE RAT

T. Soukup

Institute of Physiology
Czechoslovak Academy of Sciences
Vídeňská 1083, 142 20 Prague 4, Czechoslovakia

INTRODUCTION

Mammalian skeletal muscles possess considerable capacity
to regenerate following muscle injury (e.g. Studitsky, 1977).
Grafting of a muscle results in the degeneration of muscle fibers
due to ischemia, but within several days, their regeneration
takes place and the differentiating graft is revascularized and
reinnervated during the first three weeks after the operation
(for review see Carlson, 1983). Intrafusal fibers of muscle spindles undergo the same process of degeneration and regeneration
as the whole muscle and they do regenerate in free grafts (Carlson and Gutmann, 1975; Schmalbruch, 1977; Rogers, 1982). The
standard technique of free grafting has a great drawback, as the
onset and extent of regeneration of the neuromuscular pathways
disrupted at the time of grafting is difficult to control. To
overcome this disadvantage, nerve-intact grafts (Rogers and Carlson, 1981), bupivacaine treatment (Milburn, 1976) or other techniques, e.g. temporary ischemia (Diwan and Milburn, 1986; Scott
et al., 1988) were introduced for the study of spindle regeneration. These techniques leave the innervation pathways in the muscle intact, thus giving a better chance for regeneration than
can be expected after standard grafting, where the reinnervation
is likely to be more random. Because of the possible practical
utilization, we returned to the classical model of standard free
grafting, but we allowed a sufficiently long time for muscle regeneration. In the present paper we focussed on the histochemical
reactions of intrafusal fibers after autotransplantation. The
preliminary results were published previously (Soukup, 1981).

The experiments were performed on six 2-month-old female
rats of the Wistar strain. Under Nembutal anesthesia, the right
EDL muscle was excised, the nerve and blood supply were removed
and the muscle was sutured back to its original place. Both
transplanted and control left EDL were examined 1, 3 and 10
months later.

HISTOCHEMICAL REACTIONS OF REGENERATED INTRAFUSAL FIBERS

Cryostat sections were stained for myofibrillar adenosine-

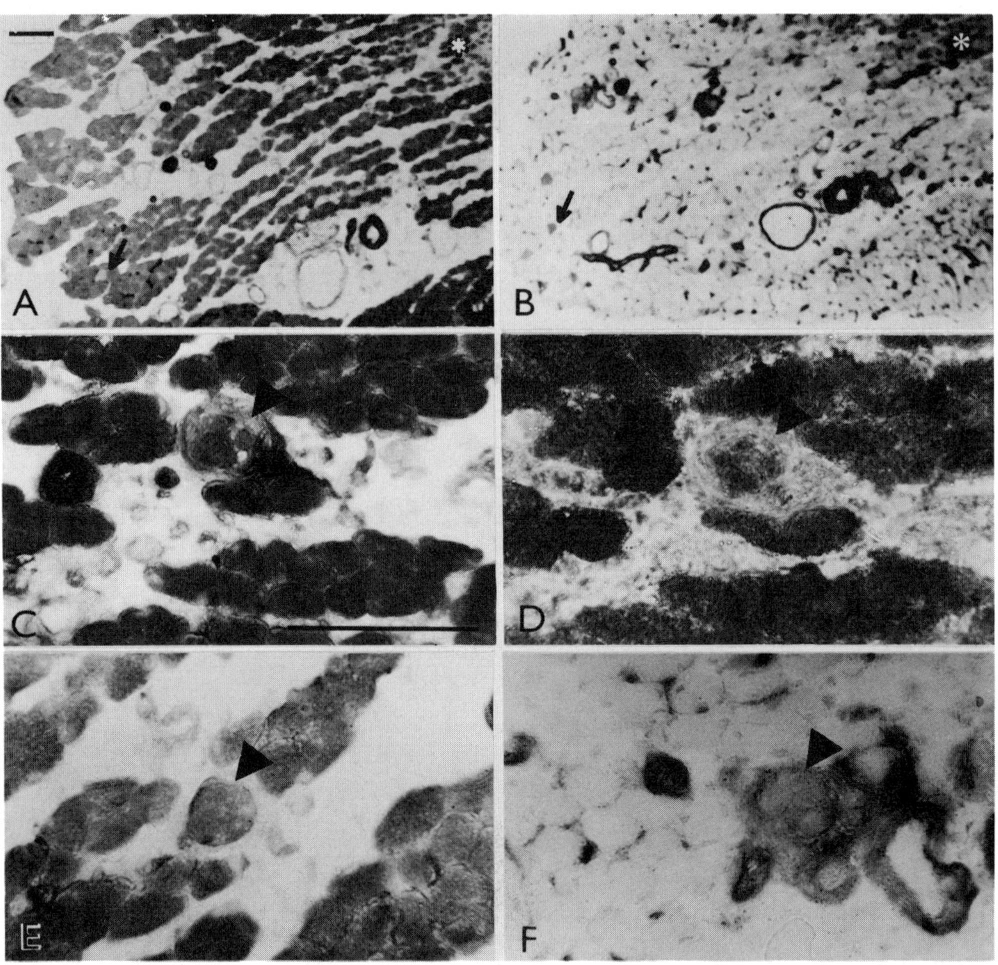

Fig. 1. Transverse sections through grafted EDL muscle stained
for alkali (A,C,E) or acid (B,F) preincubated ATPase and
SDH (D) activity, 1 month after autotransplantation in
a 2-month-old rat. Intrafusal fibers (arrowheads) are
not differentiated into histochemically distinct fiber
types (C-F). Note the peripheral rim of surviving fibers
with preserved histochemical distinctions (arrows) and
the less differentiated center of the graft (asterisks)
(A,B). Scale bars (A-B, C-F) indicate 100 μm.

triphosphatase (ATPase) activity after acid (pH 4.3 and 4.5)
and alkaline (pH 10.4) preincubations (Guth and Samaha, 1970)
and for succinic dehydrogenase (SDH) activity (Nachlas et al.,
1957).

   One month after grafting a peripheral rim of extrafusal
fibers surviving the temporary ischemia and remaining differentiated into the fiber types was clearly distinguishable from the
regenerated muscle fibers that exhibited homogeneous ATPase
(Fig. 1A, B) or SDH reactions. Regenerated spindles, found among
newly formed extrafusal muscle fibers, contained more than typical four intrafusal fibers exhibiting the same ATPase (Fig. 1C,
E, F) and SDH (Fig. 1D) activities resembling those of surround-

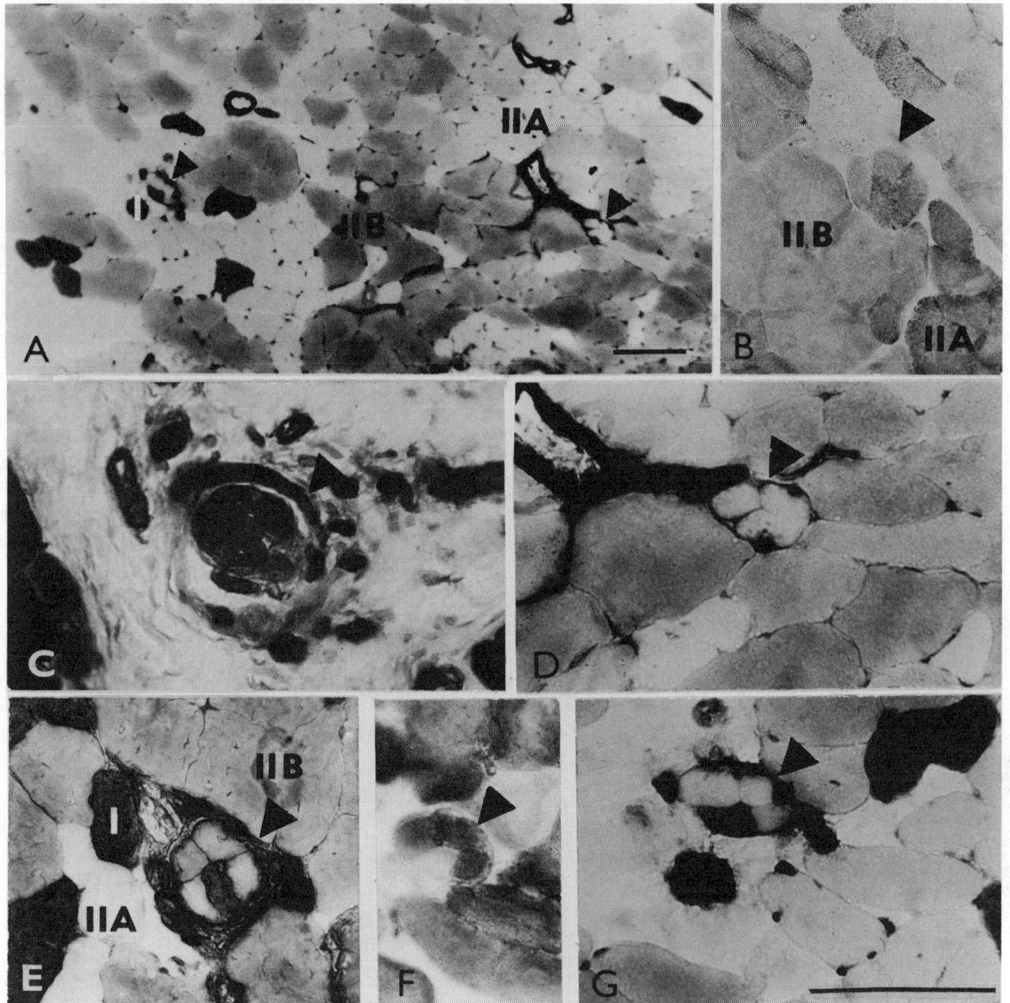

Fig. 2. Transverse sections through an EDL graft 10 months after
autotransplantation in a 2-month-old rat. Sections were
stained for alkali (C,F) and acid (A,D,E,G) preincubated
ATPase and SDH (B) activity. Both extrafusal and intrafusal muscles fibers are differentiated into histochemically distinct fiber types. Note the presence of muscle
spindles with intrafusal fibers of the fast type only
(B-D), or with fast and one slow type intrafusal fiber
(E-G). Scale bars A and G (common for B-G) indicate
100 μm.

ing extrafusal fibers. The capsule was thin and tightly bound
with the intrafusal muscle fibers.

Ten months after grafting, the muscle fiber types were
clearly differentiated. Extrafusal fibers exhibited a pattern
typical for reinnervation (Fig. 2A). The majority of muscle spindles (Fig. 2C, D) contained intrafusal fibers of only the fast
type, with alkali-stable and acid-labile ATPase activity. In
about every third spindle, one intrafusal fiber was of the slow

113

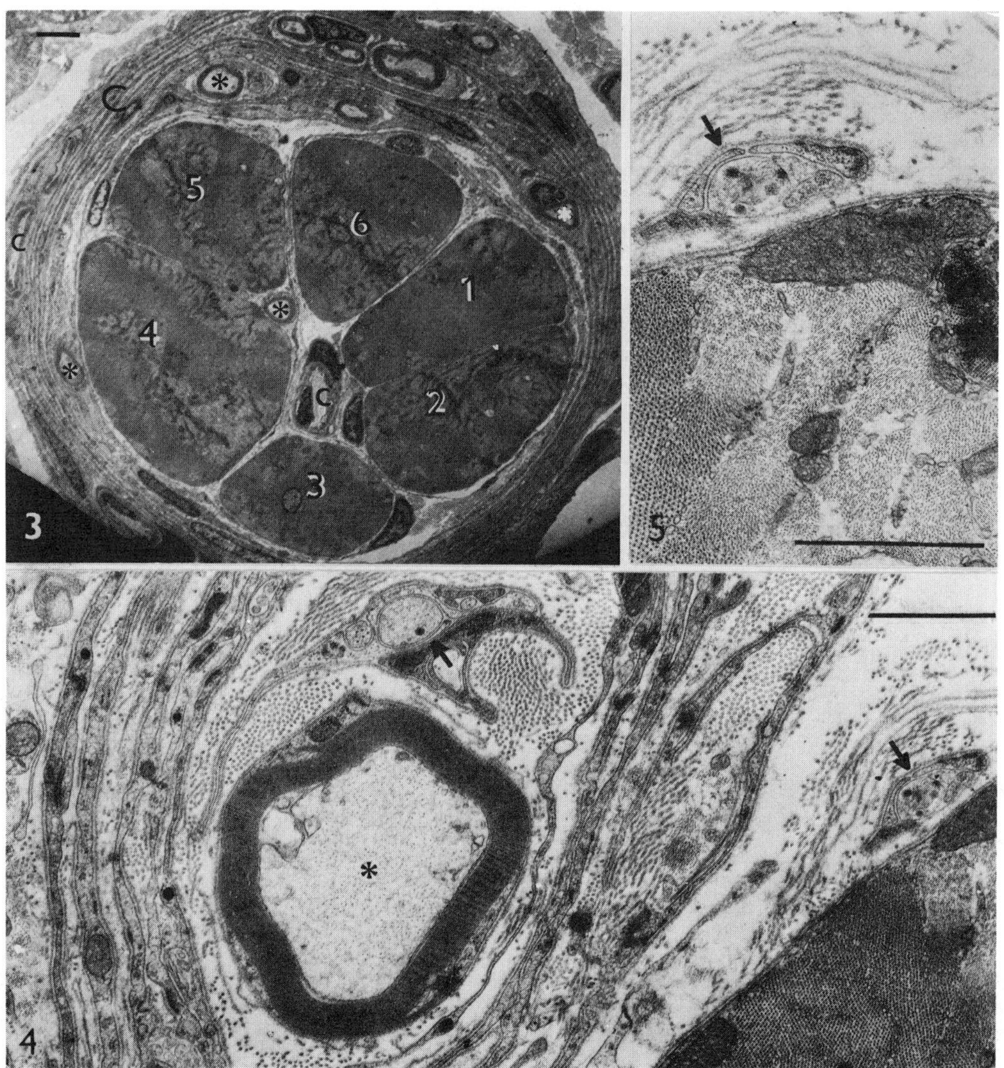

Figs. 3-5. A transverse section through a typical spindle from
the autotransplanted EDL muscle in a 2-month-old rat,
3 months after the operation. The spindle contains
many myelinated (asterisks) and unmyelinated nerve
fibers (arrows). The multilayered capsule (C) tightly
encircles the intrafusal fibers, the number of which
is increased (numerals 1-6). c - capillary. Enlargements (Figs. 4 and 5) demonstrate an area near an intrafusal fiber numeral 5. Note that an unmyelinated
fiber (arrow) is separated from the intrafusal fiber
only by basement membranes. Scale bars indicate 2.5 μm
in Fig. 3 and 1 μm in Figs. 4 and 5.

extrafusal type, with high acid and low alkali-preincubated
ATPase activity (Fig. 2E-G). About half of the spindles of the
first type and some spindles of the second type (Fig. 2E) contained two types of intrafusal fibers corresponding to fast extrafusal IIB(FG) fibers, only partially inhibited at pH 4.5 and
to IIA(FOG) fibers inhibited completely at this pH (cf. Soukup

et al., 1979). The SDH reaction (Fig. 2B) also suggested a similarity between intrafusal and extrafusal fast fibers. No spindles
with a typical equatorial characteristics were found on serial
sections cut through the whole muscle belly. The capsule became
thicker at this time (Fig. 2C, E), but it still tightly surrounded the intrafusal bundle usually without any conspicuous periaxial space.

<u>Ultrastructural characteristics</u>. In order to know more
details about the structure of regenerated spindles, two EDL
muscles were studied in the electron microscope 3 months after
autotransplantation. EDL muscles were excised under ether anesthesia, fixed in 1 % glutaraldehyde and 1 % paraformaldehyde,
postfixed in 2 % OsO_4, dehydrated and embedded in Durcupan (Flu-
ka). Ultrathin sections were cut on a Reichert OM U3 ultramicrotome, stained with 1 % uranyl acetate and 0.1 % lead citrate and
examined in a Philips 300 electron microscope. Ultrastructural
analysis of seven spindles confirmed that the number of intrafusal fibers was usually increased; some of them were separated
from each other only by their basal laminae, the inner capsule
was less developed (Fig. 3). Spindles were innervated with large
myelinated (Figs. 3 and 4) and small (0.1 - 0.3 µm) unmyelinated
nerve fibers (Figs. 4 and 5). Myelinated nerve fibers ran between
capsular lamellae and/or inside the spindle, but the nerve terminals were not observed. No accumulation of nuclei or typical
enlargement of the periaxial space or any other characteristics
of the equatorial region were found in any of the spindles examined. It was therefore not possible to distinguish intrafusal
types on cross sections. Small unmyelinated fibers accompanied
myelinated fibers not only in the capsule (Fig. 4), but they frequently penetrated into the spindle and ran along the surface of
an intrafusal fiber (Fig. 5), or in the free spaces among fibers.
The capsule was highly vascularized and a capillary was often
seen in between intrafusal fibers (Fig. 3).

DISCUSSION

In contrast to minced muscle preparations (Zelená and Sobotková, 1971), spindles in free grafts of whole muscles regenerate,
although they have altered histochemical and structural properties (Carlson and Gutmann, 1975; Schmalbruch, 1977; Rogers, 1982).
They are not differentiated into the typical three fiber types
(cf. Soukup, 1976), but their ATPase activity resembles that of
surrounding extrafusal fibers. The rat EDL is a fast muscle composed of IIA and IIB muscle fibers with some slow type I fibers.
Correspondingly, the majority of regenerated intrafusal fibers
exhibit ATPase and SDH reactions typical for IIA and IIB fast
extrafusal fibers. Even those intrafusal fibers with high acid
ATPase resemble slow extrafusal fibers, but not nuclear bag_1 or
bag_2 fibers. This suggests that spindles could be reinnervated
by branches of extrafusal motor axons although the presence of
motor terminals was not confirmed. No sensory nerve terminals
were found in regenerated muscle spindles of this series, nuclear
bags and chains were also absent, which corroborates previous
findings (Milburn, 1976; Rogers, 1982). We conclude that the morphogenetic interactions leading to the differentiation of intrafusal fibers during development (Zelená, 1976), are deficient
in spindle regeneration.

REFERENCES

Carlson, B.M., 1983, The regeneration and transplantation of
        skeletal muscle, in: "Nerve, Organ and Tissue Regenera-
        tion: Research Perspectives", F.J. Seil, ed., Academic
        Press, pp. 431-454.
Carlson, B.M., and Gutmann, E., 1975, Regeneration in free grafts
        of normal and denervated muscles in the rat: Morphology
        and histochemistry, Anat. Rec., 183:47-62.
Divan, F.H., and Milburn, A., 1986, The effects of temporary
        ischaemia on rat muscle spindles, J. Embryol. exp. Mor-
        phol., 92:223-254.
Guth, L., and Samaha, F.J., 1970, Procedure for the histochemical
        demonstration of actomyosin ATPase. Exp. Neurol., 28:
        365-367.
Milburn, A., 1976, The effect of the local anaesthetic bupiva-
        caine on the muscle spindle of rat. J. Neurocytol., 5:
        425-446.
Nachlas, M.M., Tsou, K.C., De Souza, E., Cheng, C.S., and Selig-
        man, A.M., 1957, Cytochemical demonstration of succinic
        dehydrogenase by the use of a new p-nitrophenyl substi-
        tuted ditetrasole, J. Histochem. Cytochem., 5:420-436.
Rodgers, S.L., 1982, Muscle spindle formation and differentiation
        in regenerating rat muscle grafts. Devl. Biol., 94:265-283.
Rodgers, S.L., and Carlson, B.M., 1981, A quantitative assessment
        of muscle spindle formation in reinnervated and non-rein-
        nervated grafts of the rat extensor digitorum longus
        muscle, Neuroscience, 6:87-94.
Schmalbruch, H., 1977, Regeneration of soleus muscles of rat
        autografted in toto as studied by electron microscopy.
        Cell Tiss. Res., 177:159-180.
Scott, S.J.A., Barker, D., and Berry, R.B., 1988, Muscle spindle
        recovery in orthotopic grafts of cat muscles, this volume.
Soukup, T., 1976, Intrafusal fibre types in rat limb muscle spin-
        dles. Morphological and histochemical characteristics.
        Histochemistry, 47:43-57.
Soukup, T., 1981, Histochemical differentiation of intrafusal
        and extrafusal muscle fibres after autotransplantation,
        Physiol. bohemoslov., 30:457.
Soukup, T., Vydra, J., and Černý, M., 1979, Changes in ATPase
        and SDH reactions of the rat extrafusal and intrafusal
        muscle fibres after preincubations at different pH, Histo-
        chemistry, 60:71-84.
Studitsky, A.N., 1977, "Muscle Transplantation in Animals",
        Medgiz, Moscow.
Zelená, J., 1976, The role of sensory innervation in the develop-
        ment of mechanoreceptors, Progr. Brain Res., 43:59-64.
Zelená, J., and Sobotková, M., 1971, Absence of muscle spindles
        in regenerated muscles of the rat, Physiol. bohemoslov.,
        20:433-439.

ACKNOWLEDGEMENTS

    The author wishes to thank Dr. J. Zelená and Dr. P. Hník
for critical comments during the preparation of the manuscript
and Mrs. M. Krupková and Mr. M. Doubek for their skilful technical assistance.

MUSCLE-SPINDLE RECOVERY IN ORTHOTOPIC GRAFTS OF CAT MUSCLES

J.J.A. Scott, D. Barker and R.B. Berry*

Department of Zoology, University of Durham, Durham
DH1 3LE, U.K., and *Shotley General Hospital
Consett, DH8 0NB, U.K.

INTRODUCTION

During a muscle-graft operation the grafted muscle undergoes a period
of ischaemia before it is revascularized by the new vascular pedicle. In
these experiments, the effects of this ischaemia on the subsequent recovery
of cat muscle spindles has been examined in orthotopic grafts of the extensor digitorum longus (EDL) muscle.

Some preliminary results of this study have been published previously
(Barker et al., 1987).

METHODS

The experiments were performed on 11 adult cats (average weight,
2.4kg) which were anaesthetized with sodium pentobarbitone (45mg/kg, i.p.,
supplemented as required) during both the preliminary surgery and acute
recording procedures.

In the first group of three cats, which provided a control for the
study, the muscle nerve to EDL was cut close to muscle entry and the ends
joined with 10/0 epineurial sutures. The proximal and distal tendons were
cut and repaired with 8/0 sutures. The blood supply was left intact.

In the first ischaemic group (5 cats) the nerve and tendons were
likewise cut and sutured and the main blood vessels were occluded with
Acland clamps for 1.5h, any small vessels being divided with a bipolar
cautery. The second ischaemic group (3 cats) was identical except that
the occlusion was maintained for 4.5h while the muscle was reflected into
a bath of mammalian Ringer at 4^{0}C. In each animal, the effectiveness of
both the occlusion and subsequent restoration of the blood supply was
demonstrated by the absence and subsequent presence of fluorescence by
the muscle under u.v. light following injection of fluorescein (15mg/kg
(i.v.).

After a recovery period of 13 weeks, the animals were prepared for
single-unit recording from EDL muscle-spindle afferents isolated in the
dorsal (L 6 and 7) roots. Ramp-and-hold stretches of the muscle of 2mm
amplitude were applied at 2.5, 5 and 10mms^{-1} with a hold of 1 or 3s. At

the termination of each experiment, the EDL muscle was removed and processed for the production of teased, silver preparations of the spindles
according to the method of Barker & Ip (1963) as modified by Barker et al.,
(1985).

RESULTS

Histological Analysis

Both primary (Ia) and secondary (II) afferents participate in the reinnervation of muscle spindles after nerve section, as well as afferents
(Ib) that formerly terminated in tendon organs (Banks et al, 1984). All
form annulospiral endings, and the success of the reinnervation achieved
may be quantified by ascertaining the proportion of the spindle population
that receive annulospiral endings.

In the control animals 63% of the spindles (n=135) received annulospiral innervation. In the ischaemic animals the proportion was similar,
being 61% (n=101) after 1.5h ischaemia and 66% (n=122) after 4.5h.

Physiological Analysis

Following the 13-week recovery period, a number of the regenerated
afferents responded abnormally to ramp-and-hold stretch. The most common
abnormality was the failure of the afferent to maintain firing throughout
the stretch cycle (Fig. 1). Such afferents either ceased firing at the peak
of the stretch (Fig. 1C), or fired throughout the stretch itself and for
part, but not all, of the hold phase (Fig. 1D). In the control group 75%
of the isolated afferents (n=107) responded normally, compared with 65%
(n=171) and 79% (n=104) after 1.5 and 4.5h ischaemia, respectively.

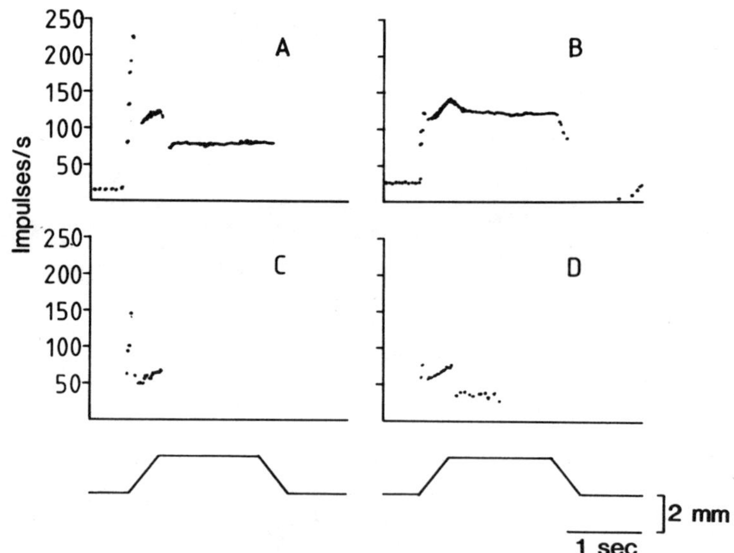

Fig. 1.   Instantaneous frequency records of the responses to stretch of
          EDL spindle afferents. A, normal Ia-like response; B, normal
          II-like response; C and D, abnormally responding afferents from
          a 1.5h ischaemic animal.

118

The mean peak firing rates (measured at the peak of a 10mms^{-1} stretch) of the afferent populations in each group were significantly lower than in normal animals (P<0.05, Student's t test) being 71.7±3.9, 63.9±3.3 and 70.6±4.5 impulses/s in the control, 1.5h and 4.5h animals, respectively, compared with 89.5±3.6 in normal animals. There was no significant difference between the three operated groups. This reduction compared with normal can be attributed to the presence of the abnormally-responding afferents which displayed reduced firing rates. The mean peak firing rates of the populations of normally-responding afferents were all similar to normal.

The afferents were also identified as being Ia-like (Fig. 1A) or II-like (Fig. 1B) according to the criteria of Barker et al (1986) based on the patterns of response. In normal animals 46% of the afferents gave Ia-like responses to the ramp-and-hold stretch. In the control animals 36% were identified as being Ia-like, falling to 27% and 17% after 1.5h and 4.5h ischaemia, respectively. The ratios of group I:II afferents, however, did not vary. In the normal animals 44% of the afferents conducted at more than 70ms^{-1} compared with 47%, 50% and 48% in the control, 1.5 and 4.5h groups, respectively (conduction velocity (CV) was measured central to the lesion site).

The CV distributions of the Ia-like and II-like afferents are shown in Fig. 2. In the normal animals there was an overlap of the distributions between 66 and 73ms^{-1} (Fig. 2A). In the control animals the degree of overlap was much greater (Fig. 2B), the CV ranges being 46-115ms^{-1} and 19-115ms^{-1} for the Ia-like and II-like afferents, respectively. Similar degrees of overlap were observed in the two ischaemic groups (Fig. 2C, D). Thus the CVs of the II-like afferents encompassed the full range of spindle-afferent CVs. The CVs of the Ia-like afferents, however, lay predominantly above 70ms^{-1}. In the control animals 67% of the afferents that gave a Ia-like response conducted at more than 70ms^{-1}, the corresponding proportions being 79% and 93% in the two ischaemic groups (Fig. 2).

In the three 4.5h ischaemic animals, the ventral roots were also divided to test for the presence of functional γ efferents. Totals of 26 γs and 6 γd efferents were isolated which activated one or more of the afferents, such that 39 γs and 10 γd actions were recorded. In each case, the nature of the activation was clearly identifiable from the afferents' responses to stretch in the presence and absence of efferent stimulation, and, in those cases where an efferent activated more than one afferent, the type of activation was consistent between afferents.

DISCUSSION

The main finding from these experiments is that the periods of ischaemia of up to 4.5h duration had no profound effects on either the reinnervation or the functional recovery of the muscle spindles in the orthotopic grafts. The overall level of spindle-afferent reinnervation, the proportion of normally-responding spindle afferents, and the mean firing rates were not significantly different in either of the ischaemic groups compared with the control animals.

Periods of ischaemia of up to five hours result in widespread ultra-structural changes in both extrafusal (Karpati et al., 1974) and intra-fusal (Diwan & Milburn, 1986) muscle fibres in rats, the earliest effects becoming apparent after 0.5h. In the intrafusal fibres these degenerative changes initially affect the mitochondria, followed by disintegration of the plasmalemma and loss of the Z-lines (Diwan & Milburn, 1986).

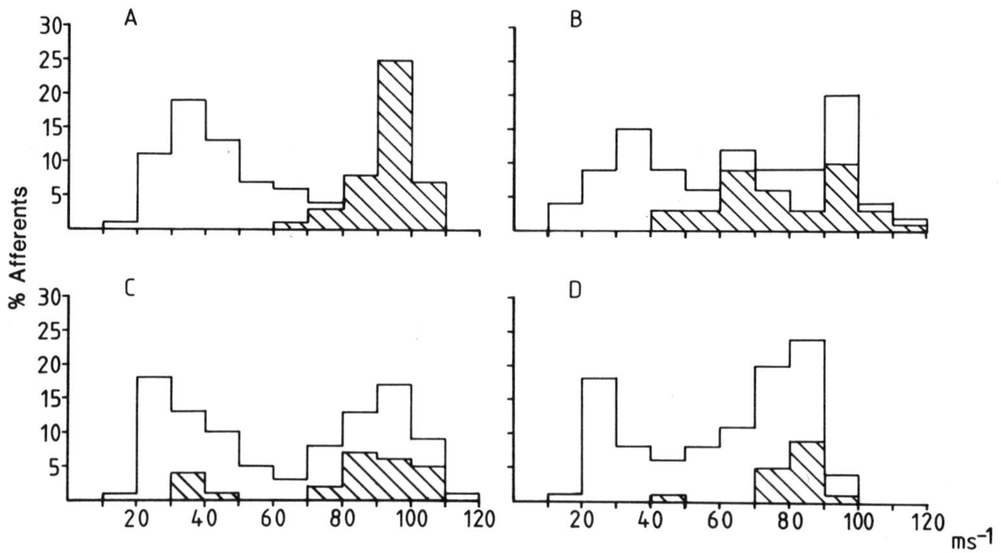

Fig. 2. Conduction velocity distributions of the Ia-like (hatched areas) and II-like spindle afferents from grafted EDL muscles. A, normal muscles; B, control muscles (tendons and nerves sectioned and sutured, blood supply left intact); C, orthotopic grafts with 1.5h ischaemia; D, grafts with 4.5h ischaemia.

Following reopening of the blood vessels after the 1.5 or 4.5h ischaemia, revascularization was immediate, as demonstrated by the infiltration of the muscle with fluorescein. Since the vascular network of the muscle would be relatively undisturbed, repair would be much faster than after free grafting of the muscle (Jennische, 1986) and might be completed prior to reinnervation by the regenerating muscle nerve.

The ratio of Ia-like:II-like responses recorded from each group of animals showed a progressive change which occurred in the absence of any change in the ratio of group I:II afferents sampled. Normally, 46% of the afferents show Ia-like responses to stretch, and 44% of the spindle-afferent population conduct at more than 70ms^{-1}. The proportion of Ia-like responses was 36% in the control animals, falling to 27% and 17% in the two ischaemic groups, though the proportion of afferents with CVs above 70ms^{-1} was unchanged at 50% and 48%, respectively. An increasing proportion of fast-conducting afferents were therefore giving II-like responses to muscle stretch. In the control animals such II-like responses could be generated by Ib (Banks, 1987) or Ia afferents reinnervating the secondary region of the spindles, or reinnervating the primary region and failing to form sufficient terminations on the bag$_1$ fibre (Banks & Barker, 1983). It is notable that relatively few of the Ia-like responses were generated by slowly-conducting afferents (Fig. 2). This suggests that bag$_1$ reinnervation in the primary site is more likely to be effected by group I rather than group II axons.

The periods of ischaemia applied in combination with section of the muscle nerve further reduced the proportion of Ia-like responses and it is possible that the ischaemia induced a long-lasting change in the nature of the bag$_1$ fibres, thereby altering the dynamic sensitivity of the af-ferent response to stretch. Further experiments are in hand to attempt to resolve this question.

A similar pattern of a reduced proportion of Ia-like responses was reported by Barker et al.,(1986) following long-term denervation, and they attributed this change to alterations in the mechanical properties of the bag_1 fibres.   In the present experiments, the periods of denervation were insufficient to give rise to such changes since these only became apparent after the spindles had been denervated for 12 weeks or more (Barker et al., 1986).

REFERENCES

Banks, R.W., 1987, Responses to small-amplitude sinusoidal stretching of cat peroneus brevis muscle spindles reinnervated after nerve section, J. Physiol., In Press.

Banks, R.W., and Barker, D., 1983, Recovery of cat hindlimb muscle afferents following nerve section, J. Physiol., 345:97P.

Banks, R.W., Barker, D.,and Stacey, M.J., 1984, Reinnervation of cat muscle spindles by foreign afferents after nerve section, J. Physiol., 357:21P.

Barker, D., Berry, R.B.,and Scott, J.J.A., 1987, Recovery of cat muscle spindles in orthotopic muscle grafts, J. Physiol., In Press.

Barker, D.,and Ip, M.C., 1963, A silver method for demonstrating the innervation of mammalian muscle in teased preparations, J. Physiol., 169-74P.

Barker, D., Scott, J.J.A.,and Stacey, M.J., 1985, Sensory reinnervation of cat peroneus brevis muscle spindles after nerve crush, Brain Res., 333:131-138.

Barker, D., Scott, J.J.A.,and Stacey, M.J., 1986, The reinnervation and recovery of muscle receptors after long periods of denervation, Exp. Neurol., 94:184-202.

Diwan, F.H.,and Milburn, A., 1986, The effects of temporary ischaemia on rat muscle spindles, J. Embryol. exp. Morphol., 92:223-254.

Jennische, E., 1986, Rapid regeneration in post-ischaemic skeletal muscle with undisturbed microcirculation, Acta Physiol. Scand., 128:409-414.

Karpati, G., Carpenter, S. Melmed, C.,and Eisen, S.S., 1974, Experimental ischaemic myopathy, J. Neurol. Sci., 23:129-161.

INVESTIGATION OF MECHANORECEPTORS IN THE SKELETAL MUSCLES OF
RATS UNDER DIFFERENT EXPERIMENTAL CONDITIONS

A.N. Studitsky, R.P. Zhenevskaya, M.M. Umnova,
I.L. Novoselova and T.P. Seene

Severtzov Institute of Evolutionary Animal
Morphology and Ecology
USSR Academy of Sciences
Moscow, USSR

When studying the receptors of skeletal muscles, most
attention is paid to muscle spindles, other types of sensory
nerve endings receiving less attention (Zhenevskaya, 1974).
In recent years, the methods of transplantation of skeletal
muscles, minced or whole, which were first developed in the
laboratory of A.N. Studitsky (Studitsky, 1959) have been widely used. These methods have led to numerous studies on the
restitution in transplanted muscles of various types of mechanoreceptors. Less understood is the fate of mechanoreceptors
under other experimental conditions, in particular, after denervation of the skeletal muscle, after transplantation of a
minced muscle and under the effect of glucocorticoids.

In the present study, sexually mature albino rats were
used (males, body mass 120-250 g). Three series of experiments
were performed. In series I the processes of de- and regeneration of sensory endings were studied after denervation of
the gastrocnemius muscle and its subsequent spontaneous reinnervation (the denervation of the gastrocnemius muscle was
attained by cutting the tibial nerve at a distance of 1 cm
from the muscle). In series II, the restoration of mechanoreceptors was studied in the autotransplanted gastrocnemius
muscles grafted in situ after mincing to fragments of 1-2 mm
in size. In series III, intrafusal muscle fibers of the soleus
muscle were studied by electron microscopy, following intraperitoneal injections of the glucocorticoid dexamethasone
(1 mg/kg of body mass) for 10 days resulting in atrophy mainly
of glycolytic muscle fibers (Seene and Viru, 1982).

The process of degeneration and restitution of sensory
endings was studied in muscle spindles, Golgi organs, tendon
spindles and sensory colbs. As a result of denervation, the
terminals of sensory nerve fibers degenerate, the capsule of
receptors becomes thickened and the receptor bodies remain.
On spontaneous re-innervation sensory fibers grow into receptors and their innervation is recovered. The development of
additional axons was shown (Zhenevskaya and Umnova, 1965) in

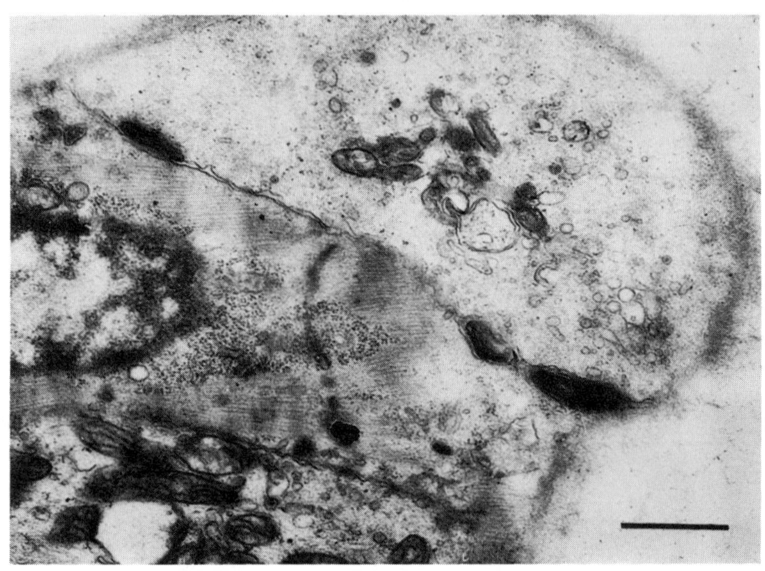

Fig. 1. A sensory ending on the bag$_2$-fiber. Myelin-like structures near synaptic cleft. Bar: 1 μm.

the sensory colbs 60 days after the nerves had been cut.

In the autografts developing from minced muscle tissue the formation of new sensory colbs or Paccinian corpuscles was recorded 3-4 weeks after transplantation. The development of de novo muscle spindles in autografts from minced muscle tissue was very rare.

The ultrastructure of nuclear bag$_{1,2}$ intrafusal muscle fibers was studied during administration of dexamethasone in longitudinal ultra-thin sections. The bag$_1$ muscle fiber was 11 μm in the widest section of the equatorial zone and was filled with densely-packed lightly stained muscle nuclei arranged in several rows and with very small nucleoli. In the equatorial region, intrafusal fibers have very few myofibrils, separating muscle nuclei from sensory endings. The sensory endings of a longitudinally-cut B$_1$-fiber were spaced at a roughly equal distance from one another, which was 4-8 μm on either side of the muscle fiber. The bag$_2$-fiber situated in the spindle portion adjoining to the equatorial zone was larger in size. It was 16 μm in the widest section and 7 μm in the narrowest part. In this region the fiber contained a large number of densely-packed myofibrils. These were absent in the center of the fiber, where several round muscle nuclei were arranged in a chain pattern, some of them having small granular nucleoli. The nuclei were surrounded by sarcoplasm containing mitochondria, ribosomes, granular sarcoplasmic reticulum. The distance between sensory endings was 7-17 um. In the sensory region of the bag$_2$-fiber a small leptomere was occasionally found 0.6 μm long and 0.25 μm wide, lying between Z-lines transversely to the axis of myofibrils. In the polar intracapsular zone several motor presumably gamma endings (Fig.2) were observed on the bag$_2$-fiber. Basement membrane intervened

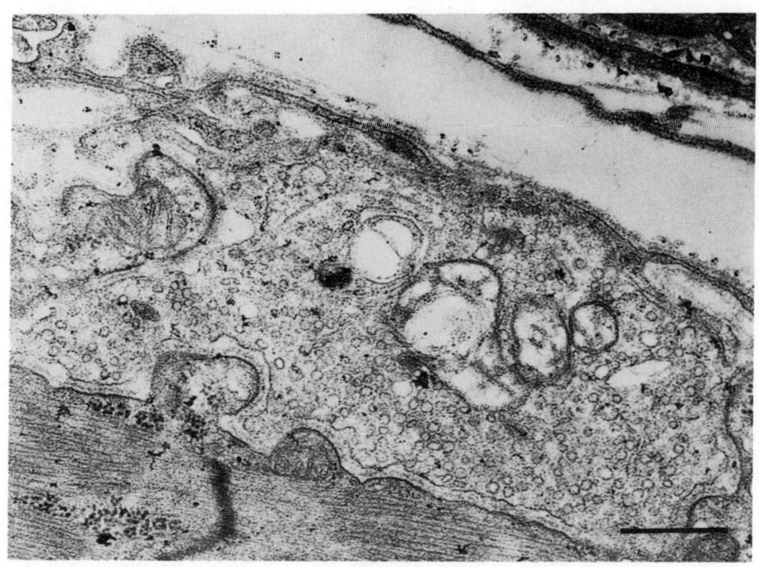

Fig. 2. Presumably gamma ending on a bag$_2$-fiber. Bar: 0.5 μm.

between the pre- and postsynaptic membranes. The motor terminal contained numerous light vesicles and a few mitochondria.
The postsynaptic membrane formed a few wide and shallow folds.
Neurofilaments were absent in the motor terminals. Motor endings on the bag$_2$-fiber had a very small endplate. During administration of dexamethasone we did not observe degenerative
changes of myofilaments in bag$_{1,2}$-fibers. We noted the formation of myelin-like structures near the synaptic cleft(Fig.1)of
sensory endings, a local broadening of the synaptic cleft of
sensory endings, destruction of some mitochondria, the presence of lysosomes of heterogenous structure in the sarcoplasm
of intrafusal fibers. In fact, symptoms that apparently can
underlie the disturbance of synaptic transduction in these
regions. It is known that after injection of glucocorticoids,
no atrophy of intrafusal muscle fibers occurs in the slow
soleus muscle (Stern and Hannapel, 1973), while the reaction
of the muscle spindles to stretch changes in the fast gastrocnemius muscle (Botterman et al., 1981). Thus, the results
obtained demonstrate the lability of the response of the sensory link of innervation of skeletal muscles to various experimental conditions.

REFERENCES

Botterman, B.R., Eldred, E., and Edgerton, V.R., 1981, Spindle
        discharge in glucocorticoid - induced muscle atrophy.,
        Experimental Neurology, 72:25.
Seene, T., and Viru, A., 1982, The catabolic effect of gluco-
        corticoids on different types of skeletal muscle fibres
        and its dependence upon muscle activity and interaction
        with anabolic steroids, J. Steroid Biochemistry, 16:349.
Stern, L.Z., and Hannapel, L.K., 1973, The muscle spindle cor-
        tisone-induced muscle atrophy, Experimental Neurology,
        40:484.

Studitsky, A.N., 1959, "Experimental Studies of Muscle Surgery",
    Nauka, Moscow.
Zhenevskaya, R.P., 1974, "Neurotrophic Regulation of the Plastic
    Activity of Muscle Tissue", Nauka, Moscow.
Zhenevskaya, R.P., and Umnova, M.M., 1965, Degeneration and
    restoration of sensory nerve endings in the skeletal
    muscles, Arkhiv Anat., Histolog., Embryol., 49:3.

SENSORY NERVE ENDINGS IN MINCED MUSCLE OF YOUNG AND OLD RATS

N.V. Bulyakova

A.N. Severtzov Institute of Evolutionary Animal
Morphology and Ecology
Moscow, USSR

Muscle and tendon sensory innervation plays an important
role not only in the fine coordination of motions (Cooper,
1960), but also in the success of skeletal muscle regeneration
(Zhenevskaya, 1974; Studitsky, 1977). In the embryogenesis of
muscular tissue, the formation of mechanoreceptors is induced
by sensory nerve fibers. Their differentiation is completed
two to three weeks after the rat´s birth (Zelená, 1964; Milburn, 1973; Zelená and Soukup, 1977). Thus, sensory neurons
exert a morphogenetic influence upon muscle at that age. Later,
the transmission of impulses from the periphery to the nerve
centers is improved in sensory nerve fibers. It was shown that
after nerve section performed in newborn rats, muscle mechanoreceptors disintegrated rapidly and did not reappear (Zelená
and Hník, 1960). Some tendency to restoration of mechanoreceptors is observed only in two to three-week-old rats (Schiaffino
and Pierobon Bormioli, 1976).

The present experiments were designed to determine the
mechanoreceptor restoration in minced muscle autografts of
newborn, 1-week-old, 1-month-old and 24 to 30-month-old rats.
The regeneration process of muscle tissue after mincing undergoes changes similar to its embryogenesis. All transplants
were analysed 60 days after grafting, fixed in 10 % neutral
formol, histological sections were stained by the silver impregnation method of Bielschowsky - Gros.

The experimental data showed that free terminal plexuses
of sensory nerve fibers and non-free arbor-like receptors were
restored in minced muscle grafts of young and old rats. Generally, they were found near blood vessels and in connective
tissue. Simple encapsulated receptors were represented by modified sensory bulbs. They had some glial cells, one argyro-
phil terminal and were covered with a thin connective capsule
(Fig. 1). Modified sensory bulbs were less frequently found in
old rat grafts than in young ones. More complex mechanoreceptors such as Golgi tendon organs and muscle spindles were not
found. Only in neonatal and 1-week-old rats modified tendinous spindles could be seen in connective tissue of the
grafts. They had some glial cells and short terminals

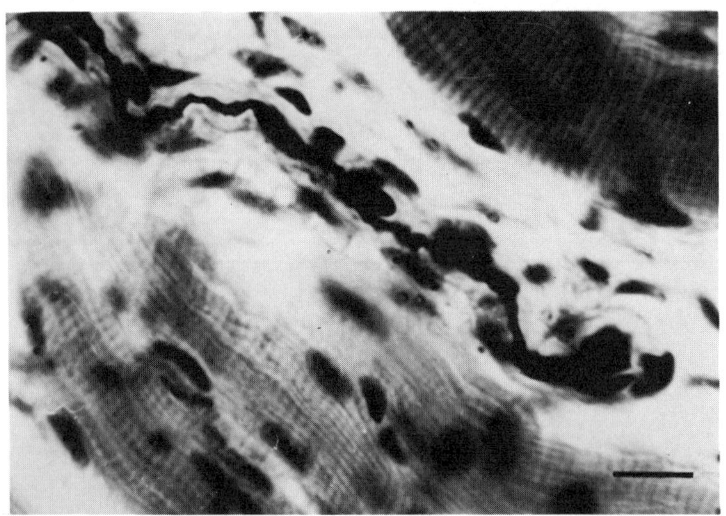

Fig. 1. A modified sensory bulb in minced gastrocnemius muscle
of a neonatal rat. 60-day graft. Bar: 20 μm.

formed by an ingrowing sensory nerve fiber (Fig. 2). Regions
similar to the formation of muscle spindles were also sometimes
found in minced muscle grafts in rats of the same age. They
were represented as muscle fibers with axial nuclear continuous
and discontinuous chains. Growing sensory axons ran parallel
to these fibers and formed one or two loops around them. Such
structures did not have a connective tissue capsule, bag intrafusal muscle fibers and motor endings at the polar zones
(Fig. 3).

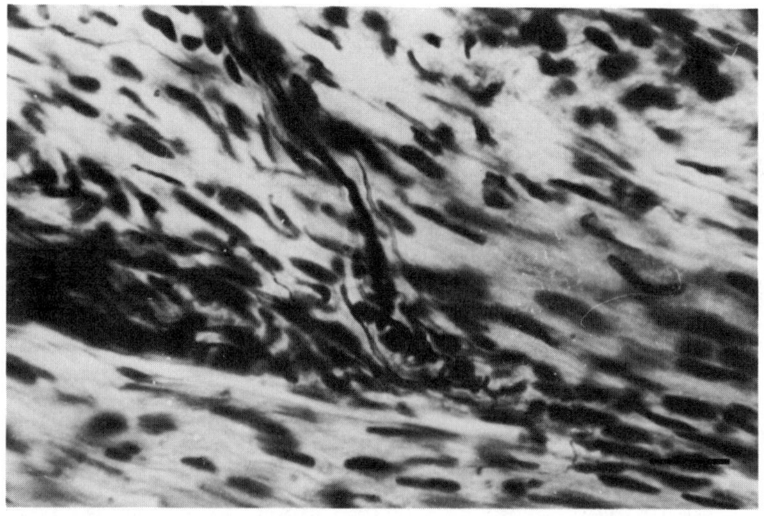

Fig. 2. An atypical tendon organ in minced gastrocnemius muscle
of a neonatal rat. 60-day graft. Bar: 20 μm.

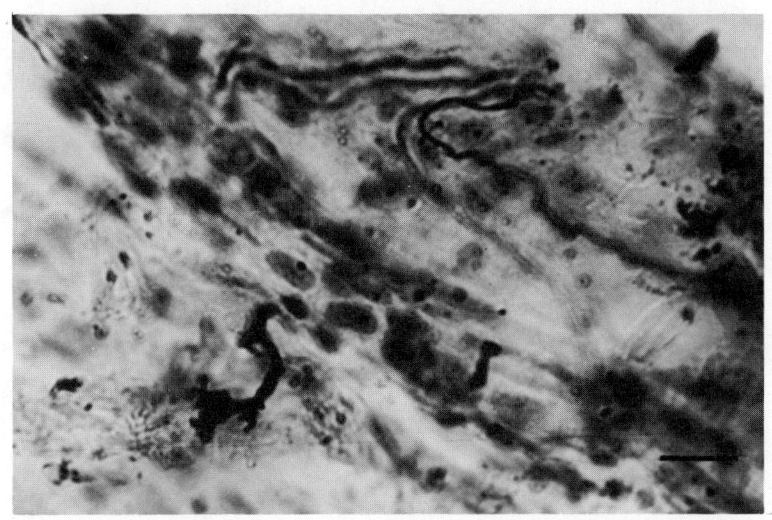

Fig. 3. An atypical muscle spindle in minced gastrocnemius
muscle of a neonatal rat. 60-day graft. Bar: 20 μm.

Complex encapsulated mechanoreceptors were not discovered
in the minced muscle grafts of adult rats, as demonstrated by
Zhenevskaya (1974), Zelená and Sobotková (1971), Carlson
(1973). Comparing our experimental data with the literature
it may be suggested that some tendency to recovery of muscle
spindles and tendon organs in newborn and 1-week-old rat
grafts is rather the result of the morphogenic effect of sensory nerve upon target cells in skeletal muscles at that ontogenetic period. Insufficient restoration of the sensory innervation in grafts can reduce the plastic activity of minced
muscle tissue in young and old rats.

REFERENCES

Carlson, B.M., 1973, The regeneration of skeletal muscle, Am.
     J. Anat., 137:119-150.
Cooper, S., 1960, Muscle spindles and other muscle receptors,
     in: "Structure and Function of Muscle", G.H. Bourne,
     ed., Acad. Press, New York-London, pp. 381-420.
Milburn, A., 1973, The early development of muscle spindles in
     the rat, J. Cell Sci., 12:175-195.
Schiaffino, S., and Pierobon Bormioli, S., 1976, Morphogenesis
     of rat muscle spindles after nerve lesion during early
     postnatal development, J. Neurocytol., 5:319-336.
Studitsky, A.N., 1977, "Muscle Transplantation in Animals",
     Medgiz, Moscow.
Zelená, J., 1964, Development, degeneration and regeneration
     of receptor organs, in: "Progress in Brain Research",
     M. Singer, J.P. Schade, eds., Elsevier Publ. Company,
     Amsterdam, 13:175-213.
Zelená, J., and Hník, P., 1960, Absence of spindles in muscle
     of rats reinnervated during development, Physiol. bo-
     hemoslov., 9:373-381.

Zelená, J., and Sobotková, M., 1971, Absence of muscle spindles
        in regenerated muscle of the rat, <u>Physiol. bohemoslov.</u>,
        20:433-439.
Zelená, J., and Soukup, T., 1977, The development of Golgi
        tendon organs, <u>J. Neurocytol.</u>, 6:171-194.
Zhenevskaya, R.P., 1974, "Neuro-trophic regulation of plastic
        activity in muscle tissue", Nauka, Moscow.

MECHANORECEPTORS IN IRRADIATED TRAUMATIZED SKELETAL MUSCLE

UNDER STIMULATED REGENERATION

M.F. Popova, R.P. Zhenevskaya and V.S. Azarova

A.N. Severtzov Institute of Evolutionary Animal
Morphology and Ecology
Moscow, USSR

The recovery process in the skeletal muscle after injury
provides information about tissue interactions in the whole
organism. As a result of disturbances of tissue regulations
in muscles irradiated with X-rays the process of posttraumatic
regeneration is suppressed considerably. It was shown in the
laboratory of A.N. Studitsky that the condition of skeletal
muscle plays an important role for increasing its regenerative
ability and radioresistance (Studitsky and Popova, 1962; Popova, 1963, 1964, 1971, 1984). The role of sensory innervation
in the regulation of plastic processes in the muscle is considerable (Zhenevskaya, 1974).

In our experiments, the recovery of muscle spindles was
examined in irradiated (20 Gy) and transected gastrocnemius
muscle of the rat under conditions stimulating regeneration by:
(1) implantation of minced muscle tissue into the defect in an
irradiated muscle; (2) denervation of the muscle 3 weeks before irradiation and operation. The irradiated and transected
muscles served as control. This experiment was conducted on
male albino rats. The gastrocnemius muscles were fixed in 10 %
neutral formol 1, 2, 3 weeks and 1, 2, 4 months after irradiation and the operation. Histological sections were stained by
the silver impregnation method of Bielschowsky.

The results of control series showed that the regeneration process in the irradiated traumatized skeletal muscles is
impaired. The inflammatory reaction is suppressed and fibrin
and necrotized material are therefore accumulated in the area
of the trauma for a long time. The proliferative myoblastic
activity of muscle fibers is very low so that the defect becomes filled with wide layers of dense connective and fatty
tissue. By one week after irradiation and the operation, swelling and widening of the spindle capsule occur in all of the
muscle spindles in the proximal muscle stump (Fig. 1). The
muscle spindles in the distal stump do not contain any nerve
terminals. Intrafusal muscle fibers have clouded sarcoplasm and
are slightly dedifferentiated. By the end of the first month
of regeneration many muscle spindles in the proximal stump are
still swollen. In the distal stump, the majority of spindles

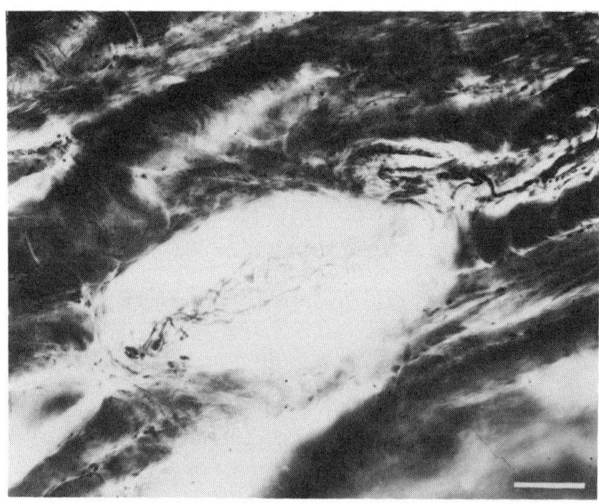

Fig. 1. The swelling of muscle spindle capsule in the proximal
        muscle stump of irradiated and transected muscle
        one week after irradiation and the operation. Bar:50 µm.

have no terminals, the connective tissue is proliferating and
the capsule is thickening. Two months after irradiation and
operation reinnervated spindles can be seen in the proximal
part. They have small motor endplates at the poles and sensory
terminals in the equatorial region. The spindles in the distal
part are denervated and after 4 months there are no spindles
in the distal muscle portion.

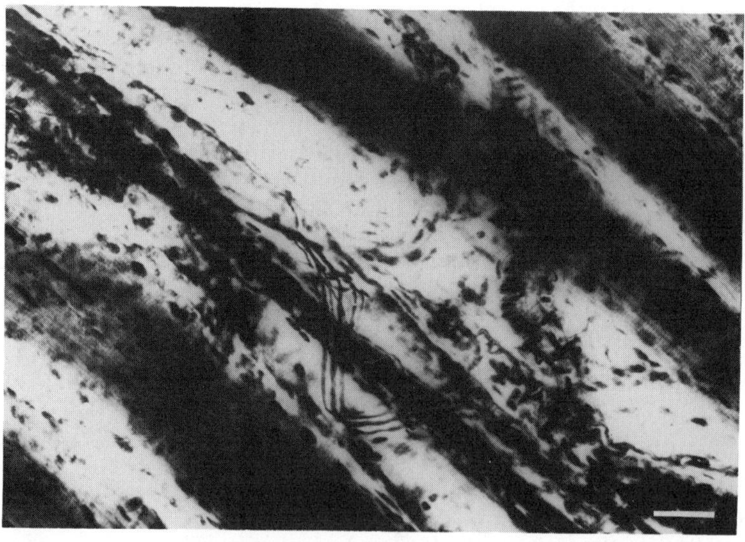

Fig. 2. Recovery of sensory innervation in muscle spindle
        2 months after irradiation, cross section and im-
        plantation of minced muscle into defect (proximal
        stump). Bar: 20 µm.

In experimental series 1 and 2 with stimulation of the
regeneration process in irradiated skeletal muscles initial
stages of reinnervation of muscle spindle poles can be found
in the proximal stump by the end of one week. By 3 weeks the
muscle spindles are reinnervated and the capsule is normal in
size. Motor axons form motor endplates on the poles, sensory
axons form plexuses around the intrafusal muscle fibers in
the equatorial region. By 2 months the spindles in the proximal stump have normal motor and sensory innervation (Fig. 2).

Thus, preliminary muscle denervation before irradiation
and the operation, or the implantation of minced muscle
into the defect of irradiated and transected gastrocnemius
muscle assist the recovery of tissue regulation and regeneration in the skeletal muscle.

REFERENCES

Popova, M.F., 1963, Influence of muscle regeneration state on
        its radioresistance, Reports of Acad. of Sciences of
        USSR, 148:223-226.
Popova, M.F., 1964, in: "Restoration processes by irradiation
        defects", Moscow, pp. 231-236.
Popova, M.F., 1971, Influence of regenerating muscle tissue
        on colony-forming bone marrow ability of irradiated
        mice. Reports of Acad. of Sciences of USSR, 199:457-459.
Popova, M.F., 1984, Radioresistance and stimulating properties
        of the regenerating mammalian tissues, Nauka, Moscow.
Studitsky, A.N. and Popova, M.F., 1962, Biological protection
        of the skeletal muscle tissue against ionizing radia-
        tion, Reports of Acad. of Sciences of USSR, 145:198-201.
Zhenevskaya, R.P., 1974, Neuro-trophic regulation of plastic
        activity of the muscle tissue, Nauka, Moscow.

PART IV

REINNERVATION

LOSS OF SENSORY AND MOTOR NEURONS AFTER NERVE INJURY IN YOUNG AND ADULT RATS

Henning Schmalbruch

Institute of Neurophysiology, University of Copenhagen

Panum Institute, Blegdamsvej 3C, DK-2200 København N

INTRODUCTION

The concept of Wallerian degeneration implies that the distal part of a cut axon is lost whereas the neuron survives. There are, however, exceptions from this rule, and in particular immature neurons may die after axotomy (Gudden, 1870; Lieberman, 1974). Hence, functional deficits after nerve repair may not only be due to aberrant reinnervation but also to neuron death. Axotomy-induced neuron death has also attracted interest because it relates to the neuronotrophic action of the target.

Loss of spinal ganglion cells after nerve injury was first observed by Marinesco (1892) in human amputees and has been demonstrated experimentally in young and adult mammals (for refs. Schmalbruch, 1987b); the quantitative results for several reasons need confirmation. The number of afferent axons was unknown and not related to the number of lost neurons. The counts of neurons in normal ganglia vary extensively and in general are underestimates (Schmalbruch, 1987a). Despite variable contributions by the individual ganglia, with few exceptions (Bondok and Sansone, 1984; Arvidsson et al., 1986), only one of the ganglia supplying a nerve was investigated. Studies of motoneuron loss after nerve injury are scarce. This review describes the results of own studies on neuronal loss after sectioning the sciatic nerve in rats aged 0-4 weeks (Schmalbruch, 1984, 1986a,b, 1987a,b).

METHODS

The sciatic nerve of Wistar rats was cut at midthigh 0-4 weeks after birth; in some 4-week-old rats, nerve regeneration was prevented by ligating the proximal stump. The anatomy of the spinal roots and ganglia and of the nerve was investigated in adult rats. Myelinated afferent and efferent axons were identified by de-efferentiation, and afferent and efferent unmyelinated fibers were distinguished by sympathectomy. De-afferentiation was technically not possible and in all cases caused severe motor deficits. The tissues were fixed by vascular perfusion with glutaraldehyde, osmicated, and embedded in epoxy resin as for electron microscopy. For light microscopy, sections 3 μm thick were stained with p-phenylenediamine. Some rats were perfused with formaldehyde, and paraffin sections of the spinal cord and ganglia were stained with cresyl violet. Horseradish peroxidase (HRP) was injected into the proximal segment of regenerated nerves to label the motoneurons that still projected to the site of injury.

Myelinated axons were counted on light micrographs 1,500- 2,000x; all axons were counted in small nerves (less than 1,000 fibers), while, in large nerves, the number of axons was calculated from their density and the total area of the nerve. Unmyelinated axons were counted in electron micrographs 11,000x, and the total number was calculated from the ratio unmyelinated-myelinated fibers and the total number of myelinated fibers. The number of dorsal root ganglion (DRG) cells was determined by counting their nucleoli in series of 3-μm epoxy sections and of 5-μm and 8-μm paraffin sections; errors due to split and multiple nucleoli were corrected by applying empirically determined correction factors (Coggeshall et al., 1984). Unilateral loss of motoneurons after nerve injury was evaluated in cross sections of the spinal cord and was ascertained by injecting HRP into the proximal segment of the nerve. The deficit of motoneurons was quantitated by counting myelinated fibers in the ventral roots. The contralateral ganglia and ventral roots served as controls to determine the deficit after nerve injury. Detailed and critical descriptions of the methods are given in the original articles (Schmalbruch, 1984, 1986a, 1987a,b).

NORMAL ANATOMY

The sciatic nerve at midthigh contained 19,000 sensory and 1,600 myelinated motor axons (Table 1). It connects to the lumbar segments L4-6. The mean number of neurons in the three dorsal root ganglia (DRG) was 41,000 when 3-μm epoxy sections were used. The data obtained from 5-μm paraffin sections were 23% smaller, and those from 8-μm paraffin sections were still smaller and not reproducible. This was due to the fact that not all nucleoli were seen because they projected into the densely stained cytoplasm. It was also found, that calculated correction factors caused an underestimate of the number of neurons; the number of fragments of each nucleolus that were visible was smaller than assumed for the various formulas. The three ventral roots together contained 3,865 myelinated fibers, i.e. about 40% of the motor axons leaving the three segments reached the nerve at midthigh.

REGENERATION OF THE SECTIONED SCIATIC NERVE

The continuity of the nerve was always restored unless this was intentionally prevented. Neonatally operated rats in contrast to rats operated at age 4 weeks never developed a neuroma. The neonatally sectioned

Table 1. Normal anatomy of the sciatic nerve
and its spinal supply

	motor	sensory myel.	unmyel.	sympathetic	total	rats number
sciatic n., mean number of axons	1,600	6,000	13,000	6,500	27,000	6-7

	L4	L5	L6	total	rats number
DRG cells, range	10,700 -16,000	13,200 -17,200	13,400 -15,100	38,500 -46,600	5
ventral root fibers, mean(S.D.)	1,518(205)	1,629(201)	718(219)	3,865	30

nerves were thinner than normal, both proximally and distally to the site of
injury. The distal nerves contained myelinated axons but there were much
fewer of them in neonatally operated rats than in rats operated at 1-4
weeks. The muscles of neonatally operated rats were rarely reinnervated,
and there were never more than 100 normal-sized or hypertrophic fibers in
a muscle. The muscles denervated at 4 weeks had become reinnervated unless
this had been prevented. Small groups of atrophic fibers or type grouping
sometimes were the only signs of previous denervation.

LOSS OF SENSORY NEURONS AFTER NERVE SECTION

The total number of neurons in right and left DRG L4-6 in three adult
rats differed by 0.05-8%. Four rats were investigated 39-89 weeks after
nerve section at birth. The deficits were 32-49% or 9,000-17,000 neurons.
This indicates that most but not all of the 19,000 axotomized neurons had
died. Six rats were investigated at age 23-96 weeks after the nerve had
been cut at 4 weeks. In 5 rats the side differences were 18-33% or 7,000
-11,500 neurons. In one rat, however, no deficit at all was found. This rat
did not show exceptionally good reinnervation of the muscles, and it was
not possible to relate the extent of cell loss to the extent of nerve
regeneration.It should, however, be noted that muscle reinnervation is an
indirect indicator for the regeneration of sensory axons. Size histograms
of the surviving DRG cells did not reveal selective loss of large or small
neurons, independent of whether the nerve had been cut at birth or at age
4 weeks.

The histological appearance of the ganglia after nerve section differed. Rats operated at 4 weeks showed signs of neuronal degeneration
(peripheral nuclei, nuclear indentations, chromatolysis, vacuolization of
the nuclei or the cytoplasm, deposition of large lipid droplets, proliferation of satellite cells), whereas such changes were not seen in rats in
which the nerve had been cut at birth. The presence of degenerative
changes in neurons axotomized months ago suggests that injured neurons
were still present at the time of study. Neurons axotomized at birth
probably had all died unless they had reached a target. This assumption is
supported by the fact that the neuronal deficit was larger after neonatal
nerve section than after nerve section at 4 weeks.

The largest ganglion cell deficit in most rats occurred in the ganglia
L4 and L5, but, in 4 of 10 rats, large deficits were also found in L6
ganglia, and in two rats the deficit at L6 was even larger than at L5. This
probably reflected the variable contribution of the ganglia to the nerve.

MOTONEURON LOSS AFTER NERVE SECTION

The number of contralateral ventral root fibers L4-6 in normal rats
differed by less than 1%. Six rats were studied 5-30 weeks after neonatal
nerve section. The deficit on the operated side was 1,696 (S.D. 172).
Considering the number of motor axons at the site of injury (1,600) this
finding suggests that almost all motoneurons were lost. Less than 1% of the
normal number of motoneurons were labelled by HRP 3 months after the operation. The deficit of ventral root fibers, 7-16 weeks after nerve section at
1 week, was 1,031 (S.D. 148; 6 rats). HRP labelling showed a distinct reduction of the number of labelled cells. Three rats were operated at 2 weeks;
the deficit after 13 weeks was 424 (S.D. 52). When the nerve was sectioned
at 4 weeks and was studied 12-42 weeks later, no consistent loss of ventral
root fibers was found. Also lack of muscle reinnervation in rats in
which the nerve had been ligated did not cause any loss of motoneurons. HRP
labelling demonstrated somewhat shrunken but otherwise normal motor nuclei.

The findings in neonatally operated rats were at variance with reports
of muscle reinnervation (Zelená and Hník, 1960; Engel and Karpati, 1968;
McArdle and Sansone, 1977). Lowrie et al.(1982) even found that the localization and the size of the motor nuclei in the spinal cord was restored,
i.e. that each muscle was reinnervated by all its original motoneurons,
while in adult rats the connection between the muscles and the spinal cord
is altered after denervation-reinnervation (Brushart and Mesulam, 1980).
Most authors crushed rather than cut the nerve. In order to evaluate the
effect of crushing instead of cutting, the sciatic nerve of 4 newborn rats
was crushed for 2 min with a watchmaker's forceps. The deficit of ventral
root fibers after 5 weeks was 1,231 (S.D. 229), i.e. large but considerably
smaller than after nerve section. Nevertheless, electron microscopy of the
freshly crushed nerve revealed that many of the small (less than 0.5 μm) and
still unmyelinated axons had survived, i.e. that axotomy was incomplete.

The axonal deficits for the individual roots in 6 rats operated at
birth were as follows. L4: 615 (S.D. 179); L5: 749 (S.D. 169); L6 337
(S.D. 189). Thus, most motor axons were derived from the segments L4 and
L5, but there was also a sizable contribution by the segment L6.

DISCUSSION

Axotomy in neonatal rats causes almost complete loss of motoneurons
and extensive but incomplete loss of sensory neurons. The sensitivity of
motoneurons decreases during the first postnatal week, and at age 4 weeks
all motoneurons survive, independent of whether or not they re-establish
target contact. In contrast, many sensory neurons die when they are axotomized at age 4 weeks.

The number of DRG cells in normal rats (L4-6: 41,000) was larger than
in previous studies, but it agreed with the number of sensory axons in the
nerve (19,000). Less than half of the DRG cells at L4-5 project into the
nerve beyond midthigh (Devor et al., 1985). Bondok and Sansone (1984)
counted only 14,800 neurons at L4-6, while Arvidsson et al. (1986) found
13,000-29,200 neurons. In the ganglion L5, Feringa et al. (1985) counted
2,000 (S.D. 230) and Arvidsson et al. (1986) 5,100-11,400 neurons. The
differences between these studies, and also the interindividual variations,
suggest that the traditional method to count nucleoli in rather thick serial
sections is not without problems (Schmalbruch, 1987a). The percentage of
L4-6 ventral root fibers which entered the nerve was 40%; this also suggests
that the total number of DRG cells is 40,000 rather than 20,000 or even less.

Loss of immature spinal motoneurons after axotomy was first described
by Romanes (1946) in newborn mice, but these results received little
attention and were not discussed in connection with studies on muscle
reinnervation in newborn rats. Romanes (1946), at a time when labelling
methods were not yet available, overestimated the number of sectioned motor
axons and concluded that only 50% of the neurons died. The absolute loss of
anterior horn cells he found agrees well with recent counts of motoneurons
in mice (Baulac and Meininger, 1983), and it seems fair to conclude that
also in newborn mice close to 100% of the axotomized motoneurons die.

The loss of axotomized neurons probably relates to the lack of neuronotrophic influences. The target dependency of motoneurons was studied in
chick (Chu-Wang and Oppenheim, 1978; Oppenheim, 1981) and amphibian embryos
(Prestige, 1970; Lamb, 1981). Nevertheless, the axons in these models were
cut during the period of natural (developmental) motoneuron death, and
induced and natural motoneuron death were viewed as part of the same
phenomenon. Natural motoneuron death in rat is probably completed before
birth (Bennett et al., 1986), and the conditions in mammals may be

different. Kuno et al. (1986) sectioned the gastrocnemius nerve of newborn rats close to the muscle and found that the majority of motoneurons survived when they were allowed to reinnervate the muscle, but that they died when this was prevented. A neuron injured far from its muscle may die before its axon reaches a target. This explains why it is possible to cross-innervate neonatal rat muscles (Brown et al., 1976; Gerding et al., 1977). Motoneurons to intercostal muscles which at birth are more mature than hind-limb motoneurons regenerate new axons, but no endplates are formed (Dennis and Harris, 1980). The failure to form neuromuscular junctions may have contributed to the scarcity of muscle reinnervation in the present study. The death of sensory neurons, at least in neonatally operated rats, is probably due to the lack of nerve growth factor (NGF). Exogeneous NGF prevents DRG cell loss in newborn rats (Yip et al.,1984) but not in adult rats (Johnson et al., 1986). There is circumstantial evidence that adult DRG cells survive when the nerve is crushed, i.e. when the conditions for axonal regeneration are optimal (Toft et al., this volume).

REFERENCES

Arvidsson, J., Ygge, J., and Grant, G., 1986, Cell loss in lumbar dorsal root ganglia and transganglionic degeneration after sciatic nerve resection in the rat, Brain Res., 373: 15-21.

Baulac, M., and Meininger, V., 1983, Postnatal development and cell death in the sciatic motor nucleus of the mouse, Exp.Brain Res., 50: 107-116.

Bennett, M.R., Abbott, M., Everett, A.W., and Lavidis, N.A., 1986, Motoneurone numbers in rat neonatal lateral columns: use of antibody to choline acetyltransferase to identify motoneurones, Neurosci.Lett., 63: 101-105.

Bondok, A.A., and Sansone, F.M., 1984, Retrograde and transganglionic degeneration of sensory neurons after a peripheral nerve lesion at birth, Exp. Neurol., 86: 322-330.

Brown, M.C., Jansen, J.K.S., and Van Essen, D., 1976, Polyneural innervation of skeletal muscle in new-born rats and its elimination during maturation, J.Physiol. (London), 261: 387-422.

Brushart, T.M., and Mesulam, M.-M., 1980, Alteration in connections between muscle and anterior horn motoneurons after peripheral nerve repair, Science, 208: 603-605.

Chu-Wang, I.-W., and Oppenheim, R.W., 1978, Cell death of motoneurons in the chick embryo spinal cord. II. A quantitative and qualitative analysis of degeneration in the ventral root, including evidence for axon outgrowth and limb innervation prior to cell death, J.Comp.Neurol., 177: 59-86.

Coggeshall, R.E., Chung, K., Greenwood, D., and Hulsebosch, C.E., 1984, An empirical method for converting nucleolar counts to neuronal numbers, J.Neurosci.Meth., 12: 125-132.

Dennis, M.J., and Harris, A.J., 1980, Transient inability of neonatal rat motoneurons to reinnervate muscle, Dev.Biol., 74: 173-183.

Devor, M., Govrin-Lippmann, R., Frank, I., and Raber, P., 1985, Proliferation of primary sensory neurons in adult rat dorsal root ganglion and the kinetics of retrograde cell loss after sciatic nerve section. Somatosensory Res., 3: 139-167.

Engel, W.K., and Karpati, G., 1968, Impaired skeletal muscle maturation following neonatal neurectomy, Dev.Biol., 17: 713-723.

Feringa, E.R., Lee, G.W., Vahlsing, H.L., and Gilbertie, W.J., 1985, Cell death in the adult rat dorsal root ganglion after hind limb amputation, spinal cord transection, or both operations, Exp.Neurol., 87: 349-357.

Gerding, R., Robbins, N., and Antosiak, J., 1977, Efficiency of
    reinnervation of neonatal rat muscle by original and foreign
    nerves, Dev.Biol., 61: 177-183.
Gudden, B., 1870, Experimentaluntersuchungen über das peripherische
    und centrale Nervensystem, Arch.Psychiatr.Nervenkrankh., 2:
    693-723.
Johnson, E.M., Rich, K.M., and Yip, H.K., 1986, The role of NGF in
    sensory neurons in vivo, Trends Neurosci., 9: 33-37.
Kuno, M., Kashihara, Y., and Miyata, Y., 1986, Motoneuron death induced
    by axotomy in neonatal rats depends upon the timing of muscle
    reinnervation, Muscle Nerve, 9 (Suppl.): 112.
Lamb, A.H., 1981, Axon regeneration by developing limb motoneurones in
    Xenopus laevis, Brain Res., 209: 315-323.
Lieberman, A.R., 1974, Some factors affecting retrograde neuronal
    responses to axonal lesions, in: "Essays on the Nervous System",
    R. Bellairs and E.G. gray eds., Clarendon Press, Oxford, pp. 71-105.
Lowrie, M.B., Krishnan, S., and Vrbova, G., 1982, Recovery of slow and
    fast twitch muscle following nerve injury during early post-natal
    development in the rat, J.Physiol. (London), 331: 51-66.
McArdle, J.J., and Sansone, F.M., 1977, Re-innervation of fast and
    slow twitch muscle following nerve crush at birth,
    J.Physiol. (London), 271: 567-586.
Marinesco, G., 1892, Ueber Veränderung der Nerven und des Rückenmarks
    nach Amputationen; ein Beitrag zur Nerventrophik,
    Neurol.Centralbl., 11: 463-467.
Oppenheim, R.W., 1981, Neuronal cell death and some related regressive
    phenomena during neurogenesis, in: "Studies in Developmental
    Neurobiology. Essays in Honor of Viktor Hamburger", M.W. Cowan
    ed., Oxford University Press, New York, pp. 74-133.
Prestige, M.C., 1970, Differentiation, degeneration, and the role of the
    periphery: quantitative considerations, in: "The Neurosciences,
    2nd Study Program", F.O. Schmitt, ed., The Rockefeller University
    Press, New York, pp. 73-82.
Romanes, G.J., 1946, Motor localization and the effects of nerve injury
    on the ventral horn cells of the spinal cord, J.Anat., 80:
    117-131.
Schmalbruch, H., 1984, Motoneuron death after sciatic nerve section in
    newborn rats, J.Comp.Neurol., 224: 252-258.
Schmalbruch, H., 1986a, Fiber composition of the rat sciatic nerve,
    Anat.Rec., 215: 71-81.
Schmalbruch, H., 1986b, The anatomy and fiber composition of the
    normal rat sciatic nerve, Clinical Neuropathology, 5: 115.
Schmalbruch, H., 1987a, The number of neurons in dorsal root ganglia
    L4-6 of rat, Anat.Rec., in press.
Schmalbruch, H., 1987b, Loss of sensory neurons after sciatic
    nerve section in rat, Anat.Rec., in press.
Yip, H.K., Rich, K.M., Lampe, P.A., and Johnson, E.M., 1984, The effects
    of nerve growth factor and its antiserum on the postnatal
    development and survival after injury of sensory neurons in rat
    dorsal root ganglia, J.Neurosci., 4: 2986-2992.
Zelená, J., and Hník, P., 1960, Absence of spindles in muscles of rat
    reinnervated during development, Physiol. Bohemoslov., 9:
    373-381.

Acknowledgements- The technical help of Mrs. M. Bjærg, and the
    financial support provided by the Danish Medical Research Council,
    the Foundation for Experimental Research in Neurology, and the
    Foundation for Progress in Medicine are gratefully acknowledged.

SENSORY INNERVATION OF ATYPICAL SPINDLES AFTER NERVE CRUSH IN
NEWBORN RATS

T. Soukup and J. Zelená

Institute of Physiology, Czechoslovak Academy of
Sciences
Vídeňská 1083, 142 20 Prague 4

INTRODUCTION

Denervation experiments (Zelená and Hník, 1963a; Zelená,
1976) and ultrastructural studies indicate (Landon, 1972; Milburn, 1973; 1984; Kozeka and Ontell, 1981) that sensory axons
trigger the development of muscle spindles by inducing the formation of nuclear bags in selected primary myotubes and by initiating the encapsulation of nascent spindles. Sensory axons are
also mainly responsible for further differentiation and maintenance of the muscle spindle structure (Zelená and Soukup, 1973;
1974; Kucera, 1987), whereas fusimotor axons contribute to a full
maturation of myosin profiles of intrafusal fiber types (teKronnie et al., 1982). If skeletal muscles are denervated in newborn
(Zelená and Hník, 1960; 1963b) or 4-day-old rats (Schiaffino and
Pierobon Bormioli, 1976), most muscle spindles disintegrate during the transient denervation period due to the loss of innervation; only occasional spindles survive with a reduced number of
intrafusal fibers and can be reinnervated again when the muscle
nerves regenerate.

Several months after crushing the sciatic nerve in newborn
rats and subsequent muscle reinnervation, muscle spindles in the
extensor digitorum longus, anterior tibial and soleus muscles
are drastically decreased in number to 1-3 per muscle and significantly reduced in size by 30-50 %; their structure is atypical;
although intrafusal fibers remain thin in diameter, they lack
nuclear accumulations and their number per spindle does not ex-
ceed 2, whereas their mean number in normal control spindles is
3.8 (Zelená and Hník, 1960; 1963b; Hník and Zelená, 1961). Atypical spindles formed after nerve crush apparently receive both
sensory and fusimotor innervation (Zelená and Hník, 1963a,b) and,
despite all their structural shortcomings, respond to stretch by
increasing the rate of discharge in a similar way as do normal
muscle spindles (Hník, 1964).

After nerve section and suture performed in 4-day-old rats,
however, muscle spindles from reinnervated extensor digitorum
longus and soleus muscles lack sensory innervation, as has been
revealed by electron microscopy; they have no periaxial space,

and the only characteristic in which the "spindle" fibers differ
from extrafusal fibers is their encapsulation (Schiaffino and
Pierobon Bormioli, 1976). In view of these ultrastructural findings, we decided to reexamine, by electron microscopy, atypical
spindles reinnervated after nerve crush, to ascertain the ultrastructural characteristics of their sensory innervation and of
their intrafusal fibers.

MATERIALS AND METHODS

The right sciatic nerve in new-born Long Evans rats was
crushed with an adapted pair of forceps at the mid-thigh level
5-6 h after birth. After 3-9 months, the electrophysiological
examination of calf muscles was performed (Paleček et al., this
volume); in four rats, the soleus muscle was then excised, extended to its resting length and fixed with a solution of 1 %
paraformaldehyde and 1 % glutaraldehyde in a 0.12 mol.l^{-1} phosphate buffer (pH 7.3), dehydrated and embedded in Durcupan.
Serial semithin section were cut through the whole muscle block,
and if a spindle was discovered, alternating semithin and ultrathin sections were prepared throughout the whole spindle length.
The ultrathin sections were stained with 1 % uranyl acetate and
0.1 % lead citrate and examined in a JEM 100 B electron microscope.

RESULTS

The examination of the sample of 6 muscle spindles removed
from reinnervated soleus muscles 3-9 months after nerve crush
confirmed and extended previous results obtained by light-micro-
scopical methods (Zelená and Hník, 1963a,b). The atypical spindles consisted of only 1 or 2 intrafusal fibers enclosed in a
spindle-shaped capsule, all had a distinct periaxial space enlarged at their mid-portion, and all were innervated by sensory
axons which formed terminals on intrafusal fibers up to several
hundred micrometers of their length. Four spindles contained a
single intrafusal fiber, one spindle had two fibers and in one
spindle, two intrafusal fibers fused together at one of the
poles. All intrafusal fibers examined lacked nuclear accumulation
and were thin, with diameters of 6-9 μm; the fused fiber had a
diameter transiently enlarged to 14 μm, which is still within
the range of normal diameters of intrafusal fibers of adult rats
(Zelená and Hník, 1963a,b; Soukup, 1976). These atypical fibers
had centrally localized nuclei throughout the zone innervated
with sensory terminals, and peripheral nuclei in polar regions;
otherwise their whole profile was filled with tightly packed
myofibrils, except for central rings and islets of sarcoplasm
with sarcoplasmic organelles to be found around the nuclear
poles or in between the nuclei (Figs. 1 and 2). Occasionally, a
large granular dense body was found in the axial region, and
leptomeres occurred near the fiber circumference (Fig. 2). Intrafusal satellite cells were infrequent.

One to 4 myelinated sensory axons with diameters of 1.5-4 μm
were found to supply the atypical spindles. After entering the
capsule, the axons or their myelinated branches often ran in parallel with the intrafusal fibers towards the poles, and repeatedly sent out short unmyelinated branches to the surface of intrafusal fibers; each of the branches formed a separate set of

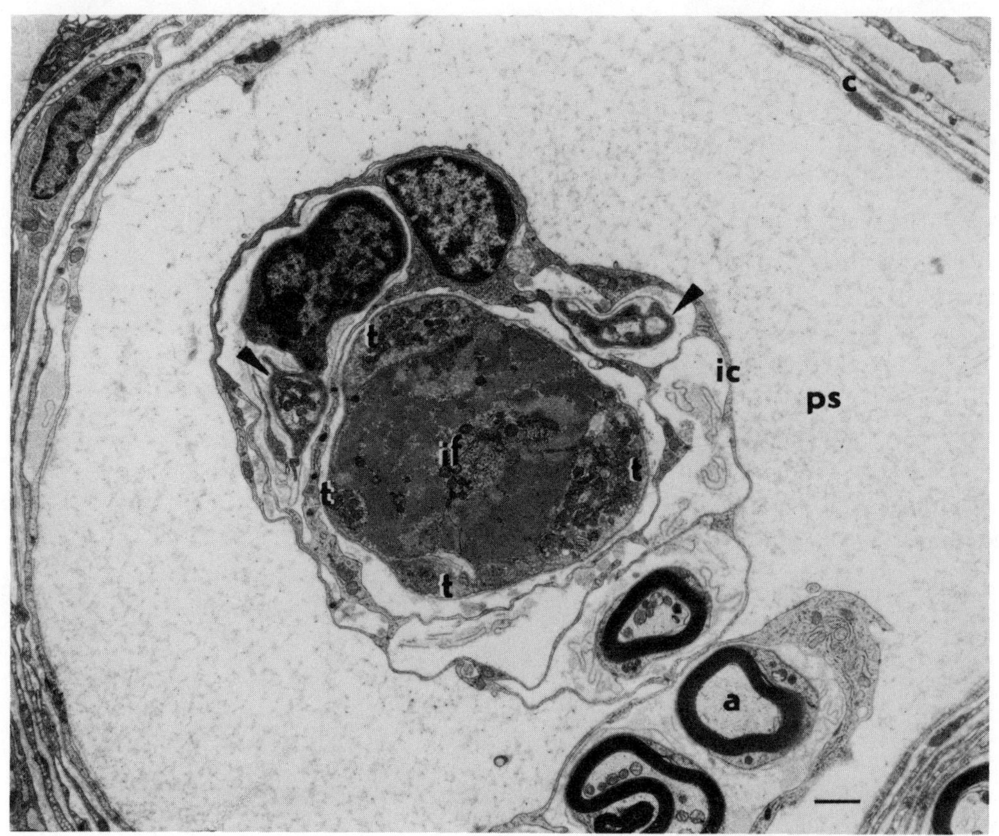

Fig. 1. Transverse section of an atypical spindle from rat soleus
muscle 3 months after neonatal crush of the sciatic
nerve. The spindle consists of a single intrafusal fiber
(if) encircled by an inner capsule (ic), a periaxial
space (ps) and the outer capsule (c). Four profiles of
sensory terminals (t) are located on the surface of the
intrafusal fiber. Myelinated axons (a) and their unmyelinated branches (arrowheads) approach the intrafusal fiber.
Bar indicates 1 μm.

terminal coils. In most instances, axonal branches shed their
Schwann cell ensheathment before penetrating the basal lamina of
the fiber and forming the terminal. The terminals lay beneath
the basal lamina in shallow depressions on the fiber surface. On
transverse sections, they appeared as oval (Fig. 1) or horseshoe-
like profiles (Fig. 2). They were filled mainly with mitochondria
and contained groups of clear vesicles at their margins. Occasionally, a Schwann cell process extended over the terminal and
covered part of its surface beneath the basal lamina. The inner
capsule was composed of a few attenuated layers of fibroblast-
like cells and surrounded the intrafusal fiber with the adjacent
axons and Schwann cells throughout the zone of sensory innervation (Fig. 1). The free space around the fiber and between the
layers of the inner capsule often contained duplicated and folded
basal laminae, and collagen and elastic fibrils (Figs. 1 and 2).
The periaxial space between the inner and outer capsule was well
developed in all the spindles examined (Fig. 1), but the maximal

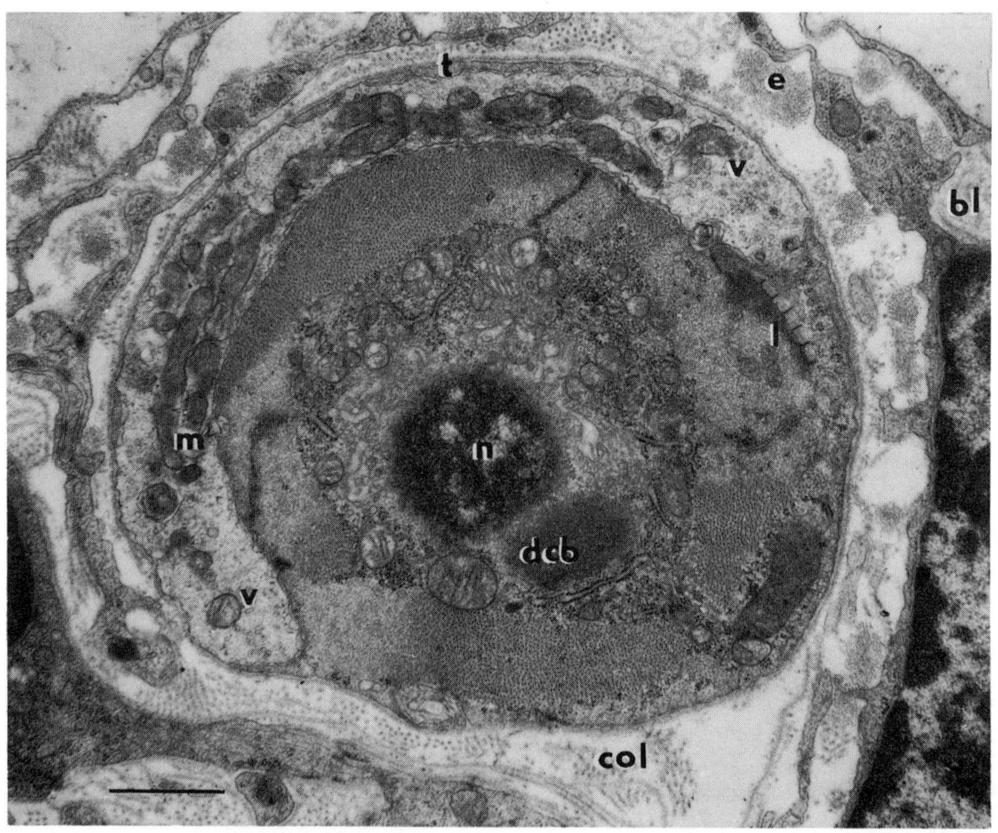

Fig. 2. Transverse section of an intrafusal fiber 3 months after
nerve crush. A sensory terminal (t) filled with mitochondria (m) and vesicles (v) encloses half of the fiber
circumference. The profiles of a nucleus (n) and a dense
cytoplasmic body (dcb) are seen in the axial region surrounded by a ring of sarcoplasm with sarcoplasmic orga-
nels. A leptomere (l) is seen at the fiber circumference.
col - collagen; e - elastin; bl - basal lamina. Bar indicates 1 µm.

spindle diameter was 25-30 µm, which is approximately half the
control value. The capsule consisted of 3-5 layers of capsular
cells with narrow interspaces, and had a normal ultrastructure
(Fig. 1).

CONCLUDING REMARKS

     Our results confirm and extend previous findings concern-
ing the structure and innervation of atypical spindles that develop in rat muscles after neonatal nerve crush (Hník and Zelená,
1961; Zelená and Hník, 1963a,b). The pattern of reinnervation
of atypical spindles by sensory axons is irregular. In most instances, atypical spindles have an increased number of intracapsular myelinated axons in their midportion, and their spiral
endings are often discontinuous and consist of a number of separate loops. However, the ultrastructure of regenerated sensory
terminals appears normal. We could not discern, in atypical spindles, Ia and II sensory axons and terminals, since all regene-

rated axons have greatly reduced diameters and the innervation
pattern is altered.

The ultrastructure of atypical spindles differs substantially from that of anomalous spindles found in leg muscles after
perinatal nerve section (Schiaffino and Pierobon Bormioli, 1976),
which is apparently a consequence of a different inductive influence of sensory and motor axons upon surviving spindle remnants. Conceivably, the phenotype of intrafusal fibers is not
yet fully determined early after birth and is still pliable at
the onset of reinnervation. After nerve section, regenerating
sensory axons do not find their way to the surviving spindles;
their intrafusal fibers become reinnervated only by motor axons
and therefore redifferentiate into encapsulated extrafusal muscle fibers of a slow or fast type (Schiaffino and Pierobon Bormioli, 1976). After nerve crush, the continuity of the nerve
with the periphery is maintained; regenerating sensory axons
grow along the preserved peripheral pathways and reinnervate
the spindle remnants. Sensory endings then exert a decisive morphogenetic influence upon spindle differentiation, even when
the spindles also receive fusimotor innervation; the reinnervated
spindle fibers differentiate as intrafusal fibers of small diameter, the periaxial space is built up beneath the capsule and
the spindles become functional (Hník, 1964; Paleček et al., this
volume). However, the atypical intrafusal fibers lack nuclear
accumulation, and their number remains restricted to one or two
fibers, which may be due to a reduced morphogenetic capacity of
primary sensory neurons or, more probably, to a failure of intrafusal fibers to respond to sensory induction, because histo-
differentiation is advanced and muscle spindles cannot generate
any more myotubes or incorporate myoblasts. In relation to the
altered structure, response to stretch of atypical spindles also
remains subnormal (Paleček et al., this volume).

In most rat muscles studied so far, no new intrafusal fibers and no new spindles are formed following reinnervation after
neonatal nerve crush. The only exception is the gastrocnemius
muscle, in which a normal or variable number of muscle spindles
can be recovered after reinnervation (Zelená and Hník, 1963a;
Werner, 1973; Sekiya et al., 1986) and additional atypical muscle spindles can differentiate anew (Zelená and Hník, 1963a;
Zelená, 1964) when sprouting of regenerating axons is enhanced
after neonatal nerve crush by exogenous (Sekiya et al., 1986)
or, presumably, by endogenous nerve growth factor. Why uncomitted
myotubes and myoblasts are available for spindle neoformation in
the gastrocnemius and not in other muscles, remains to be clarified.

REFERENCES

Hník, P., 1964, Functional characteristics of free nerve endings
        and atypical spindles after muscle reinnervation in very
        young rats, Physiol. bohemoslov., 13:216-219.
Hník, P., and Zelená, J., 1961, Atypical spindles in reinnervated
        rat muscles, J. Embryol. exp. Morphol., 9:456-467.
Kozeka, K., and Ontell, M., 1981, The three-dimensional cyto-
        architecture of developing murine muscle spindles, Devl.
        Biol., 87:133-147.
Kronnie, G. te, Donselaar, Y., Soukup, T., and Zelená, J., 1982,
        Development of immunohistochemical characteristics of

intrafusal fibres in normal and de-efferented rat muscle spindles, Histochemistry, 74:355-366.

Kucera, J., 1987, Postnatal maturation of deafferented muscle spindles in the rat, J. Physiol., 382:167P.

Landon, D.N., 1972, The fine structure of the equatorial regions of developing muscle spindles in the rat, J. Neurocytol., 1:189-210.

Milburn, A., 1973, The early development of muscle spindles in the rat, J. Cell Science, 12:175-195.

Milburn, A., 1984, Stages in the development of cat muscle spindles, J. Embryol. exp. Morphol., 82:177-216.

Paleček, J., Vejsada, R., Hník, P., Soukup, T., Vlachová, V., and Asmussen, G., 1988, Functional properties of atypical muscle spindles after nerve crush in newborn rats, in: "Mechanoreceptors - development, structure and function", P. Hník, T. Soukup, R. Vejsada and J. Zelená, eds., Plenum Press, New York, this volume.

Schiaffino,S., Pierobon Bormioli, S., 1976, Morphogenesis of rat muscle spindles after nerve lesion during early postnatal development, J. Neurocytol., 5:319-336.

Sekiya, S., Homma, S., Miyata, Y., and Kuno, M., 1986, Effects of nerve growth factor on differentiation of muscle spindles following nerve lesion in neonatal rats, J. Neurosci., 6:2019-2025.

Soukup, T., 1976, Intrafusal fibre types in rat hind limb muscle spindles. Morphological and histochemical characteristics, Histochemistry, 47:43-57.

Werner, J.K., 1973, Mixed intra- and extrafusal muscle fibers produced by temporary denervation in new-born rats, J. Comp. Neurol., 150:279-302.

Zelená, J., 1964, Development, degeneration and regeneration of receptor organs, Prog. Brain Res., 13:175-213.

Zelená, J., 1976, The role of sensory innervation in the development of mechanoreceptors, Prog. Brain Res., 43:59-64.

Zelená, J., and Hník, P., 1960, Absence of spindles in muscles of rats reinnervated during development, Physiol. bohemoslov., 9:373-381.

Zelená, J., and Hník, P., 1963a,Effect of innervation on the development of muscle receptors, in: "The Effect of Use and Disuse on Neuromuscular Functions", E. Gutmann and P. Hník, eds., Academia, Prague, pp. 95-105.

Zelená, J., and Hník, P., 1963b, Motor and receptor units in the soleus muscle after nerve regeneration in young rats, Physiol. bohemoslov., 12:277-290.

Zelená, J., and Soukup, T., 1973, Development of muscle spindles deprived of fusimotor innervation, Z. Zellforsch., 144:435-452.

Zelená, J., and Soukup, T., 1974, The differentiation of intrafusal fibre types in rat muscle spindles after motor denervation, Cell Tiss. Res., 153:115-136.

ACKNOWLEDGEMENTS

The expert technical help of Mrs. M. Krupková and Mr. M. Kunz is gratefully acknowledged.

FUNCTIONAL PROPERTIES OF ATYPICAL MUSCLE SPINDLES AFTER NERVE
CRUSH IN NEWBORN RATS

J. Paleček, R. Vejsada, P. Hník, T. Soukup, Viktorie
Vlachová and G. Asmussen[+]

Institute of Physiology, Czechoslovak Academy of
Sciences
Vídeňská 1083, 142 20 Prague 4, Czechoslovakia
[+]Carl-Ludwig-Institute of Physiology
Karl-Marx-University Leipzig, GDR

INTRODUCTION

When regeneration of the peripheral nerve takes place at
an early stage of development, e.g. after nerve crush in newborn rats, the result of reinnervation is very poor. About 50 %
of the neurons the axons of which had been crushed do not survive whether they be spinal motoneurons, or spinal ganglion
cells (Romanes, 1946; Zelená and Hník, 1963; Schmalbruch, 1984;
Jenq et al., 1987). In the reinnervated muscles, the number of
muscle spindles is dramatically reduced and their structure is
permanently altered (Hník and Zelená, 1961; Zelená and Hník,
1963; Soukup and Zelená, this volume).

The activity of the atypical spindles was studied by Hník
(1964) who reported that they reacted to maintained stretch by
a slowly adapting response. In the present study we examined
the discharge pattern of these dwarfish spindles in more detail
and attempted to correlate our findings with the morphological
data obtained, on the same material, in the light and electron
microscope (Soukup and Zelená, this volume).

METHODS

We used 17 Wistar rats of both sexes with an average body
weight of 230 g for the electrophysiological experiments. The
sciatic nerve of these rats had been crushed in mid-thigh under
cold anesthesia with a fine pair of forceps 6 - 8 h after birth.
Three to nine months after the nerve crush, laminectomy L2 to
L6 was performed (under Sagatal anesthesia, 60 mg/kg i.p.). The
rats were fixed on a heated table, the tendon of the soleus
muscle was connected to a stepping motor puller operated by a
digitimer. When we could not obtain stretch-induced responses
from the soleus muscle (probably due to the sparsity of muscle
receptors), we recorded activity from the lateral head of the
gastrocnemius muscle. The soleus nerve of the operated hindlimb

Table 1. Number of Muscle Spindles and Intrafusal Fibers in
         Reinnervated Soleus and Lateral Gastrocnemius Muscles
         (Mean ± S.E.M.)

	No. of spindles per muscle	No. of intrafusal fibers per spindle
Soleus muscle	1.6 ± 0.2	1.6 ± 0.3
Lat. gastroc. muscle	6.3 ± 1.5	1.7 ± 0.2

was dissected free, while the other branches of the sciatic and
femoral nerves were transected. The dorsal roots L4 and L5 were
teased in order to obtain unit activity. A standard series of
1, 2, 3 and in some cases also 4 mm stretches were applied to
the soleus or lateral gastrocnemius muscles (stretch velocity
7 mm/s). The initial length of the muscle was set just subthresh-
old to stretch-induced activity. The records of spindle discharges were stored on a FM tape recorder (Bell-Howell) and
afterwards evaluated on a PDP 11 computer.

Out of the 17 muscles employed in our experiments, four
soleus muscles were processed for and examined in the electron
microscope (see Soukup and Zelená, this volume). Five other
muscles (soleus or lateral gastrocnemius) were fixed in Bouin´s
fixative, embedded in paraffin or paraplast and complete series
of 5 μm thick sections were examined in the light microscope
and searched for muscle spindles. The rest of the muscles studied electrophysiologically were not evaluated histologically.

RESULTS

In fifteen preparations, we obtained stretch - induced responses from either dorsal root filaments or the muscle nerve.
The majority of atypical spindles responded only phasically to
a 1 mm stretch of the muscle; however, slowly adapting responses
were regularly seen at larger stretches (Fig. 1). When tested
for vibration sensitivity, the atypical spindles exhibited a
relatively high threshold for phase-locked firing. Moreover,
they usually failed to be driven 1 : 1 above 50 Hz by sinusoidal stretch of several hundreds of micrometers. The discharge frequencies from 16 single spindle units recorded from
9 rats were evaluated using the PDP 11 programme during 1, 2,
3 and 4 mm stretches (Fig. 2). It is evident from this figure
that the firing rates of atypical spindles are lower by more
than 50 % as compared to the control values taken from the
paper by Hník and Lessler (1973). There was a large scatter of
individual values at the end of the ramp phase of stretch (f_0
in Fig. 2), but this diminished considerably during the hold
phase of stretch ($f_{0.5}$ and f_4 in Fig. 2).

The conduction velocity of most units was in the range from
10 to 40 m/s which is almost 50 % lower as compared to normal

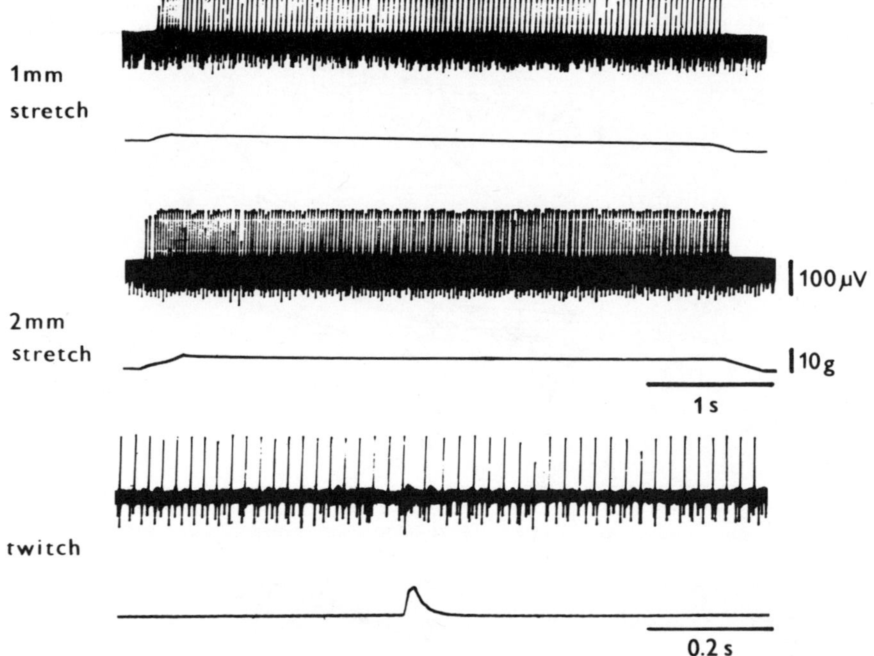

Fig. 1. Examples of activity of atypical muscle spindles from
        lateral gastrocnemius muscle, in response to 1 and 2 mm
        stretches. Upper traces, sensory activity in dorsal
        root L5 filament; lower traces, myogram. Bottom set of
        records - pause during muscle contraction. Amplitude
        calibration (for unit activity and myogram) common for
        all the records.

values of group I and II of muscle afferents in adult rats (cf.
Hník and Lessler, 1973). This reduction of conduction velocities
corresponds to the decrease in the fiber caliber spectrum found
in regenerated sensory nerve fibers following early nerve crush
(Zelená and Hník, 1963).

The histological examinations revealed that each soleus
muscle examined by light or electron microscopy contained one
or two atypical spindles with one or two intrafusal fibers
without nuclear accumulations (Table 1). In the three lateral
gastrocnemius muscles, 4, 6 and 9 spindles per muscle were
found respectively, of which 17 had only one or two intrafusal
fibers (Table 1).

In one of the soleus muscles, two units were recorded which
responded phasically at 1 and 2 mm stretches. During a 3 mm
stretch their discharge rates of the response were relatively
low. On electron microscopical evaluation of this muscle we
found two atypical muscle spindles with one intrafusal fiber
only. On the contrary, a unit from a lateral gastrocnemius
(Fig. 2 - asterisk) fired at a frequency almost within the
range of normal values and had a relatively high conduction

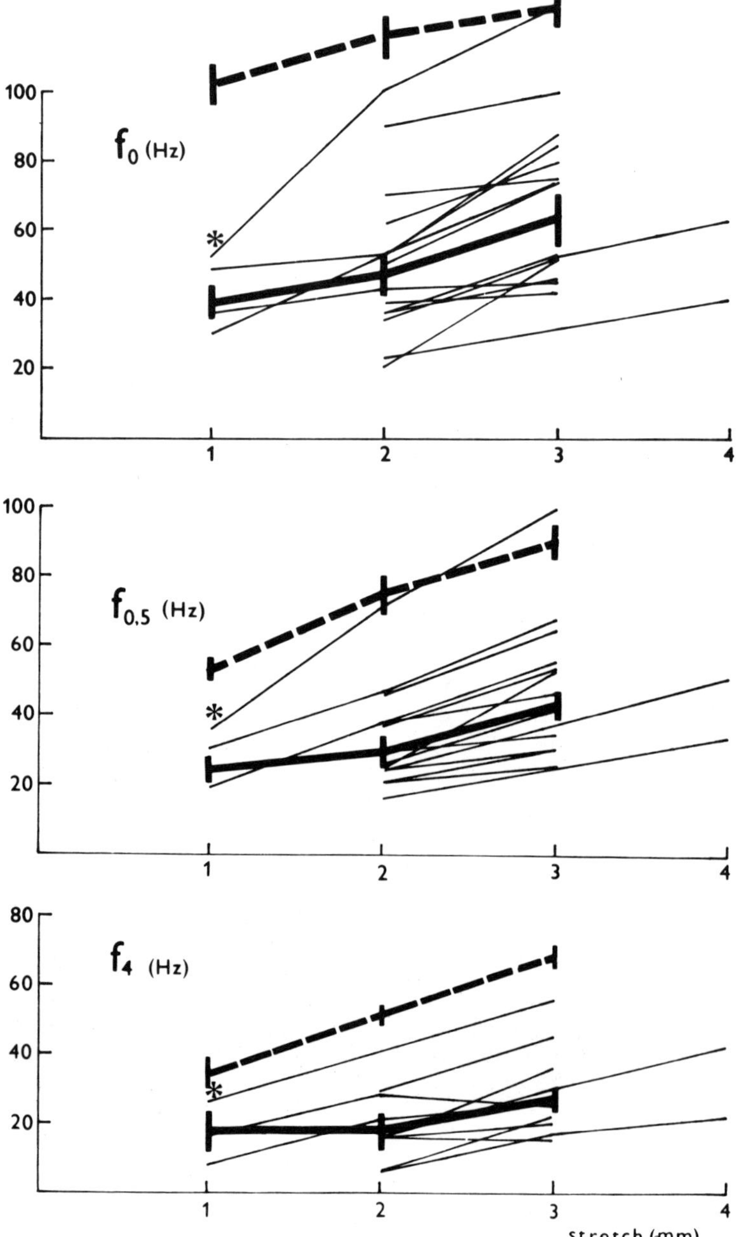

Fig. 2. Discharge frequencies of atypical muscle spindles at
the end of the ramp phase of stretch (f_0), half a second later ($f_{0.5}$) and four seconds after the beginning
of the hold phase of 1, 2, 3 and 4 mm ramp-and-hold
stretches. Normal values in adult rats (Hník and Lessler, 1973) are depicted by the thick broken line, the
present values from individual spindles in muscles reinnervated at an early postnatal age are drawn with
the thin solid lines, the thick solid line represents
the mean value ± S.E.M.

velocity (49 m/s) as compared to the others. The histological
analysis of this lateral gastrocnemius muscle revealed four
muscle spindles one of which contained a normal number of intrafusal muscle fibers.

DISCUSSION

     Muscle spindles which survive sciatic nerve crush at birth
in the rat are dwarfish and atypical in structure with only one
or two intrafusal fibers without specific nuclear accumulation,
as compared with the four fully differentiated intrafusal fibers
to be found in normal rat spindles (Hník and Zelená, 1961; Soukup and Zelená, this volume; Zelená and Hník, 1963). In spite
of this deficiency in structure, these spindles can generate a
slowly adapting response to stretch (Hník, 1964) which is characterized by a prominent dynamic and a well sustained static
component (the present study). The only functional deficit to
be found in the passive response to stretch in atypical spindles is the great reduction in the discharge frequency at any
point of the ramp-and-hold stretch, as compared to values in
normal spindles. It is possible that this reduction in firing
rates was, in part, due to the limited functional capacity of
spindle afferent nerve fibers which were found to conduct slowly
and may thus be expected to impede transmission of higher-fre-
quency trains of impulses. If this factor were the only one in
play, the greatest reduction of firing rate would occur at the
dynamic peak of the response; our data, however, rather suggest
a proportional decrease in both the dynamic and static response.
We therefore assume that other factors are involved in the subnormal responsiveness of these atypical spindles. The sensory
axon terminals make contact with one or two intrafusal fibers
only and the innervation pattern is altered in that the sensory
terminals do not form typical annulo-spiral endings (Soukup and
Zelená, this volume). The two factors apparently lead to a considerable reduction of the area of the active sensory membrane
which is directly involved in the generation of the receptor
potential. Furthermore, since the intrafusal fibers of atypical
spindles have no nuclear accumulation, the equatorial zone may
be less pliable, as compared to normal intrafusal fibers.

     It seemed of special interest to ascertain whether these
sporadic atypical spindles with a reduced capacity of monitoring the dynamic and static changes of muscle length can sub-
stitute normal spindles in their complex function during locomotion and basic spinal reflexes. In order to elucidate this
question we evaluated the EMG activity from muscles reinnervated at an early postnatal age (Hník et al., this volume).

REFERENCES

Hník, P., 1964, Functional characteristics of free nerve end-
        ings and atypical spindles after muscle reinnervation
        in very young rats, Physiol. bohemoslov., 13:216-219.
Hník, P., and Lessler, M.J., 1973, Changes in muscle spindle
        activity of the chronically de-efferented gastrocnemius
        of the rat, Pflugers Arch., 341:155-170.
Hník, P., Vejsada, R., Navarrete, R., Paleček, J., Payne, R.,
        and Borecka, U., Late effects of early hind-limb de-
        nervation and reinnervation in rats. An EMG study (this
        volume).

Hník, P., and Zelená, J., 1961, Atypical spindles in reinner-
     vated rat muscles, J. Embryol. exp. Morph., 9:456-467.
Jenq, C., Jenq, L.L., and Coggeshall, R.E., 1987, Numerical
     patterns of axon regeneration that follow sciatic nerve
     crush in the neonatal rat, Exp. Neurol., 95:492-499.
Romanes, G.J., 1946, Motor localization and the effects of
     nerve injury on the ventral horn cells of the spinal
     cord, J. Anat., 80:117-131.
Schmalbruch, H., 1984, Motoneuron death after sciatic nerve
     section in newborn rats, J. Comp. Neurol., 224:252-258.
Soukup, T., and Zelená, J., Sensory innervation of atypical
     spindles after nerve crush in newborn rats (this volume).
Zelená, J., and Hník, P., 1963, Motor and receptor units in
     the soleus muscle after nerve regeneration in very young
     rats, Physiol. bohemoslov., 12:277-290.

LATE EFFECTS OF EARLY HIND-LIMB DENERVATION AND REINNERVATION
IN RATS: AN EMG STUDY

P. Hník[1], R. Vejsada[1], R. Navarrete[2], J. Paleček[1],
R. Payne[3] and U. Borecka[4]

[1]Institute of Physiology, Czechoslovak Academy of
Sciences, 142 20 Prague, Czechoslovakia
[2]University College, London
[3]University of Aberdeen, Aberdeen
[4]Institute of Biocybernetics and Biomedical Engineering, Polish Academy of Sciences, Warsaw

INTRODUCTION

Crushing of the sciatic nerve in new-born rats leads subsequently to a permanent decrease in the number and size of both
sensory and motor nerve fibers. The number of motor axons and
muscle fibers in the soleus is decreased by more than a half;
the mean diameter of motor axons is reduced by 25 %, but that
of muscle fibers is increased by 20 %. The number of encapsulated muscle receptors is dramatically reduced in the soleus
(SOL), tibialis anterior (TA) and extensor digitorum longus
(EDL) after neonatal nerve crush (Zelená and Hník, 1963). This
is in contrast to the peripheral nerve regeneration following
nerve crush in adult animals. The recovery of muscle weight
after neonatal nerve crush was also found to be rather poor as
compared with the result of muscle reinnervation in adult animals.

It appeared to be of interest to assess the function of
these handicapped muscles.

MATERIALS AND METHODS

The sciatic nerve was crushed unilaterally in 15 rats
approximately 6 h after birth. Three to six months later an
electrode array (Hník et al., 1978) was implanted for recording muscle EMG activity in the SOL and TA on both sides. We
evaluated resting EMG activity, myotatic EMG responses to
stretch and the EMG locomotor pattern with the rat in a ro-
tating drum.

RESULTS AND DISCUSSION

1. Both the reinnervated SOL and TA exhibited spontaneous
EMG activity when the rats were sitting in the cage. This con-

tinuous activity in the TA is abnormal, since flexor muscles
in the rat hind limb do not exhibit background activity under
normal conditions (Hník et al., 1981) and it is apparently
analogous to that found in the EDL (also a flexor muscle) by
Navarrete and Vrbová (1984).

2. The SOL and TA also respond to manual stretch, although
they either contain no spindles, or the few (1 - 3) which are
present, are small atypical spindles with one or two intrafusal
fibers (Hník and Zelená, 1961; Soukup and Zelená, this volume).
The myotatic EMG activity and response to stretch in the reinnervated SOL is most likely due to an afferent input from
the synergistic gastrocnemius (which is relatively rich in
atypical spindles, in spite of early denervation) onto soleus
motoneurons.

3. During ambulation, EMG activity in the reinnervated
TA was characterized by a "double-rhythm", i.e. there was an
EMG burst both during the swing and stance phases, while the
SOL contracted according to its normal locomotor pattern.

4. An unusual phenomenon was occasionally noted in the
TA (and sometimes also in the SOL). Both muscles habitually
exhibited tonic, often single motor unit activity fluctuating
in amplitude. This phenomenon could indicate that the safety
factor at the terminal branchings for discrete muscle fibers
is lower in these muscles than normal.

5. Apparently, peripheral as well as spinal cord mechanisms contribute to the abnormal situation arising from early
postnatal peripheral nerve trauma.

REFERENCES

Hník, P., and Zelená, J., 1961, Atypical spindles in reinner-
     vated rat muscles. J. Embryol. exp. Morph., 9:456-467.
Hník, P., Kasicki, S., Afelt, Z., Vejsada, R., and Krekule,
     I., 1978, Chronic polyelectromyography in awake, un-
     restrained animals. Physiol. bohemoslov., 27:485-492.
Hník, P., Vejsada, R., and Kasicki, S., 1981, Reflex and loco-
     motor changes following unilateral deafferentation of
     rat hind limb assessed by chronic electromyography,
     Neuroscience, 6:195-203.
Navarrete, R., and Vrbová, G., 1984, Differential effect of
     nerve injury at birth on the activity pattern of re-
     innervated slow and fast muscles of the rat, J. Phy-
     siol. (London), 351:675-685.
Soukup, T., and Zelená, J., Sensory innervation of atypical
     spindles after nerve crush in newborn rats (this volume).
Zelená, J., and Hník, P., 1963, Motor and receptor units in
     the soleus muscle after nerve regeneration in very
     young rats, Physiol. bohemoslov., 12:277-290.

BRANCHING OF MYELINATED AND UNMYELINATED FIBERS DURING NERVE REGENERATION

Peter Toft, Kåre Fugleholm, and Henning Schmalbruch

Institute of Neurophysiology, University of Copenhagen

Panum Institute, Blegdamsvej 3C, DK-2200 København N

INTRODUCTION

Nerve cells projecting into peripheral nerves are usually depicted as having an axon which runs straight towards its target. In certain conditions, however, the axons branch within the nerve trunk. Ramon Y Cajal (1928) observed extensive branching of nerve fibers in regenerated nerves. The branchings occurred at the site of injury, and repeated branchings were produced when the nerve was sectioned at multiple sites. Electron microscopic data are scarce. Bray and Aguayo (1974) counted unmyelinated axons in the crushed cervical trunk of rat at different levels and found that, distal to the crush, the number of axons transiently increased by several 100% but then returned to normal. In the present study branching of myelinated and unmyelinated fibers in crushed peroneal and soleus nerves was investigated. The advantage of using the soleus nerve was that all fibers could be assessed which eliminated the error inherent to sampling procedures.

MATERIAL AND METHODS

In 28 adult rats the peroneal or soleus nerves were for 2 min crushed with a smooth-tipped forceps 1 mm wide. The peroneal nerve was crushed 5 mm proximal to the caput fibulae and the soleus nerve was crushed just distal to the last branch to the lateral gastrocnemius muscle. The soleus nerve in two rats was simultaneously crushed at two sites to investigate the effect of a distal crush lesion on the outgrowing axons. The myelinated and unmyelinated axons were counted by light and electron microscopy.

RESULTS

<u>Peroneal nerve:</u> The number of myelinated and unmyelinated axons were the same in the proximal and distal segments of normal nerves, and in the proximal segment of crushed nerves. Thirty days after a crush lesion, the peroneal nerves distally contained 1,932 myelinated axons (SD 49; 4 rats); the mean difference as compared to the proximal segments of the same nerves was +1.8%. One nerve after 30 days distally contained twice as many unmyelinated axons than proximally (10,011 vs. 4,907). The regenerated nerves, 100 days after crushing, contained 2,362 myelinated (SD 241; 5 rats) and 4,846 unmyelinated (1 rat) axons; the relative difference as compared to the proximal segments was +9% (SD 5%) and +50%, respectively.

<u>Soleus nerve</u>: Three normal soleus nerves proximally contained an average of 131 myelinated and 208 unmyelinated axons. When the axon number in the proximal nerves was set to 100%, the distal segments contained 98-102% myelinated and 103-113% unmyelinated axons. Crushed nerves proximal to the crush site, 1-19 weeks after the operation, contained an average of 128 myelinated and 217 unmyelinated axons, i.e. the number in the proximal nerve did not change after crushing. The number of regenerating axons of myelinated origin initially increased and after 4 weeks exceeded that of myelinated proximal axons by up to 20%. The sprouts were still unmyelinated after 1 week; after 2 weeks half of them and after 4 weeks all were myelinated. The number of eventually myelinated distal axons in most nerves became normal, but in some up to 21% more myelinated fibers persisted. In all soleus nerves, the number of axons was determined just distal to the crush and close to the muscle to determine whether or not branching occurred in the crush region only. There was a slight distal excess after 4 weeks (+15%), but thereafter the same number of myelinated axons was found at both levels distal to the crush. The number of unmyelinated axons after 2-4 weeks was 80-240% larger than normal; thereafter their number declined but did not reach a normal level, and an excess of 18-60% persisted. They were initially more numerous close to the muscle than just distal to the crush (10-40%), but also this excess had a tendency to decline. Two nerves at the same time were crushed twice in order to determine whether an additional crush of the distal nerve influenced sprouting at the proximal crush, and whether or not it induced extra-branching of the outgrowing axons when they passed the distal site of injury. There was, however, no evidence of any effect of injury to the distal nerve segment.

DISCUSSION

The results indicate that following a crush lesion all axons branch, but only extra branches of unmyelinated fibers are permanent. Already Gutmann and Sanders (1943) had shown that the number of myelinated fibers in a crushed nerve returns to normal. In contrast to the present results, Bray and Aguayo (1974) did not find a persisting surplus of unmyelinated axons in the crushed cervical trunk. Axonal branching is not restricted to the region of the crush. It is tentatively suggested that regenerating axons branch when searching for a target and that reestablished target contacts prevent additional branching and eliminate redundant branches. Myelinated axons grow into existing Schwann cell strands and are guided whereas unmyelinated axons do not follow predetermined pathways; this may explain their greater tendency to form permanent branches. We found no fiber loss in the proximal nerve segment although nerve section in adult rats caused the death of up to half of the sensory ganglion cells (Schmalbruch, this volume). The rapid regeneration of the crushed nerves possibly prevented neuronal death.

REFERENCES

Bray, G.M., and Aguayo, A.J., 1974, Regeneration of peripheral unmyelinated
        nerves. Fate of the axonal sprouts which develop after injury,
        <u>J. Anat.</u>, 117:517-529.
Gutmann, E., and F.K. Sanders, 1943, Recovery of fibre numbers and
        diameters in the regeneration of peripheral nerves, <u>J.Physiol.</u>
        (London), 101:489-518.
Ramon Y Cajal, S., 1928 (reprinted 1959), Degeneration and Regeneration
        of the Nervous System, Vol 1., Hafner, New York.

<u>Acknowledgments</u>- The technical help of Mrs. M. Bjærg, and the financial support by the Danish MRC, the Foundation for Experimental Research in Neurology, and Jacob and Olga Madsens Fond are gratefully acknowledged.

ABNORMALITIES OF CUTANEOUS SENSORY RECEPTORS FOLLOWING PERIPHERAL NERVE REGENERATION

Bryce L. Munger

The Milton S. Hershey Medical Center
The Pennsylvania State University, P.O.Box 850
Hershey, Pennsylvania 17033

Injuries to peripheral nerves involve complex problems in clinical management as reviewed in detail by Sunderland (1978). Sensory as well as motor sequelae often leave patients significantly impaired needing intensive rehabilitation as described by Dellon (1981). The nature of the problems encountered clinically have changed little from the descriptions by Head and Sherren (1905) at the turn of the century to Stopford's (1930) or Haymaker and Woodhall's (1953) reviews. While peripheral nerves are capable of regeneration, the results clinically can frequently be disappointing. In the case of surgical repair of peripheral nerve injuries considerable attention has been given to the alignment of nerve fascicles in the proximal and distal stump. While some benefits have been claimed for proper alignment, some misrouting is inevitable. The monograph edited by Gorio et al. (1981) explores this general problem in some detail as does Sunderland (1978). Problems of misrouting of sensory and motor axons are obvious sequelae even within a matched set of nerve fascicles. However, within sensory nerves alone mismatch is inevitable. Thus axons that should be directed to a muscle spindle can reach the skin and even within the skin mismatch may involve axons destined for rapidly adapting receptors reaching a Merkel terminal that is normally associated with slowly adapting axons. This raises the intriguing question as to what would happen to the adaptive properties of axons if they were indeed mismatched at the terminal. Would axons previously associated with Merkel terminals be rapidly adapting in a Pacinian corpuscle? While this question can not be answered rigorously at the present time it needs to be kept in mind during the following discussion that deals with anatomically defined sensory receptors that appear to be specific.

Our interest in regeneration parallels an ongoing interest in the normal development of sensory axons (Bressler and Munger, 1983; Dell and Munger, 1986 and Zahm and Munger 1983a,b; Munger and Rice, 1986). The tentative conclusion reached by Bressler and Munger (1983) that sensory axons were present prior to hair placode formation in monkey facial skin where only guard hairs were present has been substantiated by Dell and Munger (1986). They observed that sensory axons are present prior to formation of papillary ridges in monkey glabrous skin. Moore and Munger (1987) have been able to trace the onset of cutaneous innervation to an even earlier time point in human embryonic glabrous skin. These observations dealing with glabrous skin provide a probable link between neural development and cutaneous development and thus a reasonable explanation for the dermatoglyphic abnormalities that are so commonly found in hereditary neurological disorders (Dell and Munger, 1986).

But more germane to the present discussion is the integration of timing during differentiation. The temporal and spatial factors that are involved in the formation of all cutaneous derivatives can be summed as the concept of successive waves of differentiation of afferent neurons (Munger and Rice, 1986). This latter concept appears to be a generalization applicable to all developing sensory systems. As we shall discuss subsequently, this concept may be an important factor limiting successful regeneration of sensory axons in specific sensory receptors.

The classical paradigm for the study of regeneration of peripheral nerves is to compare transection of the nerve with crushing the nerve between forceps. Numerous abnormalities have been described in axons following a transection injury but rarely have abnormalities been found following crush. The excellent study of Terzis and Dykes (1980) and Dykes and Terzis (1979) are one such pair of studies in the literature. Norch and his colleagues have also made important contributions to this type of study (Horch, 1981; Horch and Lisney, 1981a; 1981b). Misdirected axons have been clearly identified by Hallin and Wiesenfeld-Hallin (1984) in human subjects that can communicate where the touched spot should be localized. All of these cited studies agree that following transection the regenerated axons are smaller and conduct more slowly. The difference between transection and crush in the success of regeneration is presumably due to the continuity of basal laminae that would be maintained in a crushed but not transected nerve.

Several studies from Ide's laboratory (Ide et al., 1983; Ide, 1983; Osawa et al., 1986) have documented the importance of residual basal laminae in the regeneration of peripheral nerves. The regrowing axons clearly prefer the inside of the old basal lamina tubes as the growth cones enter the distal segment. Thus in a complete transection of a peripheral nerve a gap or lack of continuity of basal lamina is always present that must be crossed by growth cones to reach an empty basal lamina tube. The gap so present will be filled in with the connective tissue compartment of the surrounding tissue.

The first study from the present author's laboratory on peripheral nerve regeneration involved the analysis of complex system selected for a high density of sensory axons. We chose rat mystacial vibrissae due to the highly predictable distribution of a variety of sensory receptors including lanceolate, Ruffini, Merkel, and free nerve endings (FNE's) as described by Renehan and Munger (1986a). Rice, Mance and Munger (1986) used Renehan's electron microscopic characterizations of sensory terminals and further concluded that sensory terminals are present in 5 discrete regions of rat vibrissae to include 1) FNE's in the dermal papilla, 2) Ruffini, FNE and scattered lanceolate terminals along the shaft of the hair at the level of the cavernous sinus below the Ringwulst, 3) a regular array of lanceolate and Merkel terminals above the Ringwulst at the level of the ring sinus, 4) Ruffini, circular lanceolate and FNE terminals in the dense connective tissue of the conus region and 5) Merkel terminals in the encircling rete ridge collar at the level of the skin surface.

These numerous sensory receptors are derived from two separate parent nerves: the principal nerve trunk to vibrissae referred to as the vibrissal nerve using the terminology of Dorfl (1985) enters the follicle near the base and innervates all receptors from the base or dermal papillae to the level of the ring sinus. The conus and rete ridge collar are innervated by a separate nerve that contributes to the innervation of the superficial dermal nerve plexus and innervates the intervibrissal pelage as well (Rice and Munger, 1986). Thus the sensory innervation of any one vibrissa is duplex. The important point is that the conus innervation is derived from a different source than the main vibrissal nerve.

The complex arrangement of sensory terminals within a single vibrissa is seriously perturbed following nerve transection but not appreciably different following nerve crush (Renehan and Munger, 1986b). The major abnormality in vibrissal innervation following transection is the fact that axons from the main vibrissal nerve fail to restrict their innervation to the portion of the vibrissa below the conus and instead enter the conus admixing with axons from the superficial dermal nerve net. Thus the array of lanceolate terminals and Merkel endings at the level of the ring sinus is no longer regular and geometrically arranged. The nerve that was transected in this study was the infraorbital branch of the trigeminal, a purely sensory nerve. The motor innervation to the mystacial pad is derived from branches of the facial nerve and the motor components of the trigeminal innervate deep muscles of mastication. The abnormality is thus an example of apparent misrouting of axons into the wrong nerve fascicle. In this study we could not be certain of the appropriate innervation of Merkel as compared to lanceolate terminals.

Based on the findings of abnormalities within vibrissae the present author has undertaken a collaborative study with Dykes group at Montreal and published in preliminary form with Samulack (Samulack et al., 1986). The experimental model studied by Samulack and Dykes consists of baboons who have extensive allografts of skin and associated neurovascular pedicles. The entire finger is degloved to the bone and a pedicle of nerve and artery grafted onto the same finger as a transplant control or to another animal as an allograft. The animals were treated with cyclosporin to immunosupress them and permit tolerance of the allograft. Dykes and Samulack subsequently studied the receptor properties of individual mechanoreceptive afferents and were able to find both slowly and rapidly adapting receptors in both glabrous and hairy skin. The skin from the digits was removed and the tissue sent to the present author's laboratory for processing. We have studied serial section sets of tissue stained with the Sevier-Munger (1965) silver method.

These studies are still being analyzed but several important conclusions have been derived to date. First, sensory receptors can be identified in glabrous skin although the number of Meissner corpuscles that can be recognized as such is drastically

reduced. Many Meissner corpuscles are abnormal in location in
that they are not restricted to the apex of the dermal papillae,
but rather can be found near the base of the dermal papillae.
Similar changes have been observed in reinnervated human skin
studied in collaboration with Dellon (Dellon and Munger, 1983).
But the important point to note is that in skin where Meissner
corpuscles could not be recognized as such, rapidly adapting
receptors could still be found although with difficulty.

Atrophic non-reinnervated Meissner corpuscles can also be
identified in all specimens of human and baboon reinnervated
skin studied to date with both Dykes and Dellon (see also Ide,
1982a,b). Atrophic Meissner corpuscles consist of a few remnants
of lamellar cells and compacted and condensed basal lamina. The
results indicate that the lamellar cells retract their processes
and the basal lamina collapses. The collapsed basal lamina can
be identified in PAS-stained sections and the same sections will
also contain examples of almost normal Meissner corpuscles with
thin sheets of basal lamina between the lamellae of the corpuscle. Thus some Meissner corpuscles can undergo regeneration of
the lamellar cell whereas others seemingly are unable to do so.

Pacinian corpuscles were frequently reinnervated and by
more than one axon as has been described by Zelená (1984) suggesting a profound trophic influence of the interior of a Pacinian corpuscle on regenerating axons. In fact where we could
scarcely find a Meissner corpuscle implying poor regeneration
we could frequently find reinnervated Pacinian corpuscles. In
most cases we could trace the axons to determine that the multiple axons were branches of a single parent axon. Recently Ide
et al. (1987) have found that such branching of axons in the
inner core of Pacinian corpuscles is normal in both human and
opossum digital skin. In fact they have one example from a 2
year old subject that had a supernumary digit removed surgically
and the resulting tissue clearly could not have been subject to
prior trauma and thus the end result of nerve regeneration. We
conclude that the inner core of Pacinian corpuscles must be a
most unique microenvironment promoting sprouting of normal axons
as well as regenerating axons. But we note that the inner core
does not regenerate and the numerous inner core lamellae are
absent in reinnervated Pacinian corpuscles.

We have not been able to identify Merkel cells in our allografted skin to date (Samulack et al., 1986) and have no logical
explanation. We consider the timing premature to conclude that
they could be destroyed by the immunological environment, but if
confirmed, it is further evidence that Merkel cells are not necessary for the adaptive properties of the axon. While Merkel
cells are at least infrequent in reinnervated allografted skin,
the numerous slowly adapting receptors that have been identified
physiologically would render the suggestion that the Merkel cells
are necessary for slowly adapting properties simply untenable.
Considering the other sensory receptors that have slowly adapting characteristics such as Ruffini corpuscles (Biemesderfer et
al., 1978) as well as Golgi tendon organs (Matthews, 1974).

But the most surprising findings from our collaborative
study with Dykes is the absence of any sign of normal reinnervation of hairy skin. We have consistently been unable to identify
lanceolate receptors in all specimens studied to date. This observation prompted us to re-evaluate the hairy skin between the

vibrissae in Renehan´s material and Munger and Renehan (1987)
have found equivalent abnormalities in guard hairs in the rat
mystacial pad. The normal collar of lanceolate terminals (Munger,
1971, 1982) is simply not present in any reinnervated guard hairs
following nerve transection. But even more surprising is the
absence of normal reinnervation following nerve crush. The proportion of guard hairs that have a normal collar of lanceolate
terminals is reduced by half as compared to normal. To express
the relationship a different way, in the normal adult rat almost
100 % of innervated guard hairs have a typical piloneural complex. Approximately 5 % (a liberal interpretation as the percentage may be even smaller) of guard hairs that have nerves associated with them lack the typical piloneural complex and instead
have what are tentatively identified as FNE´s. Following infraorbital nerve crush the percentage of guard hairs with only
FNE´s increases dramatically to be almost 50 % of innervated
hairs.

These findings on Meissner, Pacinian, and lanceolate terminals are all internally consistent with the hypothesis that the
specialized cells of sensory receptors become atrophic following
denervation as described by Zelená (1984). We have clear evidence
of such atrophy in Meissner corpuscles as noted above. The
lanceolate terminals on denervated or reinnervated hairs have
not been described to date and such studies are currently under-
way in the present author´s laboratory. Based on these lines of
argument we propose that the limits of regeneration are based
on the normal events of differentiation. Spatial and temporal
factors that appear to play such an important role in normal development simply can not be recapitulated during regeneration in
the adult. The die has been cast and time can not be repeated.
In the absence of an axon the specialized cells of specific sensory receptors begin to undergo atrophy and resultant collapse
of the basal laminae. Basal laminae can exist for long periods
of time resisting digestion by other scavenging cells. Thus the
long term survival of atrophic Meissner corpuscles as little
scars. But for the patient with a nerve transection, the peculiar
sensations clearly begin at the level of the sensory receptors.

REFERENCES

Biemesderfer, D., Munger, B.L., Bink, J., and Dubner, R., 1978,
        The pilo-Ruffini complex: A non-sinus hair and associated
        slowly-adapting mechanoreceptor in primate facial skin,
        Brain Res., 142:197-222.
Bressler, M., and Munger, B.L., 1983, Embryonic maturation of
        sensory terminals of primate facial hairs, J. Invest.
        Dermatol., 80:245-260.
Dell, D.A., and Munger, B.L., 1986, The early embryogenesis of
        papillary (sweat duct) ridges in primate glabrous skin:
        The dermatotopic map of cutaneous mechanoreceptors and
        dermatoglyphics, J. Comp. Neurol., 244:511-532.
Dellon, A.L., 1981, "Evaluation of Sensibility and Re-education
        of Sensation in the Hand", Williams and Wilkins, Balti-
        more.
Dellon, A.L., and Munger, B.L., 1983, Correlation of histology
        and sensibility after nerve repair, J. Hand Surg., 8:
        871-875.
Dorfl, J., 1985, The innervation of the mystacial region of the
        white mouse. A topographical study, J. Anat., 142:173-184.

Dykes, R.W., and Terzis, J.K., 1979, Reinnervation of glabrous
    skin in baboons: Properties of cutaneous mechanoreceptors
    subsequent to nerve crush, J. Neurophysiol., 42:1461-1478.
Gorio, A., Millesi, H., and Mingrino, S., eds., 1981, "Posttrau-
    matic Peripheral Nerve Regeneration", Raven Press, New
    York.
Hallin, R.G., and Wiesenfeld-Hallin, Z., 1984, Sensory abnorma-
    lities after peripheral nerve repair and their relation-
    ship to the properties of reinnervated mechanoreceptors
    in human glabrous skin, pp. 229-240, in: "Sensory Receptor
    Mechanisms", W. Hamann and A. Iggo (eds), World Scientific
    Publishing Company, Singapore.
Haymaker, W., and Woodhall, B., 1953, "Peripheral Nerve Injuries"
    W.B. Saunders, Philadelphia.
Head, H., and Sherren, J., 1905, The consequences of injury to
    the peripheral nerves in man, Brain, 28:116-338.
Horch, K., 1981, Absence of functional collateral sprouting of
    mechanoreceptor axons into denervated areas of mammalian
    skin, Exp. Neurol., 74:313-317.
Horch, K.W., and Lisney, S.J.W., 1981a, On the number and nature
    of regenerating myelinated axons after lesions of cuta-
    neous nerves in the cat, J. Physiol., 313:275-286.
Horch, K.W., and Lisney, S.J.W., 1981b, Changes in primate af-
    ferent depolarization of sensory neurons in the cat,
    J. Physiol., 313:287-299.
Ide, C., 1982a, Degeneration of mouse digital corpuscles, Am. J.
    Anat., 163:59-72.
Ide, C., 1982b, Regeneration of mouse digital corpuscles, Am. J.
    Anat., 163:73-85.
Ide, C., 1983, Nerve regeneration and Schwann cell basal lamina:
    observations of the long-term regeneration, Arch. Histol.
    Jap., 46:243-257.
Ide, C., 1984, Nerve regeneration through the basal lamina scaf-
    fold of the skeletal muscle, Neurosci. Res., 1:379-391.
Ide, C., Nitatori, T., and Munger, B.L., 1987, The cytology of
    human Pacinian corpuscles: evidence for sprouting of the
    central axon, Arch. Histol. Jap., in press.
Ide, C., Tohyama, K., Yokota, R., Nitatori T., and Onodera, S.,
    1983, Schwann cell basal lamina and nerve regeneration,
    Brain Res., 288:61-75.
Matthews, P.B.C., 1974, Receptors in Muscles and joints, in:
    "The Peripheral Nervous System", J.I. Hubbard (ed.),
    Plenum Press, New York.
Moore, S.J., and Munger, B.L., 1987, The early ontogeny of the
    afferent nerves and papillary ridges in human digital
    glabrous skin, submitted.
Munger, B.L., 1971, Patterns of organization of peripheral sen-
    sory receptors, pp. 523-556, in: "Handbook of Sensory
    Physiology", Vol. 1, W.R. Lowenstein (ed.), Springer-
    Verlag, New York.
Munger, B.L., 1982, Multiple afferent innervation of primate
    facial hairs - Henry Head and Max von Frey revisited,
    Brain Res., 4:1-43.
Munger, B.L., and Renehan, W.E., 1987, Abnormal sensory rein-
    nervation of rat guard hairs following transection or
    crush of branches of the trigeminal nerve, J. Comp. Neu-
    rol., submitted.
Munger, B.L., and Rice, F.L., 1986, Successive waves of diffe-
    rentiation of cutaneous afferents in rat mystacial skin,
    J. Comp. Neurol., 252:404-414.

Osawa, T., Ide, C., and Tohyama, K., 1986, Nerve regeneration through allogenic nerve grafts in mice, Arch. Histol. Jap., 49:69-81.
Renehan, W., and Munger, B.L., 1986a, Degeneration and regeneration of peripheral nerve in the rat trigeminal system. I. Identification and characterization of the multiple afferent innervation of mystacial vibrissa, J. Comp. Neurol., 246:129-145.
Renehan, W., and Munger, B.L., 1986b, Degeneration and regeneration of peripheral nerve in the rat trigeminal system. II. Acute response to nerve lesion, J. Comp. Neurol., in press.
Rice, F.L., Mance, A., and Munger, B.L., 1986, A comparative light microscopic analysis of the sensory innervation of the mystacial pad in the hamster, mouse, rat, gerbil, rabbit, guinea pig, and cat. I. Innervation of vibrissal follicle-sinus complexes, J. Comp. Neurol., 252:154-174.
Rice, F.L., and Munger, B.L., 1986, A comparative light microscopic analysis of the sensory innervation of the mystacial pad in the hamster, mouse, rat, gerbil, rabbit, guinea pig, and cat. II. The common fur between the vibrissae, J. Comp. Neurol., 252:186-205.
Samulack, D.D., Munger, B.L., Dykes, R.W., and Daniel, R.K., 1986, Neuroanatomical evidence of reinnervation in primate allografted (transplanted) skin during cyclosporine immunosuppression, Neurosci. Lett., 72:1-6.
Sevier, E.A., and Munger, B.L., 1965, A silver method for paraffin sections of neural tissue, J. Neuropathol. Exp. Neurol., 24:130-135.
Stopford, J.S.B., 1930, "Sensation and the Sensory Pathway," Longmans, Green and Co., London.
Sunderland, S., 1978, "Nerves and Nerve Injuries", Churchill, Livingstone, Edinburgh, London and New York.
Terzis, J.K., and Dykes, R.W., 1980, Reinnervation of glabrous skin in baboons: properties of cutaneous mechanoreceptors subsequent to nerve transection, J. Neurophysiol., 44: 1214-1225.
Zahm, D.S., and Munger, B.L., 1983a, Fetal development of primate chemosensory corpuscles. I. Synaptic relationship in late gestation, J. Comp. Neurol., 213:146-162.
Zahm, D.S., and Munger, B.L., 1983b, Fetal development of primate chemosensory corpuscles. II. Synaptic relationships in early gestation, J. Comp. Neurol., 219:36-50.
Zelená, J., 1984, Multiple axon terminals in reinnervated Pacinian corpuscles of adult rat, J. Neurocytol., 13:665-684.

REINNERVATION OF CUTANEOUS MECHANORECEPTORS

H. Aldskogius[1], Z. Wiesenfeld-Hallin[2], E. Kinnman, and
J. Persson[1]

1 Dept. of Anatomy Karolinska Institute, Stockholm, Sweden
2 Dept. of Clinical Neurophysiology, Karolinska Institute
  Huddinge Hospital, Huddinge, Sweden

INTRODUCTION

Following injury to peripheral sensory nerves residual functional
defects and abnormal sensations are common consequences (Sunderland, 1978).
An understanding of the mechanism(s) for these disturbances requires a
detailed knowledge of how these nerve lesions affect the various components
in the somatosensory system from receptor to cerebral cortex.

Two different "peripheral" mechanisms can contribute to recovery of
sensory function in the skin after peripheral nerve injury. In most situations, the injured nerve has the capability in a more or less adequate and
accurate manner to regenerate and reinnervate the denervated skin area. In
other situations, axons normally innervating areas adjacent to the denervated skin could extend branches into this skin area. This mechanism -
collateral reinnervation - could provide partial sensory recovery to a
denervated area in cases where the injured nerve does not regenerate, as
well as temporarily provide a denervated skin area with some innervation
while the injured nerve regenerates.

In the present series of studies we have examined sensory nerves and
nerve endings in the skin of the hindpaw of the rat after two different
types of sciatic nerve injuries. One, in which the transected nerve was
kept chronically injured, and another one where the nerve was repaired and
allowed to regenerate to the original territory. In the first case, the
skin of the hindpaw could only be reinnervated by collateral extension from
the neighboring saphenous nerve. The possibility of collateral reinnervation
was investigated under two different experimental situations, namely with
the saphenous nerve intact or with the saphenous nerve reinnervating the
denervated skin after a crush injury.

COLLATERAL REINNERVATION BY INTACT SENSORY AXONS

The hindpaw was chronically denervated by transection and ligation of
the sciatic nerve in female adult (200 g b.wt.) or newborn (both sexes)
Sprague-Dawley rats.

Two-24 months later the possible extension of saphenous nerve axons
into the sciatic nerve territory was examined morphologically by using

the anterograde wheat germ agglutinin-horseradish peroxidase conjugate
(WGA-HRP) tracing technique (Aldskogius et al., 1986; Kinnman and
Aldskogius, 1986). Briefly, dorsal root ganglia L3 and L4 (which are the
major ganglia for saphenous nerve dorsal root ganglion cells) were exposed
and 1 µl 2% WGA-HRP was injected in each ganglion. Twenty-four hours later
the animals were perfused with an aldehyde fixative and frozen serial
sections from the sole of the foot processed for demonstration of HRP activity, using the tetramethyl benzidine (TMB) method.

Another group of animals was subjected to physiological experiments
several months after the sciatic nerve injury. The saphenous nerve was exposed bilaterally and placed on a pair of platinum hook electrodes for
recording electrical activity following light mechanical stimulation of the
plantar, as well as the dorsal skin of the hindpaw.

In the morphological study, the normal saphenous nerve was found to
have a small number of labeled fibers in a narrow strip medially and mainly
proximally on glabrous skin on the sole of the foot, on an average 0.7 mm
from the medial edge (Kinnman and Aldskogius, 1986). Labeled fibers ending
in a fashion resembling Meissner corpuscles and Merkel cell-neurite complexes were rarely seen, while free nerve endings were more commonly observed. Many of the latter were of a very fine caliber and reached the lower
portions of the epidermis. Fiber labeling in digit I showed a similar
appearance, while only occasional fibers were present in digit II.

Following chronic sciatic nerve injury in adult rats only scattered
fine caliber labeled axons were present outside the normal saphenous nerve
territory (Kinnman and Aldskogius, 1986). The lateral extension of these
fibers varied in different experiments from 0.3-2.9 mm. No signs of
labeled fibers were found in the three lateral digits. With a neonatal
chronic sciatic nerve lesion, scattered coarse and fine calibered labeled
fibers were present outside the normal saphenous nerve territory. Their
lateral extension ranged from 1.2-3.0 mm. Labeled fibers were present on
the third digit, which we always found to completely lack innervation
normally by the saphenous nerve. It should be noted that no signs of re-
appearance of Meissner corpuscles or Merkel cell-neurite complexes were
found. The coarse labeled fibers seemed to end without any terminal specializations and usually followed a course parallell to the skin surface
just beneath the epidermis. Control experiments where dorsal root ganglia
L4 and L5 were injected in combination with saphenous nerve cut, did not
show any labeled fibers, demonstrating that sciatic nerve sensory axons
had not reinnervated the plantar skin.

In the physiological experiments the glabrous, as well as hairy, skin
of the hindpaw were examined. In control animals no activity could be recorded in the saphenous nerve after light mechanical stimulation of the
glabrous skin. On the dorsum of the foot activity could be recorded when
digits I, II and the proximal phalanx of digit III, as well as the proximal skin were stimulated. An overlap with the sciatic nerve was always
present on digit III and sometimes on digit II as well.

Chronic sciatic nerve lesion did not change that pattern. This is unlikely to be due to methodological problems. Activity could easily be
recorded when the normal saphenous nerve area was stimulated. The border-
line to the denervated area was always very marked, and even a strong,
non-nociceptive stimulus of this area did not evole a response in the
saphenous nerve. Furthermore, in the same experiments evidence was found
for some extent of collateral reinnervation by saphenous nerve C-fibers
by using the Evans blue - plasmaextravasation technique.

From these experiments we conclude that significant functional

collateral reinnervation by intact low threshold mechanoreceptive axons
does not occur in the hindpaw. A further inference from our findings would
also be that the labeled fibers outside the saphenous nerve area subserve
other sensory functions or - in some cases - may even be non-functional.

With regard to adult animals, our results are in agreement with previous observations in the cat (Horch, 1981) rabbit (Jackson and Diamond,
1983) and rat (Jackson and Diamond, 1984). However, results from previous
studies indicate that low threshold mechanoreceptor axons of the rat
undergo sprouting within a dermatome of the trunk after neonatal lesion
(Jackson and Diamond, 1984). The discrepancy between these and our findings
could have a number of different explanations. First, different strains of
rats have been used - Wistar vs. Sprague-Dawley. Second, different regions
have been examined - trunk vs. hindpaw. Third, the "shape" of the normally
innervated skin area differed - isolated "island" with the entire circumference bordering denervated skin (Jackson and Diamond, 1984) vs. longitudinal "strip" with basically only two borderlines towards the denervated
skin. A fourth factor which can perhaps not be excluded is whether some
sensory axons in the trunk could still be in the process of growing into
the skin. Conceivably, such axons could enter denervated skin which is not
part of the normal territory (cf. Jackson and Diamond, 1984 and below).

Nevertheless, our negative findings might be somewhat surprising in
view of the clinical (Leonard, 1973) and experimental (Guth, 1956) observations that cutaneous sensation returns to an area deprived of its original innervation to a much greater degree in very young compared to adult
individuals. Possibly, the basis for this relatively extensive functional
recovery resides in the central nervous system.

COLLATERAL REINNERVATION BY REGENERATING SENSORY AXONS

In order to examine the possibility that axons in the saphenous nerve
subserving low threshold mechanoreceptors might be induced to extend
branches into "foreign" territory if they were set in a growing mode, the
saphenous nerve was crushed at the same time as the sciatic nerve was
chronically injured.

Using the tracing method described above after combined sciatic-
saphenous lesions in adult rats, fine and coarse calibered axons were
present 0.9-3.5 mm lateral to the normal saphenous-sciatic nerve boundary
(Kinnman and Aldskogius, 1986). Labeled fibers were regularly present in
the plantar skin of digit III and in some cases in digit IV as well. The
density of labeling was much greater than in the single lesion experiments
described above and almost approached the normal situation. Similarly, the
qualitative appearance of the labeled nerve fibers and endings resembled
that of normal skin more closely that in single nerve lesion experiments.
Thus, occasional profiles with the appearance of Meissner corpuscles and
Merkel cell-neurite complexes were found.

In physiological experiments made as described above, no activity
could be recorded in the saphenous nerve outside its normal innervation
territory with light mechanical stimulation of the glabrous and hairy skin
of the hindpaw. This was the case even when the nerve lesions were made at
the neonatal stage. Saphenous nerve C-fibers, on the other hand, showed
extensive reinnervation of sciatic nerve territory after combined sciatic-
saphenous nerve lesions in adults as well as neonates.

These results indicate that even if the saphenous nerve is injured,
its regenerating low threshold mechanoreceptive axons do not extend
functionally effective branches into the denervated glabrous or hairy skin.

The morphological observations in glabrous skin indicate that coarse
fibers and some specialized nerve endings were present. Since no activity
could be recorded in the saphenous nerve upon light mechanical stimulation
of this skin, it would seem that these nerve fibers and endings subserve
other functions or are non-functional. Evidence for functionally effective
collateral reinnervation by saphenous nerve C-fiber afferent fibers were,
indeed, found (Wiesenfeld-Hallin and Kinnman, unpubl. obs.). These observations are in contrast to those of Jackson and Diamond (1984), who found
evidence for collateral reinnervation by regenerating low threshold
mechanoreceptive axons in the trunk. Perhaps a major reason for this discrepancy is the difference in the site of nerve crush, which seems to be
rather close to the skin in their experiments. Since there is an overpro-
duction of axonal branches at the lesion site, this might allow some of
the otherwise eliminated branches to find an appropriate target. Regional,
strain and/or other methodological differences in their study compared to
ours can perhaps not be excluded either.

REGENERATIVE REINNERVATION BY THE "ORIGINAL" SENSORY AXONS

In order to obtain information about the temporal course and morphological aspects of cutaneous reinnervation by the original nerve, we have
initiated a series of experiments to follow regenerative outgrowth of
sensory axons in the injured sciatic nerve. The sciatic nerve was transected at midthigh level and repaired with epineurial sutures. Following
various postoperative survival times, sensory axons in the sciatic nerve
were labeled by injections of WGA-HRP in the L4-L5 dorsal root ganglia.
The saphenous nerve was cut simultaneously to prevent any spurious labeling via its axons. The tibial and sural nerves, as well as plantar and
dorsal skin of the hindpaw, were processed for the demonstration of HRP
activity.

The examination showed that sciatic nerve fibers had reached the
distal portion of the tibial and sural nerve within four weeks after nerve
injury. The first fibers to reappear in the skin of the hindpaw were found
four-six weeks postoperatively. These fibers were of varying caliber and
appeared to terminate in the upper portion of the epidermis in an uncha-
racteristic manner. During the subsequent six-eight weeks a gradual increase in the density of labeled fibers occurred, some of which were thin
and extending into the epidermis. Fifteen weeks after nerve lesion, the
density of labeled fibers still appeared to be reduced and there was a
paucity particularly with regard to the occurrence of labeled Merkel cell-
neurite complexes.

These findings indicate that morphological recovery of cutaneous
mechanoreceptor nerve endings is incomplete even several months after
nerve transection. This is in line with clinical experience that sensory
recovery following nerve transection is usually deficient in adult individuals (Sunderland, 1978) as well as with previous physiological observations that functional reinnervation of the hindpaw in the rat is still
subnormal two-three months after crush injury of the sciatic (Bisby and
Keen, 1986) or saphenous (Wiesenfeld-Hallin and Kinnman, unpubl. obs.)
nerve.

SUMMARY AND CONCLUSIONS

The present study has examined the potential for mechanoreceptive
axons to reinnervate the hindpaw of the rat after two different types of
sciatic nerve lesions. One, a chronic lesion required the denervated skin
to be innervated by collateral extension from the neighboring saphenous

nerve, which was either left intact or induced to regenerate by a crush. Two, a transection followed by suture of the injured nerve, allowed this nerve to reinnervate the denervated skin. In the first case, the sciatic nerve was injured at the neonatal or adult stage. The possible presence of saphenous nerve sensory axons in the sciatic nerve territory was examined with the anterograde WGA-HRP tracing method and with electrophysiological recordings from the saphenous nerve after light mechanical stimulation of the hindpaw. The morphological studies suggested some extension of coarse sensory saphenous axons into sciatic nerve territory after neonatal injury and combined sciatic cut/saphenous crush. However, there was no physiological evidence for collateral reinnervation by low threshold mechanoreceptive axons.

Only adult animals were subjected to transection and suture of the sciatic nerve. The outgrowth sequence of regenerating sensory axons was followed with the WGA-HRP technique. By 15 weeks postlesion survival time, the number of labeled sensory axons in the hindpaw was still less than normal. This was true e.g. for Merkel cell-neurite complexes.

These findings indicate, first, that low threshold mechanoreceptive axons do not undergo functionally effective collateral reinnervation in the hindpaw and, second, that restoration of low threshold mechanoreceptive nerve endings is incomplete in the hindpaw following nerve transection with subsequent regeneration.

ACKNOWLEDGEMENTS

Expert secretarial assistance was given by Ms. Marianne Rapp. The work was supported by grants from the Swedish Medical Research Council, proj.nos. 5420 and 7913, the Folksam Insurance company and the Karolinska Institute.

REFERENCES

Aldskogius, H., Kinnman, E., and Persson, J., 1986, Labeling of cutaneous sensory nerve endings with axonally transported horseradish peroxidase and wheat germ agglutinin-horseradish peroxidase conjugate: a methodical study in the rat, J. Neurosci. Meth., 15:281-294.

Bisby, K.A., and Keen, P., 1986, Regeneration of primary afferent neurons containing substance P-like immunoreactivity. Brain Res., 365:85-95.

Guth, L., 1956, Regeneration in the mammalian peripheral nervous system. Physiol. Rev., 36:441-478.

Horch, K., 1981, Absence of functional collateral sprouting of mechanoreceptor axons into denervated areas of mammalian skin, Exp. Neurol., 74:313-318.

Jackson, P.C., and Diamond, J., 1983, Failure of intact cutaneous mechanosensory axons to sprout functional collaterals in skin of adult rabbits, Brain Res., 273:277-283.

Jackson, P.C., and Diamond, J., 1984, Temporal and spatial constraints of the collateral sprouting of low-threshold mechanosensory nerves in the skin of rats, J. Comp. Neurol., 226:336-345.

Kinnman, E., and Aldskogius, H., 1986, Collateral sprouting of sensory axons in the glabrous skin of the hindpaw after chronic sciatic nerve lesion in adult and neonatal rats: a morphological study, Brain Res., 377:73-82.

Leonard, M.H., 1973, Return of skin sensation in children without repair of nerves, Clin. Orthoped. Rel. Res., 95:273-277.

Sunderland, S., 1978, "Nerves and Nerve Injuries," 2nd Ed. Churchill Livingstone, Edinburgh, London, New York.

HYPERINNERVATION OF RAT PACINIAN CORPUSCLES IN A TOXIC DISTAL AXONOPATHY

I. Jirmanová

Institute of Physiology, Czechoslovak Academy of
Sciences
Vídeňská 1083, Prague 4, Czechoslovakia

During nerve regeneration following surgical injury, axonal
sprouts are formed by regenerating axons at the site of nerve
lesion and in the periphery. As a consequence, peripheral target
tissues may become hyperinnervated (McArdle, 1975; Rotshenker
and McMahan, 1976; Mira, 1981; Gorio et al., 1983; Zelená,
1984). Nerve regeneration subsequent to degeneration also occurs in individuals affected by various types of toxic axono-
pathies both after recovery from the intoxication and during
continuing intoxication. In our experiments, we examined whether
in neuropathic animals peripheral target tissues will also become reinnervated and if so, whether their innervation pattern
will be altered. For morphological study, Pacinian corpuscles
appeared to be the most suitable model since each of these rapidly adapting mechanoreceptors, which consists of an inner
core and a capsule, is normally innervated by only one myelinated axon and one axon terminal.

Adult Wistar rats were exposed to carbondisulphide (CS_2)
vapour inducing distal axonopathy characterized by continuous
nerve degeneration and regeneration (Jirmanová and Lukáš, 1984).
After six months of intoxication, the animals were perfused
with a fixative, the interosseous nerve with Pacinian corpuscles
attached to it was dissected out and the material was processed
for electron microscopy as described before (Jirmanová, 1987).

Both degenerative and regenerative changes were observed
in axons and Pacinian corpuscles of the diseased animals. In a
sample of corpuscles examined, 30 % were denervated and about
60 % were reinnervated. Some of the reinnervated corpuscles
were supplied by unmyelinated axons. In others, one to three
myelinated axons were already present at the nerve entry. The
axons branched and formed three to eight terminals in the original inner core (Fig. 1). Due to continuous intoxication, some
regenerated terminals were again undergoing degeneration. Lateral processes and membrane specializations characteristic for
Pacinian endings were usually absent.

It can be concluded that Pacinian corpuscles of rats with
CS_2 neuropathy became multiply reinnervated in their majority.

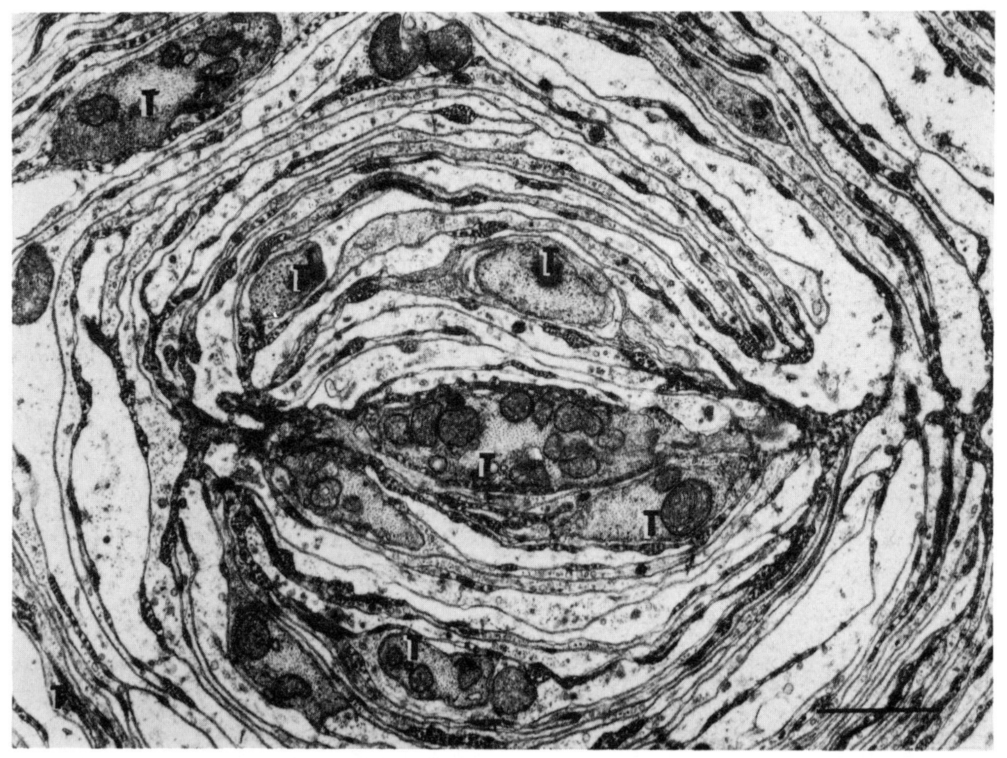

Fig. 1. An inner core is reinnervated by eight axon terminals
        (T). Bar: 1 μm.

We assume that a postdenervation hyperinnervation of Pacinian
corpuscles and other end-organs also occur in other types of
neuropathies, if not during the poisoning, then at least after
its cessation.

REFERENCES

Gorio, A., Carmignoto, G., Finesso, M., Polato, P., and Nunzi,
        M.G., 1983, Muscle reinnervation - II. Sprouting, synapse
        formation and repression, Neurosci., 8:403-416.
Jirmanová, I., 1987, Pacinian corpuscles in rats with carbon-
        disulphide neuropathy, Acta Neuropathol., 72:341-348.
Jirmanová, I., and Lukáš, E., 1984, Ultrastructure of carbon
        disulphide neuropathy, Acta Neuropathol., 63:255-263.
Mira, J.C., 1981, Degeneration and regeneration of peripheral
        nerves: ultrastructural and electrophysiological obser-
        varions, quantitative aspects and muscle changes during
        reinnervation, J. Microsurg., 3:102-120.
McArdle, J.J., 1975, Complex end-plate potentials at the rege-
        nerating neuromuscular junction of the rat, Exp. Neurol.,
        49:629-638.
Rotshenker, S., and McMahan, U.J., 1976, Altered patterns of
        innervation in frog muscle after denervation, J. Neuro-
        cytol., 5:719-730.
Zelená, J., 1984, Multiple axon terminals in reinnervated
        Pacinian corpuscles of adult rat, J. Neurocytol., 13:
        665-684.

THE EFFECT OF OPIOID PEPTIDES ON THE FUNCTIONAL RECOVERY OF DAMAGED NEURONAL STRUCTURES

G.N. Akoev, O.B. Ilyinsky[x], L.I. Kolosova,
M.I. Titov, and O.G. Trofimova

USSR Cardiology Research Center
Academy of Medical Sciences
Moscow, USSR

INTRODUCTION

The discovery of endogenous opioid peptides (OP) - endorphins and enkephalins (Hughes, 1975; Hughes et al., 1975, Kosterlitz and Hugnes, 1975) which play a conspicuous role in the nervous system function is one of the most important achievements of modern neurobiology (Krieger, 1983; Akil et al., 1984; Olson et al., 1984). The abundance of OP in nervous tissue, and the multiplicity of their functions (Atwen, 1983; North and Williams, 1983; Wood, 1983; Woolverton and Schuster, 1983) as well as the regulatory influence of endorphines on RNA synthesis (Lindvall and Bjorklund, 1974) have considerably stimulated our interest in the study of effects of OP on nervous tissue growth. Recently (Ilyinsky et al., 1985a; 1985b; 1986a) it has been shown that OP stimulate nervous tissue growth in culture. This effect is twofold: in neurons the peptides stimulate the outgrowth of neurites and their survival, while in glial cells they change the rate of their migration and, probably, their proliferation. It has been also shown that OP stimulate the processes of regeneration and growth of the sympathetic nervous tissue in vivo (Ilyinsky et al., 1986b). It was assumed that endogenous OP, released in the nervous system as a result of activation of the antinociceptive structures are playing an important role in the repair of damaged neural structures (Ilyinsky and Titov, 1985; Ilyinsky et al., 1986c). Proceeding from these facts the investigation of the OP effect on the regeneration processes of peripheral parts of the somatic nervous system as well as on the recovery of function of some mechanoreceptors was undertaken.

MATERIALS AND METHODS

One of the synthetic analogues (AE) of leu-enkephalin

---

[x]Address for correspondence: Prof. O.B. Ilyinsky, Central Institute of Medico-Biological Problems of Sports, Elizavetinsky proezd, 10, 107005 Moscow, USSR

(Tyr-D-Ala-Gly-Phe-Leu-Arg) was used. As we have shown earlier
(Ilyinsky et al., 1985b; Ilyinsky et al., 1986a) it had significant nerve-growth action. The investigation of regeneration
processes was performed on 30 adult male rats. The right sciatic nerve was cut in the middle of the thigh above the nerve
division into the main branches. The ends of the cut nerve
were connected under the microscope by means of microsurgical
epi-perineural sutures. Thirteen animals received an intramuscular injection of AE for 7 post-operative days (10 µg/kg).

Reinnervation of foot skin by sensory fibers was studied
by recording impulse activity of sciatic nerve microfasciculi
proximally to the site of injury to mechanical stimulation of
foot skin surface. The initial stage of foot muscle reinnervation with motor fibers was also estimated. The vestibular test
(Gutmann et al., 1942; Berenberg et al., 1977) for observing

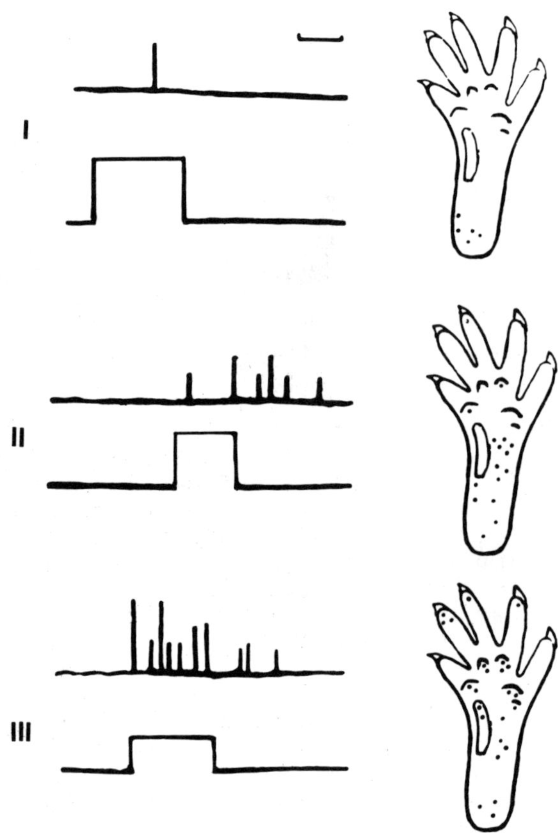

Fig. 1. Impulse response of nerve fibers of the sciatic nerve
to mechanical stimulation of the foot under AE. Left
- oscillograms of responses. Upper curves - impulse
activity, lower - stimulation point. Right - localization of receptive fields. I, II, III - 15, 30 and 45
days after the operation, respectively, Calibration -
20 ms.

the reflex spreading of the hindlimb toes in response to
a sharp lowering of the animal was also employed.

RESULTS

Fifteen days after cutting the sciatic nerve it is possible to record impulse activity in response to strong mechanical stimulating the skin of the foot in afferent nerve fibers
of rats treated with OP. The first responses were recorded
from the receptive field limited to a small area of the heel
skin surface (Fig. 1). As a rule the response consisted of
one, rarely two impulses to an on or off stimulus (Fig. 1).
It was poorly reproducible in the course of repeated stimulation. The response was observed in all three animals tested
at that period, but the number of activated nerve fibers present was rather small: 5-7 from 50 in each experiment. Con-

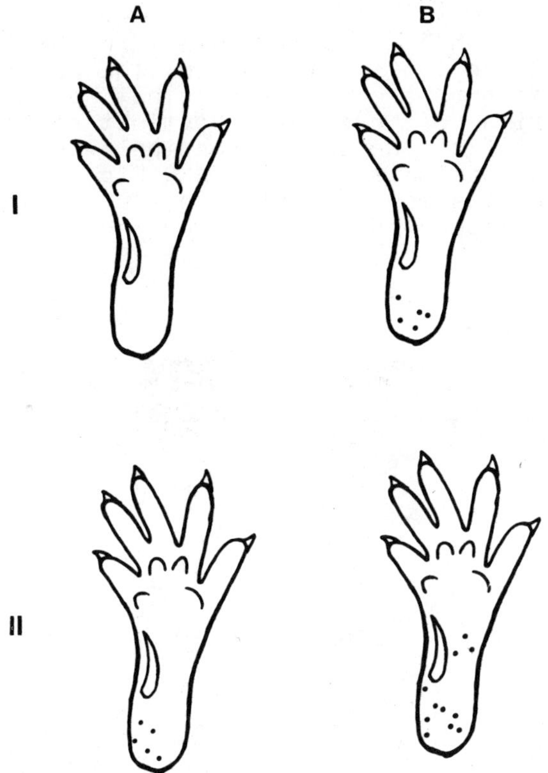

Fig. 2. AE influence on the localization of receptive fields
        on mechanical foot stimulation. A - control; B - AE
        effect. Foot areas during stimulation of which an im-
        pulse response was recorded are marked with dots.
        I - 15, II - 20 days after operation.

trol animals showed similar responses, but they appeared only
in half the animals (2 out of 4) and only on the 20th day after
the operation. The presence of impulse activity in a small part
of the tested nerve fibers (about 10 %), the character of responses and high thresholds - all these indices showed foot reinnervation is just beginning. Afferent activity at this period
should apparently be associated with the beginning of forming
of the simplest high-threshold mechanoreceptors or polyfunction-
al receptors, including the nociceptors which have been known
to regenerate first (Lundborg et al., 1982).

The receptive field of responses to mechanical stimulation
broadens gradually and by the 30th day embraces the whole foot
surface of the tested animals (by the 40th day in the controls).
As a rule, evoked impulse activity is recorded in response to
stimulation of the heel and the center of the foot, and sometimes also from the pad (Fig. 2). Forty-five days after the
operation, impulse activity in the tested animals was recorded
not only to strong mechanical stimulation but also to a light
touch. In 50 % of tested microfascicles there are fibers responding to mechanical stimulation of the pad (Fig. 2) whose
high sensitivity is well known (Sander and Zimmerman, 1986).
In control animals, responses to weak mechanical stimulation
and concentration of the receptive field in the pad were firstly observed 2 months after the operation.

The study of the vestibular reflex in these rats has shown
that, after cutting the sciatic nerve, the paw is always held
lower and the fingers are bent and do not spread in response
to sudden lowering of the animal. Distinct spreading of intact
paw fingers testified the correct manipulation of the lowering
procedure. In the animals injected AE (Fig. 3), the first signs
of reflex rehabilitation (leg flexion and abduction of the 4th
and 5th finger) appeared 6 days after the operation and

A
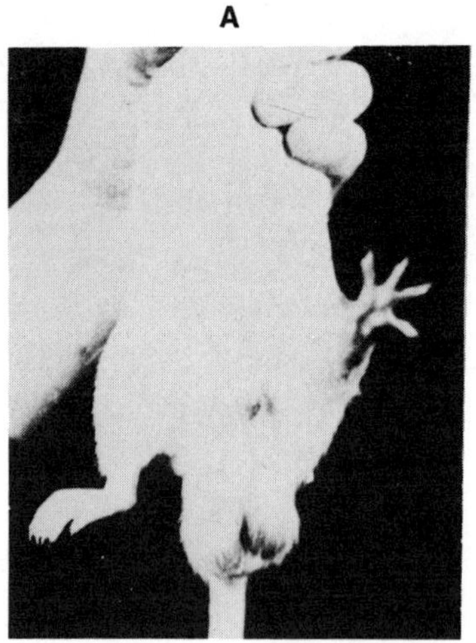 **B**
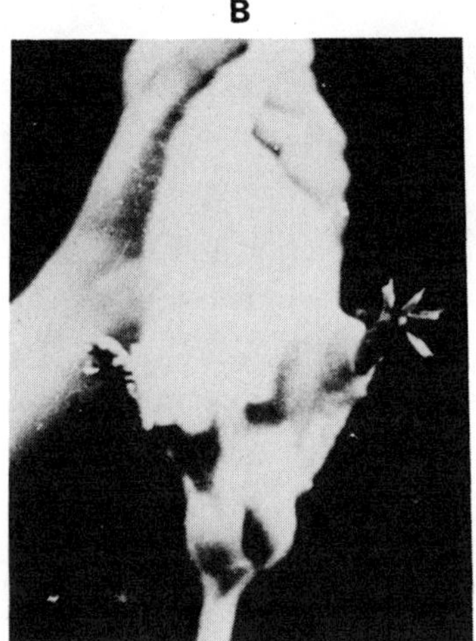

Fig. 3. AE effect on rat vestibular reflex. A - control,
        B - AE effect. Operated leg is on the left, 6 days
        after the operation.

spreading of all the five fingers after 8 days. In animals
which were not subject to peptide therapy the first signs of
reflex finger abduction were not observed until 8 days after
the operation and the spreading of all fingers until after 10
days. During 1-1.5 months the vestibular reflex was reduced
in all animals. The process of its restoration was oscillatory
in character and it was difficult to make definite conclusions
from the further observations. Probably, the vestibular test
may clearly reflect only the initial period of paw reinnervation by motor nerve fibers and this assumption coincides
with the published data (Berenberg et al., 1977).

Thus, AE stimulates sensory and motor fiber regeneration
(the time of regeneration was shortened by 25 %). The mechanism of AE action is not yet clear. It is interesting that the
half-time of degradation of the AE in the organism is short -
only two minutes (Isakova and Sepetov, 1985). Probably, AE has
a trigger mechanism of action (AE "switches on" some processes
of reparation).

OP are apparently an interesting class of substances which
also participate in enhancing the regeneration of damaged neural structures.

REFERENCES

Akil, H., Watson, S.J., Young, E., Lewis, M.E., Khachaturian,
        H., and Walker, J.M., 1984, Endogenous opioids: biology
        and function, Ann. Rev. Neurosci., 7:223.
Atwen, S.F., 1983, Characterization and distribution of brain
        opiate receptors and endogenous opioid peptides, in:
        "The Neurobiology of Opiate Reward Processes", J.E.
        Smith and J.D. Lane, eds., Elsevier, Amsterdam, N.Y.,
        Oxford.
Berenberg, R.A., Forman, D.S., Wood, D.R., Desilva, A., and
        Demarel, J., 1977, Recovery of peripheral nerve function
        after axotomy: effect triodothyronine, Exp. neurol.,
        57:349.
Gutman, E., Gutman, L., Medawar, P.B., and Young, J.Z., 1942,
        The rate of regeneration of nerve, J. Exp. Biol., 12:14.
Hughes, J., 1975, Isolation of an endogenous compound from the
        brain with pharmacological properties similar to mor-
        phine, Brain Res., 88:295.
Hughes, J., Smith, T.W., Kosterlitz, H.W., Fothergill, L.A.,
        Morgan, B.A., and Morris, H., 1975, Identification of
        two related pentapeptides from the brain with patent
        opiate agonist activity, Nature (London), 258:577.
Ilyinsky, O.B., and Titov, M.I., 1985, Nociception and struc-
        tural homeostasis, in: "Modern Physiological Problems
        of the Nervous and Muscular Systems", V.M. Okudzava,
        ed., Tbilisi University Pbl., Tbilisi.
Ilyinsky, O.B., Kozlova, M.V., Kondrikova, E.S., and Kalenchuk,
        V.U., 1985a, Effect of opioid peptides on growth and
        regeneration of the nervous tissue of rat, J. evol.
        biochem. physiol., 21:511.
Ilyinsky, O.B., Kozlova, M.V., Kondrikova, E.S., Titov, M.I.,
        Bespalova, Zh.D., Yarygin, K.N., and Yurchenko, N.N.,
        1985b, Effect of opioid peptides on nerve tissue in
        vitro, Neurophysiologia, 17:550.

Ilyinsky, O.B., Kozlova, M.V., Kondrikova, E.S., Kalenchuk,
    V.U., Titov, M.I., and Bespalova, Zh.D., 1986a, Pecu-
    liarities of opioid peptides and naloxone action on
    central and peripheral nerve system tissue in culture,
    Neurophysiologia, 18:227.
Ilyinsky, O.B., Spevak, S.E., Kargina-Terentieva, R.A., Shva-
    lev, V.N., Anaskin, A.N., Parshikova, M.V., Solovieva,
    A.I., and Titov, M.I., 1986b, Effect of the leucine-
    enkephalin analog dalargin on sympathetic reinnerva-
    tion of rat heart and skeletal muscles, Pathol. physio-
    logy, exper. therap., 5:40.
Ilyinsky, O.B., Kondrikova, E.S., Spevak, S.E., and Solovieva,
    A.I., 1986c, Participation of the brain opioidergic
    structures in healing processes, Proc. Acad. Sci. USSR,
    289:240.
Isakova, O.L., and Sepetov, N.F., 1985, Studies of peptides
    degradation in biological systems by method of nuclear
    magnetic resonance, in: "Neuropeptides in Physiology
    and Pathology", V.D. Slepushkin, ed., Cardiology Res.
    Center Pbl., Tomsk.
Kosterlitz, H.W., and Hughes, J., 1975, Some thought on the
    significance of enkephalin, the endogenous ligand, Life
    Sci., 17:91.
Krieger, D.T., Brain Peptides: what, where and why?, Science,
    222:975.
Lindvall, O., and Bjorklund, A., 1974, The glyoxylic acid
    fluorescence histochemical method: a detailed account
    of the methodology for the visualization of central
    catecholamine neurons, Histochemistry, 39:97.
Lundborg, G., Dahlin, L.B., Danielsen, N., Gelberman, R.H.,
    Longo, F.M., Powell, H.L., and Varon, S., 1982, Nerve
    regeneration in silicone chambers: influence of gap
    length and distal stump components. Exp. Neurol., 76:
    361.
North, R.A., and Williams, J.T., 1983, Neurophysiology of opi-
    ates and opioid peptides, in: "The Neurobiology of Opi-
    ate Reward Processes", J.E. Smith and J.D. Lane, eds.,
    Elsevier, Amsterdam, N.Y., Oxford.
Olson, J.A., Olson, R.D., and Kastin, A.J., 1984, Endogenous
    opiates: 1982, Peptides 4:563.
Sander, R.H., and Zimmerman, M., 1986, Mechanoreceptors in rat
    glabrous skin: development of function after nerve
    crush, J. Neurophysiol., 55:644.
Wood, P.L., 1983, Opioid regulation of CNS dopaminergic path-
    ways: a review of methodology, receptor types, regional
    variations and species differences, Peptides, 4:595.
Woolverton, W.L., and Schuster, C.R., 1983, Behavioral and
    pharmacological aspects of opioid dependence: mixed
    agonist-antagonists, Pharm. Rev., 35:33.

PART V

INTRINSIC MECHANISMS
OF MECHANORECEPTOR FUNCTION

DENSE CORED VESICLES IN SAI MERKEL CELLS AND THEIR ROLE IN

MECHANO-ELECTRIC TRANSDUCTION

A. Iggo, E. G. Pacitti and G. S. Findlater

Department of Preclinical Veterinary Sciences
University of Edinburgh, Summerhall
Edinburgh, EH9 1QH, U.K.

The SAI (slowly-adapting type I) mammalian mechanoreceptor is a well
characterised and distinctive cutaneous sensory receptor. In hairy skin
of the cat it is present as a dome-like epidermal structure innervated by
a myelinated afferent fibre that ends in expanded disk-like flattened
terminals each of which is associated with a Merkel cell (Iggo and Muir,
1969). Several studies have reported that the normal high sensitivity of
the receptor to mechanical stimulation and the characteristic irregu-
larities of the afferent discharge during sustained activity is dependent
on the presence of innervated Merkel cells (Brown and Iggo, 1963; Burgess
and Horch, 1973). The actual role of the Merkel cells in the normal
function of the SAI is not yet, however, established and views range from
the suggestion that they are the actual mechanoelectric transducers (Iggo
and Muir, 1969; Horch et al., 1974) to the opposite extreme that they can
be removed without affecting the functional properties of the sensory

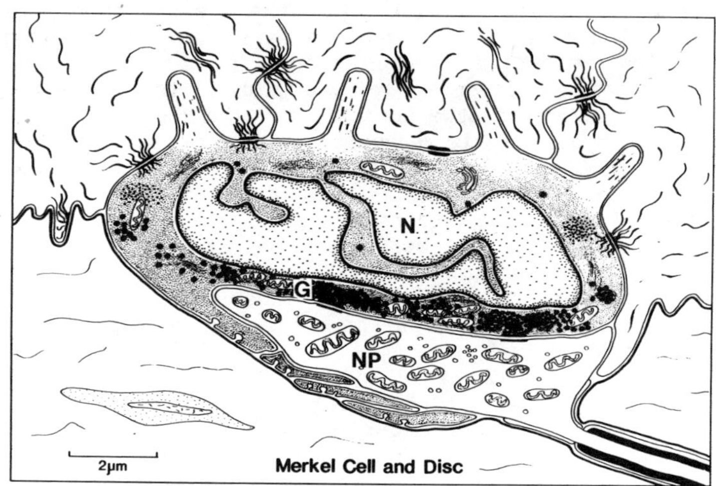

Fig. 1. Merkel cell-neurite complex, showing high
        density of granular vesicles (G) in Merkel
        cell cytoplasm between the cell's nucleus (N)
        and the expanded terminal of the nerve (NP).

receptor (Diamond, pers. comm.) or at least that they are not the transducers (Gottschaldt and Vahle-Hinz, 1982).

Any satisfactory account of the mechanism of the SAI mechanoreceptor must, however, provide an explanation for the functions of the various components that make up its normal structure (Fig. 1). These include, for example, the typical location of Merkel cells at the dermo-epidermal boundary, the additional layer(s) of epidermal cells that overly the Merkel cells, the various morphological features of the Merkel cells, the existence of special junctions between the Merkel cells and the subjacent expanded nerve terminal and so on. The present report deals with one particular aspect of the Merkel cells, namely the presence of a large number of encapsulated dense-cored vesicles in their cytoplasm. These vesicles are most abundant between the poly-lobulated nucleus of the Merkel cell and the subjacent nerve terminal (Fig. 1) and may be concentrated close to 'synapse-like' specialisations of the Merkel cell and nerve ending. If the vesicles were directly involved in the transfer of excitation between the Merkel cell and nerve-ending, that is, if the Merkel cell itself is the transducer then it might be expected that sustained mechanical stimulation of an SAI receptor would lead to depletion of granules. This has not, however, been reported and normally the granular vesicle content of the Merkel cells is apparantly unchanged when assessed by electron-microscopical examination of skin fixed immediately after prolonged mechanical stimulation. The SAI receptors, in fact, are capable of sustaining a more or less constant response to repeated intermittent mechanical stimulation (Findlater et al., 1987). Lack of information about the rate of formation and/or stability of the granular vesicles is at present an obstacle to reaching a conclusion, however, since if the rate of formation of vesicles was matched to their rate of release from the cell, or if only small numbers of vesicles were released during activity, the morphological assessment might be insufficiently sensitive to detect the changes.

In recent experiments on the metabolic requirements of the SAI receptors, in which their dependence on aerobic metabolism was examined (Findlater et al., 1987) it was found during severe hypoxia that the receptors became mechanically inexcitable at a time when the afferent fibre was still conducting normally. In these experiments, using anaesthetised cats and monkeys, SAI receptors were stimulated using quantitatively controlled mechanical indentations in conditions in which the atmosphere surrounding a severely hypoxic limb could be changed from 100% O_2 to 100% N_2. Exposure of the receptor to 100% N_2 progressively reduced the discharge in response to mechanical stimulation and finally extinguished it, and the process could be reversed by replacing the N_2 with O_2 (Fig. 2). The Merkel cells were examined electron-microscopically, in different preparations, at different stages of the cycle of O_2/N_2 replacement. In order to make valid comparisons, advantage was taken of the fact that SAI afferent fibres often innervate more than one cluster of Merkel cells. By choosing for study a unit that innervated at least two clusters or domes it was possible to compare mechanically stimulated and exhausted receptors with unstimulated receptors in otherwise identical conditions. The results obtained are tabulated in Table 1 and show clearly that there was a statistically significant reduction in the number of vesicles in the Merkel cells in exhausted receptors in hypoxic conditions when compared both with normal Merkel cells and with unstimulated cells in hypoxic conditions. This is the first time that it has been shown that the granular vesicle content of Merkel cells is labile, and although the results do not establish a causal relationship between the loss of vesicles and failure of mechanical excitability of the SAI receptors they certainly merit further investigation of the possibility that the Merkel cells are involved in transduction, whether directly in the series of

Table 1. Numerical density of granular vesicles in
         cytoplasm of Merkel cells from feline SAI
         receptors in different states of oxygenation

Experimental State	Vesicle no./μm^2 cytoplasm
Normal	14.6 ± 0.97 (n=17)
Hypoxic skin	
Not mechanically stimulated	8.3 ± 0.7 (n=16)
N_2 atmosphere, mech.stim.*	4.3 ± 0.48 (n=20)
O_2 atmosphere, mech.stim.*	7.9 ± 1.1 (n=7)

* = the receptors had ceased to respond to mechanical
    stimulation.

events leading to the discharge of afferent impulses or indirectly in some
modulatory fashion.

Calcium channel blockers and SAI receptors

The results of the experiments using hypoxic conditions were further
explored by examining the effects of calcium channel blockers on the
mechanical excitability of SAI receptors and on the vesicle content of
Merkel cells. The underlying reason for this approach was the fact that
exocytosis of neurotransmitters requires the entry of calcium ions (Rubin,
1970; Baker, 1981) and therefore if indeed the release of dense-cored
vesicles is an integral step in transduction then prevention of Ca^{2+} entry
into the Merkel cell would reduce or abolish the response of the receptor
to mechanical stimulation. In another situation where a secondary cell is

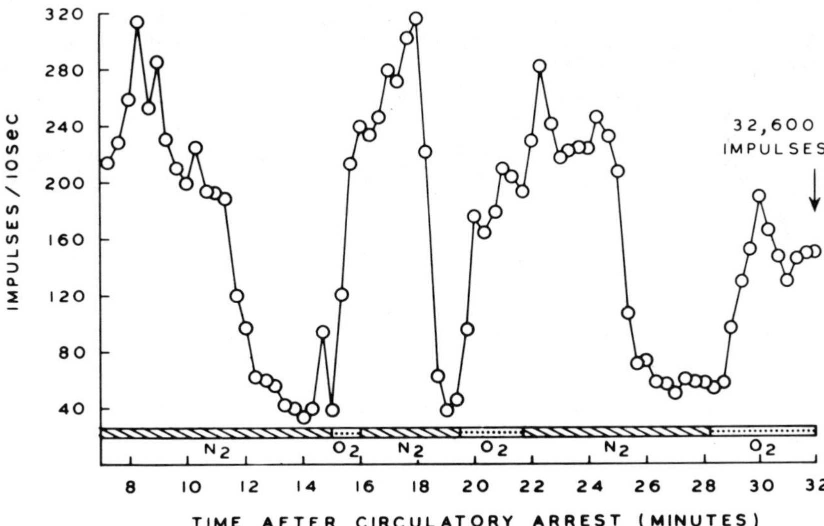

Fig. 2. Effect of hypoxia on mechanical excitability of an
        SAI mechanoreceptor in an anaesthetised monkey,
        during continuous intermittent mechanical stimul-
        ation. Graph begins 7 minutes after circulatory
        arrest. An N_2 atmosphere was formed around the limb
        at 7 minutes as indicated by the cross-hatched bar
        and periodically replaced by O_2.

Table 2. Numerical density and number of 'synaptic-like'
          structures in feline Merkel cells after exposure
          to normal saline, cobalt chloride or verapamil
          hydrochloride in the perfused hind limb circulation

Drug	No. of Vesicles/ μm^2	No. of "Synapses"/ Cell Section
Saline control	12.1 ± 1.5 (n=13)	0.6 ± 0.25 (n=13)
+CoCl$_2$ (2mM)	** 4.3 ± 0.8 (n=17)	*2.1 ± 0.40 (n=17)
++Verapamil (0.03mM)	* 6.8 ± 0.8 (n=18)	

* P<0.01   ** P<0.001 (t-test)
+ specimens examined after afferent response to mechanical
  stimulation was abolished
++specimens taken when afferent response to mechanical stimulation
  was 25% of control value

known to act as the transducer, the hair cell of the cochlea and lateral
line system, Ohmori (1988) has established that calcium ions enter the
apex of the hair cell of the chick cochlea when it is stimulated mechanically. These calcium ions then presumably lead to the release of a
transmitter from the base of the hair cell and thereby cause excitation of
the afferent nerve fibre. If a similar situation holds for Merkel cells,
then block of calcium entry using a calcium channel blocker would be
expected to prevent transfer of excitation from the Merkel cell to the
afferent nerve terminal. Three kinds of calcium channel blocker were
investigated - cobalt ions, cadmium ions and verapamil. The agents were
either introduced into the isolated circulation of the cat hind limb
(Findlater et al., 1987), injected intradermally under individual touch
domes in anaesthetised cats or added to the perfusing solution in an
isolated superfused rat skin preparation. At appropriate concentrations of
the channel blockers the responses of SAI mechanoreceptors to mechanical
stimulation in all three preparations were significantly reduced below
control levels or even abolished. In contrast the electrical excitability
of the afferent nerve fibres was unchanged. Ogawa and Yamashita (1988)
report a similar reduction in mechanical excitability of frog slowly
adapting cutaneous mechanoreceptors. The open question is whether the
effect of the channel blockers was a direct action on the nerve terminal,
or whether it was, at least in part, by an action on the Merkel cell. In
Ogawo's experiments on frog skin it was possible to compare the effects of
calcium channel blockers on both classes of slowly-adapting receptors,
those corresponding to the SAI and SAII of mammals. Both kinds were
affected, but to a greater degree and at lower concentrations of blocking
agents for the SAI analogs. It was therefore of considerable interest to
examine the Merkel cells in the mammals at a time when the responses to
mechanical stimulation were reduced or even abolished. Contrary to expec-
tations that the block of calcium entry would prevent the exocytosis of
dense-cored vesicles there was actually a statistically significant
reduction in the number of granules in Merkel cells removed at receptor
failure in the presence of cobalt chloride and verapamil chloride (Table
2). At the same time there was an increase in the number of synaptic-like
junctions between the Merkel cells and the nerve endings. Just as with
hypoxia there was a correlation between the concentration of vesicles in
the Merkel cells and the failure of mechanical excitability of the SAI
receptors, in that at receptor failure there was a reduction in granule
content of Merkel cells, but the results do not establish a causal
relationship.

The exposure of the SAI receptors to cobalt chloride led initially to
the spontaneous discharge of action potentials in the SAI afferent fibres,
as well as an increased response to mechanical stimlulation, and it is
thus possible that the cobalt ions were themselves responsble for the
release of granules with consequent excitation of the afferent nerve
terminal. The complexity of calcium actions in cells might indeed be such
that the approach used in the present experiments is too indirect to give
a clear-cut result. If, for example, calcium is required for the transport of vesicles from their site of formation in the Golgi apparatus, or
even for the formation of the vesicles themselves (Lavoie and Bennet,
1983), then the reduction in vesicle numbers after exposure of the
receptors to the channel blockers could be the consequence of both their
reduced formation as well as the initial release of stored vesicles on
exposure to cobalt chloride. In this case the reduced mechanical excitability of the receptors in the presence of channel blockers such as
cobalt may result not only from lack of calcium entry limiting exocytosis
but also from a diminished availability of dense-cored vesicles.

The results of the present investigation provide new evidence for an
association between the presence/availability of dense cored vesicles in
Merkel cells and the mechanical sensitivity of SAI cutaneuous mechanoreceptors in mammalian skin but do not provide conclusive evidence for the
direct involvement of the Merkel cells in transduction or for the specific
function of the dense cored vesicles.

References

Baker, P. F., and Knight, D. E., 1981, Calcium control of exocytosis
    and endocytosis in bovine adrenal medullary cells, Phil. Trans.
    R. Soc. Lond., B 296, 83-103

Brown, A. G., and Iggo, A., 1963, The structure and function of
    cutaneous 'touch corpuscles' after nerve crush, J. Phyisol., 165,
    28-29P.

Burgess, P. R., and Horch, K. W., 1973, Specific regeneration of
    cutaneous fibers in the cat, J. Neurophysiol., 36, 101-114.

Findlater, G. S., Cooksey, E. J., Anand, A., Paintal, A. S., and Iggo,
    A., 1987, The effects of hypoxia on slowly adapting type I (SAI)
    cutaneous mechanoreceptors in the cat and rat, Somatosensory Res.,
    5, 1-17.

Gottschaldt, K. M., and Vahle-Hinz, C., 1982, Evidence against
    transmitter function of met-enkaphalin and chemosynaptic impulse
    generation in Merkel cell mechanoreceptors, Exp. Brain Res., 45,
    459-463.

Horch, K. W., Whitehorn, D., and Burgess, P. R., 1974, Impulse
    generation in type I cutaneous mechanoreceptors, J. Neuro-
    physiol., 37, 268-281.

Iggo, A., and Muir, A. R., 1969, Structure and function of a
    slowly adapting touch corpuscle in hairy skin, J. Physiol.
    200, 763-769.

Lavoie, P. A., and Bennet, G., 1983, Accumulation of $^3H_2$ fucose-
    labelled glycoprotein in the Golgi apparatus of dorsal root
    ganglion neurons during inhibition of fast-axonal transport
    caused by exposure of the ganglion to Co^{2+}-containing or Ca^{2+}-
    free medium, Neurosci, 8, 351-362.

Ogawa, H., and Yamashita, Y., 1988, Mechanoelectric transduction
    in the slowly cutaneous afferent units of frogs, Progress in
    Brain Research, in press.

Ohmari, H., 1988, Mechanoelectric transduction of chick hair
    cell. Progress in Brain Res., in press.

Rubin, R. P., 1970, The role of calcium in the release of neuro-
    transmitter substances and hormones, Pharmacol. Rev., 22, 389-428.

ELECTROPHYSIOLOGICAL STUDIES ON MERKEL CELLS ISOLATED FROM RAT VIBRISSAL

MECHANORECEPTORS

Colin Nurse[1] and Ellis Cooper[2]

[1]Department of Biology
 McMaster University
 Hamilton, Ontario, Canada   L8S 4K1

[2]Department of Physiology
 McGill University
 Montreal, Quebec, Canada   H3G 1Y6

INTRODUCTION

Tactile sensation, an important mechanism animals use to explore their
environment, is conveyed by various mechanoreceptors in skin (Iggo and
Andres, 1982). One of the earliest to evolve in vertebrates and one with a
ubiquitous distribution in the integument is a complex comprising a
specialized epidermal cell, the Merkel cell, and the terminal(s) of a large
myelinated sensory nerve (Halata, 1975; Diamond, 1979). Though the cell
was first recognized over 100 years ago by F. Merkel (1875), its role in
mechanosensory function is still in dispute in part because the relative
inaccessibility of the complex in situ has prevented direct physiological
recording. Recent evidence based on recordings well away from the transduction site has favoured the sensory nerve ending rather than the Merkel cell
as the mechanotransducer in both the mammal and lower vertebrate (Gottschaldt
and Vahle-Hinz, 1981; Mills et al., 1985; Diamond et al., 1986). These findings raise questions about the actual physiological role of the Merkel cell
in mechanoreception especially in view of the occurrence of synaptic features
between the nerve ending and the Merkel cell (e.g., Diamond et al., 1986).
To address this issue and to test further the possibility of mechanosensitivity in Merkel cells we have initiated studies aimed at understanding the
physiological properties of the cell using patch clamp recording techniques
(Hamill et al., 1981). The preparation we used was the neonatal rat vibrissa which contains numerous Merkel cells that can be readily isolated
and identified among dispersed cells of the outer root sheath by their
ability to take up selectively the fluorescent dye, quinacrine (Nurse et
al., 1983). Further, the more recent finding that these same quinacrine
fluorescent Merkel cells could also be vitally stained with neutral red
(Nurse, 1985), proved more convenient for these studies since recordings
could be done on identified cells under a conventional (inverted) light
microscope, thus eliminating the need for fluorescence illumination.

Cell Preparation and Recording Conditions

Dissociated cells from the outer root sheaths of 1-5 day old rat

vibrissae were obtained by combined enzymatic and mechanical dissociation
of the isolated follicles as previously described (Nurse et al., 1983).
The cells were allowed to attach to a cover slip coated with poly-L-lysine
for 1 - 48 hr at 37°C in a humidified atmosphere of 95% air - 5% CO_2. Just
before recording they were incubated in medium containing 0.01 mg/ml neutral
red for 30-45 min to stain the Merkel cells which occurred with a frequency
of ca. 1 in 10 - 15 cells (Nurse, 1985). The recording methods were similar
to those reported elsewhere (Cooper and Shrier, 1985).

Whole Cell Electrophysiological Studies

A total of 60 neutral red-stained Merkel cells were investigated in
the cell-attached, outside-out patch, and whole-cell configurations. In
current-clamp conditions depolarizing pulses produced no regenerative membrane responses (Fig. 1A), even when the membrane was held at hyperpolarized
potentials (-75 to -100 mV) to remove any possible steady state inactivation; this suggests that these cells do not generate action potentials.
The membrane, however, does show rectification in response to depolarizing
currents (Fig. 1A). The input resistance and capacitance of the cells
estimated from these and other measurements were in the range 5 - 15 G Ω
and 1.5 - 2.5 pF respectively; from these capacitance values and assuming
a specific membrane capacitance of 1 $\mu F/cm^2$ we estimate the cells were 8 -
12 μm in diameter, a size range typical of Merkel cells (Halata, 1975;
Nurse et al., 1983). The resting potential could not be reliably determined
due to rapid internal dialysis of the whole cell; our best estimate for
this value is ca. -30 mV. A further complication resulting in an unstable
membrane potential results from the fact that a single channel opening can
cause a substantial variation in membrane potential due to the high input
resistance of the cell.

Under voltage clamp depolarizing steps from a holding potential of
-50 mV produced only outward membrane currents (Fig. 1B). The current-
voltage relation for the whole cell was non-linear and it appeared that
there was an increased probability of channel opening with depolarizing
steps up to +20 mV (Fig. 1D; squares). In general the probability of channel opening appeared highest shortly after the voltage step and tended to
decrease with time (not shown). With prolonged depolarizations (several
seconds) it was possible to resolve single channel currents (Fig. 1C),
which showed an approximately linear current-voltage relation and had a
conductance of ca. 80 pS (Fig. 1D; circles). Extrapolation of this graph
gives a tentative reversal potential of ca. -45 mV.

Ion Species Involved in Channel Gating

Indirect evidence from recording single channels in the cell-attached
mode suggested that ion channels in Merkel cells are non-selectively permeable to cations. To test directly whether Na^+ and K^+ movements through the
channel could exert a major control over the cell's membrane potential we
examined the effects of rapidly changing the external solution around the
cell while recording in the whole cell configuration. In Fig. 2, brief
"puffs" of low Na^+ or high K^+ solutions were ejected intermittently (via a
solenoid valve) under slight positive pressure (ca. 1 psi) from the tip of
a broken pipette placed 0 - 20 μm from the cell surface, while the culture
was continuously perfused with a different or identical solution. This procedure effectively swamped the immediate exterior of the cell with the
pipette solution for the period the solenoid valve remained open. As shown
in Fig. 2A, puffs of high K^+ solution (130 mM) applied into a low K^+ perfusate (5.4 mM), at a constant level of extracellular Na^+ (10 mM), caused a
marked potential change whose magnitude and direction depended on the initial holding potential. In Fig. 2B, puffs of low Na^+ solution (14 mM)
applied into a high Na^+ perfusate (140 mM), at a constant level of extra-

cellular K$^+$ (5.4 mM) resulted in potential changes in the opposite direction. Since only one class of ion channels has been detected on Merkel cells these results suggest that the channels are permeable to both Na$^+$ and K$^+$.

<u>Tests for Mechanosensitivity in Merkel Cells</u>

Several tests suggest that these isolated Merkel cells lack mechanosensitivity. For example, deforming the cell membrane either by moving the pipette up and down several μm while attached to the cell, or by first lifting the cell from the culture surface with the pipette attached and deforming it on the bottom of the dish, did not produce currents that could be described as mechanosensitive. Though the magnitude of the deformation was not quantifiable it is likely that it was far in excess of the threshold for exciting single Merkel cell–neurite touch receptors <u>in situ</u> (see Diamond et al., 1986). In addition, membrane stretch produced by application of

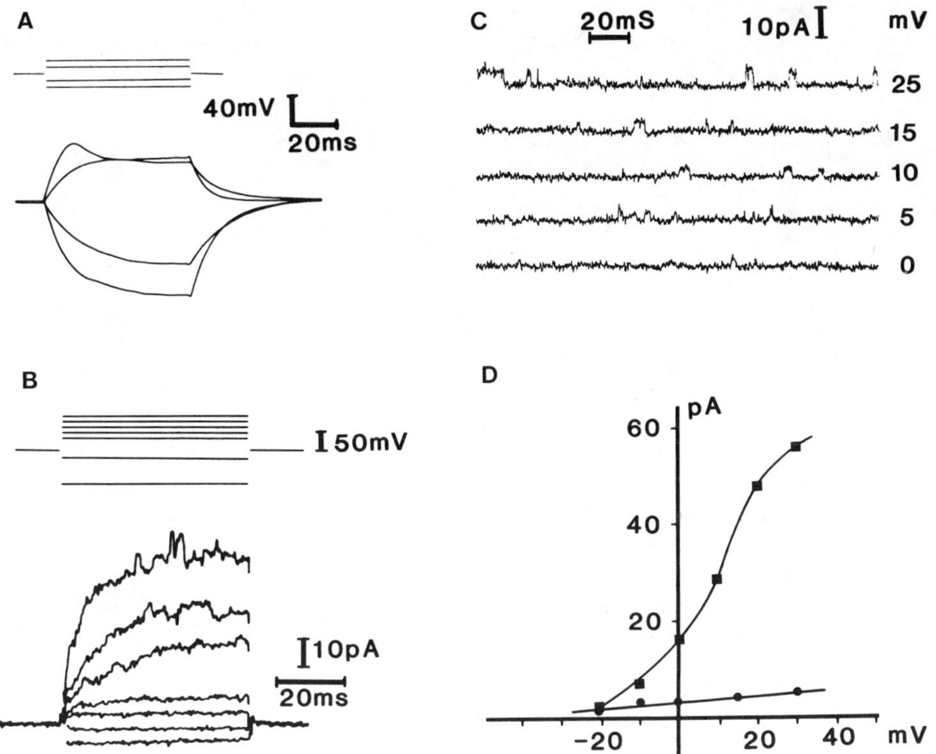

Fig. 1. Whole cell recordings from Merkel cells. <u>A</u>. Current clamp. Lower traces show change in membrane potential in response to depolarizing and hyperpolarizing currents (upper traces). <u>B</u>. Voltage clamp. The family of traces (lower records) show membrane currents in response to voltage steps (upper records) from an initial holding potential of –50 mV. <u>C</u>. Single channels. Outward currents (upward deflections) recorded when the cell (same as in <u>B</u>) was held at the 5 depolarized levels indicated on right of figure. <u>D</u>. Current-voltage or I-V curves. The I-V relation for the whole cell (same as in B) is non-linear (squares) and suggests increased probability of channel opening at depolarized levels between 0 and +30 mV. The single channels (resolved after the outward currents had largely inactivated) had a linear I-V relation (circles) and a conductance of ca. 80 pS.

pipette suction while recording from cell attached patches did not alter
gating properties (cf. Guharay and Sachs, 1984). Furthermore, application
of puffs of perfusate (Fig. 2C) sufficiently strong to dislodge or move
loosely-attached debris on the culture surface produced no change in membrane potential, consistent with a lack of mechanosensitivity in the cell.

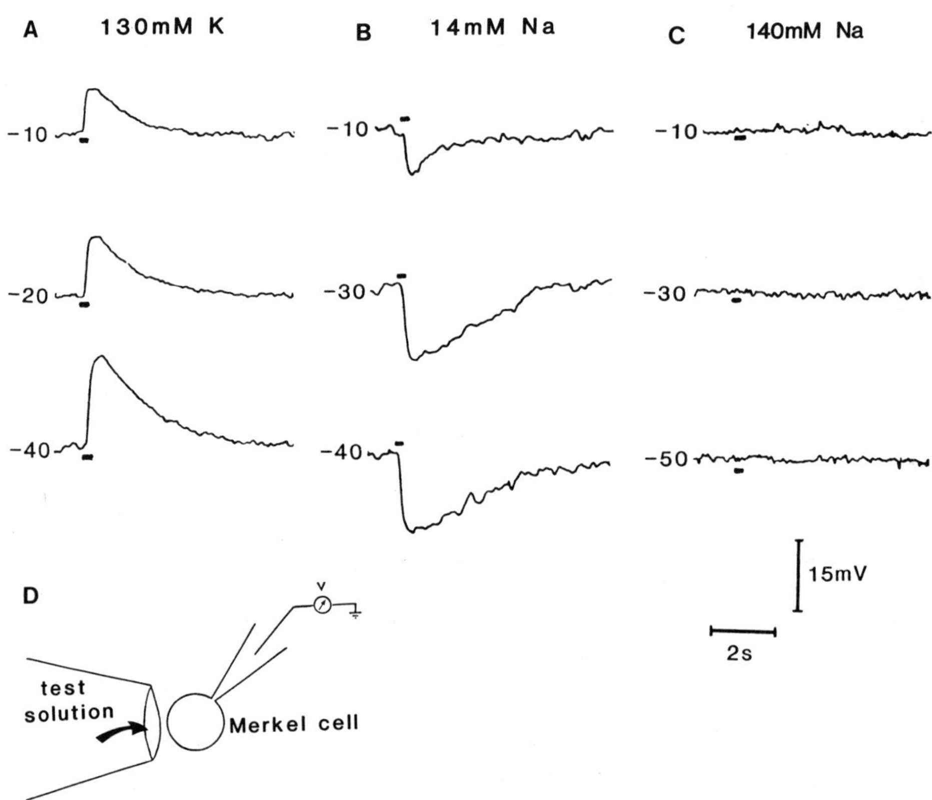

Fig. 2.   Control of membrane potential by cations. This figure shows
           the results of rapidly changing the external solution as a
           function of the initial holding potential. In A, B and C
           test solutions were ejected by pressure pulses (gated by a
           solenoid valve) from a broken pipette placed 0–20 μm from
           the Merkel cell surface (Fig. D), while the culture was
           perfused with a different or identical solution. The black
           bar under or over each trace represents the duration that the
           solenoid valve was open; ion concentrations above figures
           refer to the test solutions. A. High K$^+$. The perfusate
           contained (mM): NaCl, 10; KCl, 5; sucrose 250; CaCl$_2$, 2.8;
           HEPES/NaOH buffer, 10. The test solution was the same as
           the perfusate except that the sucrose was replaced by 130
           mM KCl. B. Low Na$^+$. A different cell from A, where the
           perfusate was (mM): NaCl, 140; KCl, 5.4; CaCl$_2$, 2.8; MgCl$_2$,
           0.18; HEPES/NaOH buffer, 10. The test solution was similar
           to the perfusate except that most of Na was replaced by
           sucrose for a final concentration of 250 mM sucrose and
           14 mM NaCl. C. Regular Na$^+$ control. In this case both the
           test solution and the perfusate were of the same composition
           (see perfusate in B).

DISCUSSION

     The main findings obtained from these initial electrophysiological
recordings from isolated vibrissal Merkel cells are: (i) the cells do not
appear to generate action potentials; (ii) they appear to lack mechanosensitive properties; and (iii) they contain predominantly a cation channel
(conductance ca. 80 pS) that discriminates poorly between Na^+ and K^+ and
regulates the membrane potential. It is unlikely that staining of the cells
with the basic dye, neutral red, in some way altered their physiological
properties since this and related dyes do not affect the Merkel cell-
neurite touch receptors in the isolated salamander skin (Diamond et al.,
1986 and unpublished observations) nor the physiology of similarly-
stained neurons in other preparations (Stuart et al., 1974).

     The finding that the isolated Merkel cells failed to respond to
various mechanical stimuli was of interest since it supported the view that
the cells lack a transducer function. This conclusion was reached in other
studies in the mammal and lower vertebrate, where one of the main arguments
was based on the demonstration of a non-involvement of chemical transmission
in the initiation of the mechanosensory response (Gottschaldt and Vahle-
Hinz, 1981; Diamond et al., 1986). Though our findings add further support
to the idea that Merkel cells are not mechanotransducers, we cannot presently exclude the possibilities that in our studies this negative finding
was due to: (i) the use of young cells that had not yet acquired mechanosensitivity; (ii) the enzymatic procedure which may have removed or disrupted important elements required for mechanosensitivity; and (iii) wash
out of essential cytoplasmic constituents by dialysis under whole-cell
clamp.

<u>Possible Physiological Role of Cation Channels in Merkel Cells</u>

     The presence of voltage-gated, cation channels in Merkel cells is a
novel finding though its significance is presently only speculative. The
result at least makes it unlikely that Merkel cells act merely as "passive
abutments for the nerve endings" (Gottschaldt and Vahle-Hinz, 1981). An
attractive and testable hypothesis concerning the possible function of these
cation channels in Merkel cells derives from studies on a variety of other
mechano- or vibration sensitive receptors (e.g., Pacinian corpuscles, hair
cells), which are generally surrounded by a K^+-rich extracellular fluid
(Illyinsky et al., 1976; Hudspeth, 1983). The effect of high K^+ may be
exerted via lowering the resting potential or it may facilitate opening of
the tranduction channels (see Guharay and Sachs, 1984). Thus, a possible
physiological function of the Merkel cell is to maintain a high-K^+ milieu
around the apposed mechanosensitive nerve endings via these cation channels
and hence to increase receptor sensitivity. These channels would be
expected to facilitate $Na^+ - K^+$ exchange and result in a steady state K^+
accumulation around the abuting nerve endings. This reasoning implies that
in the absence of Merkel cells, the mechanosensitive nerve endings should
still function but at higher-than-normal mechanical thresholds since the
high K^+ microenvironment would not be generated. Indeed in a variety of
preparations disruption of the coupling between nerve ending and Merkel cell
leads to a reduction in mechanosensitivity or an increase in receptor
threshold (e.g., Mearow and Diamond, 1983; Diamond et al., 1986; see also
Mills et al., 1985).

     At present we do not know whether the cation channel we describe in
Merkel cells is also permeable to calcium ions. If so, this could provide
a basis for release of the contents of the cytoplasmic dense cored granules.
Notwithstanding the evidence supporting a target function for the Merkel
cell (Diamond, 1979) it is expected that a more detailed electrophysiological characterization of the cell by the methods described in this study

will help elucidate its physiological role in mechanoreception.

<u>Acknowledgements</u>

This study was supported by N.S.E.R.C. and M.R.C. Canada.

REFERENCES

Cooper, E., and Shrier, A., 1985, Single channel analysis of the fast
transient potassium currents from rat nodose neurons, <u>J. Physiol.
London</u>, 369: 199–208.

Diamond, J., The regulation of nerve sprouting by extrinsic influences,
<u>in</u>: The Neurosciences: 4th Study Program, F.O. Schmitt and
F.G. Worden, eds., pp. 937–955, MIT Press, Cambridge, Mass. (1979).

Diamond, J., Holmes, M., and Nurse, C.A., 1986, Are Merkel cell–neurite
reciprocal synapses involved in the initiation of tactile responses
in salamander skin? <u>J. Physiol. Lond.</u>, 376: 101–120.

Gottschaldt, K.-M., and Vahle-Hinz, C., 1981, Merkel cell receptors:
Structure and transducer function, <u>Science</u>, 214: 183–186.

Halata, Z., 1975, The mechanoreceptors of the mammalian skin. Ultrastructure and morphological classification, <u>Adv. in Anat.</u>, 50: 1–77.

Guharay, F., and Sachs, F., 1984, Stretch-activated single ion channel
currents in tissue-cultured embryonic chick skeletal muscle, <u>J.
Physiol. Lond.</u>, 352: 685–701.

Hamill, O.P., Marty, A., Neher, E., Sakmann, B., and Sigworth, F.J., 1981,
Improved patch-clamp techniques for high resolution current recording
from cells and cell-free membrane patches, <u>Pflugers Arch.</u>, 391: 85–100.

Hudspeth, A.J., 1983, Mechanoelectrical transduction by hair cells in the
acousticolateralis sensory system, <u>Ann. Rev. Neurosci.</u>, 6: 187–215.

Iggo, A., and Andres, K.H., 1982, Morphology of cutaneous receptors, <u>Ann.
Rev. Neurosci.</u>, 5: 1–31.

Illyinsky, O.B., Akoev, G.N., Krasnikova, T.L., and Elman, S.I., 1976, K^+
and Na^+ ion content in the Pacinian corpuscle fluid and its role in
the activity of receptors, <u>Pflugers Arch.</u>, 361: 279–285.

Mearow, K.M., and Diamond, J., 1983, The development of mechanosensory
function and synaptic morphology when regenerating axons arrive at
nerve-free Merkel cells in Xenopus skin, <u>Soc. Neurosci. Abst.</u>, 9:777.

Merkel, F., 1875, Tastzellen und Tastkorperchen bei den Haustieren und
beim Menschen, <u>Arc. Microsck, Anat. Entwmech.</u>, 11: 636–652.

Mills, L., Mearow, K., Visheau, B., and Diamond, J., 1985, Are Merkel cells
the mechanosensory transducers in rat touch domes and Xenopus touch
spots? <u>Soc. Neurosci. Abstr.</u>, 11: 119.

Nurse, C.A., Mearow, K.M., Holmes, M., Visheau, B., and Diamond, J., 1983,
Merkel cell distribution in the epidermis as determined by quinacrine
fluorescence, <u>Cell and Tissue Res.</u>, 278: 511–524.

Nurse, C.A., 1985, Novel staining properties of epidermal Merkel cells
and their mechanosensory innervation, <u>Soc. Neurosci. Abstr.</u>, 11: 438.

Stuart, A.E., Hudspeth, A.J., and Hall, Z.W., 1974, Vital staining of
specific monoamine-containing cells in leech nervous system, <u>Cell
Tiss. Res.</u>, 153: 55–61.

EFFECTS OF INTRACELLULAR Ca^{2+} ON THE FROG MUSCLE SPINDLE IN RELATION TO CYCLIC AMP ACTION

Fumio Ito, Masahiro Sokabe, Falih Hassan Diwan
Noriaki Fujitsuka and Atsushi Yoshimura

Department of Physiology, Nagoya University School of
Medicine, Nagoya 466, Japan

INTRODUCTION

In sensory systems, cyclic nucleotides or calcium ions (Ca^{2+}) have been proposed as intracellular messengers for coupling the light activation of the photopigment rhodopsin to channel activity and thus modulating light-sensitive conductance (Lipton et al.1977; Miller and Nicol, 1979; Fesenko et al., 1985), or for coupling between the odorant effect and olfactory transduction (Pace et al., 1985). Recently we have found that application of dibutyryl cyclic AMP, forskoline (adenylate cyclase activator) or isobutylmethylxanthine (phosphodiesterase inhibitor) increased the rate of spontaneous discharges and decreased the responsiveness to stretch in decapsulated frog muscle spindles (Sokabe et al.,1987). Similar changes in the afferent discharges were observed with application of carbonyl cyanide chlorophenylhydrazone which would increase intraaxonal Ca^{2+} through uncoupling mitochondria. Application of quercetin (Ca^{2+} pump inhibitor) resulted in a prolonged increase in the discharge rate, lasting 15 - 20 s after the release of stretch. These results suggest a certain cascade between cAMP and intracellular Ca^{2+} in the frog muscle spindle. The present work deals with the effects of modulators of intracellular Ca^{2+} in order to confirm the contribution of intracellular Ca^{2+} to the mechano-sensory encoding.

MATERIALS AND METHODS

Muscle spindles were isolated from the semitendinosus muscles of the frog <u>Rana catesbeiana</u>. The preparation, recording and stretching methods are described in detail elsewhere (Ito et al., 1982). After individual action potentials and responses to a ramp-and-hold stretch were recorded by means of an air-gap method, the Ringer's solution in which the spindle was immersed was replaced by a Ca^{2+}-free (without any Ca^{2+} cilater) Ringer's solution containing 0.2 % collagenase. The capsule was digested by incubation with the enzyme for 30 min at 34°C in such a condition that the muscle spindle was set in the air-gap recording chamber. After digestion, the collagenase solution was replaced with normal Ringer's solution at room temperature (21 - 25°C). Chemicals used were calcium ionophore A 23187 (A 23187, Sigma) and a polymeric derivative of prostaglandin B_1 (prostaglandin Bx; PGBx, kindly supplied by Dr.T.Ohnishi of Hahnemann Medical College). A-23187 was diluted with dimethyl sulfoxide to be at 0.1 % or less in the final concentration, because the application of the solvent at concentration of more than 1 % resulted in a reduction of discharge rates of muscle spindles during slack and stretch.

RESULTS

Calcium ionophore A 23187

At first a classical calcium ionophore A 23187 was used to know the effect of intracellular Ca^{2+} on the muscle spindle. Application of a Ringer's solution containing 10 μM A 23187 produced a slight increase in the rate of spontaneous discharges in slackened spindle receptor. When the ionophore was added into a Ringer's solution containing 10 mM $CaCl_2$, a distinct increase in the rate was observed, as shown with a dotted line in Fig.1A. The rate decreased rapidly after the preparation was washed with normal Ringer's solution.

Although the response to stretch did not change significantly during perfusion of 5 – 10 μM A 23187 in normal Ringer's solution, a significant decrease in stretch responses was observed at higher concentrations of Ca^{2+} (20 mM in Fig.1D). It, however, is obscure whether the increase in spontaneous discharges and the deterioration of the responsiveness arose from intra- or extra-cellular Ca^{2+} increase. Therefore, the effect of extracellular Ca^{2+} is examined below.

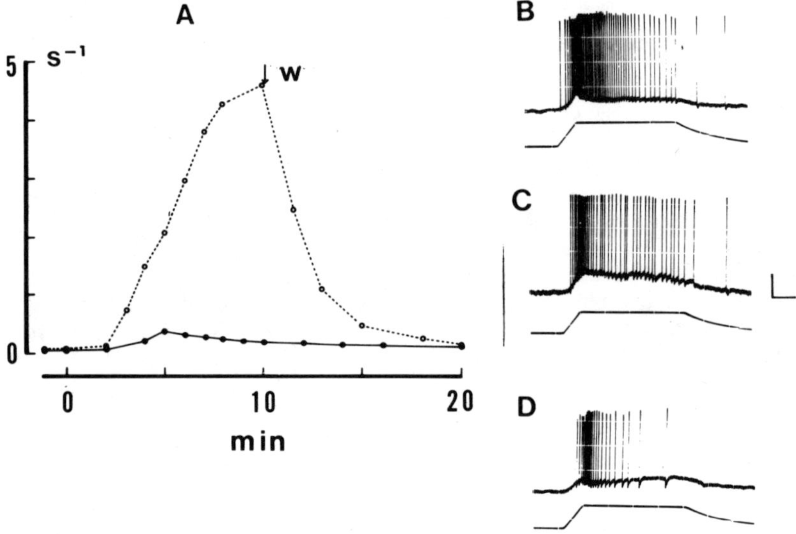

Fig.1. Effects of A 23187 on a decapsulated muscle spindle. A: the rate of spontaneous discharges in a slackened condition was plotted with time after application of the drug. At 10 μM the rate increased slightly (solid line). Application of 10 μM A23187 in addition to 10 mM $CaCl_2$ (dotted line) resulted in a steep increase in the rate until washing (W). B – D: Simultaneous record of responses (upper traces) of a muscle spindle during a ramp-and-hold stretch (lower traces) in normal Ringer's solution (B), in a Ringer's solution containing 5 μM A 23187 (C), and in both 20 mM $CaCl_2$ and 5 μM A 23187 (D). Calibration; 1 mV; 0.5 s.

196

<u>Calcium</u>

The rate of spontaneous discharges was decreased by increasing the extra-axonal calcium concentration. At 20 mM external calcium the spontaneous discharge disappeared in the in situ length. In order to reveal the rate change, the muscle spindle was slightly stretched so that the discharge rate was approximately 30 impulses/s in normal Ringer's solution, in the experiments of Fig.2A.

The increase in external calcium resulted in a decay of the number of spike responses during a ramp-and-hold stretch from the in situ length. When the muscle spindle was perfused with a Ringer's solution containing 20 mM calcium, a few discharges occurred only during the ramp phase of stretch (see Fig.2E). The stretch responses could be recovered by replacing the high calcium solution with normal Ringer's solution. The application of 5 - 10 mM calcium resulted in an increase in the amplitude of spindle potential but abolished the spontaneous afferent discharges and minimized responses during stretch (Fig.2C & D). This implies that the afferent discharges can not be triggered by the spindle potential, and supports our hypothesis that the spindle potential has to be discriminated from a pure receptor potential in the frog muscle spindle (Ito & Fujitsuka, 1983 ; Ito et al., 1985). To summarize, the increase of extracellular Ca^{2+} tends to inhibit both spontaneous discharges and responses to stretch. Thus the above described effects of A 23187 at higher Ca^{2+} concentration might include both extra- and intra-cellular Ca^{2+} effects.

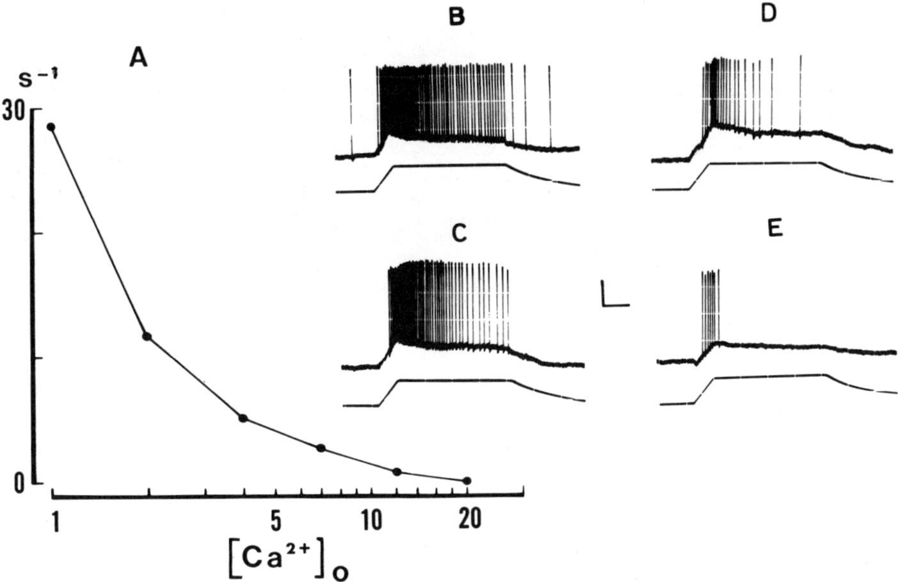

Fig.2. Effects of external calcium concentration on decapsulated muscle spindles. A: the rate of spontaneous discharges in a stretched state were plotted with the concentration of external calcium (abscissa). B - E: Simultaneous records of responses (upper traces) of a muscle spindle during a ramp-and-hold stretch (lower traces) in normal Ringer's solution (B), in a Ringer's solution containing 5 mM (C), 10 mM (D), and 20 mM $CaCl_2$ (E). Calibration; 1 mV and 0.5 s.

<u>Prostaglandin Bx (PGBx)</u>

PGBx has been synthesized from prostaglandin B_1 (Polis et al., 1979) and has a molecular weight of approximately 24000. This drug has been suggested to make a complex with Ca^{2+} and to act as a potent Ca^{2+} ionophore (Ohnishi & Devlin,1979). The rate of spontaneous discharges from a slackened muscle spindle began to increase approximately 2 min after the receptor was perfused with a Ringer's solution containing 2 μM PGBx (Fig.3A). The rate attained the maximum 5 min after the perfusion, and then decayed slowly. Increase in the concentration of PGBx did not enhance the rate further more.

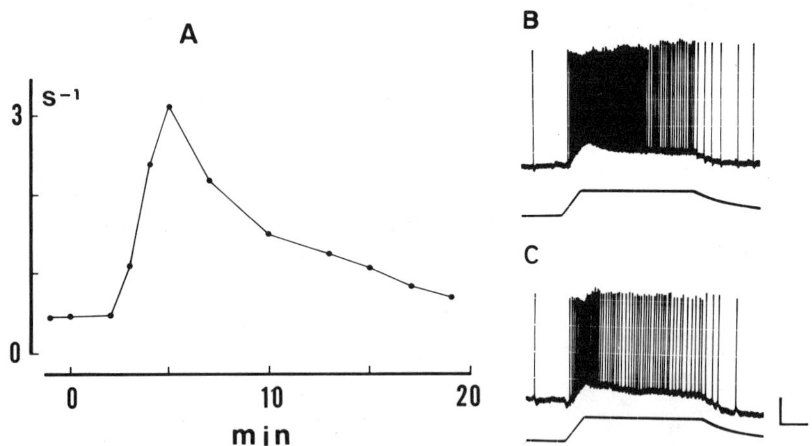

Fig.3. Effects of prostaglandin Bx on decapsulated muscle spindles. A: The rates of spontaneous discharges in a slackened condition were plotted with time after application of the drug (2 μM). B and C: Simultaneous records of responses (upper traces) of a muscle spindle during a ramp-and-hold stretch (lower traces) in normal Ringer's solution (B) and in a Ringer's solution containing 10 μM PGB (C). Calibration; 1 mV and 0.5 s.

A slight decrease in the rate of afferent impulses during stretch, without significant changes in the amplitude of spindle potential, was observed during the period of increased rate of spontaneous discharges by application of 2 μM PGBx (Fig.3C in comparison with the control Fig.3B). These results together with the preceding ones, strongly suggest that increased intracellular Ca^{2+} induced both an elevation of spontaneous discharge rate and deterioration of stretch responses. Such dual effects of intracellular Ca^{2+} well coincide with those of cAMP. While relatively low external Ca^{2+} (0.2 - 5 mM) antagonized these effect (Sokabe et al.,1987), high Ca^{2+} abolished both spontaneous discharges and stretch responses of the muscle spindles.

DISCUSSION

The results of Ca^{2+} ionophores showed that the increase in intra-axonal Ca^{2+} induces an elevation of spontaneous discharge rate and a slight deterioration of stretch responses of frog muscle spindles. The actions well mimic those of dibutyryl-cAMP applied on the spindles extracellularly. Thus our hypothesis that the cAMP actions on the mechano-electric encoding processes of the frog muscle spindle are mediated by intracellular Ca^{2+} is supported by the present experiments. The fact that increasing the external calcium results in a decrease of the rate of spontaneous discharges (Fig.2B) suggests that the rate is determined by a balance between external and internal Ca^{2+} activities across the sensory terminal membrane. A possible explanation that Ca^{2+} may affect the rate of discharges through altering the membrane potential of the sensory terminal by screening or binding membrane surface charges will be examined in future.

ACKNOWLEDGEMENT

This work was supported in part by the Grant-in-Aid for Scientific Research from the Ministry of Education, Science and Culture of Japan (61480106).

REFERENCES

Fesenko,E.E., Kolensnikov,S.S., and Lyubarsky,A.L., 1985, Induction by cyclic GMP of cationic conductance in plasma membrane of retinal rod outer segment, Nature, 313:310-313.

Ito,F. and Fujitsuka,N., 1983, Electrical threshold of the sensory nerve terminal of the frog muscle spindle: a role of spindle potential for generating afferent impulses, Neurosci. Lett., 37:233-237.

Ito,F., Fujitsuka,N., and Fan,X.L., 1985, Reversal of the static component of spindle potential by imposed depolarizing current in the frog muscle spindle, Brain Research, 326:107-116.

Ito,F., Komatsu,Y., and Fujitsuka,N., 1982, $G_{K(Ca)}$-dependent cyclic potential changes in the sensory nerve terminal of frog muscle spindle, Brain Research, 252:39-50.

Lipton,S.A., Rasmussen,H., and Dowling,J.E., 1977, Electrical and adaptive properties of rod photoreceptors in Bufo marinus, II. Effects of cyclic nucleotides and prostaglandins, J.Gen.Physiol., 70:771-791.

Miller,W.H., and Nicol,G.S., 1979, Evidence that cyclic GMP regulates membrane potential in rod photoreceptors, Nature, 280:64-66.

Ohnishi,S.T. and Devlin,T.M., 1979, Calcium ionophore activity of a prostaglandin B_1 derivative (PGBx), Biochem. Biophys. Res. Commun. 89:240-245.

Pace,U., Hanski,E., Salomon,Y., and Lancet,D., 1985, Odorant-sensitive adenylate cyclase may mediate olfactory reception, Nature, 316: 255-258.

Polis,B.D., Kwong,S., Polis,E., Nelson,G. and Shmukler,H.W., 1979, Studies on PGBx, a polymeric derivative of prostaglandin B_1: I. Synthesis and purification of PGBx, Physiol. Chem. Physics, 11: 109-123.

Sokabe,M., Fujitsuka,N., Kori,A.A. and Ito,F., 1987, Effects of cyclic nucleotides and calcium on encoding processes in frog muscle spindle, Brain Research, in submitting.

TRANSDUCTION MECHANISMS IN PACINIAN CORPUSCLES

Stanley J. Bolanowski, Jr.

Department of Physiology, Univ. of Rochester Medical School
Rochester, N.Y. 14642

Institute for Sensory Research, Syracuse University
Syracuse, N.Y. 13210

Since their detection by Lehman in 1741 (cited by Pallie, et al., 1970), a wealth of information has been obtained on the Pacinian corpuscle, its morphology, chemistry, physiology and the role that it plays in tactile perception. While ostensibly acting as the substrate for the P channel of the somatosensory system (Verrillo, 1966, and Bolanowski and Verrillo, 1982), the corpuscle is also found in the mesentery of cat where it is possible to manipulate them directly. This accessibility has made the corpuscle prototypical for the understanding of mechanotransduction, especially in reference to non-ciliated mechanoreceptors. The relationship among the stimulus, receptor potential and the neural impulse that it generates has been defined, but the actual mechanism underlying transduction and its neural locus are still unknown. The present report focuses on the form of the receptor-potential's intensity characteristic and the possible neural structures producing it.

The relationship between a stimulus and the corpuscle's receptor potential has been determined through the use of mechanical deformations in the form of pulses (e.g., Diamond, et al., 1958; Loewenstein, 1961), triangles (Il'inskii, 1965) and sinusoids (Sato, 1961; Bolanowski and Zwislocki, 1984a). Those studies using pulses and ramps have been the basis for several mathematical models for transduction. For example, Loewenstein (1961) proposed a relationship between the peak receptor potential (V) and the fraction of membrane that becomes excited (x):

$$V=(Ebx)/(1+bx)$$

where E is the resting potential and b is a constant of proportionality such that b/(1+b) equals the ratio between the potential achieved at maximum excitation and the resting potential. Zwislocki (1973), on the other hand, has formulated an equation which implies that the corpuscle first linearly integrates stimulus energy and the effective energy of the intrinsic noise, and then performs a square-root transformation on the result. Saturation of the response is approximated by an exponential function. Indeed, many other mathematical formulations have been used to model receptor potentials including linear, power, logarithmic, log tan and exponential forms, although Lipetz (1971) has shown that many of them can be reduced to the form first proposed by Loewenstein.

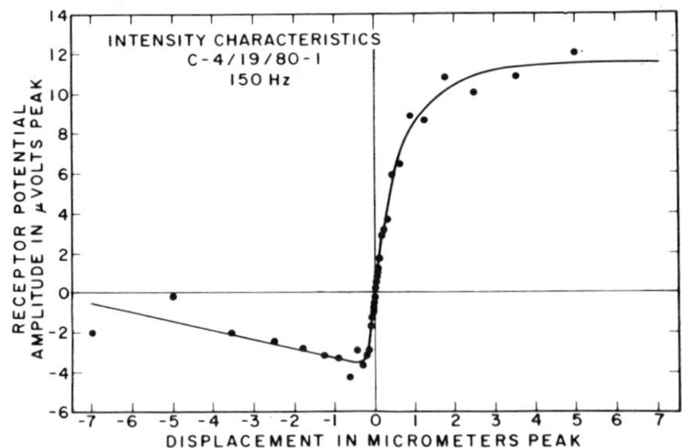

Fig. 1 Intensity characteristic obtained on a Pacinian corpuscle relating receptor-potential amplitude to stimulus displacement for sinusoidal displacements at 150 Hz. The smooth curves in this and Figs. 2 and 3 have been fitted by eye (see Bolanowski and Zwislocki, 1984a).

Stimuli in the form of pulses and ramps typically do not permit both compression into and withdrawal away from the receptor under identical stimulating conditions. Only reversal of the stimulus polarity (Gray and Sato, 1953) and rotation of the capsule (Il'inskii, 1965), after previous stimulation, will allow this. The results of such studies suggested that different receptor potentials to different stimulus directions can be produced. Using sinusoids, which if presented around some static indentation will produce both compression and withdrawal within the same cycle of stimulation, Bolanowski and Zwislocki (1984a) have delineated two, heretofore unknown, electrophysiological properties. The first property is that rather than having a receptor potential that is "half-wave rectified", as found for all other vertebrate mechanoreceptors (e.g. Hudspeth and Corey, 1977), the corpuscle's input-output characteristic can approach, asymmetrically, a full-wave rectification. Figure 1 shows the relationship between receptor potential amplitude and stimulus displacement as measured with sinusoidal stimulation. The static indentation (110 μm) of the stimulus is given by 0 on the abscissa, with compressions and withdrawals produced around this value. Positive values on the ordinate signify depolarization of the receptive membrane while negative values signify hyperpolarization. The use of monophasic pulses or ramp displacements would only reveal the positive part (upper right quadrant) of the intensity characteristic. This part of the characteristic, of course, can be approximated by either Loewenstein's or Zwislocki's model. However, the negative portion of the characteristic (lower left quadrant) shows that a release from hyperpolarization can occur on the opposite phase of stimulation. This release from hyperpolarization causes the corpuscle to produce two impulses/cycle for large stimulus displacements (Bolanowski and Zwislocki, 1984b). The degree of this release from hyperpolarization can be exceedingly large, actually producing depolarizations, as shown in Fig. 2, or virtually nonexistent, as shown in Fig. 3. The insets in both Figs. 2 and 3 show the actual time course of the signal-averaged receptor potential. Thus, corpuscles can either display half-wave rectification (Fig. 3), full-wave rectification (Fig. 2) or, more typically, something in between (asymmetric rectification, Fig. 1). The significance of this is that the response characteristics cannot be modeled by either Loewenstein's or Zwislocki's formulations.

The second property found by Bolanowski and Zwislocki (1984a) is that there appear to be two functionally distinct populations of corpuscles, one population more sensitive to stimulus compression, the other more sensitive

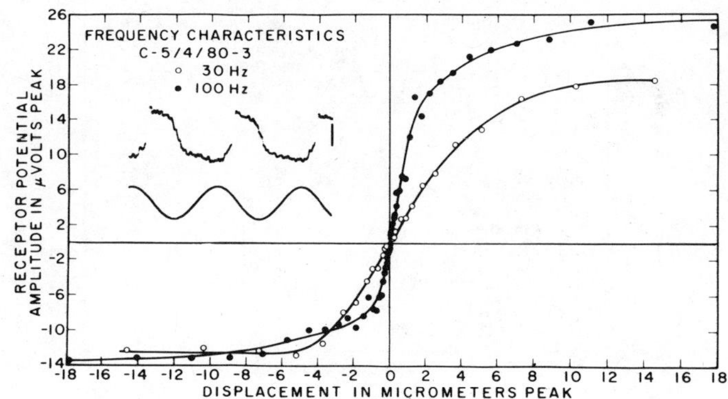

Fig. 2   Intensity characteristics obtained on a corpuscle at a stimulus frequency of 400 Hz (Bolanowski and Zwislocki, 1984a).

to stimulus withdrawal. Figure 4 shows the phase-frequency relationship between maximum depolarization and peak stimulus compression for eight corpuscles. Two types of responses occur. One type has a phase response which varies from -235 to -535°, over the frequency range of 20 to 1000 Hz. The other type undergoes the same 300° phase change from the lowest to highest frequency, but starts at a phase of about -55°. Thus, the phase difference between the two groups is 180°. In other words, some corpuscles respond preferentially to stimulus compression while other corpuscles respond to stimulus withdrawal.

Il'inskii (1965) proposed a model based on "membrane-stretching" and the fact that a major portion of the terminal axon is elliptical in cross-section (Il'inskii, et al., 1968). He proposed that the elliptical shape of the unmyelinated portion of the axon was responsible for the 180° phase change in the corpuscle's responses when it was rotated by 90° (Il'inskii, 1965). Compression along the minor axis would, for a constant cross-sectional area, increase the circumference and produce depolarization by stretching the membrane. Compression along the major axis (90° rotation of the corpuscle) would produce hyperpolarization because of the decrease in the circumference. He did not, however, actually calculate how the circumference would change for various amounts of deformation. If his model

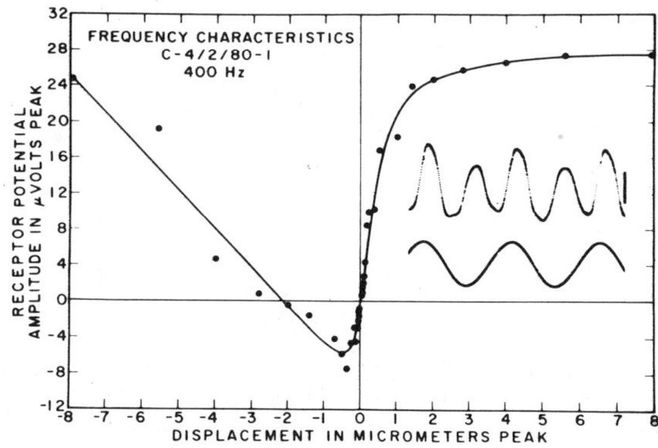

Fig. 3   Intensity characteristic obtained on a corpuscle at stimulus frequencies of 30 and 100 Hz (Bolanowski and Zwislocki, 1984a)

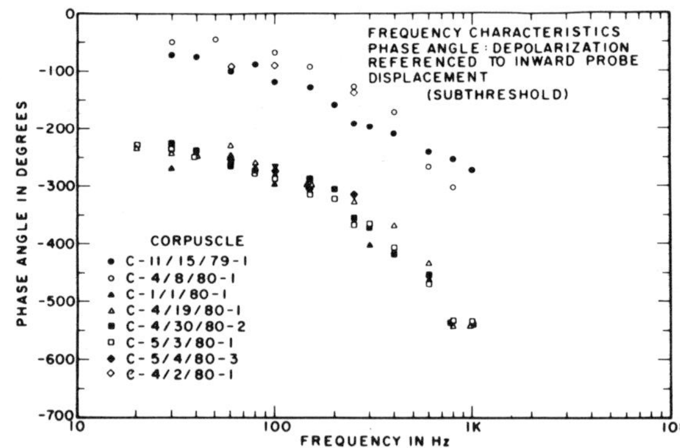

Fig. 4   Frequency characteristic relating the phase angle between receptor-potential depolarization and inward displacement for eight Pacinian corpuscles (Bolanowski and Zwislocki, 1984a).

is correct, then the results of such a calculation should match the physiological data presented in Figs. 1-3. Using the values of cross-sectional area and the static ratios of major to minor axes given by Il'inskii, et al. (1968), I have made calculations to see how the circumference of an ellipse changes with the length of the minor axes, assuming constant cross-sectional area. In the calculations, the dynamic ratio between the major and minor axes changes with amplitude of the stimulus, the situation being analogous to the one used to generate the intensity characteristics of the receptor potential. The results are shown by the solid line given in Fig. 5. The shape of the curve does not correspond to any of the characteristics shown in Figs. 1-3. The model is particularly inadequate when the degree of asymmetry in the rectification found for various corpuscles is considered. The other curve in Fig. 5 shows that even axial dimensions at the limit of Il'inski et al.,'s (1968) measurements, the function (----) changes only slightly. Thus, the actual shape and dimension of the neurite probably does not determine the shape of the intensity characteristic.

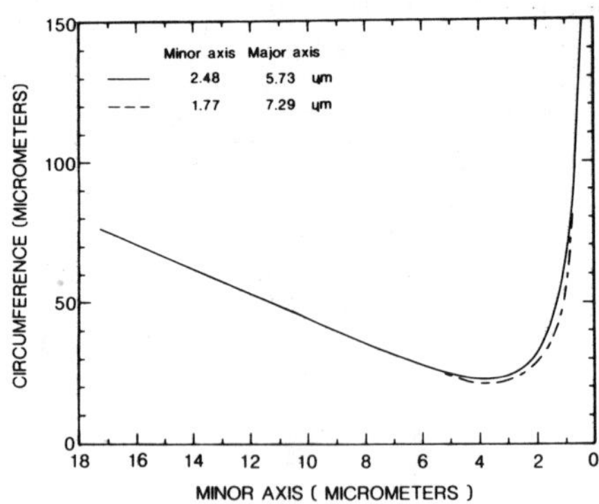

Fig. 5   Plots of the circumference (ordinate) of an ellipse for changes in the dimension of the minor axis when the area is held constant. Values used to determine the area and circumference were obtained from Il'inskii, et al. (1968). They were: minor axis, 2.48 ± 0.77 μm and major axis, 5.73 ± 1.56 μm.

Any model intending to describe the characteristics of the receptor potential must acccount, not only for variations in the asymmetry of rectification, but also for the fact that there are two classes of corpuscles, one that produces depolarization on compression and the other that produces it on withdrawal. These constraints can be fulfilled by proposing that there are two, effectively functional yet structurally distinct groups of cytoplasmic components contributing to the receptor potential. One group would be more sensitive to stimulus compression with the other more sensitive to stimulus withdrawal. Each group, however, would otherwise respond similarly, the electrode monitoring the integrated response of both groups. The type of response in terms of the degree of rectification and the phase relationship to the stimulus would depend on the relative contribution of the two groups. Fig. 6 shows the proposed scheme where the response characteristics of the two groups (-----, compressive sensitive;—-—; withdrawal sensitive) are summated (———). The fundamental shape of the response characteristic for each group has been taken from Fig. 3, since this receptor-potential characteristic shows only half-wave rectification. The implication of this is that the contribution of one of the cytoplasmic groups is negligibly small. The graphical discription reflecting this is shown in Fig. 6, panel C. The actual strength of the contibution of each group to the overall characteristic is implemented by scaling the magnitude of the individual contributions. For example, in panel B of Fig. 6, the withdrawal sensitive population is 0.8 times less effective than the compressive sensitive group. The summated response in this panel should be compared to the characteristic shown in Fig. 1. Similarly, if one group is only 0.25 less effective, then the curve shown in panel A of Fig. 6 is obtained (Compare this curve to that shown in Fig. 2). A 180° phase shift in the response can be obtained if the withdrawal sensitive group contributes more to the summated response. This can be seen in panels E and F of Fig. 6. Panel D of the figure shows what happens when the two groups contribute equally to the receptor potential.

The two groups of cytoplasmic elements alluded to above are probably the cytoplasmic extensions which emanate from the polar regions of the unmyelinated axon and insert into the hemilammellar clefts of the capsule (Spencer and Schaumberg, 1973, Chouchkov, 1976). Gottschaldt, et al. (1982) also propose such an idea for the Ruffini ending and Herbst corpuscle (see also Halata and Munger, 1980). Interestingly, these axon processes have also been seen in both dermal and epidermal endings (Chouchkov, 1974), straight and branched lanceolate terminals, Golgi tendon organs, and Golgi-Mazzoni

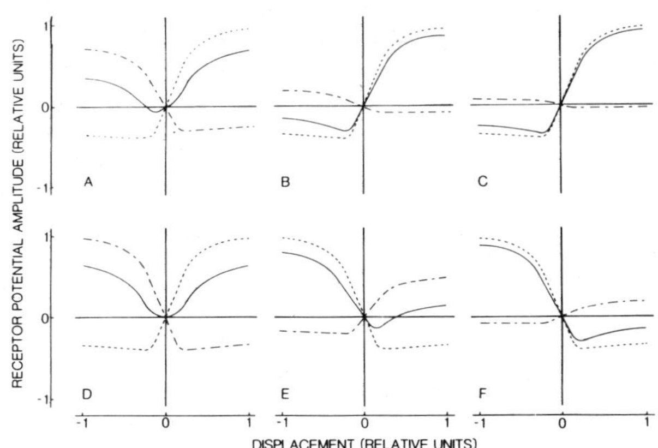

Fig. 6 Intensity characteristics relating normalized receptor-potential amplitude to normalized displacement and based on the proposed model (see text). The solid curve is a summation of the two dashed curves.

bodies (Andres and During, 1973) suggesting a common mechanism for transduction. In terms of the model, the fact that there are two groups of corpuscles becomes more a matter of which group of extensions within each corpuscle is most sensitive and therefore predominant in producing the receptor potential. Directionality of sensitivity might be imparted by their orientation within the hemilamellar clefts, by their shape or size, or by a morphological polarity, the latter requiring that a given process be bent in a particular direction for transduction to take place. Future experiments correlating the shape of receptor-potential characteristics with the corpuscle's morphology should eventually determine the manner in which the polarity of the elements is imparted. (Supported by NINCDS, NS23933).

REFERENCES

Andres, K.H., and Düring, M. von., 1973, Morphology of cutaneous systems, in: "Handbook of Sensory Physiology, Vol. II," A. Iggo, ed., Springer-Verlag, Berlin, Pp.1-28.

Bolanowski, S.J., Jr., and Verrillo, R.T., 1982, Temperature and criterion effects in a somatosensory subsystem: a neurophysiological and psychophysical study, J. Neurophysiol.,48:836-855.

Bolanowski, S.J., Jr., and Zwislocki, J.J., 1984a, Intensity and frequency characteristics of Pacinian corpuscles. II. Receptor potentials, J. Neurophysiol., 51:812-830.

Bolanowski, S.J., Jr., and Zwislocki, J.J., 1984b, Intensity and frequency characteristics of Pacinian corpuscles. I. Action potentials, J. Neurophysiol., 51:793-811.

Chouchkov, H.N., 1974, An electron microscopic study of the intraepidermal innervation of human glabrous skin, Acta. anat., 88:84-92.

Chouchkov, H.N., 1976, Ultrastructural differences between the preterminal nerve fibers and their endings in the mechanoreceptors, with special reference to their degeneration and mode of uptake of horseradish peroxidase, Prog. Brain Res.,43:77-87.

Diamond, J., Gray, J.A.B., and Inman, D.R., 1958, The depression of receptor potentials in Pacinian corpuscles, J. Physiol.,141: 117-131.

Gottschaldt, K.-M., Fruhstorfer, H., Schmidt, W., and Kraft, I., 1982, Thermosensitivity and its possible fine-structural basis in mechanoreceptors in the beak skin of geese, J. Comp. Neurol., 205:219-245.

Halata, Z., and Munger, B.L., 1980, The ultrastructure of Ruffini and Herbst corpuscles in the articular capsule of domestic pigeon, Anat. Rec., 198:681-692.

Hudspeth, A.J., and Corey, P.P., Sensitivity, polarity and conductance change in the response of vertebrate hair cells to controlled mechanical stimuli, Proc. Natl. Acad. Sci., 74:2407-2411.

Il'inskii, O.B., 1965, Processes of excitation and inhibition in single mechanoreceptors (Pacinian corpuscles), Nature (London)., 208:351-353.

Il'inskii, O.B., Volkova, N.K., and Cherepnov, V.L., 1968, Structure and function of Pacinian corpuscles, Fiziol. Zh. SSSR., 54:295, Translation in Neurosciences Translation.,56:637-643.

Lipetz, L.E., 1971, The relationship of physiological and psychological aspects of sensory intensity, in " Handbook of Sensory Physiology, Vol 1." W.R. Loewenstein, ed., Springer-Verlag, Berlin, Pp. 191-225.

Loewenstein, W.R., 1961, Excitation and inactivation in a receptor membrane, Ann. N.Y. Acad. Sci., 81:510-534.

Loewenstein, W.R., 1971, Mechano-electric transduction in the Pacinian corpuscle, in: "Handbook of Sensory Physiology, Vol. 1," W.R. Loewenstein, ed., Springer-Verlag, Berlin, Pp. 269-290.

Pallie, W., Nishi, K, and Oura, C., 1970, The Pacinian corpuscle, its
    vascular supply and the inner core, Acta Anat. (Basel).,77:508-520.
Sato, M., 1961, Response of Pacinian corpuscles to sinusoidal vibration. J.
    Physiol., 159:391-409.
Spencer, P. S., and Schaumberg, H.H., 1973, An ultrastructural study of the
    inner core of the Pacinian corpuscle, J. Neurocytol., 2:217-237.
Verrillo, R.T., 1966, Vibrotactile sensitivity and the frequency response of
    the Pacinian corpuscle, Psychon. Sci.,4:135-136.
Zwislocki, J.J., 1973, On intensity characteristics of sensory receptors: A
    generalized function, Kybernetik.,12:169-183.

THE ROLE OF CYTOSKELETON IN MECHANORECEPTOR ACTIVITY OF
PACINIAN CORPUSCLES

R.R. Gataullin and A.T. Zaripov

Kazan Medical Institute
Kazan, USSR

INTRODUCTION

In recent years much data has accumulated concerning the
possible role of the cytoskeleton in cell membrane excitability. It was shown that colchicine which disrupts intracellular microtubules (MT), blocks the axolemma K^+-channels (Terakawa and Watanabe, 1976), K^+, Ca^{2+} and Cl^--channels of Aplysia
neuronal membrane (Baux et al., 1981). Two articles have appeared (Moran and Varela, 1971; Schafer and Reagan, 1981)
which describe that colchicine inhibits the cockroach spine
mechanoreceptor electrical responses and this effect might be
due to specific disruption of dendritic MT. MT´s structural
integrity is necessary for generation of electric potentials
in crustacean stretch receptors (Witte, 1980). We continued
the investigation of this problem and studied the influence
of colchicine on generator potentials (GP) of single Pacinian
corpuscles isolated from the mesentery of cats.

METHODS

The preparation of receptors and extracellular recording
of GP was performed by a standard method (Chelyshev and Gataullin, 1983) with the help of a hook electrode. We used colchicine (Merck) prepared in carbogen aerated Ringer-Locke´s
solution. Lumicolchicine was prepared by prolonged irradiation
of a colchicine solution with ultraviolet light. For electron
microscopic examination, Pacinian corpuscles were fixed in
glutaraldehyde and osmium tetroxide, dehydrated and embedded
in Epon-Araldite resin.

RESULTS AND DISCUSSION

In 9 out of 11 experiments the perfusion of receptors
with 0.1 mM colchicine solution led to a depression of GP
amplitude (Fig. 1). This effect appeared 15-25 min after
switching to the drug solution. Further perfusion of colchicine completely depressed the Pacinian corpuscle´s mechano-

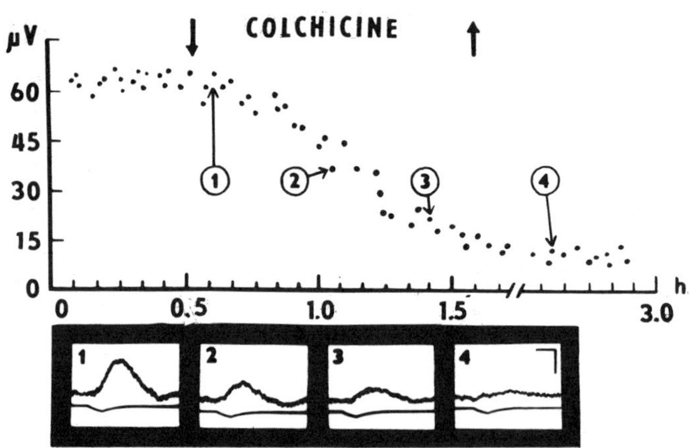

Fig. 1. Depression of GP amplitude in colchicine solution (a
record of one experiment).
Abscissa, time in hours; ordinate, GP amplitude in µV.
Lower part of the figure - electrical responses of
Pacinian corpuscles at various times of the experiment.
Upper trace is the generator potential; lower trace is
the record of the mechanical stimulus. Calibration
= 50 µV, 2 ms.

sensitivity which did not recover during subsequent prolonged
washing with normal Ringer-Locke's solution. Perfusion of mechanoreceptors with a solution containing a colchicine photo-
derivative, lumicolchicine (0.1 mM), failed to inhibit their
mechanosensitivity in all 6 experiments. Electron microscopic
examination showed that the sensitive nerve ending had lost
its microtubules after incubation in the colchicine solution.

It is known that colchicine depolymerizes MT by binding
tubulin monomers and by inhibition of polymer formation. The
doses of colchicine which are active in various species differ largely from 0.001 mM to 10 mM. It also depends on different cell membrane permeability to the alkaloid (for review
see Dustin, 1984). In our experiments 0.1 mM concentration of
colchicine was effective, it destroyed MT and decreased the
GP amplitude.

It was thus found that colchicine inhibits GP of Pacinian
corpuscles. This effect may be associated with the disintegration of axonal MT.

REFERENCES

Baux, G., Simonneau, M., and Tauc, L., 1981, Action of colchicine on membrane currents and synaptic transmission in
Aplysia ganglion cells, J. Neurobiol., 12:25.
Chelyshev, Yu.A., and Gataullin, R.R., 1983, Effect of morphine
and leucine-enkephalin on the electrophysiological

responses of the Pacinian corpuscles of the cat, <u>Neurosci. Lett.</u>, 39:165.

Dustin, P., 1984, "Microtubules", Springer Verlag, Berlin - Heidelberg - New York.

Moran, D.T., and Varela, F.G., 1971, Microtubules and sensory transduction, <u>Proc. Natl. Acad. Sci. USA</u>, 68:757.

Schafer, R., Reagan, P.D., 1981, Colchicine reversibly inhibits electrical activity in arthropod mechanoreceptors, <u>J. Neurobiol.</u>, 12:155.

Terakawa, S., and Watanabe, A., 1976, Effects of colchicine and other antimitotic drugs on the electrophysiological properties of a crayfish axonal membrane, <u>Proc. Jpn. Acad.</u>, 52:82.

Witte, H., 1980, Electrophysiologische Untersuchungen zur Beeinflussung des Transductionsverhaltens der abdominalen Streckreceptoren von Oronectes limosus durch Colchicin and Vinblastin, <u>Zool. Jb. Physiol.</u>, 84:198.

PART VI

STRUCTURE AND FUNCTION
OF MATURE MUSCLE RECEPTORS

STROBOSCOPIC CINEMATOGRAPHIC AND VIDEORECORDING OF DYNAMIC BAG_1 FIBRES
DURING RAPID STRETCHING OF ISOLATED CAT MUSCLE SPINDLES

I.A. Boyd, M.H. Gladden, D. Halliday and M. Dickson

Institute of Physiology
University of Glasgow
Glasgow, UK

INTRODUCTION

When an isolated cat muscle spindle is subjected to a ramp and hold
stretch the primary sensory annulo-spiral round the dynamic bag_1 (Db_1)
fibre is extended and then creeps back part way towards its original state
over several seconds since the poles of the fibre give way (Boyd, 1976).
This 'creep' at the final length is small, or absent, when the fibre is
inactive, and is greatly enhanced by activation of the fibre (Fig. 1; Boyd,
Gladden & Ward, 1981). The length sensitivity of the primary sensory
ending is greatly increased during the dynamic phase of stretching, but the
static length sensitivity is little increased because the creep causes
pronounced slow adaptation of the Ia discharge (Fig. 2).

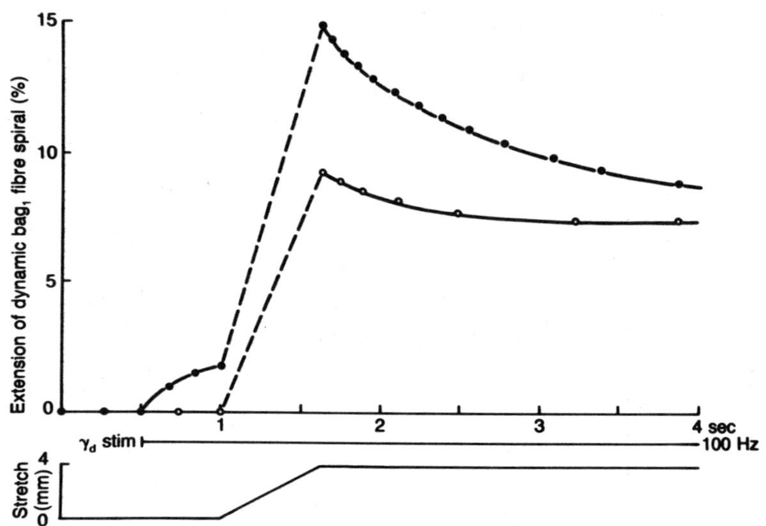

Fig. 1. The degree of extension of the annulo-spirals of the primary
sensory ending round the dynamic bag_1 (Db_1) fibre in response to a
ramp stretch in the presence (●) and absence (o) of stimulation of a
dynamic γ axon to the fibre for the period indicated by the horizontal
line (Boyd, Gladden & Ward, 1981).

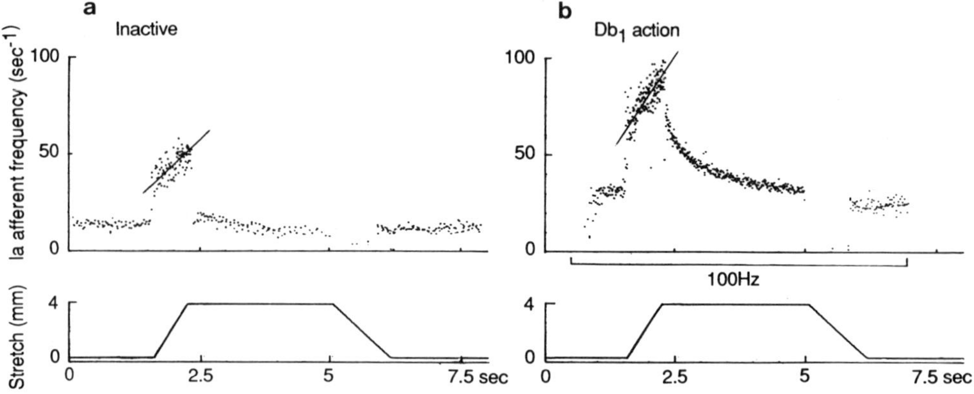

Fig. 2 Instantaneous frequencygrams of the response of a primary sensory ending to a ramp stretch and release (a), and the modification produced by stretching during stimulation of a dynamic γ axon at 100Hz for the duration of the horizontal line (b). 3 traces superimposed.

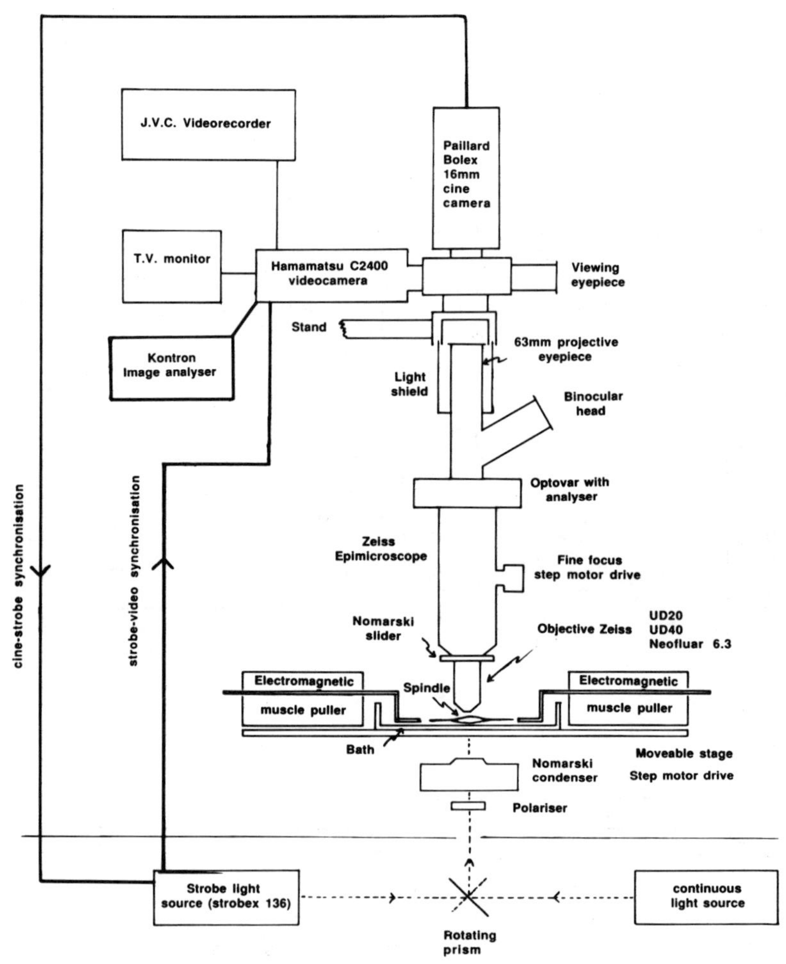

Fig. 3 Experimental arrangement for stroboscopic cine or videorecording of sarcomere length or displacement in active dynamic bag$_1$ intrafusal fibres during rapid stretches.

The creep may represent the decay of activation of the Db_1 fibre by
stretch (Boyd, 1976). Studies involving glycogen depletion, sinusoidal
stretching, or after-effects of fusimotor stimulation also suggest that
'stretch activation' of the Db_1 fibre may occur (Boyd & Gladden, 1985, Part
II review). Poppele & Quick (1981) demonstrated that, during very slow
stretches, shortening of sarcomeres in the fluid space region of the Db_1
fibre could be observed during a slow ramp stretch, associated with a rise
in intrafusal tension. It is doubtful, however, whether the Db_1 fibre was
behaving normally since it was clamped at the end of the capsular sleeve
region, excluding the extracapsular region from the stretching and possibly
damaging it.

Hitherto, we have been unable to observe intrafusal events <u>during</u>
rapid stretches (Fig. 1, broken line). The image on cine or videoframes
is blurred by the movement, the region under observation often moves
laterally out of the microscope field, and vertical displacement defocusses
the image. We have now overcome the first difficulty with high speed
stroboscopic cinephotography and videorecording, and partially overcome the
latter two by manual compensatory movements of the microscope stage and
fine focus during the stretch.

METHODS

An isolated tenuissimus muscle spindle-nerve preparation is stimulated
repetitively at 100Hz if a dynamic γ axon in the nerve has the lowest fusimotor threshold. If not, then the Db_1 fibre is activated with succinylcholine 10 $\mu g \ell^{-1}$ added to the bath. Stretch of about 10% applied to both
ends of the spindle is balanced so that the region under observation remains
stationary during the stretch. Stretch velocity is 0.05 to 0.50 resting
lengths per second.

The intrafusal bundle is observed through a Zeiss epimicroscope using
Nomarski interference contrast optics and water immersion lens. Movements
of individual intrafusal fibres or their primary sensory annulo-spirals
relative to the stationary capsule are recorded at 25, 32, or 50 frames per
sec, using a stroboscopic light source synchronised to either cine camera
or videocamera. Lateral displacement and defocussing of the image during
rapid stretch is compensated by driving stepmotors on microscope stage or
fine focus. Data about sarcomere movements and spacing are fed into a
Kontron Image Analyser for subsequent quantitative study.

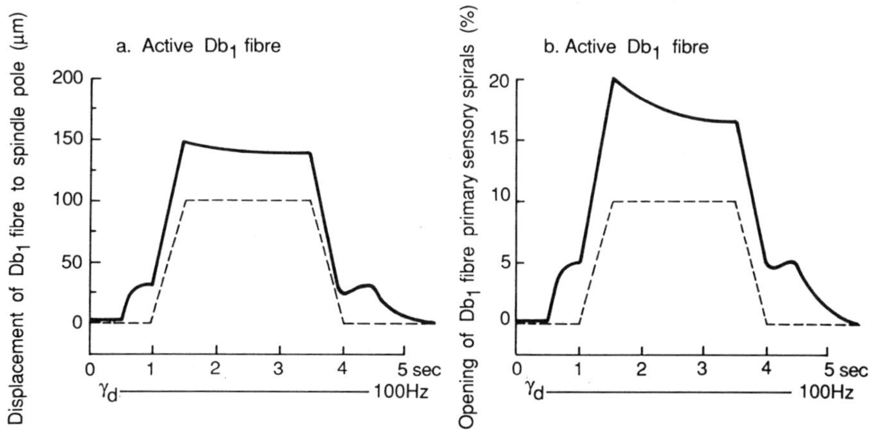

Fig. 4. Displacement of the active Db_1 fibre towards the spindle pole 1mm
from the equator (a) and the extension of its primary annulo-spiral (b)
during a 10% ramp stretch and release of an isolated spindle (full lines).

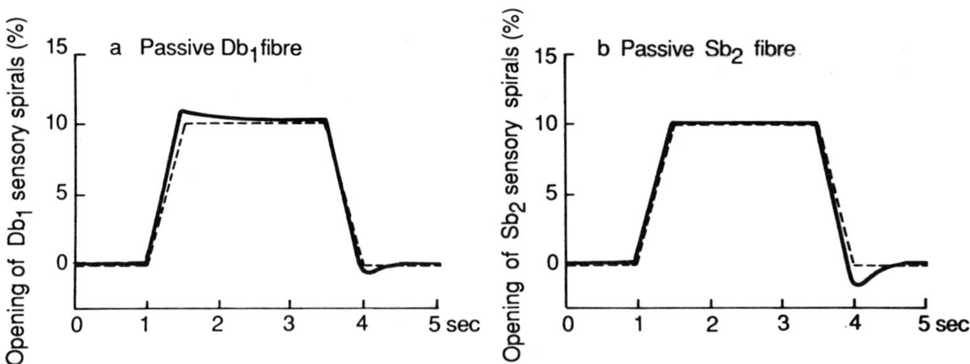

Fig. 5   The degree of extension of the primary annulo-spirals round the passive dynamic bag$_1$ (Db$_1$) fibre (a) and the passive static bag$_2$ (Sb$_2$) fibre (b) in response to a 10% stretch of an isolated spindle. Broken lines show expected changes if the fibres extended uniformly throughout their length.

RESULTS

The semi-quantitative results shown in Figs. 4, 5 and 6 summarise the observations made on ten isolated tenuissimus spindles. When a dynamic γ axon is stimulated at 100Hz maximal local sarcomere shortening of the order of 10% occurs in the capsular sleeve region close to the site of the fusi-motor end-plate(s) and sometimes also in the extracapsular region where there are usually no fusimotor endings. The Db$_1$ fibre in the fluid space region is displaced between 20 μm and 40 μm towards the pole (Fig. 4a), and its primary sensory annulo-spiral is extended about 5% (Fig. 4b).

During stretch of the active Db$_1$ fibre its sarcomeres in the fluid space and inner capsular sleeve regions move rapidly towards the spindle pole by an amount greater than the polar movement of capsule and other intrafusal fibres at the same point (Fig. 4a). During the plateau of stretch the Db$_1$ fibre creeps back towards the equator, and its annulo-spiral closes slowly (Fig. 4b) causing a phase of slow adaptation of the Ia discharge.

When the spindle is shortened, the Db$_1$ sensory spiral closes with only a very small overshortening. When stimulation ceases, the Db$_1$ fibre relaxes over a period of about 1 sec and the sensory spiral returns gradually to its initial state (Fig. 4b). The passive Db$_1$ fibre also exhibits the small overshortening but shows little or none of the stretch effect (Fig. 5a). The passive Sb$_2$ fibre, unexpectedly, shows a considerably greater degree of overshortening on release of stretch than the Db$_1$ fibre, but the stretch effect and creep are absent (Fig. 5b).

The translational movement of the Db$_1$ fibre during stretch increases with distance from the equator and may be as great as 10 sarcomeres excess movement (30 μm approx.) in the capsular sleeve region (Fig. 6). There is usually no displacement of the fibre relative to the spindle capsule at the site of maximal sarcomere shortening. It has not yet been possible to observe what happens beyond this point. Creep of the Db$_1$ fibre is greatest in the secondary sensory region and then decreases with distance from the equator so that it is absent in the capsular sleeve region (Fig. 6). Beyond the γ contraction zone, creep following stretch is towards the pole.

Shortening of sarcomeres during stretch, indicating stretch activation has not so far been observed in the fluid space region but may occur in the capsular sleeve where measurement has not yet been possible.

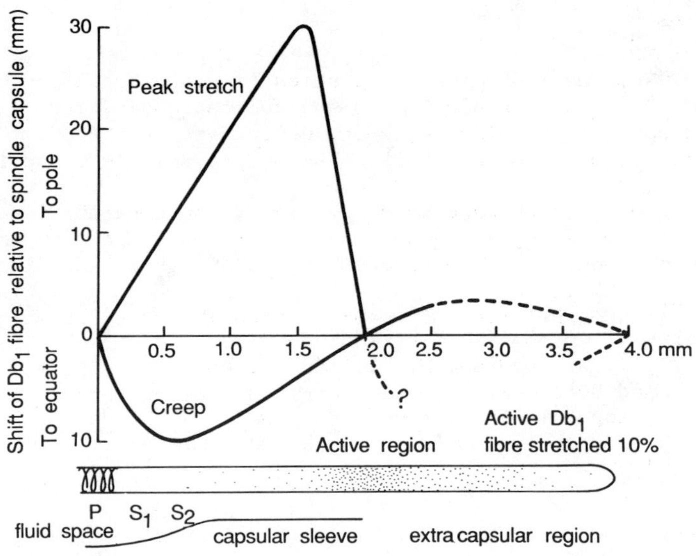

Fig. 6. Displacement of the active Db_1 fibre towards the pole in excess of the movement of capsule and other intrafusal fibres at the same point at the peak of a ramp stretch, and the amplitude of the subsequent creep during the stretch plateau, both plotted against distance from the spindle equator.

DISCUSSION

Contraction of the Db_1 fibre by fusimotor stimulation or Succinylcholine stiffens the fibre in capsular sleeve and extracapsular regions. A 10% stretch applied to the spindle then results in a proportionately greater stretch of the sensory annulo-spiral (say 15%). Uniform extension of sarcomeres in fluid space and inner capsular sleeve region (say 12%) would explain the increase in excess polar shift of the Db_1 fibre with distance from the equator (Fig. 5).

The creep phenomenon is mostly due to lengthening of sarcomeres in the fluid space region during the plateau of stretch and not of sarcomeres in the contracting region. This could be due to decay of weak stretch activation of the whole fibre, or simply to the breaking of unstable bonds. Computerised image analysis of small sarcomere length changes during stretch is being conducted to resolve this point, with simultaneous measurement of intrafusal tension.

Poppele & Quick (1981) and Poppele (1985) observed shortening of sarcomeres in the fluid space when the passive Db_1 fibre was stretched slowly. Further, when fusimotor contraction of the Db_1 fibre was produced at one pole the sarcomeres in the fluid space at the opposite pole did not extend during stretching indicating that activity must have been induced in them by the stretch. We believe that the Db_1 fibre was damaged since it was gripped at the end of the capsular sleeve and that the passive fibre was behaving atypically. Stretch activation of the active Db_1 fibre may have been a genuine effect, however, though possibly exaggerated by the exclusion of the extracapsular zone from the stretching.

REFERENCES

Boyd, I. A., 1976, The mechanical properties of dynamic nuclear bag
    fibres, static nuclear bag fibres and nuclear chain fibres in
    isolated cat muscle spindles, Prog. Brain Res., 44:33-50.
Boyd, I. A., and Gladden, M. H., 1985, "The Muscle Spindle", Macmillan,
    London.
Boyd, I. A., Gladden, M. H., and Ward, J., 1981, The contribution of
    mechanical events in the dynamic bag$_1$ intrafusal fibre in isolated
    cat muscle spindles to the form of the Ia afferent axon discharge,
    J. Physiol., 317:80-81P.
Poppele, R. E., 1985, Relation between intrafusal muscle mechanics and the
    sensitivity of spindle sensory endings, in: "The Muscle Spindle",
    pp.167-171, I. A. Boyd and M. H. Gladden, eds., Macmillan, London.
Poppele, R. E., and Quick, D. C., 1981, Stretch induced contraction of
    intrafusal fibres in the cat muscle spindle, J. Neurosci. 1:1069-1174.

ACKNOWLEDGEMENT
This work was supported by The Wellcome Trust.

    It was with great sadness of heart that all those who
knew prof. I.A. Boyd and his active role during this meeting
learned of his untimely death shortly afterwards. He will be
missed by his many friends and admirers.

FUSIMOTOR ENDINGS RESPONSIBLE FOR CHAIN FIBRE 'DRIVING' OF PRIMARY SENSORY
ENDINGS IN CAT MUSCLE SPINDLES

I.A. Boyd, F. Sutherland and E.R. Arbuthnott

Institute of Physiology
University of Glasgow
Glasgow, U.K.

The effect of stimulation of single static γ axons, in ventral spinal
root filaments, on the response of primary sensory endings from tenuissimus
muscle spindles in dorsal root filaments was correlated with the ultrastructure of the fusimotor endings of these same axons. Static γ axons
causing 'driving' of the Ia discharge were stimulated at frequencies from
0 to 150Hz with the spindle at constant length, and when subjected to ramp
stretches. Seven spindles were isolated and intrafusal events observed
during axon stimulation. The spindles were fixed with glutaraldehyde and
reconstructed from serial transverse sections. Ultrathin sections were
prepared at several levels of section through all the fusimotor endings.

Seven axons to five spindles ended in from one to four Mc plates
(Arbuthnott et al., 1982) with prominent sole plates and complex folding of
muscle sarcoplasm beneath and between the axon terminals (Fig. 2a), each
plate lying on a different chain fibre. All axons produced 1:1 driving of
the primary endings up to 50Hz at least, and usually to 100Hz or 150Hz.
Such strong driving frequently held the Ia discharge constant during muscle
shortening and sometimes suppressed the response to lengthening (Fig. 1a).
This driving action persisted at all muscle lengths tested.

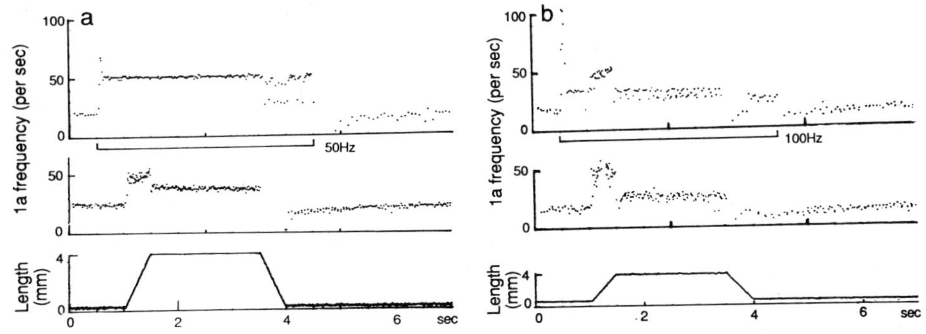

Fig. 1. The response of spindle primary sensory endings to a ramp stretch
(lowest traces) with and without stimulation of a static γ axon for the
period shown by the horizontal line (top traces). a. γ axon to one Mc
plate (Fig. 2a) on a chain fibre gives 1:1 driving suppressing the
response to stretch. b. γ axon to six Ma plates on three chain fibres;
1:3 driving does not suppress the response to length changes.

Four γ axons to four spindles ended in from two to six Ma plates, with
superficial axon terminals and subjunctional folding beneath the terminals
only or absent, on from one to three chain fibres (Fig. 2b). All the axons
produced weak driving of the primary endings, usually 1:2 up to 50Hz, and
1:3 at higher stimulation frequencies. The driving tended to be dependent
on muscle length and the Ia discharge was never held constant during muscle
lengthening or shortening (Fig. 1b).

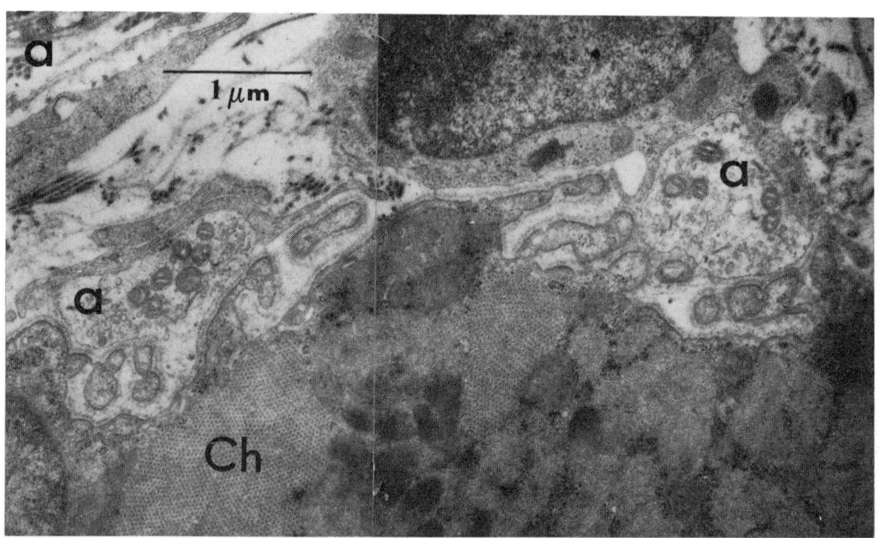

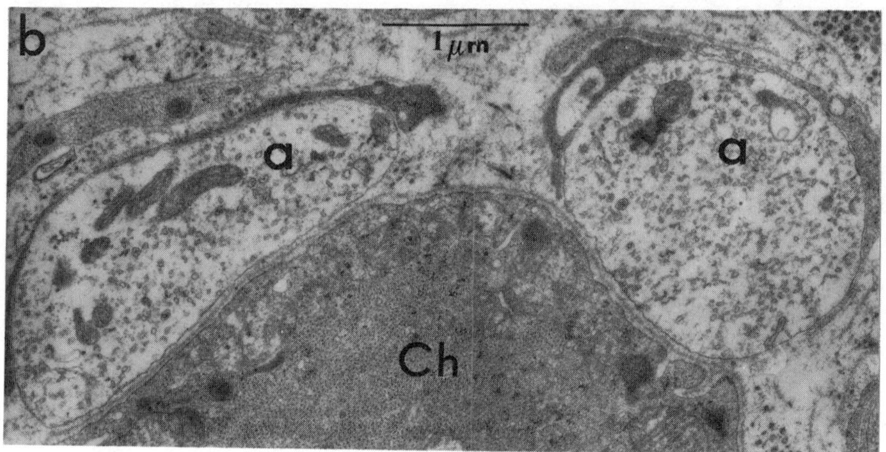

Fig. 2. Ultrathin sections through the centre of fusimotor end plates on
nuclear chain fibres previously studied functionally. _a_. Mc plate;
two axon terminals indented into finger-like folds of muscle sarcoplasm.
Strong Ia driving produced, shown in Fig. 1a. _b_. Ma plate; two
superficial axon terminals on unfolded muscle fibre membrane. Weak Ia
driving produced similar to Fig. 1b. These plates were the terminations of different static γ axons to the same spindle.

REFERENCES

E.R. Arbuthnott, K.J. Ballard, I.A. Boyd, and F.I. Sutherland (1982),
The ultrastructure of cat fusimotor endings and their relationship to foci
of sarcomere convergence in intrafusal fibres, _J. Physiol._ 331: 285-309.

LONG-CHAIN FIBRES IN SPINDLES OF CAT SUPERFICIAL LUMBRICAL
MUSCLES

L. Decorte, F. Emonet-Dénand, D.W. Harker and
Y. Laporte

Laboratoire de Neurophysiologie, College de France
75231 Paris cédex 05, France

A high incidence of long-chain fibres, i.e. of nuclear
chain fibres that extend for 1 mm or more beyond the spindle
capsule (Barker et al., 1976), was observed in spindle poles
of cat superficial lumbrical muscles which were reconstructed
over their whole length from serial transverse sections
(Decorte et al., 1987).

Out of 42 spindle poles, no less than 30 (71 %) contained
at least one long-chain fibre (1 in 17 poles, 2 in 11 poles,
3 in 2 poles). Of 246 poles of chain fibres, 45 (18 %) were
"long". In 4 spindles, in which both poles could be completely
examined, 10 long-chain fibres were observed; in 8 of these,
only one pole was "long" whereas the opposite pole ended either
intracapsularly or at a short distance outside the capsule.
Within the A region and the inner B region, the diameter of
long-chain fibres was similar to that of other chain fibres
but in the C region it was approximately equal to that of the
bag_1 fibres.

The high proportion of long-chain poles in superficial
lumbrical muscles suggests that spindles of these muscles are
extensively supplied by static β axons, since long-chain
poles are presently considered to be innervated by axons of
this type (Harker et al., 1977; Jami et al., 1978; Jami et al.,
1979, Kucera and Hughes, 1983; Kucera, 1984; Kucera et al.,
1984). However, attention should be given to several facts.
Long-chain fibres are not the only effector of static β axons
since β axons were found to supply other chain fibres among
the longest in spindle poles (Harker et al., 1977; Jami et al.,
1978; Kucera et al., 1984). Long-chain poles may be innervated
by other types of motor axon and short-poles of long-chain
fibres may be coinnervated by static gamma axons (for details
and references, see Decorte et al., 1987).

Rabbit lumbrical spindles were found to be extensively
supplied by static and dynamic β axons (Emonet-Dénand et al.,
1970) using curarizing drug for differentially blocking of
extra- and intrafusal neuromuscular junctions. However, in cat
lumbrical spindles, only a few skeletofusimotor axons could be

observed with this technique (Ellaway et al., 1971). More effective techniques for differential block being now available, it would be worthwhile, in view of the high proportion of long-chain fibres observed in cat lumbrical spindles, to re-investigate the β innervation of these spindles.

REFERENCES

Barker, D., Banks, R.W., Harker, D.W., Milburn, A., and Stacey, M.J., 1976, Studies of the histochemistry ultrastructure, motor innervation and regeneration of mammalian intrafusal muscle fibres. In: "Progress in Brain Research", edited by S. Homma. Amsterdam: Elsevier, vol. 44, p. 67-88.

Decorte, L., Emonet-Dénand, F., Harker, D.W., and Laporte, Y., 1987, High incidence of long-chain fibers in spindles of cat superficial lumbrical muscles, J. Neurophysiol., 57:1050-1059.

Ellaway, P.H., Emonet-Dénand, F., and Joffroy, M., 1971, Mise en évidence d'axones squeletto-fusimoteurs (axones β) dans le muscle premier lombrical superficiel du chat, J. Physiol. (Paris), 63:617-623.

Emonet-Dénand, F., Jankowska, E., and Laporte, Y., 1970, Skeletofusimotor fibres in the rabbit, J. Physiol. (London), 210:669-680.

Harker, D.W., Jami, L., Laporte, Y., and Petit, J., 1977, Fast-conducting skeletofusimotor axons supplying intrafusal chain fibers in the cat peroneus tertius muscle, J. Neurophysiol., 40:791-799.

Jami, L., Lan-Couton, D., Malmgren, K., and Petit, J., 1978, "Fast" and "slow" skeleto-fusimotor innervation in cat tenuissimus spindles; a study with the glycogen-depletion method, Acta Physiol. Scand., 103:284-298.

Jami, L., Lan-Couton, D., Malmgren, K., and Petit, J., 1979, Histophysiological observations on fast skeleto-fusimotor axons, Brain Res., 164:53-59.

Kucera, J., 1984, Non selective motor innervation of nuclear bag$_1$ intrafusal muscle fibers in the cat, Cell Tissue Res., 236:383-391.

Kucera, J., Hammar, K., and Meck, B., 1984, Ultra-structure of dynamic and static skeletofusimotor endings in a cat muscle spindle, Cell Tissue Res., 238:151-159.

Kucera, J., and Hughes, R., 1983, Histological study of motor innervation to long nuclear chain intrafusal fibers in the muscle spindle of the cat, Cell Tissue Res., 228: 535-547.

FUSIMOTOR INDUCED PHASE DIFFERENCES BETWEEN RESPONSES OF PRIMARY AND
SECONDARY ENDINGS FROM THE SAME MUSCLE SPINDLE

D.M. Halliday, M.H. Gladden and J.R. Rosenberg

Institute of Physiology
University of Glasgow
Glasgow, UK

INTRODUCTION

The work presented in this report forms part of a larger research
programme in which the behaviour of the muscle spindle is described under
conditions when several fusimotor inputs, and length changes are applied
simultaneously to the spindle while recording simultaneously from its
primary (Ia) and secondary (II) sensory endings. We restrict the current
presentation to a discussion of the strength of coupling or association
and time delay between the responses of primary and secondary endings from
the same tenuissimus muscle spindle during activation of single and
multiple fusimotor inputs. A description of this interaction is
particularly interesting in the light of a recent study by Edgley and
Jankowska (1987) which demonstrated the convergence of group Ia and II
afferent axons onto the same interneurones in the midlumbar segment of the
spinal cord.

EXPERIMENTAL METHODS

Single Ia and II sensory axons isolated in dorsal root filaments were
identified as belonging to the same muscle spindle by a modification of a
procedure introduced by Bessou and Laporte (1963), and illustrated in Figs.
1a and 1b. The thresholds of single Ia and II sensory axons to intramuscular stimulation of the tenuissimus nerve at different distances from
the entry of the nerve into the muscle is plotted against the distance from
the entry point of the nerve (Figs. 1a,b). Ia and II axons were assumed to
arise from the same muscle spindle when an abrupt change in their thresholds occurred simultaneously as the stimulating electrode was moved
further from the point of entry of the tenuissimus nerve. In each case
thresholds were expressed relative to that of Ia or II sensory axons passing
through the region without terminating. Fig. 1a illustrates the location
of two spindles by this method, one 4mm from nerve entry supplied by a Ia
and two II axons and another at 10mm with a Ia and a single II axon. Fig.
1b illustrates the threshold responses of the Ia and II axons of the muscle
spindle discussed in this report.

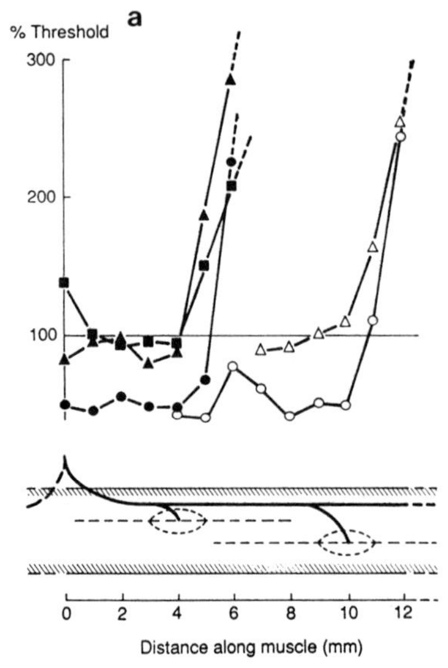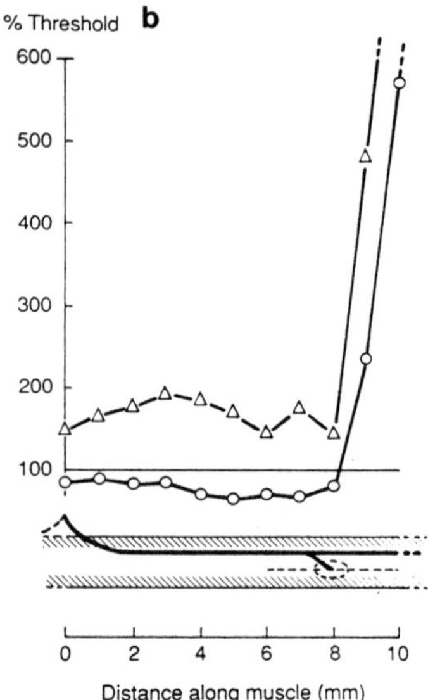

Fig. 1 Threshold of group Ia and II afferents relative to position of stimulation along muscle for (a) afferents of two adjacent spindles (b) single pair described in this study.

STATISTICAL METHODS

There are several approaches to measures of the strength of association between spike trains (Brillinger, 1975b). One measure of association, called the coherence, $|R_{ab}(\lambda)|^2$, is based on the correlation between the Fourier transforms of the two spike trains. Suppose one has records of duration T from two spike trains N and M. Set

$$d_N^T(\lambda) = \int_0^T e^{-i\lambda t}dN(t) \quad \text{and} \quad d_M^T(\lambda) = \int_0^T e^{-i\lambda t}dN(t).$$

The coherence may then be defined as

$$\text{Coherence} = |R_{NM}(\lambda)|^2 = \lim_{T\to\infty} | \text{Corr}\{d_N^T(\lambda), d_M^T(\lambda)\}|^2$$

where "Corr" denotes correlation. The coherence may be estimated in terms of the cross-spectrum between processes N and M, f_{NM}, and the autospectrum of each process, f_{NN}, f_{MM}, as

$$|R_{NM}(\lambda)|^2 = \frac{|f_{NM}|^2}{f_{NN}f_{MM}}$$

The coherence takes on values from 0 to 1, with 0 occurring in the case of independence (Brillinger, 1975a,b). The time delay between two spike trains was estimated from the phase relation derived from the cross-spectrum between the two processes (Brillinger, 1975a).

RESULTS

The behaviour of Ia and II sensory endings in response to single and multiple fusimotor inputs have been examined in four experiments. In one experiment two static γ axons were identified in ventral root filaments, each supplying the spindle static bag_2 and chain fibres. The spindle was isolated to observe its fusimotor innervation pattern, and subsequently

226

examined histologically to determine the relation of Ia and II spirals to the location of this input. Fig. 2 illustrates the pattern of fusimotor innervation in relation to the position of the Ia and II spirals. The static γ_b axon is seen to innervate both poles of the spindle, whereas the static γ_o innervates only one pole. The fusimotor axons were stimulated independently with sequences of pulses having an exponential distribution of intervals.

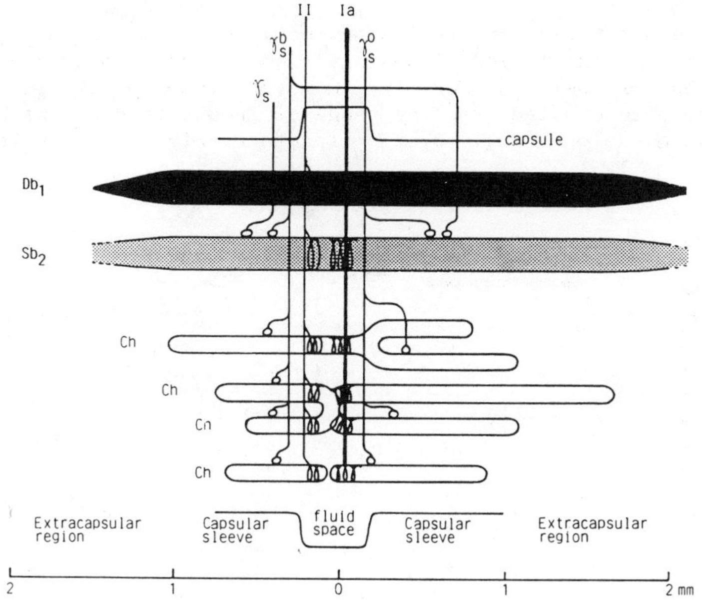

Fig. 2   Schematic diagram of fusimotor innervation.

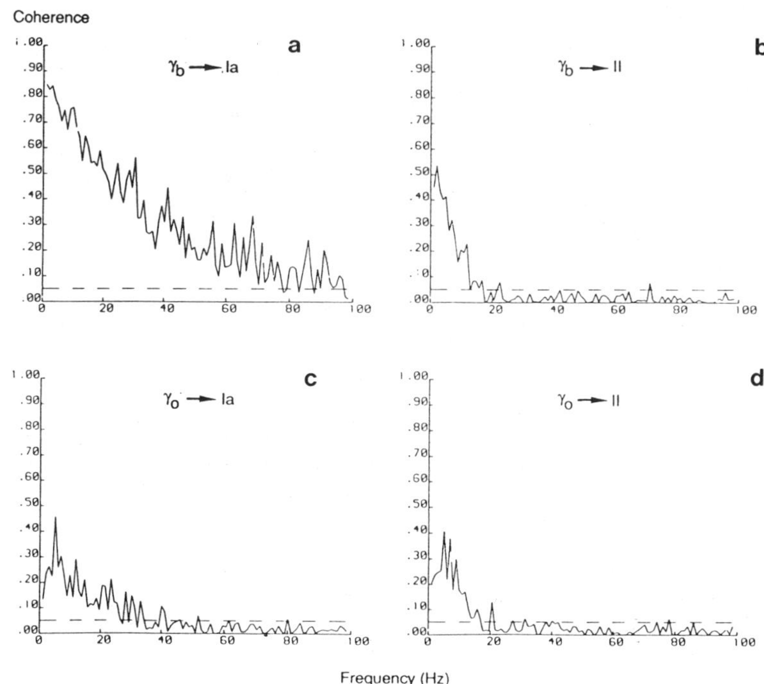

Fig. 3   Coherence between static γ_b and primary (a) and secondary (b), and between static γ_o and primary (c) and secondary (d) endings.

The coherence plots shown in Fig. 3 illustrate that the activity in each of the static axons individually is coupled to that of both the Ia and II endings. However, the strength of coupling and range of frequencies over which coupling occurs onto the Ia axon is greater for the γ_b axon than for the γ_o axon (Fig. 3a compared with Fig. 3c). In contrast, although the γ_b has a stronger effect on the II ending, than the γ_o, the range of frequencies over which each of these axons influences the II ending is the same (Fig. 3b compared with Fig. 3d).

In the absence of any fusimotor input there is no coupling between the activity in the Ia and II endings from the same spindle (Fig. 4a). In the presence of fusimotor activity, however, the activity in the Ia and II endings becomes coupled over the range of frequencies at which each of the γ static axons alone (Fig. 4b,c) or together (Fig. 4d) effects the II ending.

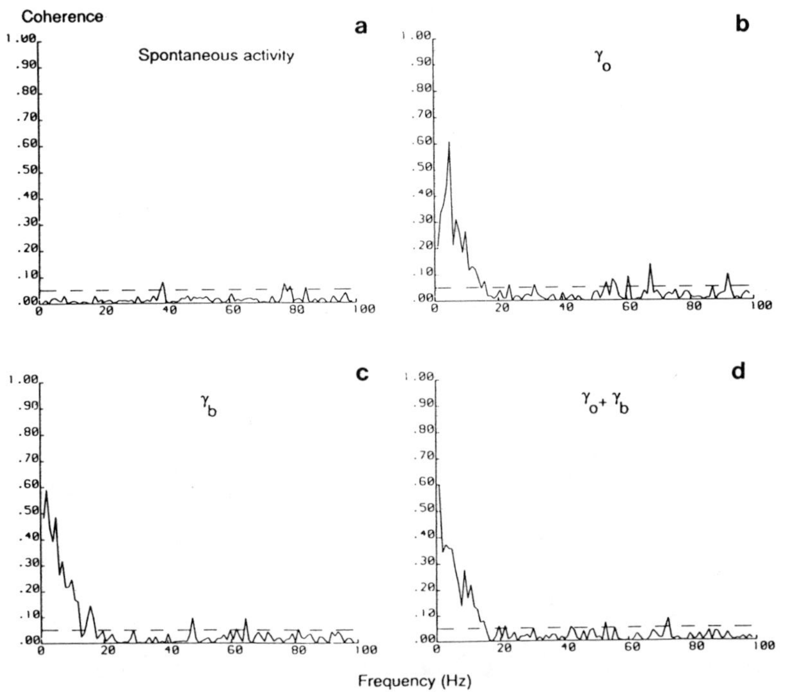

Fig. 4 Coherence between Ia and II endings with (a) no fusimotor stimulation (b) γ_o present (c) γ_b present (d) both γ_o and γ_b present.

On structural grounds alone it is not surprising that Ia and II responses become coupled in the presence of single or multiple fusimotor inputs. However, the timing relations of this coupling depends on the nature of the fusimotor input.

In the presence of γ_o stimulation alone Ia and II spikes, on average, arrived simultaneously at the dorsal roots as indicated by the phase curve shown in Fig. 5a. In the presence of γ_b stimulation there was a phase lead of 20 msec of the Ia activity over that of II (Fig. 5b), approximately 18 msec in excess of what one would expect in terms of conduction velocity differences alone. In the presence of both γ_o and γ_b, the phase lead of the Ia over the II was reduced to 13 msec.

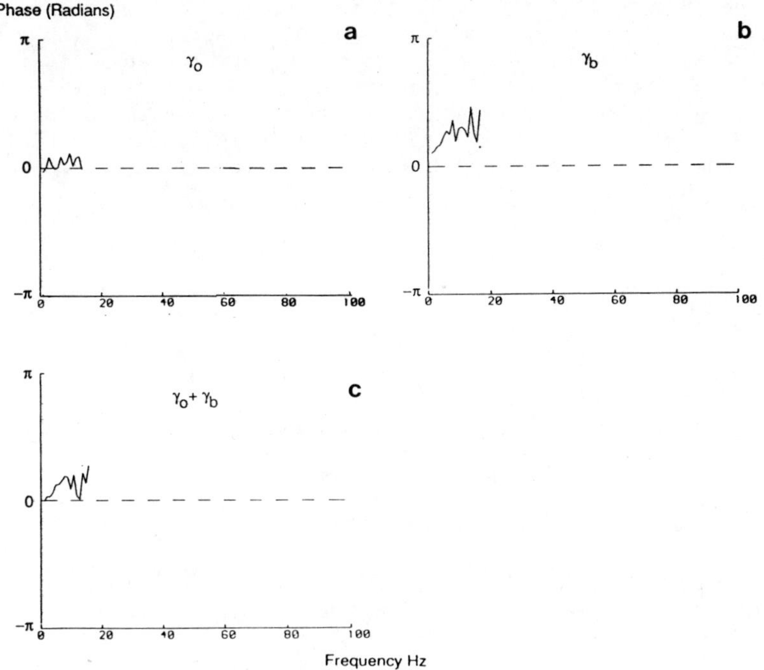

Fig. 5   Phase of Ia relative to II for fusimotor stimulation with (a) γ_o only (b) γ_b only (c) γ_o and γ_b.

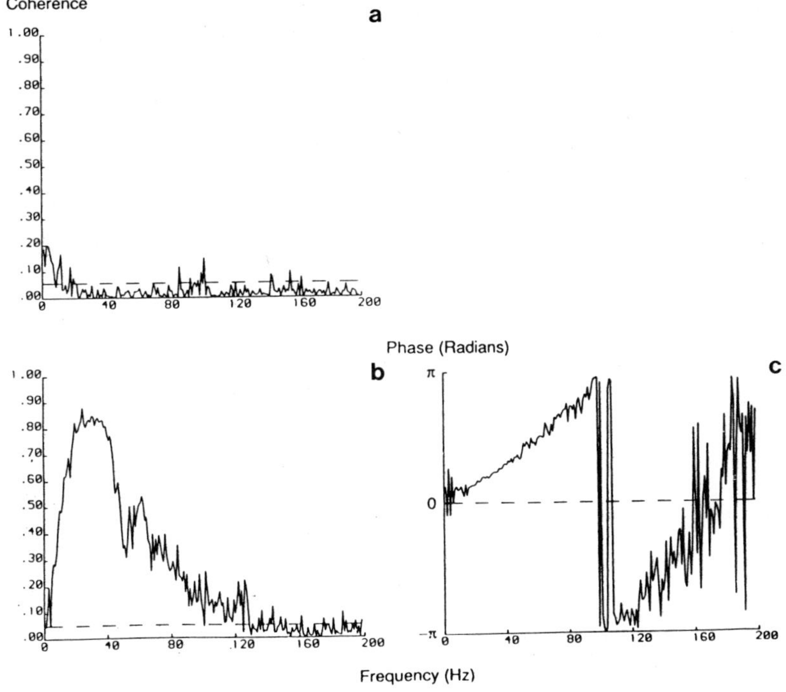

Fig. 6   Coherence between Ia and II with ventral roots left intact for (a) natural activity (b) forepaw squeezed and (c) phase of II relative to Ia for 6(b).

In one experiment the ventral roots were left intact during recordings from the Ia and II endings from the same muscle spindle. With intact ventral roots, where weak background static fusimotor activity may be expected, there was weak Ia/II coupling at low frequencies (Fig. 6a) which became strong over a broad range of frequncies during squeezing of the ipsilateral forepaw (Fig. 6b). During this reflexly induced increase in fusimotor input, a 4 msec phase lead of the II over the Ia response occurred at the level of the dorsal roots (Fig. 6c).

CONCLUDING COMMENTS

It seems likely, therefore, that Ia/II coupling would occur under natural conditions, with the earlier afferent spike acting as a conditioning pulse for the later spike when they converge onto the same central neurone. In this case, one might suggest that the timing between Ia and II spikes is a variable controlled by fusimotor input.

REFERENCES

Bessou, P., and Laporte, Y., 1963, in: "Symposium on Muscle Receptors,"
    Hong Kong University Press, pp. 106-119.
Brillinger, D. R., 1975a, Time Series, Data Analysis and Theory, Holt,
    Rinehart & Winston, New York, pp. 232-236.
Brillinger, D. R., 1975b, "The Identification of Point Process Systems,"
    Ann. Probab., 3:909-929.
Edgley, S. A., and Jankowska, E., 1987, "An interneuronal relay for group
    I and II muscle afferents in the midlumbar segments of the cat spinal
    cord," J. Physiol., 389:647-674.

ACKNOWLEDGEMENTS

This work was supported in part by grants from the M.R.C. and Wellcome Trust.

NEURAL CONTROL OF ACh SENSITIVITY OF MUSCLE SPINDLES IN CATS

R.S. Arutyunian

Sechenov Institute of Evolutionary Physiology
and Biochemistry
Leningrad, USSR

In our study a comparison was made between changes of ACh
sensitivity of muscle spindles after chronic motor denervation
of the muscle and treatment of the nerve by colchicine, a substance which disrupts the fast axonal transport (Bisby, 1976),
leaving the slow axonal transport practically unaffected
(Karlsson and Sjöstrand, 1969). The purpose of this comparison
was to clarify the possible role of axonal transport in the
regulation of ACh sensitivity of intrafusal muscle fibers.

Experiments were carried out on the extensor digitorum
longus (EDL) of cats. Standard methods were used for recording of afferent activity of muscle spindles in dorsal root filaments. Denervation of the EDL was performed by extradural
transection of ventral roots from L_5 to S_2 segments. Disruption of axonal transport was achieved by treatment of the common peroneal nerve with 20 mM colchicine for 30 min in the
lower third of the thigh. ACh was injected intraarterially
through a polyethylene catheter which was inserted into the
contralateral femoral artery so that its tip reached the bifurcation of the abdominal aorta.

ACh sensitivity of intrafusal fibers was originally shown
by Hunt (1952). After chronic motor denervation of the muscle
an increase of ACh sensitivity of intrafusal muscle fibers
occurs (Arutyunian, 1969). ACh in doses 10-50 times lower than
in control animals increased the discharge frequency during
contraction of extrafusal fibers. Responses of denervated
muscle spindles to ACh which activate both extrafusal and
intrafusal fibers resemble those of normal muscle spindles
evoked by simultaneous stimulation of alpha and gamma efferents
which were described by Hunt and Kuffler (1951).

In some experiments on normal muscle spindles we observed
a population of sensory endings which were insensitive to ACh.
After motor denervation of the muscle we also observed some
secondary endings which were insensitive to ACh and responded
by complete cessation of their activity during the whole period
of muscle contraction. We suggest that sensory endings which
responded after injection of ACh by acceleration of discharge

frequency originate from nuclear bag fibers, whereas secondary endings which were insensitive to ACh originate from nuclear chain fibers. This assumption is based on Gladden's (1976) observation on the sensitivity of nuclear bag fibers to ACh and insensitivity of nuclear chain fibers to this substance. These differences in susceptibility of nuclear bag and chain fibers to cholinergic substances allow the identification of afferents originating from nuclear bag and nuclear chain fibers.

Experiments after treatment of common peroneal nerve with colchicine were carried out after 6-8 and 20-22 days. Application of colchicine produces no changes in the ability of the nerve to control the muscle activity assessed by the magnitude of muscle contraction (Albuquerque et al., 1972). Six to eight days after application of colchicine to the nerve, ACh sensitivity of intrafusal fibers increased. Injection of 5 µg/kg of ACh induced powerful acceleration of the afferent activity from muscle spindles. In control animals such low doses of ACh have no effect on the discharge frequency of muscle spindles. Only in 3 out of 24 isolated primary endings the acceleration of discharge frequency was accompanied by contraction of extrafusal fibers. This different sensitivity of intra- and extrafusal fibers to ACh after treatment of the nerve by colchicine is due to the different sensitivity of alpha and gamma efferents to colchicine. Previously Matthews and Rushworth (1957) found different sensitivity of alpha and gamma efferents to procaine which greatly reduced the size of the stretch reflex indicating thus a selective paralyzing effect of procaine on gamma efferents. All our attempts to reinforce the increased sensitivity of intrafusal fibers to ACh by treatment of the nerve with higher concentrations of colchicine (up to 30 mM) resulted in a blockade of the conduction of impulses in afferent fibers. Conduction of impulses in efferent fibers still remained. Thus the increase of ACh sensitivity of intrafusal fibers is observed after disruption of axonal transport.

After colchicine treatment of the nerve we also observed secondary endings which were insensitive to ACh. Injection of a threshold dose of ACh (50 µg/kg) have no effect on the discharge frequency in this type of secondary endings.

Our results show that the distribution of ACh sensitivity in nuclear bag and nuclear chain fibers differs significantly. This difference which could be assessed by ACh-induced changes of afferent activity from nuclear bag and nuclear chain fibers remains both after motor denervation and treatment of the nerve with colchicine. It is concluded that the effects produced by ACh in afferents from nuclear bag fibers are due to the contraction of intrafusal nuclear bag fibers. Responses of afferents from nuclear chain fibers to ACh do not involve contraction of intrafusal fibers and are due to the direct action of ACh on sensory ending originating from nuclear chain fibers.

The mechanism of increased sensitivity of denervated intrafusal fibers is presumably the same as in extrafusal fibers.

Colchicine is known to disrupt microtubules and to interfere with axonal transport of nerve fibers (Bisby, 1976; Edstrom and Mattson, 1972). It therefore seems reasonable to

assume that ACh sensitivity of intrafusal fibers is controlled
by factors which are carried by axonal transport. After treatment with colchicine, the release of trophic materials from
the nerve is suppressed and an increase of ACh sensitivity of
intrafusal fibers is observed.

REFERENCES

Albuquerque, E.X., Warnick, J.R., Tasse, J.R., and Sansone,
      C.M., 1972, Effects of vinblastine and colchicine on
      neural regulation of the fast and slow skeletal muscle
      of the rat, Exp. Neurol., 37:607-634.
Arutyunian, R.S., 1969, Acetylcholine effects on normal and
      denervated muscle receptors, Bull. exp. Biol. and Med.
      (USSR), 68:14-17.
Bisby, M.A., 1976, Axonal transport, General Pharmacol., 7:
      387-393.
Edström, A., and Mattson, H., 1972, Fast axonal transport in
      vitro in the sciatic system of the frog, J. Neurochem.,
      19:205-221.
Gladden, M., 1976, Structural features relative to intrafusal
      muscle fibres, Progr. in Brain Res., 44:51-59.
Hunt, C., 1952, Drug effect on mammalian muscle spindles, Fed.
      Proc., 11:75.
Hunt, C., and Kuffler, S., 1951, Stretch receptors discharge
      during muscle contraction, J. Physiol., 113:298-319.
Karlsson, J., and Sjöstrand, J., 1969, The effects of colchi-
      cine on the axonal transport of protein in the optic
      nerve and tract of the rabbit, Brain Res., 13:617-619.
Matthews, P.B.C., and Rushworth, G., 1957, The relative sensi-
      tivity of muscle nerve fibres to procaine, J. Physiol.,
      135:263-269.

FLUORESCENT LABELLING OF NERVE TERMINALS IN THE LIVING ISOLATED MAMMALIAN MUSCLE SPINDLE

Y. Fukami, C.C. Hunt and J. Lichtman

Washington University School of Medicine
St. Louis, MO 63110. U.S.A.

The fluorescent dye, 4 (4-diethylamino styryl)-N-methyl pyridinium iodide (4-Di-2-ASP), stains living neuromuscular terminals without producing evident damage (Magrassi, Purves and Lichtman, J. Neuroscience, 7: 1207, 1987) and has been used to follow the structure of these endings over long periods of time (Lichtman, Magrassi and Purves, J. Neuroscience, 7: 1215, 1987).

We have exposed the isolated cat muscle spindle, decapsulated in the equatorial region, to 4-Di-2-ASP. The sensory endings, both primary and secondary are intensely stained. The endings thus visualized appear very similar to those back filled with lucifer yellow through their cut axons. Alternation between fluorescence and differential interference contrast optics permits positive identification of sensory terminals seen in Nomarski images. Motor nerve terminals in the isolated spindle may also be seen to be stained by 4-Di-2-ASP.

ACKNOWLEDGEMENT

Supported by a research grant from the U.S. Public Health Service, NS 07907.

SCANNING ELECTRON MICROSCOPIC IDENTIFICATION OF MOTOR

AND SENSORY ENDINGS ON TEASED INTRAFUSAL MUSCLE FIBERS

J.M. Schröder, H. Bodden, A. Hamacher, and C. Verres

Department of Neuropathology
RWTH, Pauwelsstraße, D-5100 Aachen
Fed. Rep. Germany

Scanning electron microscopic (SEM) studies of intrafusal muscle fibers and their various sensory and motor nerve endings are not available thus far although SEM greatly increases the resolution power for analyzing surface structures in comparison to light microscopic preparations. Progress was inhibited by the difficulty of isolating (teasing) muscle spindles as well as by problems of separating the intrafusal muscle fibers and nerve terminals from the inner and outer spindle capsules.

In the present study the method of DESAKI and UEHARA (1981) (8N HCl) was modified for teasing and clearing intrafusal muscle fibers from surrounding connective tissue elements. From a total of 68 isolated muscle spindles in lumbrical muscles of 4 female Wistar rats a rather limited number of nerve terminals on intrafusal muscle fibers was obtained.

While teasing muscle spindles, the spindle capsules and other structures are transparent so that several components such as blood vessels, bundles of nerve fibers, intrafusal muscle fibers, and nuclei can be identified.

The intrafusal muscle fibers, if not treated by certain means, are surrounded by a dense network of collagen fibrils up to 3.0 µm in thickness. Nerve fibers and blood vessels are integrated in this network. Following incomplete mechanical removal of the spindle capsule, thin collagen fibrils are seen to insert onto the surface of intrafusal muscle fibers. The latter can be clearly identified by the sarcomeric pattern causing slight indentations of the muscle fiber surface at regular intervals of about 1.5 µm in a contracted state. Collagenase treatment did not sufficiently remove the collagen fibrils. Hydrolysis, however, in 8N HCl resulted in a rather complete dissolution of collagen as well as of the outer and inner spindle capsule at the juxtaequatorial and polar region of the muscle spindles. Thus, secondary sensory and motor terminals became clearly visible. At the equatorial region, however, the outer spindle capsule persisted regularly with the exception of a single specimen where there was mechanical

disruption of the outer spindle capsule. An underlying intrafusal muscle fiber was covered by a typical annulospiral
primary sensory ending. There were 8 closely-spaced spirals
extending for more than 32 µm and tapering towards one well
preserved polar end. At the other end, this terminal was
fractured or covered by the outer spindle capsule that was
hiding any further details.

The predominant feature of the identified secondary
sensory endings was that they were surrounding intrafusal
muscle fibers completely by cylindrical, flat, or more ring-
like, or spiral processes. These terminals were characterized
by a slight elevation (approximately 0.2 µm) above the surface
of the underlying muscle fiber. There were two subsequent ring-
or cuff-shaped terminals on a single, 5.3 µm thick muscle
fiber, i. e. a chain fiber, in the S1 and S2 position of BOYD.
The cuff-like appearance was caused by the close spacing of
adjacent spirals. Some spirals were as thick as 2.5 µm, but
most of them showed diameters of 0.7-1.5 µm (Fig. 1). Ana-
stomotic connections between adjacent spirals or rings, 0.5 µm
in thickness, were occasionally seen. Some spirals were
dividing dichotomically and oriented obliquely. The preterminal
nerve fiber measured 1.3 µm in diameter. It was oriented
towards the spindle pole. The branches of the terminal originated perpendicularly.

Motor nerve terminals were only rarely encountered along
the 68 muscle spindles isolated for SEM investigation. The main
reason for this was the difficulty of freeing the intrafusal
muscle fibers from its connective tissue and perineurium-like
capsule without destroying the nerve terminals and the muscle
fibers themselves. In addition, the juxtaequatorial and polar
portions of a muscle spindle, especially if freed from stabi-
lizing connective tissue elements, were extremely fragile and
brittle. These parts could get lost during the strenuous procedure of isolating, handling, hydrolysing, dehydrating, and
sputtering spindles. Because of the smallness and fragility of
the specimens they could not be oriented intentatively towards
optimal sites or angles for SEM.

In the juxtaequatorial region, a multiterminal nerve
ending was observed, 60 µm in length. It was only partially
surrounding a muscle fiber, that measured 10 µm in diameter,
i.e. a bag fiber. It originated from a single nerve fiber that
was approaching the muscle fiber from the equator. The nerve
fiber was approximately 0.7 µm in thickness, but focally
dilated to about 1.5 µm by what was thought to represent a
Schwann cell nucleus. It terminated with multiple branches
superimposed on the muscle fiber, clearly lying above the level
the latter.

The branches were showing interconnections at various
angles. Although this multiterminal nerve ending was obviously
lying above the surface of the muscle fiber it appeared to be
covered by a basement membrane that extended to the surface of
the muscle fiber.

In the polar region two types of plate endings could be
distinguished. Small end plates were noted at the more proximal
and at the distal segment of the spindle pole. They were

238

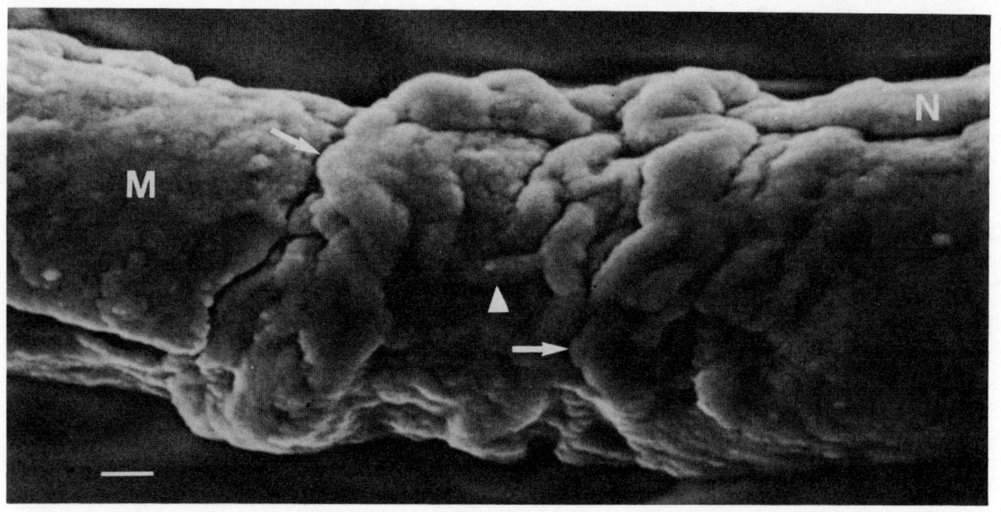

Fig. 1 Scanning electron micrograph of a secondary sensory
ending in the juxtaequatorial region (approxiamtely 150 µm from
the equator) superimposed on a nuclear chain fiber (M) that
measures 5.3 µm in diameter. The afferent nerve fiber (N) shows
several branches (arrows) which surround the underlying intrafusal muscle fiber rather completely in a meandering pattern.
Some of the branches appear to be twisted along their longitudinal axis. The branches measure 0.7 - 1.5 µm in thickness.
Anastomoses between branches occur (arrowhead). No Schwann
cells are located on the surface of this terminal.
Bar = 1 µm

indenting the surface of the muscle fiber lying approximately
within the level of the muscle fiber surface. The preterminal
axon with a diameter of 0.5-0.7 µm was seen running parallel to
the muscle fiber and indenting its surface. They were increasing in size at the terminal to a button- or plate-like
ending, 3-7 µm in length and 2-3 µm in width. The muscle fibers
underlying this type of ending measured 4-5 µm in diameter,
i.e. they were nuclear chain fibers.

At the midpolar level a larger, more irregular, complex
type of endings was noted showing small processes or projections in various directions. The processes were
of uneven thickness, branching, narrowing, and extending at
quite a number of points. The processes varied in thickness
between 0.5 and 2.0 µm. Focal distensions, possibly caused by
superimposed Schwann cell processes, 3-4 µm in thickness, were
apparent. The whole area measured approximately 20 µm in length
and 15 µm in width. The preterminal nerve fiber of this ending
was 3 µm in diameter approaching the muscle fiber from the
lateral side.

In the juxtaequatorial region there were also some thin
processes, presumably unmyelinated nerve fibers, 0.75-1.1 µm in
thickness, running parallel to and on the surface of the intrafusal muscle fiber although clearly separated from the latter.

One of the nerve fibers was branching coming from the polar
region. Several thin and short lateral branches (0.25 μm in
thickness) of this nerve fiber (0.8 μm in diameter) were seen
approaching the underlying muscle fiber more closely. Their
surface appeared to be relatively smooth. They could end
abruptly on the surface of the muscle fiber or, presumably also
due to artificial separation, free in the extracellular space.
Similar small nerve fibers were also seen in the polar region.

REFERENCES

Barker, D., Banks, RW., 1985, The muscle spindle. In: Banker, BQ.
    Engel, AG. (eds.) Textbook of myologie, Vol. I, Chapter
    10, McGraw-Hill, New York, pp.309-341.

Barker, D., Stacey, MJ. Adal, MN., 1970, Fusimotor innervation
    in the cat, Phil Trans B 258:315-346.

Boyd, IA., Gladden, MH., 1985, Review. In: Boyd, IA., Gladden,
    MH. (eds.) The Muscle Spindle. Macmillan: Basingtoke and
    London, pp.3-22.

Desaki, J., Uehara, Y., 1981, The overall morphology of neuro-
    muscular junctions as revealed by scanning electron micro-
    scopy, J. Neurocytol 10:101-110.

Kucera, J., Walro, JM., 1985, Structure and classification of
    cat intrafusal motor endings. In: Boyd, IA., Gladden, MH.
    (eds.) The Muscle Spindle. Macmillan: Basingstoke and
    London, pp.63-68.

SENSORY 'CROSS-TERMINALS' BETWEEN DYMAMIC AND

STATIC INTRAFUSAL FIBERS IN RAT MUSCLE SPINDLES

Jon M. Walro and Jan Kucera

Department of Anatomy, Northeastern Ohio Universities
College of Medicine, Rootstown, OH 44272 USA
Department of Neurology, Boston University, Boston, MA 02118

INTRODUCTION

Separate motor pathways for controlling dynamic (velocity of muscle stretch) and static (length of muscle) sensitivity are present in muscle spindles of the cat (Boyd et al., 1977). The pattern of afferent innervation parallels this separation of motor pathways in cat spindles. Each muscle spindle in the cat receives only one primary afferent which terminates on all three major types of intrafusal fiber (Banks et al., 1982). However, the input of the dynamic bag fiber is separated from that of the static bag and chain fibers by pacemakers located at terminal heminodes of branches of the primary afferent which terminate on the bag_1 fiber and at corresponding sites of branches which terminate on the bag_2 and chain fibers (Quick et al., 1980; Banks et al., 1982). Presumably the two different responses observed from primary spindle afferents result from differential excitation of pacemakers related to differences in the viscoelastic properties of intrafusal fibers.

In contrast, pathways for regulating the dynamic and static sensitivity overlap to a greater degree in muscle spindles of the rat than those of the cat. Approximately 20% of motor axons to muscle spindles coinnervate the dynamic bag fiber and static bag and/or chain fibers in spindles of the soleus, extensor digitorum longus (EDL) and hindlimb lumbrical muscles of the rat (Walro and Kucera, 1985; Walro, unpublished data). Whether the distribution of sensory terminals to dynamic and static fibers in rat muscle spindles is highly segregated analogous to the pattern present in cat spindles or whether the distribution of sensory terminals to dynamic and static fibers overlaps to some degree as does the distribution of motor axons to these types of intrafusal fibers in rat spindles is the subject of this paper.

MATERIALS AND METHODS

Four young adult female Sprague-Dawley rats (170-200g body wt.) were anesthetized with sodium pentobarbital (50 mg/kg i.p.). Two of the rats were perfused via the abdominal aorta with a solution of 2.5% glutaraldehyde-1% paraformaldehyde buffered with cacodylate at a pH of 7.2. The extensor digitorum longus muscles from both hindlimbs were

excised, cut into 10 x 1 x 1 mm strips and fixed for an additional
hour. The hindlimb lumbrical muscles were removed from two
anesthetized rats without vascular perfusion. The lumbrical muscles
together with the flexor digitorum longus (FDL) muscle and tendon were
excised and fixed in 2.5% paraformaldehyde - 1% glutaraldehyde at
resting length in a petri dish for 20 min. Entire lumbrical muscles
were freed from the tendon of the FDL and were fixed for an additional
2 h. Tissues were postfixed in OsO_4, dehydrated in an ethanol series
and embedded in epoxy.

Skip-serial sectioning at a thickness of 1-μm began in polar
regions where the three types of intrafusal fibers could easily be
identified according to established criteria (Walro and Kucera, 1985)
in sections stained with 1% toluidine blue. The sensory region of each
spindle determined by the presence of clear profiles of sensory
terminals laden with blue granules was sectioned serially into 1-μm
sections through its length. Sectioning at 1-μm stopped when a cross-
terminal, or sensory terminal shared by more than one intrafusal fiber,
was encountered and a set of ultrathin (40nm) sections was cut and
stained with uranyl acetate and lead citrate for observation under a
JEOLCO 100 electron microscope. Sectioning at 1-μm resumed until
either another cross-terminal was encountered whereby another set of
thin section was cut, or until the sensory region terminated. This
regimen permitted identification of intrafusal fibers, and
determination of the branching pattern of afferents and the
distribution of sensory terminals to different types of intrafusal
fibers.

Length of the sensory region, extent of distribution of sensory
terminals on individual intrafusal fibers, length of cross-terminals
and location of cross-terminals were calculated by summation of 1-μm
sections of spindles in both the EDL and lumbrical muscles. Lengths
were not corrected for any changes in volume incurred during fixation.
Group means for different parameters were compared by analysis of
variance (anova) with a significance level of $p = 0.05$. Differences
between individual groups indicated by anova were elucidated by a
Student-Newman-Keuls test. Data are reported as means ± standard
error.

RESULTS

Five spindles in the EDL and lumbrical muscles from three young
adult female Sprague-Dawley rats were studied. Each spindle studied in
the EDL and lumbrical muscles contained one bag_1 and one bag_2 fiber.
All five spindles in the EDL contained two chain fibers, whereas four
of five spindles in the lumbrical muscles contained only a single
chain fiber.

<u>Branching of afferents</u>

A primary afferent terminated at the equator of each spindle. A
single secondary afferent terminated in the juxtaequatorial region of
four poles (equator to tip of longest intrafusal fiber) of spindle in
the EDL muscle and two poles in the lumbrical muscle.

The initial bifurcation of both primary and secondary afferents
occurred just prior to or shortly after entry into the periaxial space.
The most frequently observed pattern of branching of the primary
afferent in both muscles was trifurcation into branches which

242

terminated predominantly on bag_2, bag_1, or chain fibers. However a
branch which terminated predominantly on bag_2 fibers occasionally
bifurcated into lower order branches which supplied one or both chain
fibers of the intrafusal bundle, and vice versa.

Four of six secondary afferents in the two muscles entered the
periaxial space together with the primary afferent near the equator of
spindles, whereas two secondary afferents penetrated the capsules
together with motor axons in more polar regions of two EDL spindles.
Secondary afferents in both muscles predominantly supplied bag_2 and
chain fibers, though not necessarily through separate branches.

Domains of afferents

The domains of primary afferents which supplied intrafusal fibers
in the EDL muscles (243 ± 20 μm) were broader than those of the
lumbrical muscles (164 ± 9 μm). The breadth of distribution of
terminals along bag_1, bag_2, and chain fibers were 185 ± 12, 164 ± 9 and
243 ± 20 for spindles in the EDL, and 154 ± 11, 140 ± 16 and 120 ± 15 μm
for spindles in the lumbrical muscles.

Secondary afferents in both muscles terminated predominantly upon
static bag_2 and chain intrafusal fibers. Breadths of domains for
secondary afferents in the EDL muscles were 3 ± 5, 29 ± 38, and $133 \pm$
59 for bag_1, bag_2, and chain fibers, and 13 ± 13, 90 ± 16, and 96 ± 32 μm
for domains on the corresponding fibers in the lumbrical muscles.

Cross-terminals

Cross-terminals, or sensory terminals shared by more than one
intrafusal fiber, were present in all five spindles in the EDL and four
of five spindles in the lumbrical muscles. Chain-chain and bag_1 -
bag_2 were the combinations of intrafusal fibers most frequently
observed to share sensory terminals in the EDL muscle (Table 1). In
contrast, most cross-terminals in lumbrical spindles involved the bag_1
and bag_2 fibers (Table 1). The majority of spindles sampled in the
lumbrical muscles had only a single chain fiber, hence chain-chain
cross-terminals were not possible in these spindles.

Cross-terminals of the primary afferent were located in the
juxtaequatorial region of spindles in both muscles. However, cross-
terminals of primary afferents in the EDL muscle were longer (7.6 ± 0.8
μm and located farther from the equator (78 ± 7 μm) than their
counterparts (4.7 ± 1.1 μm; 53 ± 8 μm) in the lumbrical muscles.
Whereas these features of cross-terminals differed between spindles in
these two muscles, length (range = $2.0 - 7.0$ μm) and location relative

Table 1. Incidence of cross-terminals of primary
and secondary (parentheses) afferents in the EDL
and lumbrical muscles of the rat. b_1 - bag_1,
b_2 - bag_2, c - chain fiber.

	EDL			Lumbrical		
	b_1	b_2	c	b_1	b_2	c
b_1	—	—	—	—	—	—
b_2	6(0)	—	—	15(0)	—	—
c	0(0)	1(5)	11(5)	2(0)	0(0)	2(0)

to the equator (range = 54 - 82 μm) were equivalent (p>0.05) for
primary afferent cross-terminals shared between different combinations
of intrafusal fibers.

Bag$_2$-chain or chain-chain were combinations of intrafusal fibers
observed to share terminals of secondary afferents in the EDL muscles.
Bag$_2$-chain cross terminals (136 ± 11 μm) were located at equivalent
distance from the equator as chain-chain cross-terminals (126 ± 8 μm),
although cross-terminals shared between chain fibers (6.8 ± 2.2 μm)
were longer than those shared by bag$_2$ and chain fibers (3.2 ± 1.0 μm)
in the EDL muscle. The two secondary afferents in the lumbrical
muscles had no cross-terminals.

Intrafusal fibers were closely apposed to each other when they
shared terminals. Processes of inner capsule cells which surrounded
each intrafusal fiber at the equator of spindles enclosed the bag
fibers in one compartment and the chain fibers in one or two other

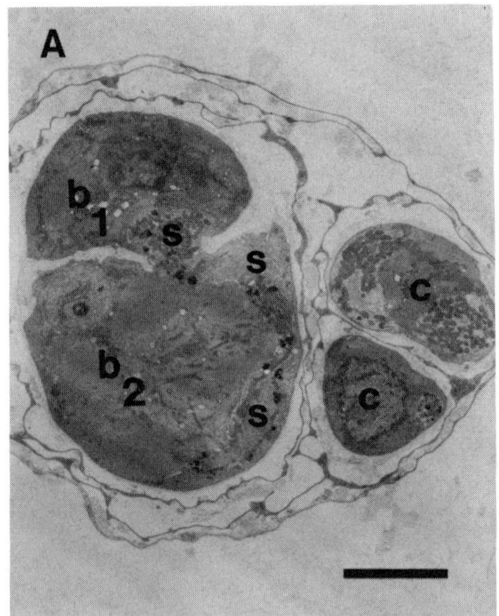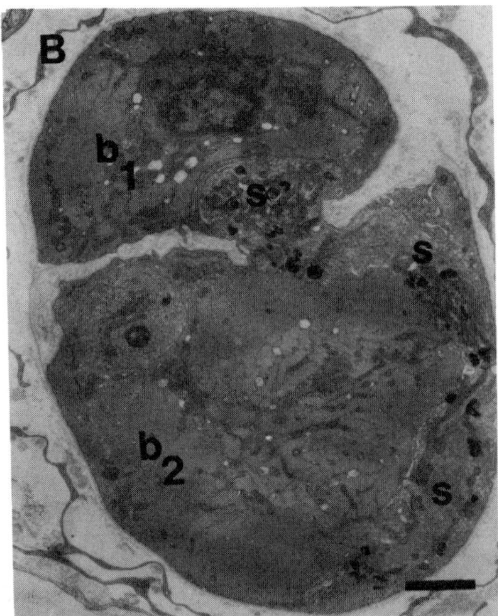

Fig. 1. A. Sensory region of a spindle in the
extensor digitorum longus muscle of a rat.
Note that the bag$_1$ (b$_1$) and bag$_2$ (b$_2$) fibers
occupy a compartment formed by inner capsule
cells which is separate from those occupied by
chain (c) fibers. A sensory cross-terminal (s)
shared by the bag$_1$ and bag$_2$ fibers is visible.
Scale bar = 5 μm. (reprinted with permission
of Elsevier Science Publishers)

B. Enlargement of Fig. 1. A. Note that the
sensory terminal streams from one bag fiber
to the other in the intrafusal bundle. A
common basal lamina encases both bag fibers
and the sensory cross-terminal in this region.
Scale bar = 2 μm.

compartments in the juxtaequatorial regions of spindles (Fig. 1A).
Sharing of cross-terminals occurred when both fibers occupied the same
compartment and were not separated by processes of inner capsule cells
at their apposed surfaces. A single external lamina enclosed both
intrafusal fibers at the sites of shared sensory terminals. The
contours of the intrafusal fibers flattened and became irregular on the
apposed surfaces, but remained rounded on their free surfaces at sites
of cross-terminals. Most of the sensory cross-terminals spanned the
gap of several microns between the apposed surfaces of the intrafusal
fiber (Fig. 1B).

DISCUSSION

The presence of sensory terminals shared by the dynamic bag_1 fiber
and static bag_2 in the EDL and lumbrical muscles analyzed in
conjunction with the incidental observation of similar cross-terminals
in the soleus muscle of rats (Diwan and Milburn, 1986) indicate that
bag_1 - bag_2 cross terminals are neither anomalous nor restricted to
fast-twitch muscles.

Features of cross-terminals in the EDL did not differ greatly from
those in the lumbrical muscles. Cross-terminals in spindles of the EDL
were located farther from the equator than those of spindles in the
lumbrical muscles. However, domains of afferents on intrafusal fibers
were also broader in the EDL than in the lumbrical spindles, hence the
difference in the mean location of cross-terminals relative to the
equator may reflect a difference in breadth of domains rather than a
difference in the location of terminals. Another difference, fewer
chain-chain cross-terminals in the lumbricals than in the EDL muscles,
is a reflection of the complement of a single chain fiber in many
spindles in the lumbrical muscle. Chain-chain cross-terminals were
present in the lumbrical spindle which contained two chain fibers.

Cross-terminals between immature intrafusal fibers are numerous in
developing spindles of both cats (Milburn, 1984) and rats (Kucera and
Walro, unpublished data), yet the incidence of cross-terminals,
particularly those between the dynamic bag_1 and static bag_2 fibers is
much greater in muscle spindles of the adult rat than in muscle
spindles of adult cats. These data suggest that many of the cross-
terminals in developing spindles are transient structures, and that
processes by which the sensory innervation to intrafusal fibers is
established differ in these two species of mammal.

The mechanism by which muscle stretch is transduced into afferent
discharge is not known. Presumably, stretch of intrafusal fibers
mechanically deforms the sensory terminals, thereby altering the ionic
permeability of the sensory terminal which in turn depolarizes the
terminal. This depolarization is hypothesized to spread
electrotonically to electrically excitable pacemakers located in the
preterminal branches of the primary afferent. Differences in the
dynamic and static responses of spindle afferents presumably result
because of differential excitation of pacemakers in branches of the
primary afferents which terminate on dynamic and static intrafusal
fibers and differences in the viscoelastic properties of the different
types of intrafusal fibers (Hulliger and Noth, 1979; Arbuthnott et al.,
1982; Banks et al., 1982). Hence activation of the dynamic bag_1 fiber,
the static bag_2 fiber, or both fibers in concert should be capable of
exciting some pacemakers in primary afferents in spindles of the rat,
but not in those of the cat.

Separate branches of the primary afferent to dynamic and static
intrafusal fibers (Banks et al., 1982; Banks, 1986), absence of cross-
terminals of the primary afferent between dynamic and static intrafusal
fibers (Banks et al., 1982), together with the paucity of motor axons
which coinnervate dynamic and static intrafusal fibers (Boyd et al.,
1977; Arbuthnott et al., 1982) support the presence of non-overlapping
systems for monitoring the dynamic and static components of stretch in
muscle spindles of the cat. This degree of separation of dynamic and
static systems is clearly not present in spindles of rats. Sharing of
sensory terminals by the dynamic bag_1 and the static bag_2 fibers,
and motor axons which coinnervate dynamic and static intrafusal fibers
are common in spindles of the rat. Motor axons which coinnervate the
bag_1 fiber together with the bag_2 and/or chain fibers are present in
spindles of monkeys (Kucera, 1985) and humans (Kucera, 1986). However,
cross-terminals of primary afferents shared by dynamic and static
fibers are absent in the extensor indicis muscle of humans (Kennedy et
al., 1975). These data suggest that the neural organization of the
dynamic and static systems of muscle spindles differs among different
taxonomic groups of mammals, ranging from complete separation of
sensory and motor innervation to dynamic and static intrafusal fibers
to a moderate degree of overlap of systems modulating the dynamic and
static components of muscle stretch.

REFERENCES

Arbuthnott, E.R., Ballard, K.J., Boyd, I.A., Gladden, M.H., and
    Sutherland, F.I., 1982, The ultrastructure of cat fusimotor
    endings and their relationship to foci of sarcomere convergence
    in intrafusal fibers, J. Physiol. (Lond.), 33:285-309.
Banks, R.W., 1986, Observations on the primary sensory ending of
    tenuissimus muscle spindles in the cat, Cell Tissue Res.,
    246:309-313.
Banks, R.W., Barker, D., and Stacey, M.J., 1982, Form and distribution
    of sensory terminals in cat hindlimb muscle spindles, Phil. Trans.
    R. Soc. Lond. B., 299:329-364.
Boyd, I.A., Gladden, M.H., McWilliam, P.N., and Ward, J., 1977, Control
    of dynamic and static nuclear bag fibres and nuclear chain
    fibres by gamma and beta axons in isolated cat muscle spindles,
    J. Physiol. (Lond.), 265:133-162.
Diwan, F.H., and Milburn, A., 1986, The effects of temporary ischemia
    on rat muscle spindles, J. Embryol. exp. Morphol., 92:223-254.
Hulliger, M., and Noth, J., 1979, Static and dynamic fusimotor inter-
    action and the possibility of multiple pacemakers operating in
    cat muscle spindles, Brain Res., 173:21-28.
Kennedy, W.R., Webster, H. deF., and Yoon, K.S., 1975, Human muscle
    spindles: fine structure of the primary sensory ending, J.
    Neurocytol., 4:675-695.
Kucera, J., 1985, Characteristics of the motor innervation of muscle
    spindles in the monkey, Am. J. Anat., 173:113-125.
Kucera, J., 1986, Reconstruction of the nerve supply to a human muscle
    spindle, Neurosci. Lett., 63:180-184.
Milburn, A., 1984, Stages in the development of cat muscle spindles,
    J. Embryol. exp. Morphol., 82:177-216.
Quick, D.C., Kennedy, W.R., and Poppele, R.E., 1980, Anatomical evidence
    for multiple sources of action potentials in the afferent fibers
    of muscle spindles, Neuroscience, 5:109-115.
Walro, J.M., and Kucera, J., 1985, Motor innervation of intrafusal
    fibers in rat muscle spindles: incomplete separation of dynamic
    and static systems, Am. J. Anat., 58:55-68.

ULTRASTRUCTURE OF ATTACHMENTS OF HUMAN INTRAFUSAL FIBERS

Vinod Sahgal, Venka Subramani and Sudarshan Sahgal[X]

Neuromuscular Studies Unit/Rehabilitation Institute
of Chicago and Westside VA Hospital[X]
Chicago, Illinois, USA

INTRODUCTION

Muscle spindles are mechanoreceptors with dynamic and static function. The dynamic function is ascribed to the Bag 1 and the static to the Bag 2 and chain fibers (Boyd and Gladden, 1985). The studies on the measurement of strain developed by the intrafusal fibers under static conditions showed that the values varied according to the region of the spindle and type of the fiber (Poppele et al., 1979).

These observations underscore the functional importance of the attachment of the intrafusal fibers. Studies to date have shown definite species variations in the attachment pattern and sites of the intrafusal fibers. In some species there are only extracapsular while in the others both extra and intracapsular attachments were seen (Bridgeman et al., 1970). In this paper we describe the site and ultrastructure of attachment of human intrafusal fibers.

MATERIALS AND METHODS

Over the last ten years muscle spindles from the biceps (6), quadriceps (12), lumbricals (10), and paraspinal muscles (12) were studied. In this group there were five whole spindles, eighteen half spindles and seventeen incomplete poles. The spindles studied consisted of 1-3 bag fibers per pole with a mean of 2.4 $\pm$ 0.5, and 3-10 chain fibers per pole with a mean of 6.3 $\pm$ 2.4. A total number of 92 bag (48 Bag 1 and 44 Bag 2) and 224 chain fibers were studied.

The muscle tissue was prepared for electron microscopy in the normal manner (Subramani et al., 1986). Thick and thin sections were cut with a LKB III ultramicrotome using glass and diamond knives. The thick sections were stained with Toluidine Blue (1%). Serial sections in sets of ten were picked up on glass slides. Bag 1, Bag 2 and chain fibers were identified according to the established criteria (Banks et al., 1977). Each spindle was divided into the three conventional regions

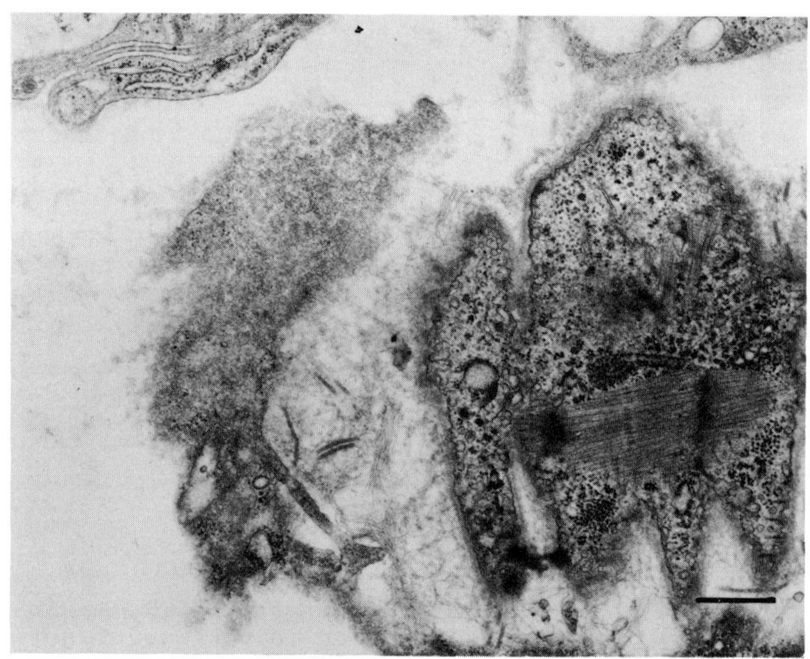

Fig. 1. Termination of the chain fiber. Note the M line, thickening of the Z band. Villous folds of the plasma membrane studded with vesicles. Basement membrane is
thickened. Attachment of the elastic fibers is visible.
Bar: 0.5 /um.

A, B and C. All intrafusal fibers were traced from the C to A
region or until they terminated. Near the termination and/or
approximation of two fibers serial thin sections were cut at
10 /um intervals.

RESULTS

In all the spindles examined the attachment of the intrafusal fibers could be classified into the following four categories:

(i) Extracapsular endomysial
(ii) Intralamellar
(iii) Periaxial
(iv) Interfiber attachments.

(i) Extracapsular Attachments: The Bag 1, Bag 2 and extracapsular portion of the chain fibers attached to the extracapsular
endomysium (C region). The attachment patterns of these fibers
were similar so they will be described together. In the C region
the intrafusal fibers passed freely through the capsular sleeve
to attach to the extrafusal endomysium. The terminating fibers
tapered for a distance of 50 to 100 /um. At their termination
the basement membrane thickened and the plasma membrane was
thrown into folds. The elastic and collagen fibers attached to
the basement membrane but never penetrated the plasma membrane.

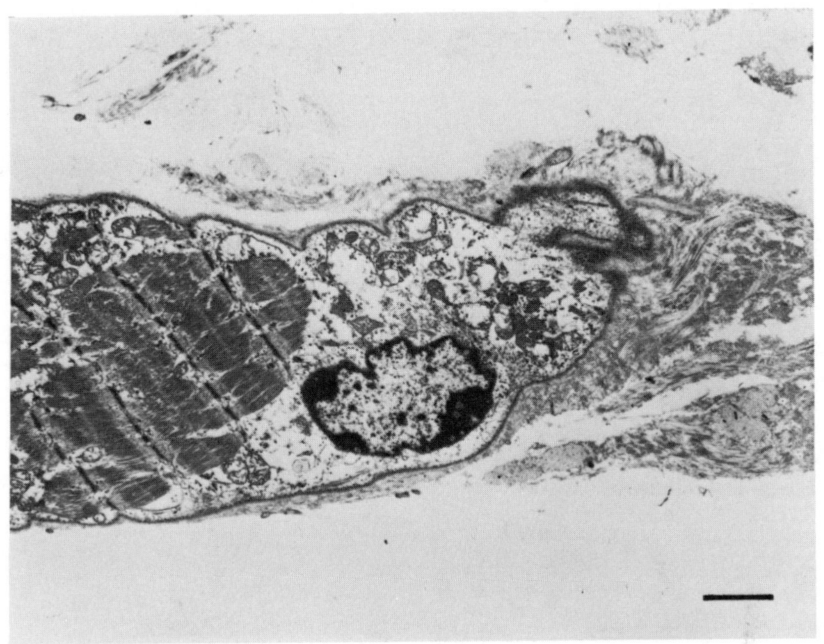

Fig. 2. Termination of the Bag 1 fiber. Note the absence of M
line. Villous folds of plasma membrane. The collagen
fibers are attaching to the basement membrane. Subsarcolemal organelles and actin filaments are visible.
Bar: 2.0 /um.

The plasma membrane was studded with vesicles. The sarcoplasm
at these attachment sites showed disarray of the sarcomere
pattern with numerous rod bodies and actin filaments which
attached directly to the plasma membrane. Direct attachment to
the fibroblastic processes was also seen (Fig. 1, 2).

(ii) Intralamellar Attachment: This type of attachment was seen
only in the B region. At this site only the short chain fibers
were attached. These fibers travelled 10-50 /um distance between
the layers of the outer capsule and attached to the collagen
fibers within the capsular wall. At the sites of binding the
plasma membrane is thrown into villous processes which attach
to the collagen fibers. These areas also show rod bodies and
leptomeres (Fig. 3). A few chain fibers pierced the outer capsule to end in the extracapsular endomysium.

     Attachment or apposition of the bag fiber to the inner
wall of outer capsule was occasionally observed. At this site
the plasma membrane was folded and showed electron dense contact
zone.

(iii) Periaxial Space: A third of the short chain fibers ended
in the periaxial space of the A region. These fibers attached
to the connective tissue within the capsule or the inner capsular process. At these attachment sites the fibers split by
longitudinal invagination of the plasma membrane, the basement

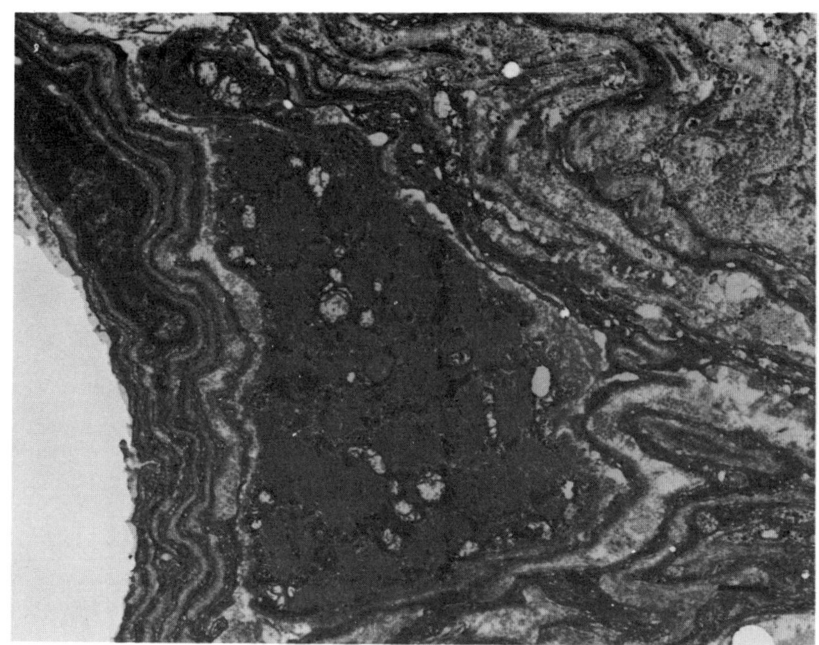

Fig. 3. Intralamellar attachment. Note chain fiber surrounded
by the layers of outer capsule. Sarcomeres are disorganized, rod bodies and streaming of Z band is visible. The
plasma membrane is folded, basement membrane is thick.
Attachment to the capsule and collagen fiber. Bar:1.0 µm.

membrane thickened, the sarcomere pattern was disorganized, and
numerous vesicle and rod bodies were seen. Within these invaginations, attachment of the collagen fibers was seen. Most of
these terminations were juxtaposed to the bag fibers. Sometimes
direct contact with the bag fiber was seen (Fig. 4).

(iv) Interfiber Attachment or Apposition: This profile was
almost exclusively seen in the A region and involved the chain
fibers in 90% of the cases and chain to bag fibers in 10% of
the cases. At the sites of attachment the plasma membrane of
the two chain fibers came together and the basement membrane
was reflected from one fiber to the other. Two basic types of
apposition were seen. First was tight junction type. These were
generally seen between the 2 chain fibers. One to six (2.8 $\pm$ 1.6)
such junctions per fiber were seen. These junctions were 100-700
µm long and 20-40 µm wide. At these junctions there was complete obliteration of the intracellular space.

The second was gap junction. This was found between the
two chain in 60% (24) and bag to chain in 20% (8) of the fibers.
There were 1-8 (3.2 $\pm$ 2) junctions between the two chain fibers
and 1-4 (2.2 $\pm$ 1.0) between the chain and bag fibers. At these
sites the basement membrane was reflected on each fiber and sometimes it was thickened and thrown into folds. The plasma membrane at these sites came close together and was separated by
a 1-16 µm wide intercellular space. The two plasma membrane
leaflets were thickened with electron dense zones. In the sub-

250

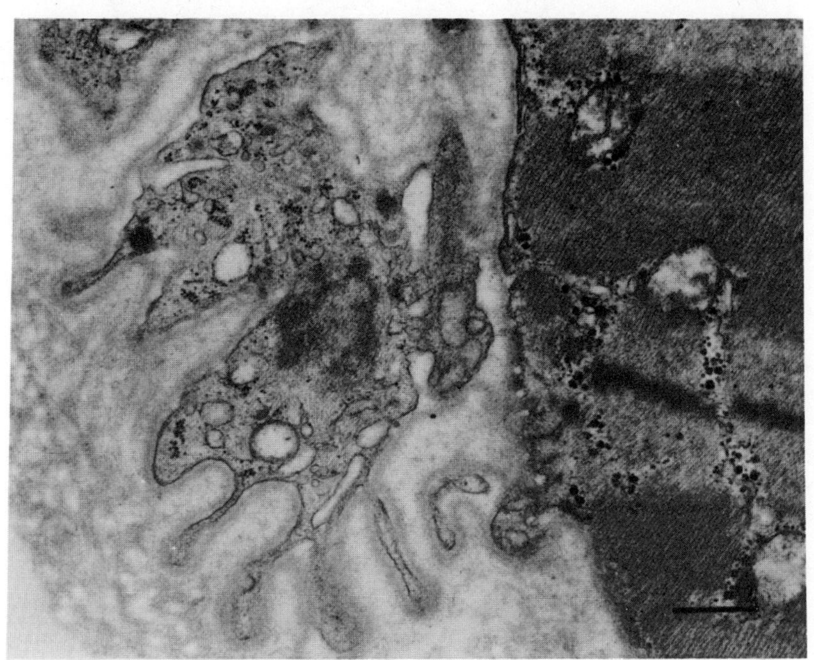

Fig. 4. Note the periaxial termination of chain fiber. The
plasma membrane is thrown into folds. Basement membrane
is thickened. Attachment of collagen fibers is seen.
In the subsarcolemal region vesicles, actin filaments
and Z band material are visible. A few terminal processes share their basement membrane with bag fiber. Note
the leptomeres and vesicles. Bar: 0.5 /um.

sarcolemal region leptomeres, dense core vesicles and actin
filaments which attached to the plasma membrane were seen
(Fig. 5, 6).

DISCUSSION

     As in the cat (Boyd, 1962), Gibbon rat (Bridgeman et al.,
1969) and Chinese hamster (Desaki et al., 1983), the human intrafusal fibers also demonstrated extracapsular, intralamellar
and interfiber attachment. In addition to these, some chain
fibers demonstrated periaxial termination. The extracapsular
and interlamellar attachments with their thickening of the
basement membrane, attachment of collagen fiber, villous projections of plasma membrane, rod bodies, leptomeres, disorganized sarcomeres, and vesicles were reminiscent of developing
myotendinous junction (Mair and Tome, 1972). This structure
would be ideally suited for the mechanical function of generating tension. Thus, these contacts were given the general
classification of mechanical contacts.

     The tight junctions with their lack of intercellular space,
reflected basement membrane, thickened apposed plasma membrane
with the leptomeres and filamentous attachment were interpreted

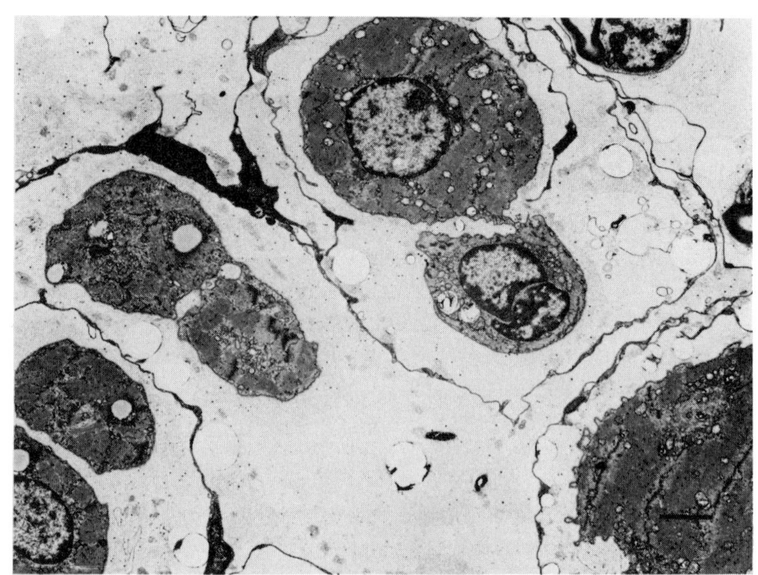

Fig. 5. Juxtaequatorial region. Note two chain fibers attaching
to each other by three villous projections. Two chain
fibers in approximation. Note also attachment of Bag 2
and chain fibers. Note a part of nuclear Bag 1 fiber.
Two myelinated axons are also seen. Bar: 2.0 /um.

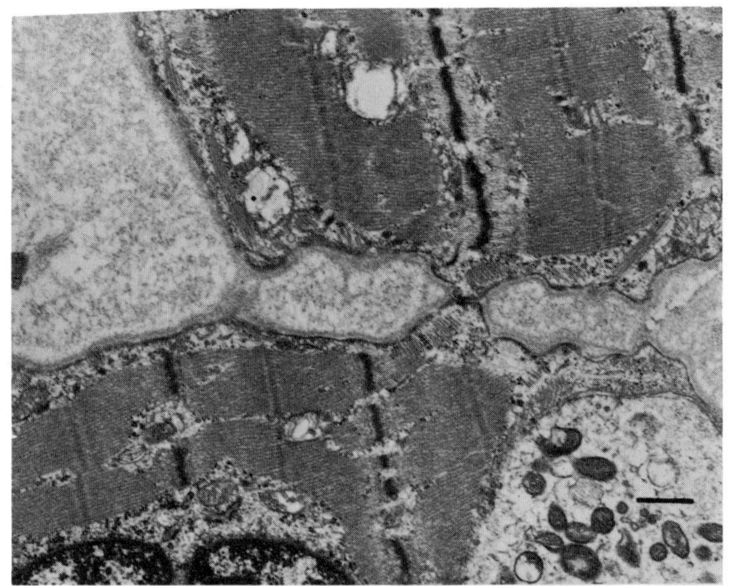

Fig. 6. Two chain fibers seen side by side. Note the sensory
ending. Junctional complex is visible. Note the reflected basement membrane at the plasma membrane process. The junctional site has two leaflets of membrane
separated by intracellular space. Site of adhesion
without intercellular space with electron dense button
like area. Another zone where basement membrane of two
fibers came in contact and are thickened. Bar: 0.5 /um.

as firm contacts which maintain anatomical relationship between
the chain fibers and may also serve as electrical low resistance
contact. The gap junctions with their intercellular space without desmosome-like structure was considered low resistance electrical contact as seen in the central nervous system. Very
occasional fully developed junctional complex consisting of gap
junction and desmosome-like area was seen.

The bag fibers (both Bag 1 and 2) in addition to the extracapsular attachment showed lateral attachment to the periaxial
collagen and elastic fibers indicating lateral anchoring of the
fibers. These findings strongly suggest that the muscle spindle
is capable of actively modifying its profile in relation to the
extrafusal fiber length. The chain fibers may play a major role.
The afferent discharge patterns, thus, need to be interpreted
in the light of these observations.

REFERENCES

Banks, R.W., Harker, D., and Stacey, M.J., 1977, A study of
        mammalian intrafusal muscle fibers using a combined
        histochemical and ultrastructural technique, J.Anat.,
        133:571.
Boyd, I.A., 1962, The structure and innervation of the nuclear
        bag muscle fibre system and the nuclear muscle fibre
        system in mammalian muscle spindles, Philos. Trans. R.
        Soc. London (Biol.), 245:81.
Boyd, I.A., and Gladden, M.H., 1985, Morphology of mammalian
        muscle spindles (A Review), in: "The Muscle Spindle",
        I.A. Boyd and M.H.Gladden, eds.,Stockton Press,New York.
Bridgeman, C.F., Shumpert, E.E., and Eldred, E., 1969, Inser-
        tions of intrafusal fibers in muscle spindles of the
        cat and other mammals, Anat. Rec., 164:391.
Desaki, J., and Uehara, Y., 1983, A fine-structural study of
        the termination of intrafusal muscle fibers in the
        Chinese hamster, Cell Tissue Res., 234:723.
Mair, W.G.P., and Tome, F.M.S., 1972, The ultrastructure of the
        adult and developing human myotendinous junction, Acta
        Neuropathol. (Berlin), 21:239.
Poppele, R.E., Kennedy, W.R., and Quick, D.C., 1979, A determi-
        nation of static mechanical properties of intrafusal
        muscle in isolated cat muscle spindles, Neuroscience,
        4:401.
Subramani, V., Sahgal, V., Sahgal, S., and Shah, A., 1986,
        Morphology and morphometry of motor endings in primate
        intrafusal fibers, Exptl. Brain Res., 64:149.

THE CAPSULAR SLEEVE OF MUSCLE SPINDLES IN MOUSE AND MAN

WITH SPECIAL REFERENCE TO THE CYTOSKELETON

W.K. Ovalle and P.R. Dow

Department of Anatomy, Faculty of Medicine

University of British Columbia, Vancouver, Canada

INTRODUCTION

Muscle spindles are enveloped by a capsular sleeve that is endowed with two cellular portions, an outer and inner capsule. Both portions of the capsule contribute to a permeability barrier which is impervious at the equator yet leaky at its distal ends to vascularly infused tracers (Dow et al., 1980). In addition to its protective role in sequestering the sensory regions of intrafusal fibers, the capsule must be designed to accommodate alterations in shape and repetitive changes in length that normally occur during muscular shortening. Bridgeman and Eldred (1984) compared the capsule to a double-ended squeeze bulb, and postulated a pressure sensitive role in response to subtle fluxes in intramuscular pressure. To test this hypothesis, the mechanical properties inherent in the capsule and its contractile ability need to be clarified.

The distribution of various components of the cytoskeleton has been characterized in many cell types by fluorescence microscopy and ultra-structural methods (Schliwa, 1986). Among the intermediate filaments, those of the vimentin variety comprise the cytoskeleton of many cells (Steinert et al., 1984). Actin filaments are also present in the cyto-plasmic matrix of non-muscle cells where they may determine cell shape, cell motility and cell-to-cell adhesion (Stossel, 1984). The aim of this study is to elucidate the cytoskeletal nature of the capsular cells of muscle spindles in mouse and man by morphologic and cytochemical methods.

MATERIALS AND METHODS

Spindles from mouse soleus and human lumbrical muscles were examined. Conventional methods of tissue processing for transmission electron micro-scopy were undertaken (Ovalle and Dow, 1983). The immunocytochemical lo-calization of vimentin was also determined on cryostat sections of mouse muscle according to a modification of Wolosewick and De Mey's (1982) pro-cedure. Mouse monoclonal antibody to vimentin was obtained from Lab- Sys-tems (Helsinki), and fluorescein-coupled goat anti-rabbit immunoglobulin-G was the secondary antibody. Both antibodies were used at a dilution of 1 μg/ml in 1% phosphate-buffered bovine serum albumin at 37°C. Control sections were incubated in the absence of primary antibody. The fluores-cent probe, NBD-phallacidin, binds specifically to filamentous F-actin in

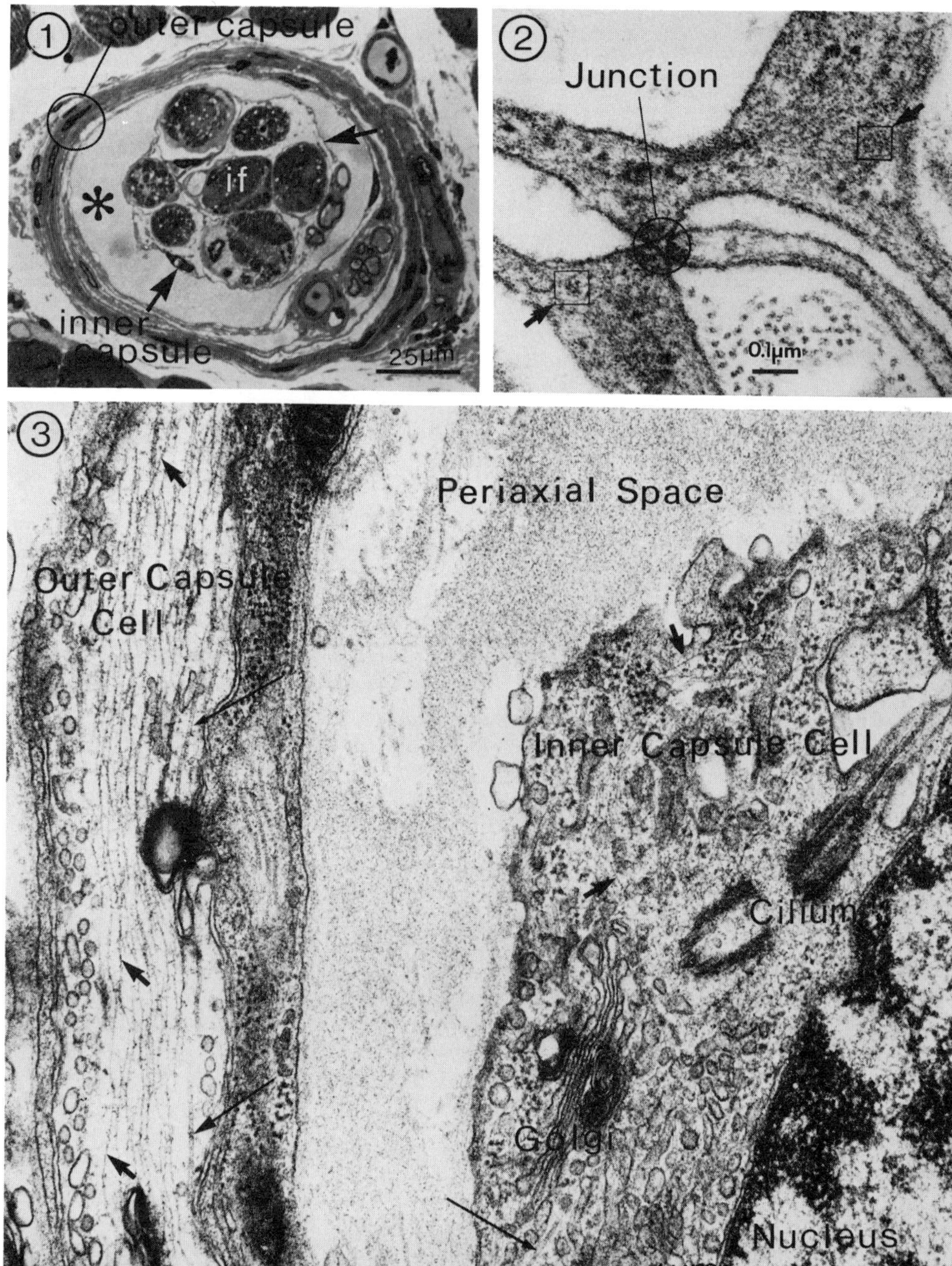

Fig. 1. The location of outer and inner capsular sleeves is evident
in this epon transverse section of a human muscle spindle.
Figs. 2-3. These electron micrographs show salient features of the
cytoplasmic matrix of capsular cells in the mouse, including
many filaments (short arrows) and microtubules (long arrows).

the cytoplasmic matrix (Barak et al., 1983). To demonstrate F-actin in
frozen sections, NBD-phallacidin (Molecular Probes) was applied at a concentration of 0.17 µM in phosphate buffered saline (PBS) to sections of
murine and human muscle fixed in 3.7% paraformaldehyde in PBS. After glycerol mounting, sections were viewed with a light microscope fitted with
epifluorescence or phase contrast optics.

OBSERVATIONS

    Light microscopic images of spindles viewed in transverse (Fig. 1) and
longitudinal (Fig. 5) planes show the arrangement of the two portions of
the mammalian spindle capsule. An equatorial periaxial space intervenes
between the two capsular layers, and the number of concentric perineurial
lamellae of the outer capsule varies, appearing multilayered in the human
(Fig. 1) and more attenuated in the mouse (Fig. 5). In addition, the inner capsular sheath consists of a delicate syncitium of interlinked and
branching cells that more closely resemble endoneurial fibroblasts.

    Electron microscopy revealed an intricate array of cytoskeletal
elements in the capsular cells of mouse (Figs. 2-3) and man (Fig. 4).
Flattened cells of the outer capsule and cell bodies of the inner capsule
contained a lattice-work of fine filaments and microtubules, dispersed
either in parallel array or in a randomly scattered pattern among the
other organelles that are typical of these cells (Fig. 3). Collections of
closely packed filaments were also apparent in the thin extensions of the
murine (Fig. 2) and human (Fig. 4) inner capsule, and measured 6-15 nm in
diameter. Bundles of filaments were either straight or slightly wavy in
the capsular cytoplasm. They were often seen in focal collections subjacent to the plasma membrane where they appeared to be closely associated
with intercellular junctions of the inner capsule (Figs. 2, 4).

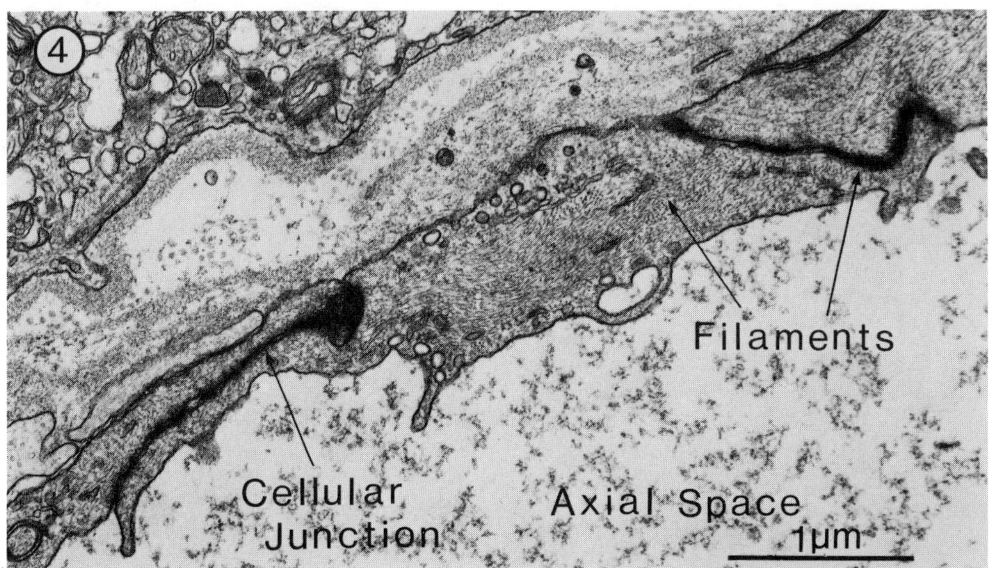

Fig. 4. Electron microscopic view of the equatorial interior of a
         muscle spindle in man. Intercellular junctions and a ple-
         thora of closely aggregated cytoplasmic filaments are promi-
         nent ultrastructural features of the inner capsular sheath.

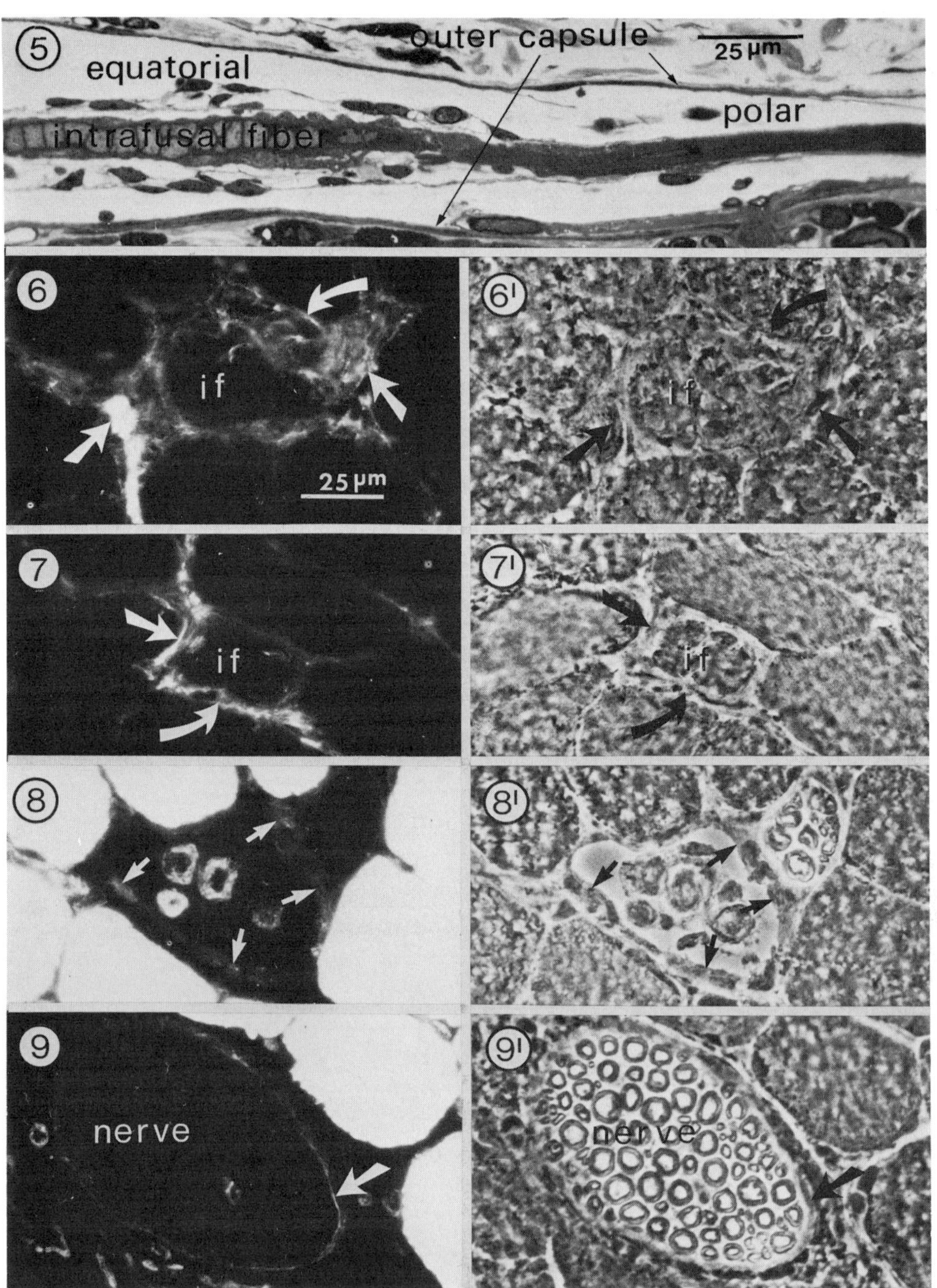

Fig. 5. Epon longitudinal section of a muscle spindle in the mouse.
Figs. 6-9. Series of fluorescence and corresponding phase contrast micrographs of mouse muscle showing the distribution of vimentin (6-7) and actin (8-9) in the spindle capsule and perineurium. Serial sections along the same spindle show co-localization of vimentin and actin in the outer capsule.

When transverse sections of murine muscle were examined with fluorescence microscopy after treatment with antiserum to vimentin, staining of the capsule in equatorial (Fig. 6) and polar (Fig. 7) regions was seen. Fluorescent labelling of the outer capsule showed vimentin-rich and vimentin-sparse regions. Immunostaining of the inner capsular sheath was generally less distinct, but the processes of inner capsule cells could often be discerned within the lumen of the receptor (Fig. 6). Antibodies to vimentin also showed affinity for the surrounding endomysium and interstitial capillaries, whereas no fluorescence was emitted either by the intrafusal fibers of spindles or by neighboring extrafusal fibers within the muscle. Moreover, no specific fluorescence was detected in control slides.

Transverse sections of murine and human muscle treated with NBD-phallacidin showed that the outer capsules of spindles in both species were labelled for actin (Figs. 8-10). The distribution of actin in murine spindle capsules exhibited a delicate and uneven lattice-like pattern with less obvious staining of the inner capsule. Actin labelling was also seen in the perineurial sleeves that surrounded the neighboring nerve fascicles (Fig. 9). In human spindles treated with NBD-phallacidin, staining for actin in the outer capsule showed a lamellated pattern where filament bundles displayed a predominantly concentric course, perpendicular to the main axis of the receptor (Fig. 10). Isolated patches of fluorescence were also seen focally within the periaxial space which presumably corresponded to cellular profiles of the inner capsule. As expected, the intrafusal fibers enclosed within the capsule and the surrounding extrafusal fibers of the muscle were both strongly labelled for F-actin (Fig. 10).

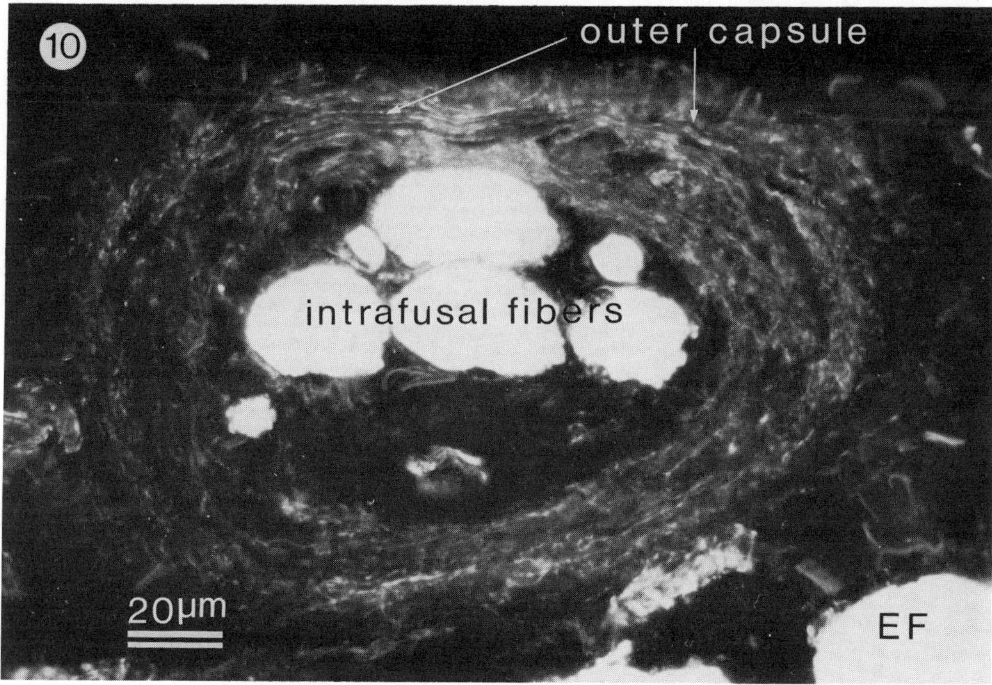

Fig. 10. Fluorescence micrograph of a human spindle treated with NBD-phallacidin to detect actin. A circumferentially disposed network of actin filament bundles comprises the multilayered outer capsule. An extrafusal fiber is also indicated (EF).

DISCUSSION

By electron microscopy, an intricate network of cytoplasmic filaments
was observed in spindle capsules of mouse and man. Parallel filament ar-
rays in cells of the outer capsule and filament aggregates in extensions
of the inner capsule were noted. With the two fluorescent microscopic
procedures applied in this study, filamentous actin and vimentin were also
detected in the spindle capsule.

Intermediate filaments of the vimentin type are characteristic of
cells of mesenchymal origin (Schliwa, 1986). While Shantha and coworkers
(1968) suggest that capsular cells of muscle spindles are ectodermally derived, Milburn (1973) contends that portions of the capsule arise from
invading fibroblasts. In our study, the presence of vimentin in cells of
both the outer and inner capsules in the mouse would seem to support a
mesenchymal origin.

Ross and Reith (1969) initially described microfilaments in the perineurium of peripheral nerve, postulating their contractile role to shorten
the nerve during body movement. A question that emerges from our study is
whether actin in the spindle capsule plays a dynamic role in the generation of contractile force. While an interaction of actin with myosin has
been implicated in the motility of non-muscle cells (Schliwa, 1986), isoforms of myosin have yet to be localized in capsular cells of the muscle
spindle. On the other hand, histochemical procedures that detect myosin
adenosine triphosphatase in skeletal muscles often localize this enzyme in
the outer capsule of the spindle as well (Ovalle and Smith, 1972).

If cells of the spindle capsule are potentially contractile, a functional relationship between these cells and paracellular elements that mediate contraction might be expected. Unmyelinated adrenergic axons and
varicosities have been observed within lamellae of the mammalian spindle
capsule (Barker and Saito, 1981; Swash and Fox, 1985) and a close relationship between mast cells and the outer capsule has been documented
(Ovalle and Dow, 1985). Moreover, an enzyme that is related to the neuropeptide, substance-P, has also been localized in the capsule (Dubovy and
Soukup, 1985; Soukup and Dubovy, 1985). While the functional implications
of these relationships have not yet been resolved, it is conceivable that
capsular contraction generated by actin filaments may alter junctional
permeability thereby affecting the capsular-blood barrier, reminiscent of
that which normally occurs between endothelial cells in the vascular wall
(Majno et al., 1969).

While the precise functions of intermediate filaments remain elusive,
they are known to be unusually stable and are thought to form a rigid and
mechanically continuous framework within cells (Steinert et al., 1984).
They may also function in determining cell shape, and have been implicated
as mechanical integrators during cell movement (Schliwa, 1986). Within
muscle spindles, the focal convergence of filament bundles at capsular
intercellular junctions may provide mechanical adhesion at these sites to
withstand deformation.

Alternatively, the cytoskeleton of the spindle capsule may provide a
supportive framework for the receptor, maintaining its overall shape and
holding the intrafusal fibers in place. Whereas some intrafusal fibers of
spindles extend beyond the ends of the outer capsule, short chain intrafusal fibers are anchored directly to the capsule at their terminal ends
(Desaki and Uehara, 1983). At these sites, the capsular cytoskeleton may
stabilize the ends of the intrafusal fibers as they undergo repeated epi-
sodes of stretch and relaxation during the movements related to the contractile activity of the surrounding muscle.

ACKNOWLEDGEMENT

This study was supported by grants from the British Columbia Health
Care Research Foundation, the Muscular Dystrophy Association of Canada,
and the Natural Sciences and Engineering Research Council of Canada.

REFERENCES

Barak, L.S., Yocum, R.R., and Watt, W.W., 1983, In vivo staining of cyto-
    skeletal actin by autointernalization of nontoxic concentrations of
    nitrobenzoxadiazole-phallacidin, J. Cell Biol., 89:368-372.
Barker, D., and Saito, M., 1981, Autonomic innervation of receptors and
    muscle fibres in cat skeletal muscle, Proc. R. Soc. Lond. B., 212:
    317-332.
Bridgeman, C.F., and Eldred, E., 1964, Hypothesis for a pressure sensitive
    mechanism in muscle spindles, Science, 143:481-482.
Desaki, J., and Uehara, Y., 1983, A fine-structural study of the termina-
    tion of intrafusal muscle fibres in the Chinese hamster, Cell Tissue
    Res., 234:723-733.
Dow, P.R., Shinn, S.L., and Ovalle, W.K., 1980, Ultrastructural study of a
    blood-muscle spindle barrier after systemic administration of horsera-
    dish peroxidase, Am. J. Anat. 157:375-388.
Dubovy, P., and Soukup, T., 1985, Histochemical localization of dipeptidyl
    peptidase IV in muscle spindles of the rat. II. Electron microscopy,
    Histochem. J., 17:582-584.
Majno, G., Shea, S.M., and Leventhal, M., 1969, Endothelial contraction in-
    duced by histamine-type mediators. An electron microscopic study, J.
    Cell Biol., 42:647-672.
Milburn, A., 1973, The early development of muscle spindles in the rat,
    J. Cell Sci., 12:175-1195.
Ovalle, W.K., and Dow, P.R., 1983, Comparative ultrastructure of the inner
    capsule of the muscle spindle and the tendon organ, Am. J. Anat., 166:
    343-357.
Ovalle, W.K., and Dow, P.R., 1985, Morphological aspects of the muscle
    spindle capsule and its functional significance, in: "The Muscle
    Spindle", I.A. Boyd and M.H. Gladden, eds., Macmillan, London, pp.
    23-28.
Ovalle, W.K., and Smith, R.S., 1972, Histochemical identification of
    three types of intrafusal muscle fibers in the cat and monkey based on
    the myosin ATPase reaction. Can. J. Physiol. Pharm., 50:195-202.
Ross, M.H., and Reith, E.J., 1964, Perineurium: evidence for contractile
    elements, Science 165:604-606.
Schliwa, M., 1986, "The Cytoskeleton", Springer-Verlag, New York.
Shantha, T.R., Golarz, M.N., and Bourne, G.H., 1968, Histological and his-
    tochemical observations on the capsule of the muscle spindle in normal
    and denervated muscle, Acta Anat., 69:632-646.
Soukup, T., and Dubovy, P., 1985, Histochemical localization of dipeptidyl
    peptidase IV in muscle spindles of the rat. I. Light microscopy, His-
    tochem. J., 17:579-581.
Steinert, P.M., Jones, J.C.R., and Goldman, R.D., 1984, Intermediate fila-
    ments, J. Cell Biol., 99:22s-27s.
Stossel, T.P., 1984, Contribution of actin to the structure of the cyto-
    plasmic matrix, J. Cell. Biol., 99:15s-21s.
Swash, M., and Fox, K.P., 1985, Adrenergic innervation of baboon and human
    muscle spindles, in: "The Muscle Spindle", I.A. Boyd and M.H. Glad-
    den, eds., Macmillan, London, pp. 121-126.
Wolosewick, J., and De Mey, J., 1982, Localization of tubulin and actin in
    polyethylene glycol embedded rat seminiferous epithelium, Biol. Cell,
    44: 85-88.

QUANTITATIVE STUDIES ON MAMMALIAN MUSCLE SPINDLES

AND THEIR SENSORY INNERVATION

Robert Banks and Michael Stacey

Department of Zoology
University of Durham
South Road, Durham DH1 3LE, U.K.

INTRODUCTION

It is not surprising, in view of the various functional roles played by skeletal muscles, that each muscle should possess a characteristic proprioceptive innervation. Muscle spindles are relatively easy to count and have been the main subject of quantitative studies. In drawing comparisons between different muscles, most authors have used the number of spindles per gram of adult muscle, or spindle density, as a measure of relative abundance. In both man and cat, where sufficient muscles have been examined, smaller muscles have been found usually to have higher spindle densities than larger muscles (reviewed by Hosokawa, 1961; Voss, 1971; and Barker, 1974). This has frequently led to the suggestion that the higher densities are functionally appropriate to small muscles involved in fine postural adjustment or manipulation, yet it has never been demonstrated that it is justifiable to relate spindle number linearly to muscle mass as a simple density.

Spindle counts actually refer to the number of separate encapsulations whose sensory complements are very varied (Barker & Banks, 1986). Each capsule usually encloses a primary sensory ending supplied to all three types of intrafusal muscle fibre (b_1b_2c unit) by a group Ia afferent axon. On either side of the primary up to 3 or 4 secondary endings may also be supplied, mainly to chain fibres, by about as many group II axons. Rarely a b_1b_2c unit may receive two Ia axons that form a double primary ending. Separate encapsulations may be linked in tandem by a continuous bag_2 fibre and, at least in the cat, some of them (b_2c units) lack a bag_1 fibre; the proportion of this type differs from one muscle to another (Bakker & Richmond, 1981; Banks et al. 1982). Their primary endings, which are rarely accompanied by secondaries, are supplied by afferents intermediate in diameter and preterminal branching pattern between those of b_1b_2c primary and S_1 secondary endings (Banks et al., 1982; Richmond et al., 1986; Kucera & Walro, 1987).

In this paper a new measure of relative spindle abundance is proposed and a comparison is made of the different provision of spindle units and their afferents in various muscles of the cat.

RELATIVE SPINDLE NUMBER

The usefulness of spindle density as a measure of relative spindle abundance relies on the assumption that absolute spindle number should be linearly related to muscle size. That this may not be so is suggested by the general occurrence of higher densities in smaller homologous muscles of different species (Table 1) as well as in the smaller muscles within a single species. Moreover, similarly sized muscles of different species may exhibit similar spindle densities: e.g. cat soleus 2.49g, 56 spindles, $23g^{-1}$ (Chin et al., 1962); human abductor pollicis brevis 2.7g, 80 spindles, $29.3g^{-1}$ (Schulze, 1955). These observations led us to question the relationship between spindle number and muscle size, and we have therefore sought evidence for its nature from published data for 75 muscles derived from rat (4 muscles), Arendt and Asmussen (1974); cat (29 muscles), Bakker and Richmond (1982), Barker (1974), Richmond and Abrahams (1975), Richmond and Stuart (1985); and man (42 muscles), Hosokawa (1961), Matthews (1972), von Hoyer (1963).

Logarithmic transformation of both spindle number and muscle weight yielded a linear relationship of the form

$$y = 1.58 + 0.32x$$

where y is $\log_{10}$ spindle number and x is $\log_{10}$ muscle weight in grams (Fig. 1). It is now possible to measure relative spindle abundance by the extent to which any muscle deviates in the richness of its spindle content from the value expected for a muscle of the same size. Some examples are given in Table 2. To those of us long conditioned in the use of spindle density it may come as a surprise to find lumbrical muscles rather poorly supplied with spindles by this measure. Dorsal neck muscles, however, retain their position as having the greatest abundance of spindles, with over five times as many as expected in intertransversarius C2-C3 of the cat.

SENSORY INNERVATION

We have previously shown that the number of secondary endings in b_1b_2c units of cat hindlimb muscles from a mixed sample dominated by tenuissimus and peroneus brevis followed a binomial distribution (Banks et al., 1982). However, it was also apparent that the average number of secondary endings per unit varied somewhat in different muscles. We have therefore analysed the sensory innervation of spindles from a variety of axial and limb muscles of the cat using teased, silver-impregnated preparations.

Table 1. Spindle-capsule Density in Homologous Muscles of Rat, Cat and Man.

Species	Capsule Density (g^{-1})		
	Lumbrical III (Hand)	Soleus	Gastrocnemius (Total)
Rat		310.5	40.0
Cat	173	23	6.5
Man	12.2	0.94	0.4

Data from Voss, 1971; Arendt & Asmussen, 1974; Barker 1974

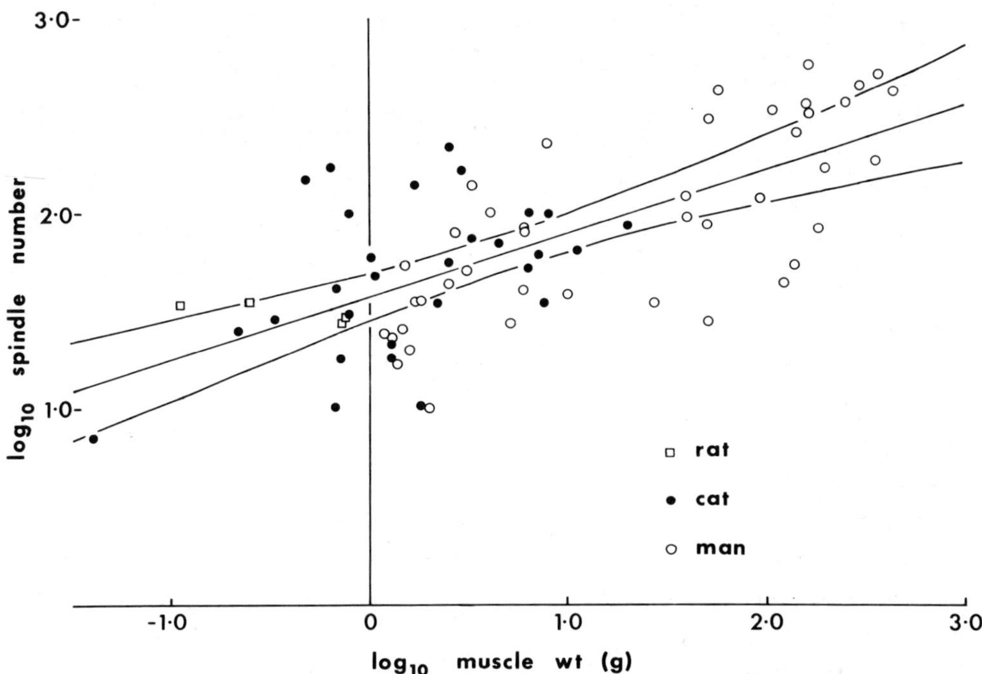

Fig. 1. Logarithmic transformations of spindle number and muscle weight
are highly correlated (r=0.69). The linear regression is shown
with 95% confidence limits.

To avoid biasing the sample, it was important whenever possible to use as
many well-stained spindles as could be obtained from individual muscles.

The composition of the sample is given in Table 3a together with data
on the proportion of all units that were of b_2c type, the proportion of
b_1b_2c units that possessed double primary endings, and the average number
($\bar{a}$) per b_1b_2c unit of afferent fibres in excess of a single Ia fibre. The
last feature is expressed in this way to take account of the double primary
endings and also a small number of II fibres that had terminals in the S_1
positions on both sides of some primary endings from extensor digitorum
longus and popliteus. Values of $\bar{a}$ ranged from 0.56 in extensor digitorum
lateralis of the forelimb to 3.5 in complexus of the neck; however, the
greatest number of afferents and endings occurred in 2 spindles from
popliteus (1 from each muscle sampled) with complements of $S_2S_1PPS_1S_2S_3$
(7 afferents) and $S_2S_1PPS_1S_2S_3S_4$ (8 afferents).

Frequency distribution of the number of afferents in excess of a
single Ia differed for each type of muscle, but when several complete
muscles of a single type (e.g. peroneus brevis) were examined the individual
distributions were all similar. For each muscle the value of $\bar{a}$ was used
to calculate the corresponding Poisson distribution and a set of binomial
distributions from which one could be selected that presented the greatest
overlap with the observed frequency distribution. Whenever there were
sufficient degrees of freedom a x^2 goodness-of-fit test was performed to
determine if the observed distribution differed from the calculated one.
The results are shown in Table 3b. In general, the binomial distributions
fitted the observed data better than did the Poisson distributions, as is
particularly clear in the case of peroneus brevis which was the type most
extensively sampled in that we were able to analyse virtually all the
spindles from 5 muscles. The complexus sample was slightly different

Table 2. Deviations of Actual Spindle-capsule Abundances from Values
Expected in Terms of Muscle Weight.

Species/Muscle	Number		Deviation	
	Expected	Actual	Linear	Log
Cat/Intertrans	32.8	176	5.4	+0.73
Man/Obl cap inf	73.6	232	3.2	+0.50
Cat/Biv cervicis	45.1	140	3.1	+0.49
Rat/Soleus	18.8	34	1.8	+0.26
Man/Bic brachii	194.3	320	1.6	+0.22
/Soleus	265.3	408	1.5	+0.19
Rat/Plantaris	24.1	35	1.5	+0.16
Cat/Rect femoris	75.0	104	1.3	+0.14
/Soleus	50.9	56	1.1	+0.04
/Int V (hand)	23.1	25	1.1	+0.03
/Gastr medialis	71.9	62	0.86	−0.06
Rat/Gastr medialis	34.0	28	0.82	−0.08
Cat/Lumb III (hand)	13.6	7	0.51	−0.29
Man/Plantaris	80.6	39	0.48	−0.32
/Lumb III (hand)	44.2	20	0.45	−0.34
Cat/Occipitoscap	33.3	11	0.33	−0.48
Man/Infraspinatus	183.5	54	0.29	−0.53
/Thyreohyoideus	47.4	12	0.25	−0.60
Cat/Infrahyoideus	45.9	6	0.13	−0.88

Abbreviations: Bic(eps); Biv(enter); Gastr(ocnemius); Int(erosseus);
Intertrans(versarius); Lumb(rical); Obl(iquus) cap(itis) inf(erior);
Occipitoscap(ularis); Rect(us).

from the best-fitting binomial distribution, but this may well have been
due to sampling bias since the number of spindles analysed was small
compared to the number present in a single muscle, largely because of the
poor staining which is a common problem with dorsal neck muscles.

We feel justified in concluding, therefore, that the observed data
are well described by binomial statistics, so that each muscle may be
classified two-dimensionally by the parameters n, p (Fig. 2). Where p is
very small (<0.1) the distributions reduce to Poisson form, but as p
increases the advantages of the binomial distribution become more apparent.
This should be borne in mind when considering interosseus ($p = 0.41$) for
which there were insufficient degrees of freedom to test the binomial
distribution, but whose observed and Poisson distributions differed only
slightly.

DISCUSSION

In attempting to interpret these results it must be recalled that
the spindle is a structure that is not entirely purpose-built from unique

Table 3. a) Composition of the Sample of Cat Muscles Used in the Analysis of Spindle Sensory Innervation, Together with Some Basic Observations. b) Comparisons of Observed Frequency Distributions of Afferent Axons in Excess of One Ia with Distributions Calculated According to Poisson and Binomial Statistics. χ^2 Goodness-of-Fit Test.

a

b

Muscle	All Units					Poisson			Binomial		
				b_1b_2c Units							
Type	Number Sampled	Total Number	$\%b_2c$ Type	% with Double P	$\bar{a}$	χ^2	df	P	χ^2	df	P
EDLat	1	32	16	0	0.56	–	0	–	–	0	–
S&DLumb	14	52	4	0	0.9	2.34	1	NS	–	0	–
PB	5	164	7	0.7	1.22	9.11	3	$0.5{>}P{>}0.1$	0.18	1	NS
Interosseus	5	39	0	0	1.23	8.68	2	$0.5{>}P{>}0.1$	–	0	–
Soleus	3	47	11	0	1.36	0.48	2	NS	1.2	1	NS
EDL	1	71	20	12	1.58	3.38	3	NS	3.82	1	NS
Tenuissimus	14	70	7	6	1.86	1.82	3	NS	1.47	2	NS
FDL	2	81	7	0	1.89	3.61	3	NS	1.93	2	NS
ECLat	2	64	10	0	2.26	7.54	4	NS	0.11	1	NS
Popliteus	2	68	13	8	2.72	4.49	3	NS	1.27	2	NS
Complexus	2	64	34	0	3.5	10.7	4	$0.5{>}P{>}0.1$	4.85	1	$0.5{>}P{>}0.1$

Abbreviations: ECLat, extensor caudae lateralis; EDLat, extensor digitorum lateralis; EDL, extensor digitorum longus; FDL, flexor digitorum longus; PB, peroneus brevis; S&DLumb, superficial and deep lumbricals.

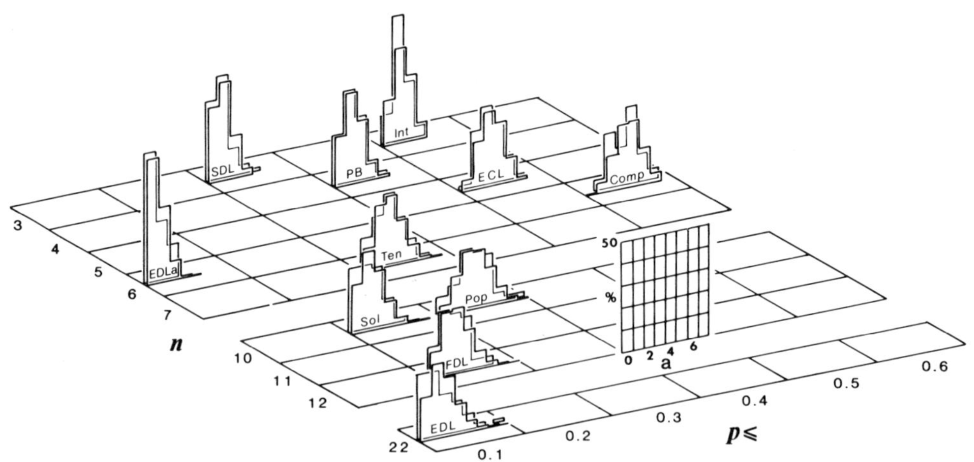

Fig. 2.   Pairs of histograms showing observed (rear) and best-fitting
          binomial (front) distributions of numbers of afferents (a) in
          excess of a single Ia.   The histograms are positioned on the
          grid according to their binomial parameters (n, p).   Abbreviations:
          Comp(lexus); ECL, Extensor caudae lateralis; EDLa, Extensor
          digitorum lateralis; Int(erosseus); Pop(liteus); Sol(eus);
          SDL Superficial and deep lumbrical; Ten(uissimus); and as Table 3.

components, but is a makeshift structure assembled from what is available.
In order to understand which features are functionally important in the
adult we must recognize that the significance of some features may relate
principally to the way that the spindle is constructed.

Our results indicate that in the case of the muscle spindle the
proprioceptive needs of functionally diverse muscles are met in two
mainly independent ways: i) variations in the abundance of spindles (and
hence Ia afferents), and ii) variations in the proportional balance of Ia
and II afferents.   The association of II with Ia afferents is clearly a
random process, but it does not seem to depend only on the average number
of afferents present.   One obvious factor that may also be important is
the period of time during which developing afferents continue to arrive
at the presumptive spindle.

The other features considered here, double primary endings and b_2c
units, might not be independently produced, but might arise as inevitable
consequences of the developmental programme in which relative timing of
events such as myotube formation, regional maturation, and neurite arrival
are crucially important (e.g. Milburn, 1984).   For example, if Ia and II
afferents differentiate only during spindle development, b_2c units would
be more likely to occur if there is high competition among neurites (either
sensory/sensory or sensory/motor) for available target sites on primary
myotubes early in development.   This can be brought about if most of the
afferents arrive at an early stage (which perhaps gives rise to a distribution with high n and low p, e.g. extensor digitorum longus); or if there
is simply a very large number of afferents (e.g. complexus).   This offers
an alternative explanation for the occurrence of b_2c units in cat muscles
to the specific functional role that has been postulated (Richmond et al.
1986), and that could just as easily be played by secondary endings.

REFERENCES

Arendt, K.-W., and Asmussen, G., 1974, Die Anzahl und die Verteilung von
    Muskelspindeln im M. triceps surae der Ratte, Anat. Anz., 136: 207-216.
Bakker, G.J., and Richmond, F.J.R., 1981, Two types of muscle spindles in
    cat neck muscles: A histochemical study of intrafusal fiber composi-
    tion, J. Neurophysiol., 45: 973-986.
Bakker, G.J., and Richmond, F.J.R., 1982, Muscle spindle complexes in
    muscles around upper cervical vertebrae in the cat, J. Neurophysiol.,
    48: 62-74.
Banks, R.W., Barker, D., and Stacey, M.J., 1982, Form and distribution of
    sensory terminals in cat hindlimb muscle spindles, Philos. Trans. Roy.
    Soc., B299: 329-364.
Barker, D., 1974, The morphology of muscle receptors, in: "Handbook of
    Sensory Physiology, III/2" Hunt, C.C., ed., Springer, Berlin.
Barker, D., and Banks, R.W., 1986, The muscle spindle, in: "Myology",
    Engel, A.G., and Banker, B.Q., eds., McGraw-Hill, New York.
Chin, N.K., Cope, M., and Pang, M., 1962, Number and distribution of spindle
    capsules in seven hindlimb muscles of the cat, in: "Symposium on
    Muscle Receptors", Barker, D., ed., Hong Kong University Press, Hong
    Kong.
Hosokawa, H., 1961, Proprioceptive innervation of striated muscles in the
    territory of cranial nerves, Tex. Rep. Biol. Med., 19: 405-464.
Kucera, J., and Walro, J.M., 1987, Heterogeneity of spindle units in the
    cat tenuissimus muscle, Am. J. Anat., 178: 269-278.
Matthews, P.B.C., 1972, "Mammalian Muscle Receptors and their Central
    Actions", Arnold, London.
Milburn, A., 1984, Stages in the development of cat muscle spindles, J.
    Embryol. Exp. Morph., 82: 177-216.
Richmond, F.J.R., and Abrahams, V.C., 1975, Morphology and distribution of
    muscle spindles in dorsal muscles of the cat neck, J. Neurophysiol.,
    38: 1322-1339.
Richmond, F.J.R., Bakker, G.J., Bakker, D.A., and Stacey, M.J., 1986, The
    innervation of tandem muscle spindles in the cat neck, J. Comp. Neurol.,
    245: 483-497.
Richmond, F.J.R., and Stuart, D.G., 1985, Distribution of sensory receptors
    in the flexor carpi radialis muscle of the cat, J. Morph., 183: 1-13.
Schulze, M.L., 1955, Die absolute und relative Zahl der Muskelspindeln in
    den kurzen Daumenmuskeln des Menschen, Anat. Anz., 102: 290-291.
von Hoyer, E., 1963, Untersuchungen über Anzahl, Länge und Verteilung der
    Muskelspindeln in der Unterschenkelmuskulatur, Anat. Anz., 113: 36-44.
Voss, H., 1971, Tabelle der absoluten und relativen Muskelspindelzahlen der
    menschlichen Skelettmuskulatur, Anat. Anz. 129: 562-572.

THREE-DIMENSIONAL DISTRIBUTION OF MUSCLE SPINDLES AND GOLGI

TENDON ORGANS IN THE HUMAN ABDUCTOR POLLICIS BREVIS*

Heather Stephens, Line Chabot, Denis Miron and
Therese Simard

Departement d'Anatomie
Faculte de Medecine
Universite de Montreal
Montreal H3C 3J7

INTRODUCTION

It is well known that there are large numbers of muscle spindles in
human muscles involved in fine movements (Matthews, 1972). However, their
spatial distribution as well as that of Golgi tendon organs has not been
fully investigated. In animals the distribution of muscle spindles is
related to the pattern of intramuscular nerve branching, to the proportion
of aerobic or anaerobic fibers, and possibly to the presence of
intramuscular tendons (Barker, 1972). We examined the 3-dimensional
distribution of muscle spindles and Golgi tendon organs, both singly and in
relation to one another as well as their position relative to the
intramuscular nerve branches, to the intramuscular tendon as well as the
distal and proximal tendinous insertions and in relation to the extrafusal
motor endplates. This study was carried out on the human abductor pollicis
brevis.

MATERIAL AND METHODS

After dissection of the abductor pollicis brevis from formalin-embedded
cadavers and paraffin embedding, 5 micron serial transverse sections were
stained with trichrome blue. Every fifth section was analysed either for
the presence of muscle spindles, Golgi tendon organs or other histological
landmarks. Their length and position were noted. Using an epidiascope the
projected image of selected sections was traced with the position of the
muscle receptors marked relative to intramuscular nerves and tendons. The
images were centred using the peripheral contours of serial sections as a
guide and the information digitalized on an IBM microcomputer using the 3-
dimensional rotation software from Bioquant (Stephens,1985). The
reconstructions were displayed with a Gould plotter equipped with 6 colour
pens to enable concomitant and separate analysis of parameters such as
the position of muscle spindles, Golgi tendon organs, intramuscular nerves,
tendons, myotendinous junctions or motor endplates. The latter two were
identified by acetylcholinesterase histochemistry performed on fresh muscles
prior to paraffin embedding.

---

*Supported by the Medical Research Council of Canada.

RESULTS AND DISCUSSION

There were an average of 90 muscle spindles per abductor pollicis
brevis, similar to the finding of Voss (1959), each with an average length
of 3.5 mm. They were concentrated at the motor nerve entry point towards
the proximal muscle insertion. They covered the vast majority of the total
territory, although they were absent from the peripheral ends. On the other
hand, the Golgi tendon organs, as expected, were restricted to the distal
and proximal insertions. However, there were significantly more of these
receptors at the distal insertion than at the proximal. This may be due to
the fact that there is a thick well-compartmentalized tendon at the distal
end as opposed to the flattish proximal webbing. There were approximately 30
Golgi tendon organs, with a mean length of 1.5 mm, per abductor pollicis
brevis. Fig. 1 illustrates a 3-dimensional reconstruction of the spatial
distribution of the muscle spindles seen from the side and from an angle.
Fig. 2 illustrates similar views of the Golgi tendon organs. These
reconstructions will permit elucidation and comparison of the territories
covered by these two sensory receptors.

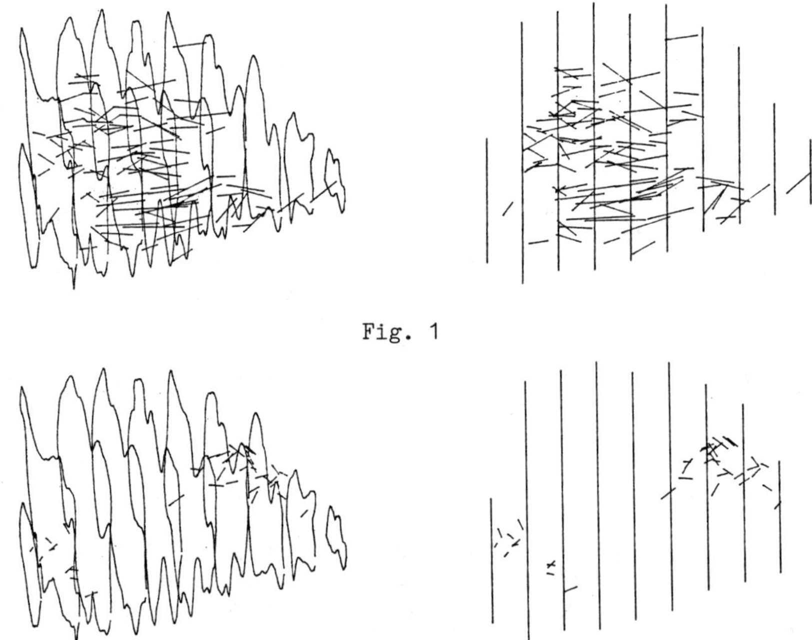

Fig. 1

Fig. 2

REFERENCES

Barker, D., 1972, The morphology of muscle receptors, in: "Handbook of
    Sensory Physiology", Vol. III/2, C.C. Hunt, ed., Springer Verlag,
    Berlin-Heidelberg-New York, pp. 1-190.
Matthews, P.B.C., 1972, "Mammalian Muscle Receptors and Their Central
    Actions", Arnold, London.
Stephens, H.R., 1985, Distribution and 3-D reconstruction of muscle
    spindles in mouse skeletal muscles, in "The Muscle Spindle",
    I. Boyd and M. Gladden, eds., Stockton Press, New York, pp. 83-88.
Voss, H., 1959, Weitere Untersuchungen über die absolute und relative
    Zahl der Muskelspindeln in verschiedenen Muskelgruppen des Men-
    schen, Anat. Anz., 107:190.

MYOFIBRILLAR AND CYTOSKELETAL PROTEINS IN HUMAN MUSCLE SPINDLES*

P-O. Eriksson[1,2], G. Butler-Browne[3], D. Fischman[4], B. K. Grove[1],
S. Schiaffino[5], I. Virtanen[6] and L-E. Thornell[1]

Depts. of [1]Anatomy and [2]Clinical Oral Physiology, University
of Umeå, Sweden; [3]Paris, France; [4]N. Y., USA; [5]Padova, Italy;
[6]Helsinki, Finland

INTRODUCTION

From differences in morphological, histochemical and mechanical properties three types of intrafusal (IF) fibre have been distinguished: "dynamic
nuclear bag_1", "static nuclear bag_2" and "nuclear chain" fibres. We have recently shown that the muscle spindles in the human masticatory muscles are
far more complicated in structure than reported before for human and animal
muscle spindles (Eriksson and Thornell, 1985, 1987). To attain further information about the structure and function of human muscle spindles we have
analysed the composition of different myofibrillar and cytoskeletal proteins
in IF fibres and spindle capsule tissue.

MATERIALS AND METHODS

Human masseter and temporal muscle spindles were examined. Serial frozen muscle cross-sections were stained either histochemically for myofibrillar ATPase activity and oxidative and glycolytic enzymes or immunohistochemically with type-specific antisera. Antibodies to myosin isozymes, M-
band proteins (M-protein, myomesin and MM-CK) and cytoskeletal proteins were
used together with indirect peroxidase-antiperoxidase (PAP) or indirect im-
munofluorescence staining.

RESULTS (Table 1, Fig. 1)

In sections stained with anti-embryonic myosin chain fibres showed
strong and bag_2 weak reactivity while bag_1 fibres were in general unreactive. With the anti-neonatal myosin chain and bag_2 fibres were strongly
stained while some bag_1 were stained, others not. Sections treated with
anti-fast and anti-slow myosin, respectively, were heterogeneous in staining
pattern. Only chain fibres stained strongly with the anti-fast serum. Some
intrafusal fibres reacted with anti-M-protein serum while some bag_1 were
non-reactive. Most of the fibres stained by the anti-M-protein were also
stained by the anti-MM-CK serum. All intrafusal fibres reacted with anti-
desmin. The spindle capsule reacted strongly with antibodies against embryonic myosin, vimentin, vinculin, fibronectin, and moderately with anti-
spectrin.

*Supported by the Swedish Medical Research Council (3934 and 6874), the K-O.
Hansson Foundation, Umeå University, and the Swedish Dental Society.

Table 1   Immunohistochemical staining activity in intrafusal fibres, capsule and extrafusal fibres of human masseter muscle.

	Embry-onic	Neo-natal	Slow tonic	M-pro-tein	Myo-mesin	Desmin	Vimentin
Bag$_1$	−	−..+	+	−..+	+	+	−/+*
Bag$_2$	+	+	+	+	+	+	
Chain	+	+	−	+	+	+	−/+*
Capsule	+	−	−	−	−	−	+
Type I	−	−..+	−	−..+	+	+	−
Type II	−	+	−	+	+	+	−
ATPase-IM	−	+	−	+	+	+	−

− unreactive, + weak to strong staining activity, * staining with some monoclonal antibodies

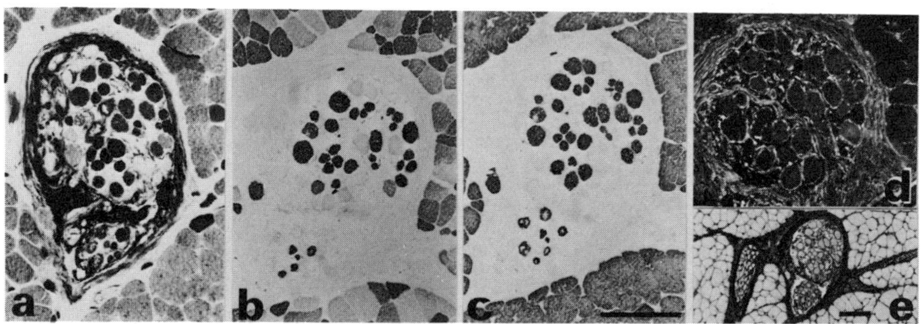

Fig. 1   Serial cross-sections of a large masseter muscle spindle stained for a)myofibrillar ATPase at pH 9.4, and antibodies against b)M-protein, c)myomesin, d)vimentin (FITC) and e)fibronectin. Bar: 100 µm.

DISCUSSION

The functional response properties of the mammalian muscle spindle are well known, however, it is not clear how these properties should be assigned to the subunits of the spindle. Our study shows that with antibodies against neonatal myosin, slow tonic myosin, MM-CK and M-protein a distinct classification of the intrafusal fibres in human muscles can be performed. Bag$_1$ fibres contain slow twitch and slow tonic myosins and lack M-protein and MM-CK in the myofibrillar M-band. Bag$_2$ fibres contain determinants recognized by antibodies against embryonic, neonatal, fast and slow tonic myosins, and all three M-band proteins. Finally, the chain fibres contain embryonic, neonatal and fast twitch myosins and all M-band proteins. The capsule contains cells which have a high amount of vimentin. As the capsule cells also show high myofibrillar ATPase activity one can conclude that they have both a myofibrillar and a cytoskeletal organization similar to smooth muscle cells or to fibroblasts of the contractile type.

REFERENCES

Eriksson, P-O. and Thornell, L. E., 1985, Heterogeneous intrafusal fibre composition of the human masseter muscle, in: "The Muscle Spindle", eds. I. A. Boyd and M. H. Gladden, MacMillan Press Ltd., London, pp. 95-100.

Eriksson, P-O. and Thornell, L. E., 1987, Relation to extrafusal fibre-type composition in muscle spindle structure and location in the human masseter muscle, Archs oral Biol, 32:483-491.

IMMUNOHISTOCHEMICAL DEMONSTRATION OF CONNECTIVE TISSUE MACROMOLECULES

AT THE EQUATOR OF CHICK MUSCLE SPINDLES

Alfred Maier and Richard Mayne

Department of Cell Biology and Anatomy
University of Alabama at Birmingham
Birmingham, AL

INTRODUCTION

Past structural studies on muscle spindles have in large part focused
on identifying intrafusal fiber types, and on characterizing sensory and
motor innervations. Connective tissue components of the receptor have received considerably less attention. Cooper and Gladden (1974) have explored
the functional relationship between intrafusal fibers and the elastic and
reticulin fibrils which surround them, and knowledge about the structure and
function of inner and outer capsules has come from the work of Shanta et al.
(1968), Kennedy and Yoon (1979), Ovalle (1976) and Hikida (1985). However,
the molecular basis of the connective tissue matrix occuring in spindles has
yet to be demonstrated. We undertook the current immunohistochemical study
to remove some of this uncertainty by evaluating the distribution of nine
connective tissue macromolecules in the external lamina of intrafusal fibers
and in the outer capsule at the equator of chick muscle spindles.

METHODS

The equatorial regions of 50 muscle spindles from tibialis anterior,
extensor digitorum longus, pars media of gastrocnemius and extensor metacar-
pi radialis of normal eight-week-old chicks were examined. Small pieces
of muscle were frozen by immersion into melting isopentane. Nine consecu-
tive cross sections were cut at a time on a freezing microtome, air-dried
(20°C) for 30 minutes and incubated with 4% bovine serum albumin (BSA).
After 10 minutes the BSA was drained off and to each of the nine sections
10 µl of a different primary monoclonal antibody was applied. The antibodies that were used are specific for collagen types I, IV, V and VI, a basal
lamina form of heparan sulfate proteoglycan (HS), fibronectin (F), laminin
(L), tenascin/brachionectin (T/B) and the stubs of chondroitin-4-sulfate (CS)
(Caterson et al., 1987). The sources, designations and concentrations of the
antibodies, besides anti-chondroitin sulfate, have been described previously
(Maier and Mayne, 1987). Sections were incubated with the primary antibodies for 15-30 minutes and then washed three times with phosphate buffered
saline (PBS). This was followed by incubation for 30 minutes with 10 µl of
the secondary antibody, fluorescein-conjugated goat anti-mouse IgG (Cappel
Laboratories, Cochranville, PA), diluted 1:30 with PBS. All incubations were
carried out in a moist chamber at 37°C. After a final three washes with PBS,
sections were mounted on glass slides with a 7:3 mixture of PBS and glycerol.

Control sections were treated as described above, except that incubation with the primary antibody was omitted. Fluorescence was observed with a Leitz Ortholux II microscope, using an I_2 filter block. The location of reaction product was verified by alternately inspecting sections with fluorescence and phase contrast microscopy.

RESULTS

Figure 1 is a schematic drawing, showing how avian intrafusal fibers and structures immediately adjacent to them appear at the equator. The drawing, based on compilation of data from several sources (James and Meek, 1973; Ovalle and Dow, 1983; Hikida, 1985; Maier, 1985), should be useful in recognizing the sites of the immunohistochemical reactions described below. Sensory terminals and intervening sensory satellite cells contact intrafusal fibers. Satellite cells, sensory terminals and intrafusal fibers are surrounded by a common external (basal) lamina. On the outside of the external lamina, and over the region of the sensory terminals, lies a crescent-shaped collagenous sheath. Inner capsule cells and their cytoplasmic extensions provide an outer wrapping for the entire intrafusal fiber complex. Extensions of inner capsule cells may invade the collagenous sheath.

Control sections incubated without primary antibody were negative (Fig. 2). The external lamina of intrafusal fibers was positive for collagen type IV, laminin and heparan sulfate. For heparan sulfate no great changes were noted in the intensity of staining around the circumference of intra-fusal fibers. Staining for collagen type IV was very weak or absent, and strong for laminin, in that portion of the external lamina which covered the sensory terminals (Figs. 3-5). Reaction product for chondroitin sulfate was only seen in the portion of the external lamina over the sensory terminals (Table 1). Collagen type VI was found just external to the external lamina. Its distribution was also not uniform, being virtually absent between the external lamina and the collagenous sheath, but appearing as a strong band elsewhere (Fig. 6; arrow). The collagenous sheath itself was strongly positive for collagen type I (Fig. 7) and moderately positive for collagen type V and laminin. Faint collagen type VI reaction product was also recognized (Figs. 4,6; Table 1). In some instances it was difficult to see whether all reaction product was restricted to the collagenous sheath, or whether some of it, especially in the outer portions of the sheath, was also located in cytoplasmic extensions and cell bodies of inner capsule cells.

The outer spindle capsule at the equator was strongly positive for collagen types IV and VI, laminin, heparan sulfate and tenascin/brachionec-tin. As a rule, dense reaction product appeared across the entire width of the capsule, making it stand out prominently against the thinner deposits around the extrafusal fibers (Figs. 8-12). Staining for collagen type V was weak and for collagen type I moderate. Fibronectin staining was also moderate, but its distribution within the outer capsule and contiguous vessels and nerves was spotty rather than uniform (Fig. 13; Table 1). Tenascin/brachionectin reaction product was only seen in the outer capsule and in the walls of nerves and vessels associated with the capsule (Fig. 12). The capsule was negative for chondroitin sulfate. Next to tenascin/brachio-nectin, the reactions for heparan sulfate and fibronectin produced the least amount of staining in the periaxial space (Figs. 10,13).

DISCUSSION

The observed reaction products most likely represent valid structural associations because the fluorescent staining was consistently identified

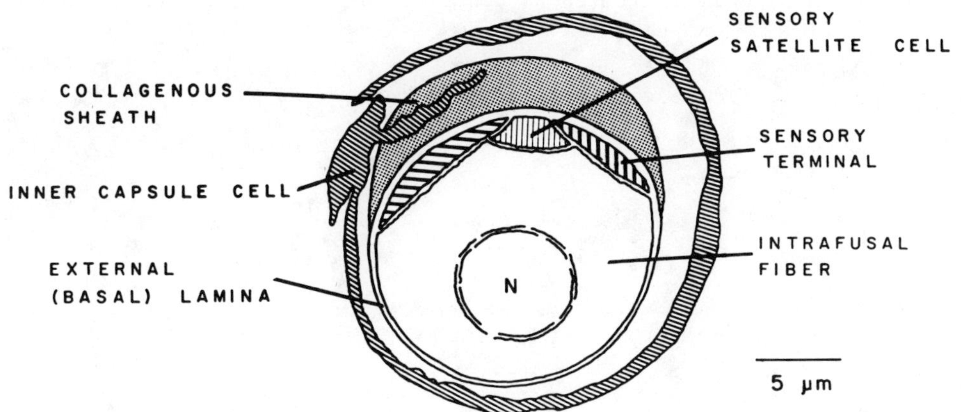

Fig. 1.   Schematic   drawing of cross section of avian intrafusal fiber and closely associated structures as they appear at the equator. N = nucleus of intrafusal fiber.

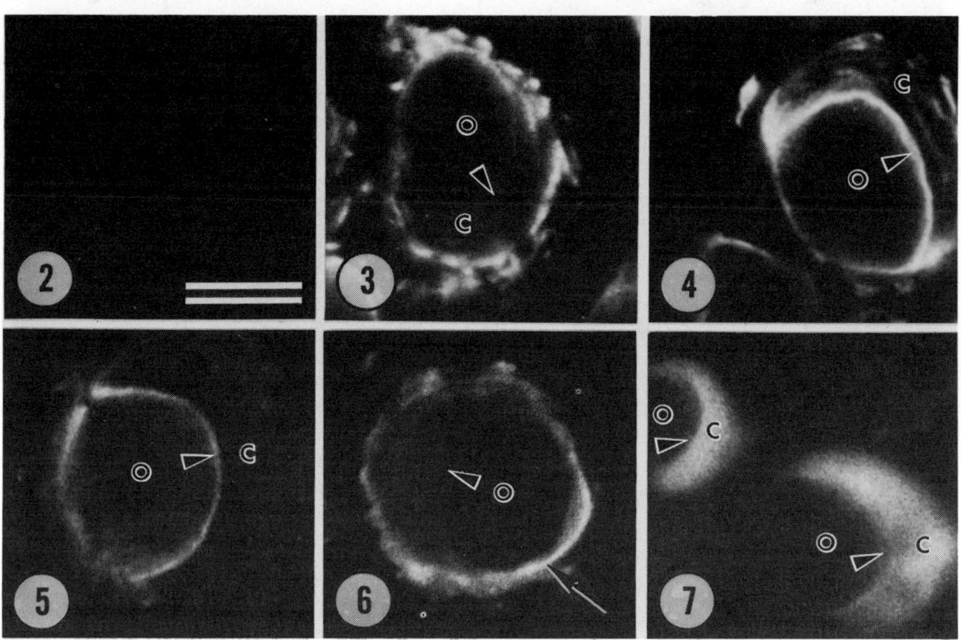

Figs. 2–7.   Cross   sections of chick intrafusal fibers at the equator incubated without primary antibody (control) (2), and with antibodies against collagen type IV (3), laminin (4), heparan sulfate proteoglycan (5), collagen type VI (6) and collagen type I (7).   Centers of intrafusal fibers are marked with circles, and location of external lamina over sensory terminal with arrowheads.   Arrow in Figure 6 points to strong band of collagen type VI opposite the sensory terminals.   Bar is 10 micra.

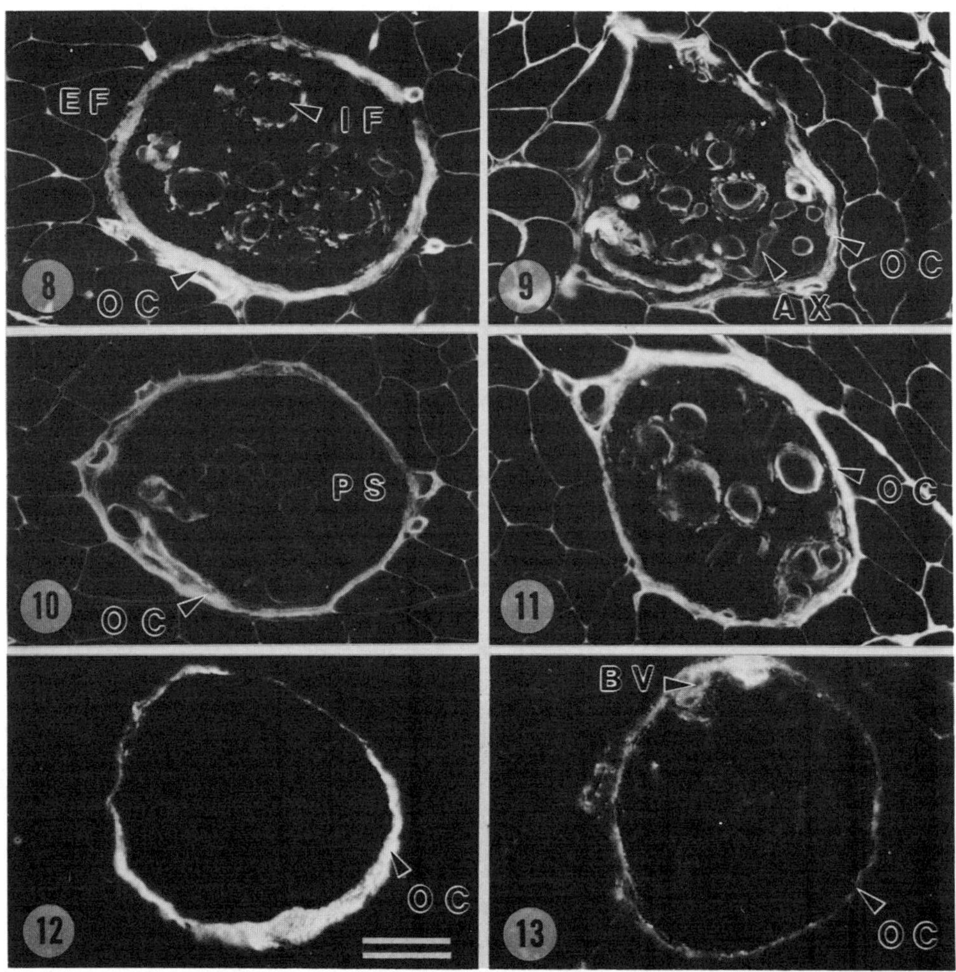

Figs. 8–13. Appearance of cross sections at the equator of chick muscle
spindles incubated with antibodies against collagen type IV
(8), laminin (9), heparan sulfate proteoglycan (10), collagen
type VI (11), tenascin/brachionectin (12) and fibronectin (13).
The more pronounced staining of the outer capsule relative to
the external lamina of extrafusal fibers is readily apparent.
Bar is 30 micra. Ax, axon; BV, blood vessel; EF, extrafusal
fiber; IF, intrafusal fiber; OC, outer spindle capsule; PS,
periaxial space.

with phase microscopy as being localized in the respective connective tissue
components of the muscle spindle (external lamina, collagenous sheath, etc.);
however, resolving of structural details must await investigation with the
electron microscope.

Specialization of the external lamina over the sensory terminals has
been recognized with the electron microscope. In pigeons the external
(basal) lamina over the sensory terminals is much thicker than elsewhere, and
it is thought to be the result of fusion of the external laminae of
intrafusal fibers, sensory terminals and sensory satellite cells (Hikida,
1985). The current study shows that in chicks the external lamina around

278

Table 1. Summary of antibody-antigen reactions

	Collagen type								
	I	IV	V	VI	L	HS	F	T/B	CS
External lamina[a]	–	d	–	–	+++	++	–	–	+++
External lamina[b]	–	++	–	e	++	++	–	–	–
Coll. sheath[c]	+++	+[f]	+	d	+	–	–	–	–
Outer capsule	+	+++	d	+++	+++	+++	+[g]	+++	–

[a] of intrafusal fibers over sensory terminals
[b] of intrafusal fibers not over sensory terminals
[c] some of the staining may be located in adjacent inner capsule
  cells
[d] weak or very weak
[e] strong reaction external to external lamina
[f] spotty or weak
[g] spotty

intrafusal fibers at the equator is not immunohistochemically homogeneous.
That portion away from the sensory terminals has immunohistochemical
properties similar to basement membranes described by Timpl and Dziadek
(1986). On the other hand, over the region of the sensory terminals,
collagen type IV is essentially absent, while chondroitin-4-sulfate and
laminin are strongly represented. The occurence of the highly negatively
charged sulfated residues of chondroitin sulfate might be of importance in
regulating the concentration of cations during depolarization of the sensory
terminals.

The question of what structures in the mammalian muscle spindle are
responsible for differences in stiffness among regions of intrafusal fibers,
and their effect on the quality of the afferent discharge, has long occupied
the thinking about the internal working of the spindle (Poppele and Quick,
1985; Gladden, 1986). Cooper and Gladden (1974) have suggested that elastic
fibers that surround intrafusal fibers are part of a mechanism that increases
stiffness in the equatorial region. In the avian spindle the presence of
collagen type I as the major component of the collagenous sheath, a type
otherwise occurring in large amounts in muscle tendons, should decrease the
extensibility of the equatorial region and the degree of deformation placed
on the sensory terminals. It is not completely clear how the sheath attaches
to the external lamina of the intrafusal fiber; however, laminin, a cell
attachment protein (Timpl and Dziadek, 1986), appears to be present at the
points where the narrow ends of the crescent meet the intrafusal fiber
(Fig. 4). Quite likely, the concentration of collagen type VI away from the
region of the sensory terminals also affects the mechanical properties of the
equatorial region.

The outer spindle capsule consists of several layers of perineural
epithelium and the associated external (basal) laminae (Shanta et al., 1968;
Ovalle, 1976), and it is believed to be an effective barrier against high
molecular weight tracers (Dow et al., 1980, Kennedy and Yoon, 1979). The
"basement membrane" proteins which are present in the outer capsule of the
chick spindle are also seen in virtually all other basement membranes (Timpl
and Dziadek, 1986). Simply the thickness of the layers of the reaction
products supports the notion that the outer capsule is an effective diffusion
barrier. A possible function for the fibronectin within it may be to link

cell membranes or external laminae to the intracellular matrix of the outer
capsule (Hantai et al., 1983). Tenascin/brachionectin is found in ther outer
capsule and at the myotendinous junction (Chiquet and Fambrough, 1984);
however, the proteins from the two sites do not appear to connect to each
other (Maier and Mayne, 1987). It has been suggested that tenascin/brachio-
nectin is important for epithelial growth during development (Chiquet-
Ehrisman et al., 1986). Its presence in the outer capsule may be related
to the initial formation of the capsule from perineural epithelium.

The results of the current study show that the distribution of connective tissue macromolecules can be readily documented with monoclonal antibodies, and that the connective tissue network of the avian spindle is
intricate. Complete documentation of this network will contribute to a
better understanding of how the sensory region functions.

REFERENCES

Caterson, B., Calabro, T., Hampton, A., 1987, Monoclonal antibodies as probes
      for elucidating proteoglycan structure and function, In Biology of Pro-
      teoglycans, Wight, T.N., Mecham, R.P., eds., Academic Press, in Press.
Chiquet-Ehrisman, R., Mackie, E.J., Pearson, C.A., and Sakakura, T. 1986,
      Tenascin: an extracellular matrix protein involved in tissue inter-
      actions during fetal development and oncogenesis, Cell 47:131-139.
Chiquet, M., and Fambrough D.M., 1984, Chick myotendinous antigen. I. A
      monoclonal antibody as a marker for tendon and muscle morphogenesis, J.
      Cell Biol., 98:1926-1936.
Cooper, S., and Gladden M.H., 1974, Elastic fibres and reticulin of mammalian
      muscle spindles and their functional significance, Quart. J. Exp.
      Physiol., 59:367-385.
Dow, P.R., Shinn, S.L., and Ovalle, W.K., 1980, Ultrastructural study of a
      blood-muscle spindle barrier after systematic administration of
      horseradish peroxidase, Am. J. Anat. 157:375-388.
Gladden, M.H., 1986, Mechanical factors affecting the sensitivity of
      mammalian muscle spindles, Trends Neuro Sci. 9:295-297.
Hantai, D, Gautron, J., and Labat-Robert, J., 1983, Immunolocalization of
      fibronectin and other macromolecules of the intracellular matrix in the
      striated muscle fiber of the adult rat, Collagen Rel. Res. 3:381-391.
Hikida, R.S., 1985, Spaced serial section analysis of the avian muscle
      spindle, Anat. Rec. 212:255-267.
James, N.T., and Meek, G.A., 1973, An electron microscopical study of avian
      muscle spindles, J. Ultrastr. Res. 43:193-204.
Kennedy, W.R., and Yoon, K.S., 1979. Permeability of muscle spindle
      capillaries and capsule, Muscle Nerve 2:101-108.
Maier, A., 1985, Ultrastructure of the equatorial region of chick forearm
      muscle spindles, Anat. Rec. 211:121A.
Maier, A., and Mayne, R., 1987, Distribution of connective tissue proteins in
      chick muscle spindles as revealed by monoclonal antibodies: A unique
      distribution of tenascin/brachionectin, Am. J. Anat., In press.
Ovalle, W.K., 1976, Fine structure of the avian muscle spindle capsule,
      Cell Tiss. Res., 166:285-298.
Ovalle, W.K., and Dow, P.R., 1983. Comparative ultrastructure of the inner
      capsule of the muscle spindle and the tendon organ, Am. J. Anat.,
      166:343-357.
Poppele, R.E., and Quick, D.C., 1985, Effect of intrafusal mechanics on
      mammalian muscle spindle sensitivity, J. Neurosci., 1881-1885.
Shanta, T.R., Golarz, M.N., and Bourne, G.H., 1968, Histological and
      histochemical observations on the capsule of the muscle spindle in
      normal and denervated muscle, Acta Anat., 69:632-646.
Timpl, R., and Dziadek, M., 1986, Structure, development and molecular
      pathology of basement membranes, Intl. Rev. Exp. Pathol., 29:1-112.

PART VII

FUNCTIONAL MORPHOLOGY

OF SENSORY RECEPTORS

WHAT IS A SENSORY CORPUSCLE?

L. Malinovský

Department of Anatomy
Medical Faculty of J.E. Purkyně University
Komenského nám. 2
662 43 Brno, Czechoslovakia

Previous authors (e.g. Krause, Merkel) described different
kinds of sensory corpuscles in the skin: Grandry corpuscles in
birds, Meissner corpuscles in mammals and club-shaped corpuscles (Kolbenkörperchen) in reptiles, birds and mammals (and
lamellar corpuscles including Pacinian corpuscles). Stilwell
(1957) described free endings in joint capsules, Ruffini endings and encapsulated endings. Zabusov and Maslov (1961) distinguished three categories of sensory structures: 1. simple
or free endings, 2. receptors with a glial component and 3. encapsulated receptors. Poláček (1966) already pointed out some
problems connected with the function of the capsule as one of
the significant aspects of receptor function. Poláček, however,
did not take into account the capsule as a significant morphological criterion. He designated only those structures as corpuscles which form a lamellar inner core. According to this
classification, Meissner corpuscles were not classified as
corpuscles and were termed "Meissner endings". The same also
concerned Ruffini formations. In Grandry corpuscles the term
"corpuscle" and "ending" was used promiscuously even though
they were classified into the second type - i.e. endings without an inner core. Biemesderfer et al. (1978) found "Ruffini
corpuscles" associated with a non-sinus hair. Chouchkov (1978)
described the Ruffini corpuscle as an encapsulated receptor
without a lamellar inner core. Krstić (1984) put forth a commonly used definition of a sensory corpuscle: "A corpuscle is
an encapsulated ending with a well-delimited round or oval
structure formed by a lamellated connective tissue capsule
around one or more peripheral nerve endings in contact with
either a group of specialized connective cells, intrafusal
muscle fibers or tendon fibers". The connection of a Merkel
cell with an adherent nerve ending is also called a Merkel
corpuscle. The definition of Ruffini´s "spray" is not sufficiently clear, as yet, (see Poláček, 1966). According to this
concept, it can be concluded that all sensory nerve formations
(endings), with the exception of free endings, are sensory
corpuscles. Menisci tactus (Merkel complexes) do not possess
any capsule at all. From this point of view, the term "sensory
corpuscle" does not say anything about the special structural

properties of certain formations and it is only a name for a
"space-structure".

Physiologists mostly classify sensory formations according to their response as slowly or rapidly adapting receptors
(Iggo and Gottschaldt, 1974). Gottschaldt (1985) divides the
rapidly adapting mechanoreceptors into two subgroups: receptors
detecting acceleration and receptors detecting velocity. Ruffini
formations, Golgi tendon organs and muscle spindles are slowly
adapting mechanoreceptors. In our classification (Malinovský,
1986) we take into account the structural, developmental and
physiological properties of these formations (see Figs. 1 to 6).
In some sensory formations, the dendritic zone is in relation
to the connective tissue or the muscle tissue, which are of the
same mesodermal origin. These formations are all characterized
physiologically by slow adaptation. In another group of sensory
formations, there is a relationship of the dendritic zone
(zones) to Schwann cells (ectodermal origin). The Schwann cells
are arranged either in parallel or they form a lamellar inner
core. These are physiologically all rapidly adapting receptors,
although there are some quantitative differences between them
(see Gottschaldt, 1985). We call these formations "sensory corpuscles". We do not include free dendritic zones, the Ruffini
formations, the Golgi tendon formations, the neuromuscular
formations and the Merkel complexes into the category of sensory corpuscles, although in the latter case we can speak about
a transition to sensory corpuscles (relation of the dendritic
zone to a special cell - the Merkel cell - in salamanders and
birds with rapid adaptation).

The existence or non-existence of a capsule is not the
most significant criterion, as the capsule is missing in simple
lamellar corpuscles (the skin of the nose in the hedgehog -
Malinovský and Páč, 1979). In typical Meissner corpuscles the
capsule is also not complete. On the contrary, the existence
of a lamellated capsule is not connected with rapid adaptation.
Very simple sensory corpuscles consisting of a small number of
Schwann cells arranged in parallel to the dendritic zone without a capsule resembling those around the hair (bulboid-lanceo-
late formations) were found in the nose skin of the hedgehog
(Malinovský and Páč, 1979).

In genital corpuscles (Poláček and Malinovský, 1971; Malinovský, 1979), we found an in-parallel arrangement of Schwann
cells in some places, in other places "lamellar complexes" resembling typical simple sensory lamellar corpuscles: the dendritic zone was surrounded by several (up to 8) lamellae of
Schwann cells, in some of them with a typical cleft on one side.
We consider these lamellar complexes as a morphological and a
physiological equivalent of a simple sensory corpuscle. The
genital corpuscle is the only formation in which the rate of
adaptation has not been described up to now. We also suppose
that this formation has rapid adaptation.

On the basis of morphological and physiological criteria
the following definition of a sensory corpuscle is proposed:
A sensory corpuscle is a morpho-physiological system consisting
of a dendritic zone ("receptor portion") of a pseudounipolar
cell process, Schwann cells or equivalent cells and usually of
a capsule. The Schwann cells or equivalent cells (Merkel cells)

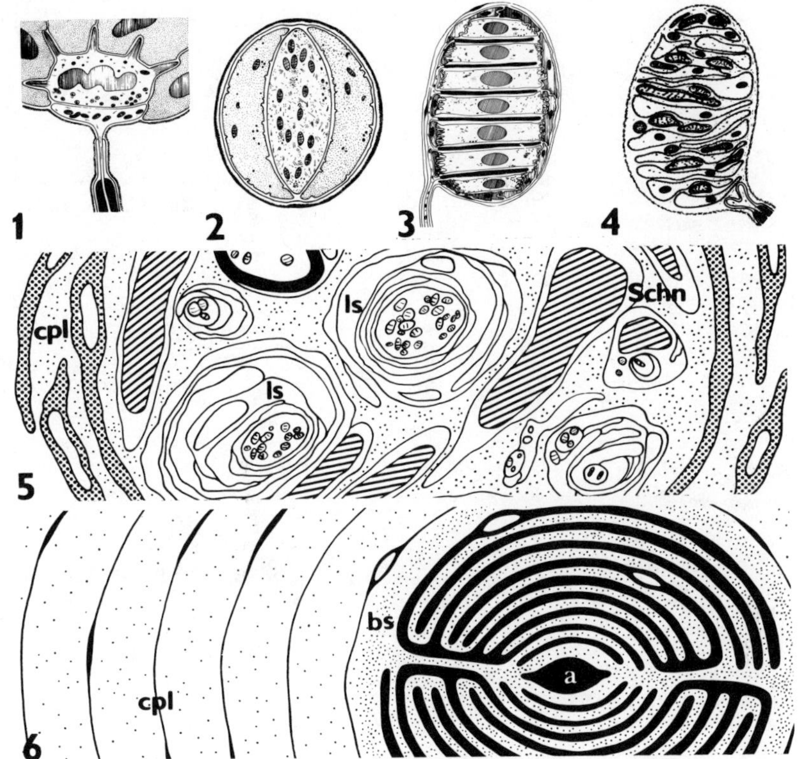

Fig. 1. Schematic structure of the Merkel complex.
Fig. 2. Schematic structure of the bulboid sensory hair unit.
Fig. 3. Schematic structure of the Grandry corpuscle.
Fig. 4. Schematic structure of the Meissner corpuscle.
Fig. 5. Schematic structure of the genital corpuscle.
ls - lamellar system, cpl - capsule lamella, Schn - Schwann cell nucleus.
Fig. 6. Schematic structure of the Pacinian corpuscle.
a - axon, ic - inner core, bs - boundary space, cpl - capsular lamellae.

are arranged either in parallel (bulboid sensory formation of hairs as the most simple sensory corpuscle, the Grandry, Merkel, Meissner and genital corpuscles) or they form a lamellar inner core (simple sensory corpuscle, the Golgi-Mazzoni corpuscle as a variety of a simple sensory corpuscle, the Herbst and Pacinian corpuscle). Physiologically the corpuscles exhibit rapid adaptation (detecting velocity or acceleration).

REFERENCES

Biemesderfer, D., Munger, B.L., Binck, J., and Dubner, R., 1978, The pilo-Ruffini complex: a nonsinus hair and associated slowly-adapted mechanoreceptor in primate facial skin, <u>Brain Res.</u>, 142:197-222.

Gottschaldt, K.M., 1985, Structure and function of avian soma-
        tosensory receptors, in: "Form and Function in Birds",
        Vol. 3, A.S. King and J. McLelland, eds., Academic
        Press, London, pp. 375-461.
Chouchkov, Ch., 1978, Cutaneous Receptors. Springer Verlag,
        Heidelberg.
Iggo, A., and Gottschaldt, K.M., 1974, Cutaneous mechanore-
        ceptors in simple and in complex sensory structures,
        in: "Mechanoreception", J. Schwartzkopff, ed., West-
        deutscher Verlag, Opladen, pp. 153-176.
Krstić, R.V., 1984, Illustrated encyclopedia of human histolo-
        gy, Springer Verlag, Heidelberg.
Malinovský, L., 1979, Lamellar complexes in human female ge-
        nital nerve endings, Folia morphol. (Prague), 27:9-10.
Malinovský, L., 1986, Classification of sensory nerve endings
        in vertebrates brought up to date, Folia morphol.
        (Prague), 34:261-264.
Malinovský, L., and Páč, L., 1979, The ultrastructure of sen-
        sory corpuscles in the hedgehog, Z. mikrosk.-anat.
        Forsch., 93:673-688.
Poláček, P., 1966, Receptors of the Joints, Acta facult. Med.
        Brno, Vol. 23.
Poláček, P., and Malinovský, L., 1971, Die Ultrastruktur der
        Genitalkörperchen in der Clitoris. Z. mikrosk.-anat.
        Forsch., 84:293-310.
Stilwell, D.L. jr., 1957, The innervation of deep structures
        of the hand. Amer. J. Anat., 101:59-74.
Zabusov, G.I., and Maslov, A.P., 1961, Opyt evolyucionno-mor-
        fologicheskoy klassifikatsii chuvstvitelnykh nervnykh
        okonchaniy, in: "Problemy morfologii i reaktivnosti
        perifericheskikh otdelov nervnoy sistemy", Kazan, pp.
        13-19.

CLASSIFICATION OF SENSORY NERVE FORMATIONS (ENDINGS)

L. Malinovský

Department of Anatomy, Medical Faculty
Purkyně University
Brno, Czechoslovakia

Sensory nerve formations are divided into 3 classes (Fig. 1):

Ist class - axonal dendritic zone is related to the tissue elements of mesodermal origin. Ia: the surrounding connective tissue elements are not organized: 1 - simple sensory formation (individual free axons), 2 - simple arborization; Ib: the connective tissue elements are organized: 3 - Ruffini spray-like formation, 4 - Golgi tendon organ, 5 - neuromuscular spindle.

IInd class is a transient class: 6 - intraepidermal or intraepithelial nerve formations, 7 - Merkel complexes.

IIIrd class - axonal dendritic zones are related to tissue elements of ectodermal origin. IIIa: the Schwann cells are arranged in parallel: 8 - bulboid (hair) sensory formations as the most simple corpuscles, 9 - Grandry corpuscles, 10 - Merkel corpuscles, 11 - Meissner corpuscles, 12 - Krause spherical corpuscles; IIIb: sensory formations with lamellar inner core: 13 - simple sensory corpuscles, 14 - Golgi-Mazzoni corpuscles, 15 - Herbst corpuscles, 16 - Pacinian corpuscles.

The formations of the Ist class are slowly adapting and those of the IIIrd class are rapidly adapting. Some of the formations belonging to the IInd class adapt rapidly in some animal species, while others exhibit slow adaptation.

Complex sensory nerve formations are divided into 3 groups (Fig. 2): I) A system consisting of sensory nerve formations of the same kind - Pinkus touch spot. II) A constant combination of sensory formations of different kinds - Eimer organ in the mole. III) A combination of sensory nerve formations with a hair or with a feather.

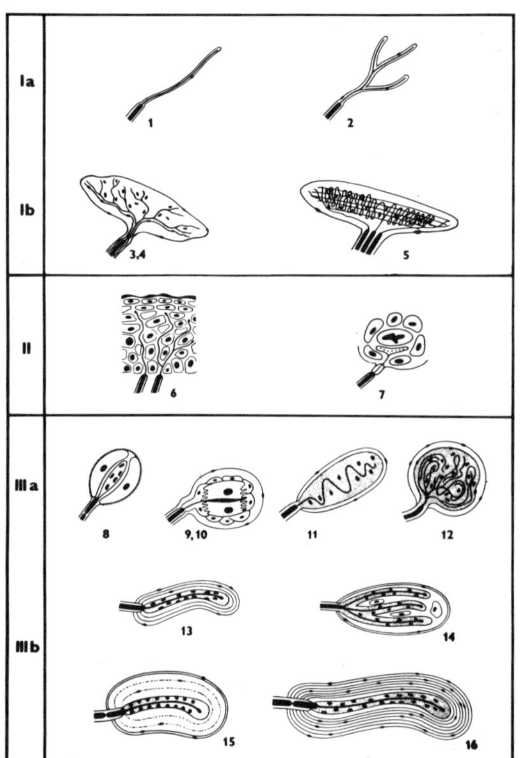

Fig. 1. Sensory nerve formations (endings)

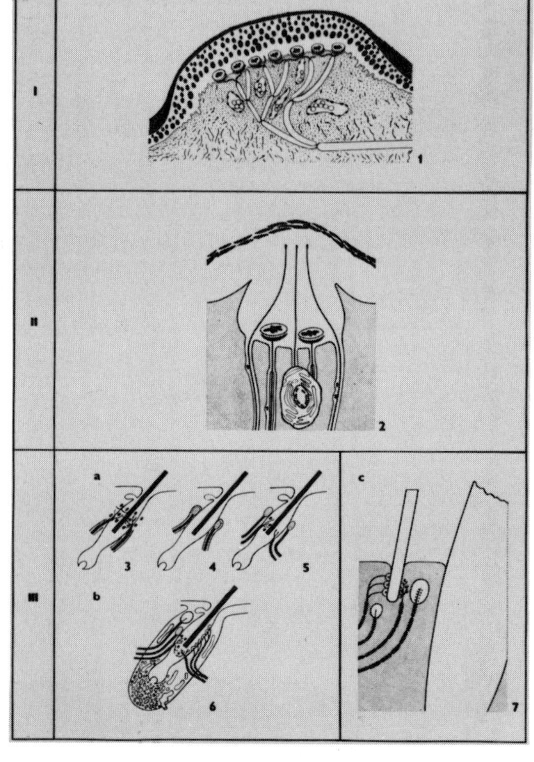

Fig. 2. Complex sensory
formations

THE EFFECTS OF AGE ON MURINE PACINIAN CORPUSCLES

Pedro B. Nava

Department of Anatomy
Loma Linda University
Loma Linda, CA

INTRODUCTION

The effects of age on the somatosensory nervous system have been observed at all of its morphological levels extending from the neuronal cytons of the dorsal root ganglia distally to specific sensory mechano-receptors (Mathewson and Nava, '85; Spencer and Ochoa, '81; Kenshalo, '77, and Cauna, '65). Documentation of age-related regression of the peripheral nervous system has come from several sources: clinical, electrophysio-logical, neuropharmacological and histopathological (Ras and Nava, '86). One of the clinical manifestations of concern is the distal blunting of vibratory sensation. Sensitivity to vibratory stimuli in the absence of detectable neuropathy is significantly reduced with advanced age in humans (Era, et al.,'86). Thresholds for all frequencies are significantly higher in the elderly. This is especially true for the lower extremities after the sixth decade (Verrillo, '80). It has been suggested that these pro-gressive losses of sensitivity to vibration stimuli may be the partial result of alterations in the morphology of the Pacinian corpuscle that occur with age (Cauna, '65). This rapidly adapting mechanoreceptor is capable of responding to sinusoidal vibratory stimuli in the range of 20-1500 Hz, with a maximum sensitivity of 200-400 Hz (Iggo, '85). These receptors are concentrated within the mesentery, subcutaneous tissues of the digits and footpads and the deep skeletal tissues (i.e., joints, ligaments, periosteum and interosseous membranes of mammals. The receptor sites are strategically clustered to facilitate transmission of extrinsic vibratory stimuli from adjacent bone to the receptors. The structure and functional properties of these corpuscles are similar for all species reported, i.e., primates, marsupials, cats and rodents. Although caution is indicated, it is plausible to extrapolate the age-related structural changes of murine crural corpuscles in this study, using silver impregna-tion and electron microscopic techniques, to the frequently observed lower extremity sensory deficits that occur in humans (Era, et al., '86, and Kenshalo, '86).

The murine crural Pacinian corpuscle is an ideal model for the study of progressive age changes in a specific mechanoreceptor. Its structure has been detailed in several morphological studies (Idé, '87; Idé and Tohyama, '85) allowing comparative correlation with other species. The development (Zelena, '78), degeneration (Zelena, '82), regeneration (Ide,

'87; Zelena, '84), transplantation (Jirmanova and Zelena, '87), the toxic effects of CS_2 (Jirmanova, '87) studies give additional credence to the establishment of this rodent receptor as a viable model for the ongoing aging and neuropathological studies of this receptor by the author.

Light Microscopy

<u>Typical Young Mouse Pacinian Corpuscles</u>. The murine crural Pacinian corpuscles 35 to 42 in number are clustered in a grape-like manner along the axis of the fibula from its midregion to the tibiofibular fusion. They are oval (350 x 115 μm) with their long axis parallel to that of the fibula. A position well suited for the detection of ground-borne vibrations. The corpuscle is innervated by a single preterminal myelinated neurite. The myelin is lost as the neurite enters the inner core. This unmyelinated portion is divided into terminal and ultraterminal segments. The ultraterminal segment ends in one or more bulbous enlargements in the distal end of the corpuscle. The inner core is made up of lamellar cells of Schwann cell origin, which is embedded in a multilayered outer core of concentric lamellae of perineurial origin.

<u>Aged Pacinian Corpuscles</u>. Silver impregnated hindlimb tissue sections from 10 age groups of mice were compared and evaluated for age changes using the Winkelmann and Schmit ('57) technique. Receptor counts and size determinations revealed no significant change in corpuscular numbers or size with age. Camera lucida tracings of impregnated Pacinian corpuscle neurites demonstrated four basic neurite pattern types (I-IV) found within the inner core of aging corpuscles. The type I profile is a simple, attenuated, pattern which does not branch and terminates in a slight bulbous enlargement. In the type II pattern, branching occurs in either the distal portion of the terminal or ultraterminal part of the neurite. There may be secondary branching of the neurite. The branches terminate as bulbous enlargements in the ultraterminal region of the receptor.In the terminal portion of the neurite,in type III, branches within the inner core and terminates in bulbous enlargements in the ultraterminal region of the receptor. Secondary branching may occur; if so, these branches also terminate as bulbous enlargements in the ultraterminal region of the corpuscle. Multiple and secondary branching of the terminal portion of the neurite characterize the type IV pattern. Some branches recur and extend proximally, giving the neurite a skein-like appearance. Some branches end in small, bulbous enlargements. These neurite patterns were confined to the inner core of the receptor. More than one pattern type may be found at any given age, but with advancing age there is a shift to more complex neurite types. A single neurite innervation of the corpuscle is retained with aging with no indication of multiple innervation.

Electron Microscopy

<u>Morphology of Typical Young Pacinian Corpuscle</u>. Each corpuscle consists of an afferent neurite, an inner core formed by lamellar cell processes of Schwann origin, a subcapsular space between the inner core and a multilayered capsule which is continuous with the perineurium. Representative ultrathin sections for each of the three regions of the receptor, as defined by Nishi, et al., ('69), were evaluated from young mice (1.5-3 months) for comparison with aged mice (12-24 months). These regions included the proximal, central and ultraterminal regions of the corpuscle which contained the preterminal myelinated, the terminal unmyelinated and the distal end of the neurite, respectively. The ultraterminal portion ends in two-to-six bulbous enlargements.

The myelinated nerve fiber (5-6 μm in diameter) in the proximal region of the receptor was enveloped by a Schwann cell. The preterminal nerve fiber had an irregular transverse outline. The basal lamina of the Schwann cell separated it from the subcapsular space. The Schwann cell had short, basal lamina-covered projections extending into the subcapsular space. Mitochondria, microtubules and small profiles of rough endoplasmic reticulum were found throughout the Schwann cell cytoplasm. The mitochondria and neurotubules were concentrated peripherally while the neurofilaments were situated centrally. As the nerve fiber approached the inner core, it became ellipsoidal in outline with the previously mentioned organelles found in the same positions in this region. The nerve fiber was no longer myelinated, but possessed an incomplete Schwann cell cytoplasmic investment. The cytoplasm of the Schwann cell contained numerous electron dense granules of various sizes, mitochondria and free ribosomes in this transitional region. A basal lamina was still associated with the Schwann cell in this region of the receptor.

The Schwann cell cytoplasmic investment was lost as the nerve fiber passed into the inner core. The nerve fiber was ellipsoidal in outline throughout its terminal course with mitochondria and neurotubules just beneath the axolemma, with neurofilaments centrally positioned. Extending from both poles of the axon into the radial-longitudinal cleft of the inner core were axon or digitate processes. These cytoplasmic processes of the axon were often branched and contained microfilaments that extended from the base of the processes into their branches. The base of these structures contained vesicles and mitochondria. The innermost cytoplasmic processes of the lamellar cells of the inner core were in close association with the nerve fiber and its axon or digitate processes.

The nerve fiber in the distal portion of the terminal region of the receptor inner core became circular in outline. The distribution of the axonal organelles was similar to the preceding portion of the nerve fiber, except for the increased number of mitochondria. It was this region that the nerve fiber characteristically divided into two-to-six branches that extended into the ultraterminal region.

The nerve fiber branches in the ultraterminal region continued for a short distance, then terminated as bulbous enlargements. These branches had an investment of inner core lamellar cell processes. The short segments leading to the bulbous enlargements, and the enlargements themselves were filled with numerous mitochondria and vesicles. Extending into the bulbous expansions were neurotubules and centrally placed neurofilaments. These bulbous enlargements were devoid of any inner core cell investments. However, these enlargements were closely associated with the lamellar cells making up the capsule of the corpuscle. The thin cytoplasmic processes from these cells separated the ultraterminal enlargements from one another and from the vascular loops penetrating the poles of the corpuscles.

Lamellar cells and their numerous concentric cytoplasmic extensions made up the inner core of the Pacinian corpuscle. They invested the axon from where the Schwann cells terminated to the ultraterminal portion of the axon where they ended prior to its bulbous enlargements. Proximally, the lamellar cell processes overlapped the Schwann cells for a short distance almost completely investing the axon. Progressing distally, the inner core increased in diameter, with this increase in size, a radial-longitudinal cleft appeared, extending from the axon towards the periphery of the inner core and disappearing as the ultraterminal region of the receptor was approached. The diameter of the inner core decreased towards the ultraterminal region. The lamellar cell processes in the polar regions of the inner core completely encircled the axon or overlapped one another. No cleft was found in these regions. The bilateral organization of the midregion of the inner core was formed from two hemilamellar stacks of thin concentric cytoplasmic processes of the inner core cells. This arrangement produced the radial-longitudinal clefts.

The nuclei of the lamellar cells in all regions were restricted to the periphery of the inner core. The majority of the mitochondria, rough endoplasmic reticulum, free ribosomes and Golgi complexes were juxta-nuclear. These cytoplasmic organelles were sparsely distributed within the thin cytoplasmic extensions of the lamellar cells. Those most frequently noted were mitochondria and free ribosomes. Microtubules and microfilaments were also seen within the cytoplasmic processes. Occasionally, electron dense granules of varying sizes were observed in these processes. A prominent feature in some regions of the inner core were vesicles which opened onto either surface of the lamellar processes. Membrane specializations occurred throughout the inner core between adjacent lamellae. A discontinuous basal lamina outlined the peripheral lamellar cells and their processes.

The capsule was composed of squamous cells that were arranged in 12-18 concentric bands or lamellae surrounding the subcapsular space, giving the receptor its characteristic onion-like appearance. The capsular cells or perineurial epithelium were spindle-shaped with their cytoplasm drawn out into thin sheets which made overlapping contacts with tight junctions between contiguous cells. Rich vesiculation and discontinuous basal lamina characterized the inner and outer surfaces of the capsular cells. Occasionally there were contacts between cells of adjacent lamellae. The lamellae were separated by fluid-filled interlamellar spaces. These spaces contained collagen fibers that were oriented both parallel and perpendicular to the long axis of the receptor. Cells of unknown origin, possibly macrophages, were present within the interlamellar spaces. The interlamellar spaces were narrower near the inner core and periphery of the receptor, with the intervening spaces the widest.

The bulbous enlargements of the nerve fiber in the ultraterminal region of the corpuscle were closely associated with the innermost layer of the capsule. This innermost layer sent cytoplasmic partitions between the enlargements compartmentalizing the ultraterminal region and separating the capillaries from the enlargements. The innermost layers of the capsule in the proximal region of the receptor separated the capillaries from the endoneurial space and the inner core of the receptor.

<u>Morphology of Aged Pacinian Corpuscles</u>. The preceding cytological analysis of the corpuscle provides an adequate baseline for evaluation of age related effects on the receptor. The sequence of change that occurs with age closely parallels those seen in the degeneration of rat Pacinian corpuscles after nerve section (Zelena, '82). The chronologies of age associated effects is more prolonged, however, with the earliest detectable ultrastructural changes in the murine receptor observed in the neurite at 12 months, followed by inner core lamellar cell, then capsular cell alterations in older animals (18-24 months).

The first neurite-associated age changes were mitochondrial. Degenerative changes ranged from swollen, vacuolated, lamellated to pleomorphic appearing structures. Randomly distributed mitochondria within the axolasm occurred in some neurites rather than the usual peripheral subaxolemmal position seen in younger mice. Large, lamellated, circular myelin-like bodies and lipofuscin accumulation of possible mitochondrial origin were noted within the neurites of 18 and 24 month old receptors. The typical ellipsoidal neurite profile of younger animals was also altered with age.

Circular profiles of diminishing circumference to absent neurites from the inner core were typical in aging corpuscles. The integrity of the radial cleft still remained in these aged receptors with lamellar cell processes and extracellular matrix filling the void left by shrinking neural element. Multiple nerve profiles, evidence of neurite branching, were observed throughout the inner core in older animals. An occasional myelinated nerve fiber could be observed at the periphery of the inner core.

Schwann and lamellar cell variations appeared subsequent to changes in the neural component of the receptor. The modifications associated with the intracorpuscular cells prior to neurite loss were separation of the basal laminae from the cell surface followed by duplication of its basal laminae. Succeeding neurite disappearance, the capsular channel, which normally holds the myelinated part of the neurite, contained a Schwann cell column composed of a few thin cell processes enclosed within a folded basal lamina tube. A notable increase in collagen was witnessed in this area. Lamellar cell processes and associated extracellular matrix occupied the void created by degenerating neurite in the terminal region of the receptor. Older animals (24 months) expressed an accruement of lipofuscin within the Schwann cells and lamellar cell processes. This accumulation preceded an increase in extracellular matrix between the lamellar cell processes. These changes were preceded by a reduction in the number of pinocytotic vesicles in the lamellar cell processes. An increase and duplication of the discontinuous basal laminae associated with the inner and outer aspects of the outer core peripheral lamellar cells were the major finding associated with this portion of the receptor with age.

CONCLUDING STATEMENTS

Presented in this report are age-related light and electron microscopic morphological findings of a specific mechanoreceptor, the Pacinian corpuscle, that may correlate with the vibratory sensitivity deficits that occur with aging.

Light microscopic results revealed that with age there is no significant loss of receptor number or change in receptor size due to age in the mouse leg. This is contrary to the findings of Cauna ('65), who reported that in human digital Pacinian corpuscles there is a decrease in number and an increase in size and complexity of these receptors. These observations indicate that the same mechanoreceptor ages differently in various species and anatomical locations.

Silver impregnation of the Pacinian neurites demonstrated an age-associated shift in neural complexity from no or little branching to extensive branching and recurring of these branches within the inner core of the corpuscle. With this shift in neural complexity the murine corpuscles maintained a single neurite innervation, in contrast to the multiple innervated receptors seen in reinnervation studies (Zelena, '84). The neural disarray witnessed in this project may account for reduction in sensitivity to high frequency vibration that occurs with age. Studies have shown that pressure directed along the Pacinian terminal neurite minor axis, which is perpendicular to the major axis, evoked maximal mechano-electrical transduction (Nishi and Sato, '68; Lowenstein, '71). Alteration of this orientation, which occurs during aging, may also contribute to loss of receptor sensitivity.

Electron microscopic findings revealed sequential age-related central to peripheral, neurite to capsular, cellular to extracellular degenerative changes within the receptor complex that could produce mechanoelectric and high-pass filter frequency deficits.

BIBLIOGRAPHY

Cauna, N., 1965, The effects of aging on the receptor organs of the human dermis, in: "Advances in Biology of Skin," Vol. 6, W. Montagna, ed., Pergamon Press, Oxford.

Era, P., Jokela, J., Suominen, H., Heikkinen, E., 1986, Correlates of vibrotactile thresholds in men of different ages, Acta Neurol. Scand., 74:210.

Idé, C., 1987, Role of extracellular matrix in the regeneration of a Pacinian corpuscle, Brain Res., 413:155.

Idé, C., and Tohyama, K., 1985, Macrophages in Pacinian corpuscles, Acta Anat. 121:110.

Iggo, A., 1985, Sensory receptors in the skin of mammals and their sensory functions, Rev. Neurol. (Paris), 141:599.

Jirmanova, I., 1987, Pacinian corpuscles in rats with carbon disulphide neuropathy, Acta Neuropathol., 72:341.

Jirmanova, I., and Zelena J., 1986, Transplantation of Pacinian corpuscles of the rat into the brain, Acta Neuropathol., 69:341.

Kenshalo, D. R., 1977, Aging effects on cutaneous and kinesthetic sensibilities, Proc. Symp. Biol. Special Senses in Aging, University of Michigan, Ann Arbor: Inst. Gerontol.

Kenshalo, D. R., 1986, Somesthetic sensitivity in young and elderly humans, J. Gerontol., 41:732.

Lowenstein, W. R., 1971, Mechano-electric transduction in the Pacinian corpuscle: Initiation of sensory impulses in mechanoreceptors, in: "Handbook of Sensory Physiology," Vol. 1, W. R. Lowenstein, ed., Springer, New York.

Mathewson, R. C. and Nava, P. B., 1985, Effects of age on Meissner corpuscles: A study of silver-impregnated neurites in mouse digital pads, J. Comp. Neurol. 231:250.

Nishi, K., and Sato, M., 1968, Depolarizing and hyperpolarizing receptor potentials in the non-myelinated nerve terminal in Pacinian corpuscles, J. Physiol. (Lond.), 199:383.

Nishi, K., Oura, C., and Pallie, W., 1969, Fine structure of Pacinian corpuscles in the mesentery of the cat, J. Cell. Biol., 43:539.

Ras, V. R., and Nava, P.B., 1986, Age-related changes of neurites in Meissner corpuscles of diabetic mice, Exp. Neurol., 91:488.

Spencer, P. S., and Ochoa, J., 1981, The mammalian peripheral nervous system in old age, in: "Aging and Cell structure," Vol. 1, J. E. Johnson, Jr., ed., Plenum Press, New York.

Verrillo, R. T., 1980, Age related changes in sensititity to vibration, J. Gerontol., 35:185.

Winklemann, R. K., and Schmit R. W., 1957, A simple silver method for nerve axoplasm, Proc. Staff Meet. Mayo Clinic, 32:217.

Zelena, J., 1978, The development of Pacinian corpuscles, J. Neurocytol., 7:71.

Zelena, J., 1982, Survival of Pacinian corpuscles after denervation in adult rats, Cell Tiss. Res., 224:673.

Zelena, J., 1984, Multiple axon terminals in reinnervated Pacinian corpuscles of adult rat, J. Neurocytol., 13:665.

SUBSTANCE P- AND SOMATOSTATIN-LIKE IMMUNOREACTIVITY IN AVIAN ENCAPSULATED MECHANORECEPTORS

C. Chouchkov and M. Davidoff

Department of Anatomy, Higher Medical Institute
6003 Stara Zagora
and Regeneration Research Laboratory
Bulgarian Academy of Sciences
1431 Sofia, Bulgaria

The identity of neurotransmitters utilized by primary sensory neurons has not been firmly established, although several candidates for neurotransmitters or modulators have been suggested (Salt and Hill, 1983). In particular, many attempts to prove histochemically any of the well-known neurotransmitters or modulators in sensory receptors have been unsuccessful. Moreover, the well documented distribution of neuropeptides in some somata of primary sensory ganglia (substance P (SP) - Katz and Karten, 1980; somatostatin (Som) - Kummer and Heym, 1986; vasoactive intestinal polypeptide (VIP) - Skofitsch et al., 1985; cholecystokinin (CCK) - Kuwayama and Stone, 1986; opioid peptides - Weihe et al., 1985) has not been generally confirmed in corresponding sensory nerve fibers of encapsulated or non-capsulated receptors. Some efforts have been made to localize in the light microscope SP-LI (like-immunoreactive) nerve fibers in human digital skin including the sensory nerves entering Meissner corpuscles (Dalsgaard et al., 1983). On the other hand, Hartschuh et al. (1979) visualized immunohistochemically a met-enkephalin-LI in Merkel cells. Recently, Björklund et al. (1986) found SP- and neurokinin A-LI only in sensory nerve fibers, whereas VIP-LI has been observed predominantly in autonomic nerves but also in some sensory nerve fibers. All known available immunohistochemical investigations examine the above mentioned neuropeptides at the light microscopical level. The aim of the present communication is to extend the knowledge concerning the existence of new peptides in bird mechanoreceptors and to define their precise cytological localization at the electron microscopical level.

MATERIALS AND METHODS

Ten 12-day-old ducks were used for the immunocytochemical studies. The animals were anaesthetized with an intraperitoneal injection of Nembutal. They were perfused via the ascending aorta, first with 80 ml sterile saline at +20 $^{\circ}$C and then with 4 % paraformaldehyde (400 ml) in 0.1 M phosphate buffer,

pH 7.4, at +4 ^{o}C. Small pieces of bill skin were excised and
immersed in the same fixative for 24 h at +4 ^{o}C. Subsequently,
the tissues were immersed for 24 h at +4 ^{o}C in the 0.1 M phosphate buffer containing 20 % sucrose. Free floated cryostat
sections 14 /um thick were prepared and immersed in 0.1 M phosphate buffer, pH 7.4, containing 20 % sucrose. Sternberger´s
peroxidase-antiperoxidase (PAP) method (Sternberger et al.,
1970) was applied using the antisera to SP, SOM, VIP, Met-ENK,
CCK, serotonin (SER) and neurotensin (Immuno Nuclear Corp.
Stillwater MN, USA). Any possible endogenous peroxidase activity was destroyed by incubating the sections in 1.2 % hydrogen peroxide in absolute methanol for 30 min. The sections
were then washed in PBS for 30 min and subsequently in PBS containing 1 % normal pig serum and 0.2 % Triton X-100 for 20 min.
The primary antisera were diluted with PBS 1:1000, and to SER
1:1100. The sections were incubated in primary antisera for
24 h at +4 ^{o}C. Following three rinses in PBS/Triton X-100, the
sections were incubated in the secondary serum (antirabbit
IgG-DAKO diluted 1:50) for 30 min at +22 ^{o}C. The developing
PAP complex (DAKO as.) was applied in a dilution of 1:200 at
room temperature for 15 min. Following a brief rinse (10 min)
the sections were transferred into the TRIS buffer, pH 7.54,
containing 0.2 mg/ml 3,3´-diaminobenzidinetetrahydrochloride
(SIGMA, St. Louis, MO, USA) and 0.02 % hydrogen peroxide.
Sections which were incubated in primary antisera preabsorbed
for 24 h at +4 ^{o}C, with the corresponding synthetic antigen at
a concentration of 20 or 200 /ug/ml antiserum at its working
dilution, served as controls.

RESULTS

The most investigated mechanoreceptors in aquatic birds
are Herbst and Grandry corpuscles. Both these receptors have
many ultrastructural similarities with the Pacinian or Merkel
corpuscles in mammals.

The basic cytological compartments of Herbst corpuscles
such as the perineural capsule cells, the subcapsular space
with fibroblasts and macrophages, the innermost Schwann receptor cells surrounding the central non-myelinated portion
of the receptor nerve fiber with its enlarged ending contain
analogous ultrastructural features as those in the Pacinian
corpuscle.

The structural organization of Grandry corpuscles is less
comparable with Merkel corpuscles but the very important scaf-
folding elements present in the Grandry and Merkel cells and
the receptor nerve fibers with their endings also have similar
ultrastructural characteristics.

The most prominent in both these cells are dense core
granules of varying size (100-300 nm). Their number and electron density are concentrated in that part of the cytoplasm
facing the nonmyelinated receptor nerve fiber and its ending.

After employing different antisera, only SP- and SOM-LI
were observed in the examined bird mechanoreceptors.

Light microscopically, SP-LI is concentrated exclusively
in Grandry cells (Fig. 1). The peroxidase reaction product is

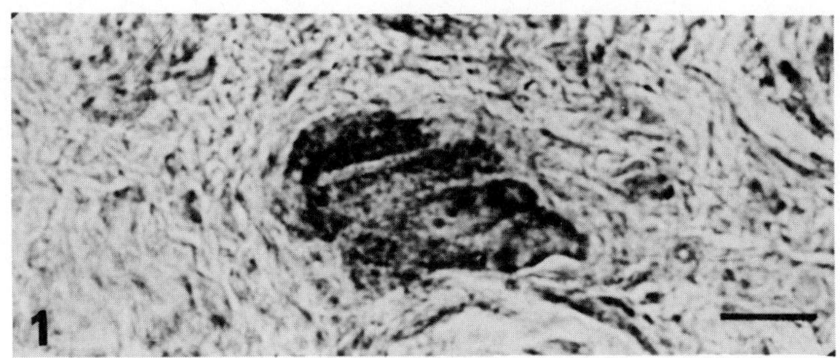

Fig. 1. SP-LI in four Grandry cells. PAP method. Unstained section. Bar: 20 /um.

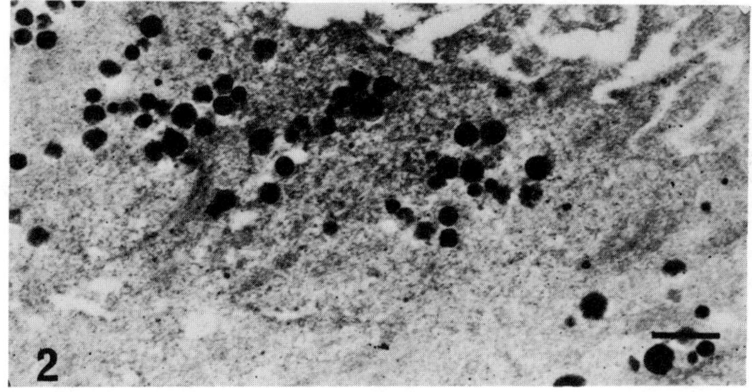

Fig. 2. Strongly positive SP-LI granules in the periphery of Grandry cell. Slightly contrasted section. Bar: 0.5 /um.

mainly polarized towards the unstained location of the receptor nerve fiber. Depending on the cell composition (two or more specialized Grandry cells), Grandry corpuscles show a different intensity of the immunohistochemical reaction.

Electron microscopically, the reaction product is localized unequivocally in the dense core granules of Grandry cells (Fig. 2). The concentration and intensity of immunoreactive grains do not depend on the size or location of the granules (Fig. 3).

SOM-LI is observed in both receptors. Histochemically, the sensory nerve fibers display SOM-LI (Fig. 4). In the electron microscope, the immunoreactivity is localized in the cytosol surrounding the clear vesicles and mitochondria near the axolemma. The cytosol of the innermost lamellae of Schwann receptor cells shows SOM-LI in some receptors.

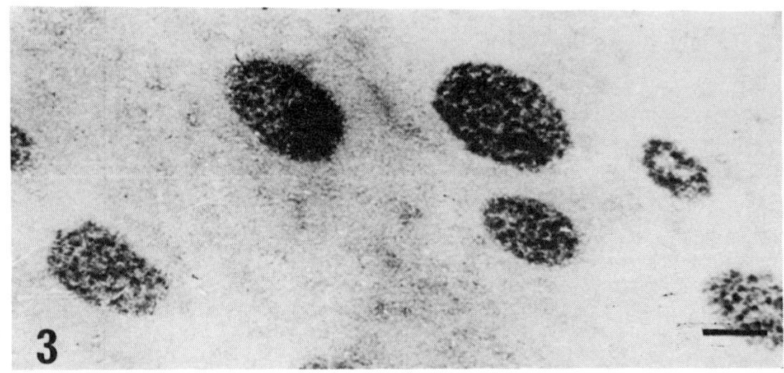

Fig. 3. SP-LI grains in uncontrasted granules of Grandry cell
cytoplasm. Uncontrasted section. Bar: 0.1 /um.

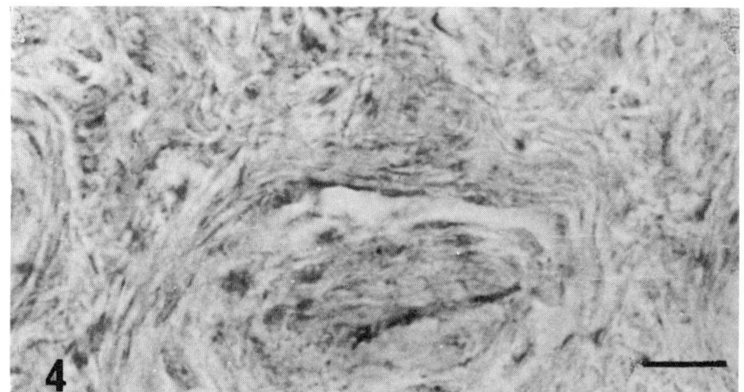

Fig. 4. SOM-LI in the receptor nerve fiber of Herbst corpuscle.
Unstained section. Bar: 20 /um.

DISCUSSION

The present immunocytochemical study implies a new im-
petus in the exploration of the adequate neurotransmitter or
modulator utilized by sensory receptors, although the physiological significance of the present findings is still not
clear. Even the putative nociceptive role of the best analysed
and most investigated SP-peptide in primary afferents is still
obscure. Moreover, we could not observe SP-LI in sensory nerve
fibers entering the corpuscles, contrary to other observations (Dalsgaard et al., 1983; Björklund et al., 1986). Only
dense core granules of Grandry cells display SP-LI. All efforts to prove the content of dense core granules in Grandry
and Merkel cells histochemically have hitherto been unsuccessful. The presented SP-like-peptide can fulfill a neurotransmitter or a modulator role, having in mind the synaptic-like
structures described by Saxod (1978), between Grandry cell
plasmalemma and adjacent axolemma, and an accumulation of dense
core granules nearby. The relationship of Grandry cells to
their afferent axons are probably more close than those of
Merkel corpuscles. Saxod (1978), Idé and Munger (1978), Pač

and Malinovsky (1985) concluded their studies of Grandry corpuscle development that Grandry cells develop in concert with
Schwann cells and afferent axons and thus originate from the
neural crest. Merkel cells apparently do not develop in this
manner. Furthermore, Grandry cells are more dependent upon the
trophic influence of sensory nerve fibers (they completely disintegrate after denervation - Chouchkov, 1978).

Regardless of the presented SP-LI, as well as the close
relationship between sensory innervation and Grandry cells, a
direct and clearcut evidence for the neurotransmitter or modulator role of dense core granules could not be established.
Also, the physiological significance of SOM-LI in receptor
nerve fibers seems more speculative at present. Our unpublished
observations show that some trigeminal neurons of ducks display
SOM-LI but whether the same neurons ensure afferent innervation
of corpuscles remains to be elucidated. Nevertheless, even if
we accept the unique role of SOM-LI trigeminal neurons in afferent receptor innervation, the physiological role of the
synthesized and transported SOM-like peptide remains obscure.
Having in mind that the list of neurotrophic substances is
growing rapidly, the presented SOM-LI in sensory receptors
evokes the option of its trophic influence in the maintenance
of both receptor nerve fibers and the surrounding cellular
medium.

REFERENCES

Björklund, H., Dalsgaard, C.-J., Jonsson, C.-E., and Hermansson,
     A., 1986, Sensory and autonomic innervation of non-hairy
     and hairy human skin. An immunohistochemical study, Cell
     Tissue Res., 243:51-57.
Chouchkov, C., 1978, Cutaneous receptors, Adv. Anat. Embr. Cell
     Biol., 54:1-62.
Dalsgaard, C.-J., Jonsson, C.-E., Hökfelt, T., and Cuello,
     A.C., 1983, Localization of substance P-immunoreactive
     nerve fibers in the human digital skin, Experientia, 39:
     1018-1020.
Hartschuh, W., Weihe, E., Buchler, M., Helmstaedter, V.,
     Feurle, G.E., and Forssmann, W.G., 1979, Met-enkephalin-
     like immunoreactivity in Merkel cell, Cell Tissue Res.,
     201:343-348.
Idé, C., and Munger, B.L., 1978, A cytologic study of Grandry
     corpuscle development in chicken tow skin, J. Comp.
     Neurol., 179:301-324.
Katz, D.M., and Karten, H.J., 1980, Substance P in the vagal
     sensory ganglia: localization in cell bodies and peri-
     cellular arborizations, J. Comp. Neurol., 193:549-564.
Kummer, W., and Heym C., 1986, Correlation of neuronal size
     and peptide immunoreactivity in the quinea-pig trige-
     minal ganglion, Cell Tissue Res., 245:657-665.
Kuwayama, Y., and Stone, R.A., 1986, Neuropeptide immunore-
     activity of pericellular baskets in the guinea-pig tri-
     geminal ganglion, Neurosci. Lett., 64:169-172.
Páč, L., and Malinovský, L., 1985, Development of the Grandry
     corpuscles in the skin of the beak of the domestic duck,
     Folia Morphologica, 33:379-384.
Salt, T.E., and Hill, R.G., 1983, Neurotransmitter candidates
     of somatosensory primary afferent fibres, Neuroscience,
     10:1083-1103.

Saxod, R., 1978, Development of cutaneous sensory receptors in
     birds, in: Handbook of Sensory Physiology, vol.9, M.
     Jacobson, ed., Springer Verlag, Berlin-Heidelberg-New
     York.
Skofitsch, G., Zamir, N., Helke, C.J., Savitt, J.M., and Jaco-
     bowitz, D.M., 1985, Corticotropin releasing factor-like
     immunoreactivity in sensory ganglion and capsaicin sen-
     sitive neurons of the rat central nervous system: co-
     localization with other neuropeptides, Peptides, 6:307-
     318.
Sternberger, L.A., Hardy, P.H., Curculis, J.J., and Meyer,
     H.G., 1970, The unlabelled antibody enzyme method of
     immunohistochemistry, J. Histochem. Cytochem., 18:
     315-333.
Weihe, E., Hartschuh, W., and Weber, E., 1985, Prodynorphin
     opioid peptides in small somatosensory primary afferents
     of guinea pig, Neurosci. Lett., 58:347-352.

EFFECTS OF NERVE INJURY AND COLCHICINE TREATMENT ON THE RECO-
VERY OF NON-SPECIFIC CHOLINESTERASE ACTIVITY IN SPECIALIZED
SCHWANN CELLS OF RAT SIMPLE LAMELLAR CORPUSCLES

Petr Dubový, Jan Hájek, Ivana Svíženská,
Lubomír Malinovský and Hana Procházková

Department of Anatomy, Medical Faculty
Purkyně University, Brno, Czechoslovakia

INTRODUCTION

The high level of non-specific cholinesterase (nCHE)
activity is a remarkable feature of sensory corpuscles. In
the light microscope, the nCHE staining persists in sensory
corpuscles for a long time after nerve lesion. However, the
electron microscopical findings suggest that the interaction
with intact axon terminals is indispensable for the synthesis
of nCHE in the sensory corpuscles (Idé, 1982a,b).

The nCHE molecules are, however, resistant to proteinase
digestion and their histochemical localization in the rough
endoplasmic reticulum (rER) is irregular like that of the
other enzymes (Novikoff, 1976; Dubový and Malinovský, 1982).
Therefore, the direct histochemical study of nCHE in sensory
corpuscles after nerve lesion is not suitable for demonstrating a neurotrophic influence on nCHE synthesis.

In the present study we explored the recovery of nCHE
staining in the sensory corpuscles of rat glabrous skin after
irreversible inhibition following injury to the sciatic and
saphenous nerve and after colchicine treatment.

MATERIAL AND METHODS

The study of nCHE recovery was investigated in the simple
lamellar corpuscles (SLC) of hindfoot glabrous skin from 10
adult and four 5-day-old Wistar rats of both sexes. Under
ether anesthesia, the right sciatic and saphenous nerves of
adult rats were ligated, crushed or transected. In the latter
operation, about 10 mm of the distal stump was removed to prevent regeneration. Cuffs prepared from an artificial fibrin
sponge (Gelaspon[R], Jenapharm) and soaked with 25 mM colchicine
dissolved in Ringer´s solution were applied around the right
sciatic and saphenous nerve. The cuffs soaked with Ringer´s
solution alone were applied around contralateral nerves for
control experiments. The 5-day-old rats were anesthetized by
being placed on ice. Their right sciatic and saphenous nerves

were crushed and transected as in adult animals.

Immediately after treatment and 5, 11 and 28 days later, the nCHE activity was irreversibly inhibited in vivo by subcutaneous injections of 6 μmol/kg b.wt. iso-OMPA (Koch Light). In preliminary experiments, we ascertained that the concentration of iso-OMPA is sufficient for total histochemical inhibition of nCHE in rat sensory corpuscles. At 3, 6, 12, 20 and 35 days after the operations, specimens of glabrous skin from the sole of the right (operated) and left (control) feet were removed and placed into 2 % paraformaldehyde, 2 % glutaraldehyde and 7.5 % saccharose in a 0.1 M cacodylate buffer (pH 7.2) for 2 hours.

For light microscopy, cryostat sections, 20 μm thick, were incubated in a medium (Karnovsky, 1964) containing butyrylthiocholine bromide and BW284c51 (1 mM) for 40 min. Unfrozen sections, 100 μm thick, were treated with the same medium for 20 min and processed for electron microscopy. The material was postfixed with 1 % OsO_4 for 2 hours, dehydrated and embedded in Durcupan. Semi-thin sections were stained with toluidine blue and examined for simple lamellar corpuscles. Ultrathin sections without lead citrate counterstaining were searched for the nCHE end-product with a Tesla BS 500 electron microscope.

RESULTS

Using the light microscope, the nCHE reactivity was demonstrated histochemically in sensory corpuscles in the dermal papillae of the skin and in some cases in the other locations beneath the epidermis. No nCHE staining occurred in the skin within 2 days after iso-OMPA injections. In the skin from control hind limbs, a weak reaction for nCHE reappeared in SLC at 3 days. A more intense reaction appeared on the sixth and subsequent days after inhibition. The sensory corpuscles of the skin from operated hind limbs were faintly restained for nCHE activity after the same period of time after application of the inhibitor as control ones, and the reaction was more intense on subsequent days.

In the electron microscope, two types of SLC were distinguished in the rat skin from the sole of the hindpaw. The axon of the sensory corpuscles in the dermal papillae was surrounded by inner core lamellae circularly arranged in the cross sections. Another type of corpuscle located elsewhere than in the dermal papillae was composed of several axonal profiles which were invested by several thin processes of Schwann cells. In both types of corpuscles the nCHE end product was confined to the extracellular spaces of the inner core cell and Schwann cell processes. Only in some cases was the end product found in rER.

Three and six days after colchicine application onto the nerves, the ultrastructure of the axonal terminals disintegrated (Fig. 2). The crushing or transection of nerves led to axon degeneration in SLC. Within 3 to 6 days after nerve injury, SLC became devoid of axons and only some axonal debris could be found (Fig. 3). A longer survival time after tran-

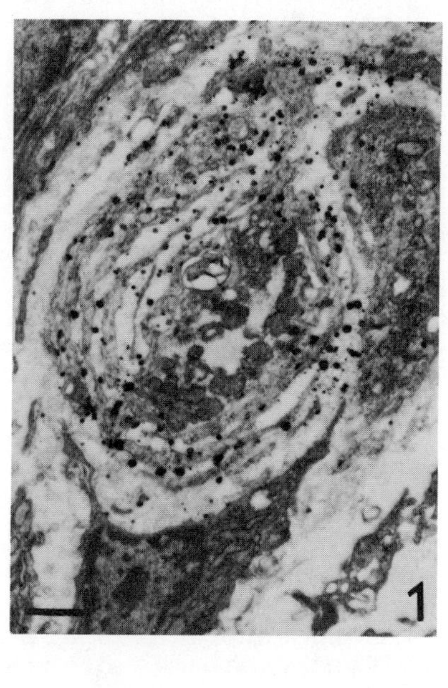
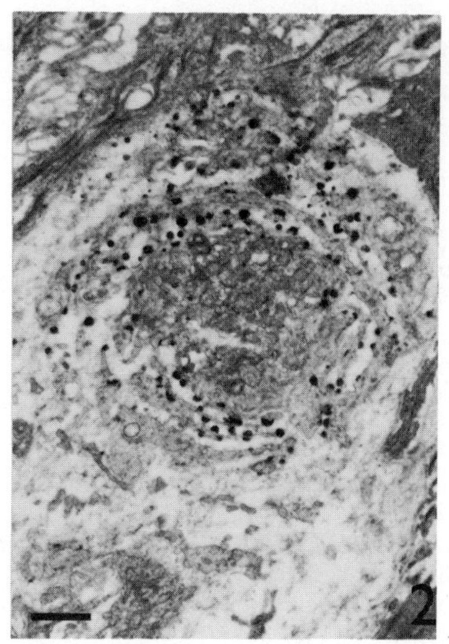
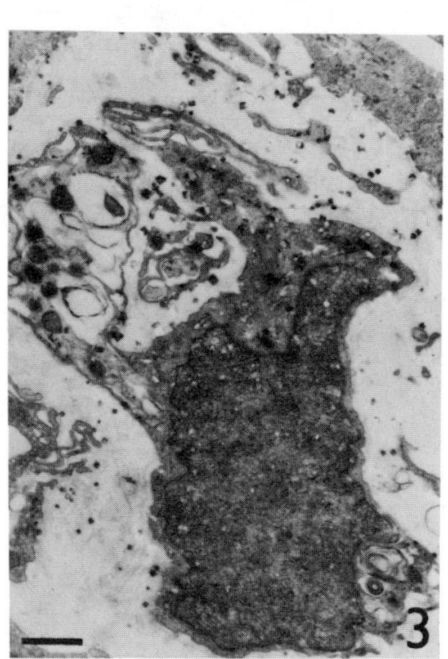
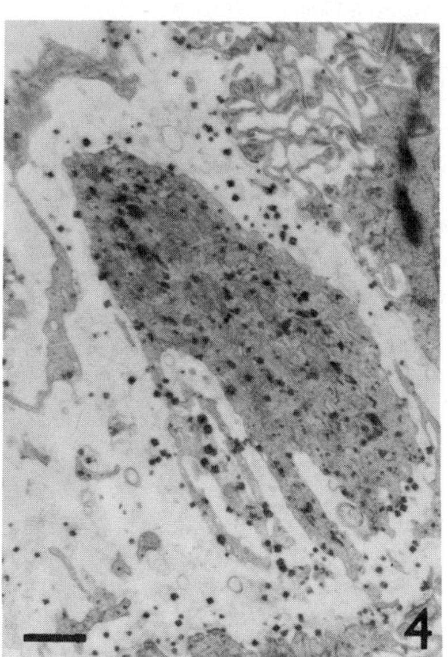

Figures 1-4. SLC of adult rat hind limb skin incubated for nCHE.
Fig. 1. Control, 6 days after nCHE inhibition.
Fig. 2. 6 days after colchicine treatment and nCHE
inhibition. Fig. 3. 6 days after crush and nCHE
inhibition. Fig. 4. 35 days after nerve transection and 7 days after nCHE inhibition. Bars =
= 0.8 μm.

section of the nerves brought about the total disappearance of
the axonal structures in cutaneous SLC of adult and 5-day-old
rats (Fig. 4). Distinct products of the nCHE reaction reappeared
in SLC of the skin from both control and operated hind limbs
after 3 and later days after inhibition (Figs. 1-4).

DISCUSSION

The molecules of nCHE are considered to be synthetized in
rER of a specialized type of Schwann cells (Idé and Saito,
1980). Besides the membrane-bound molecular forms of nCHE demonstrated in Pacinian corpuscles (Dubový, 1983), the nCHE
molecules in sensory corpuscles of the skin are released into
the interlamellar spaces of the inner core (Dubový and Černá,
1987).

The nCHE activity in the cat is almost totally inactivated
by the injection of iso-OMPA, 6 μmol/kg b.wt. (Uchida and
Koelle, 1983). The same concentration of iso-OMPA also inacti-
vates nCHE in SLC of the rat skin. The nCHE activity became
detectable by the histochemical method from the 3rd day after
inhibition. Enhanced staining of SLC on subsequent days after
the inhibition demonstrates the concomitant synthesis of new
molecules of the enzyme in the control as well as the operated
hind limbs. The sum of the findings suggests that nCHE is
indeed synthetized in SLC independently of the presence of
sensory axons or upon the transport of neurotrophic factors.

The physiological function of cutaneous mechanoreceptors
in the rat can be restored after nerve lesions (Sanders and
Zimmermann, 1986). Recently, it has been emphasized that the
basal lamina plays an important role in the regeneration of
mouse digital corpuscles (Idé, 1986). Furthermore, it has been
demonstrated that the nCHE activity persists in the interlamellar spaces up to the reinnervation of denervated sensory corpuscles (Idé, 1982b). Our present and previous findings suggest
the possible involvement of the nCHE molecules during the reinnervation of sensory corpuscles in the skin.

REFERENCES

Dubový, P., 1983, To the problem of localization of cholin-
        esterase activity in Pacinian corpuscles by electron
        microscopy, Histochem. J., 15:287-288.
Dubový, P., and Malinovský, L., 1982, Electron-microscopic lo-
        calization of cholinesterase activity in Pacinian cor-
        puscles of the cat mesentery, Z. mikrosk.-anat. Forsch.,
        96:802-816.
Dubový, P., and Černá, P., 1987, Evidence for the presence of
        different molecular forms of non-specific cholinesterase
        in Pacinian and simple sensory corpuscles. A histoche-
        mical and biochemical study, Scripta medica, 60:51-60.
Idé, C., 1982a, Degeneration of mouse digital corpuscles, Amer.
        J. Anat., 163:59-72.
Idé, C., 1982b, Regeneration of mouse digital corpuscles, Amer.
        J. Anat., 163:73-85.
Idé, C., 1986, Basal laminae and Meissner corpuscle regenera-
        tion, Brain Res., 384:311-322.

Idé, C., and Saito, T., 1980, Electron microscopic cytochemis-
    try of cholinesterase activity of mouse digital cor-
    puscle, Acta histochem. cytochem., 13:218-226.
Karnovsky, M.J., 1964, The localization of cholinesterase ac-
    tivity in rat cardiac muscle by electron microscopy,
    J. Cell Biol., 23:217-232.
Novikoff, A.B., 1976, The endoplasmic reticulum: a cyto-
    chemist's view, Proc. Natl. Acad. Sci. USA, 73:2781-2787.
Sanders, K.H., and Zimmermann, M., 1986, Mechanoreceptors in
    rat glabrous skin: redevelopment of function after nerve
    crush, J. Neurophysiol., 55:644-659.
Uchida, E., and Koelle, G.B., 1983, Identification of the pro-
    bable site of synthesis of butyrylcholinesterase in the
    superior cervical and ciliary ganglia of the cat, Proc.
    Natl. Acad. Sci. USA, 80:6723-6727.

HISTOCHEMICAL EVIDENCE OF DIPEPTIDYLPEPTIDASE IV ACTIVITY IN

THE SCHWANN CELLS SURROUNDING UNMYELINATED PORTIONS OF AXONS

Petr Dubový

Department of Anatomy
Medical Faculty, Purkyně University
Brno, Czechoslovakia

Dipeptidylpeptidase IV (DPP IV) is a serine exopeptidase
which removes X-Pro sequences from the N-terminal of natural
peptides or artificial substrates. Its biological role in peripheral nerve structures is still obscure. However, the ability of DPP IV to degrade substance P (SP) is attractive for
the explanation of the function of this enzyme in peripheral
nerve structures of the skin.

The histochemical localization of the DPP IV activity was
performed in cat glabrous skin of rhinarium and hind paw-pads
and in the skin from the 2nd, 3rd and 4th fingertips of rhesus
monkeys. Cryostat sections of fresh and paraformaldehyde-fixed
skin were incubated for light and electron microscopic localization of the DPP IV activity in peripheral nerve structures
(Dubový, in press).

In the light microscope, a high DPP IV activity was demonstrated in the blood vessels and in the perineurium of
larger nerve fascicles as described previously (Dubový and
Malinovský, 1984). The nerve fibers with a moderate azo-dye
corresponding to the DPP IV activity were found in the vicinity of blood vessels and sweat glands as well as beneath
the epidermis of glabrous skin.

Using the electron microscope, a fine electron-dense end
product was revealed in the plasma membrane both of Schwann
cell investment of unmyelinated axons and of modified Schwann
cells enveloping unmyelinated portions of sensory axons
in sensory corpuscles. In contrast to the plasma membrane of
Schwann cells, the axolemma appeared to be free of the reaction product for DPP IV.

DPP IV has the ability to cleave successively the SP(1-2),
SP(3-4) and SP(5-11) fragments from parental SP(1-11) (Heymann
and Mentlein, 1978). There is evidence that synthetic analogs
of the SP fragments corresponding to those obtained by the
action of DPP IV influence the growth of glial cells (Lindner,
1984). The mammalian skin contains numerous nerve fibers exhibiting a SP-like immunoreactivity and they also penetrate the

sensory corpuscles, e.g. Meissner corpuscles (Dalsgaard et al., 1983). These nerve fibers are located in the same loci as those in which the DPP IV reactivity was found. This suggests the functional association of the SP-containing axons with the membrane-bound DPP IV of their Schwann cells.

It has been proposed that a trophic influence of SP-containing nerve fibers on the target cells might be mediated through the SP fragments produced by the action of DPP IV (Dubový, in press). The functional relation of the SP-containing axons and the membrane-bound DPP IV might also represent one of the trophic interactions in the sensory corpuscles.

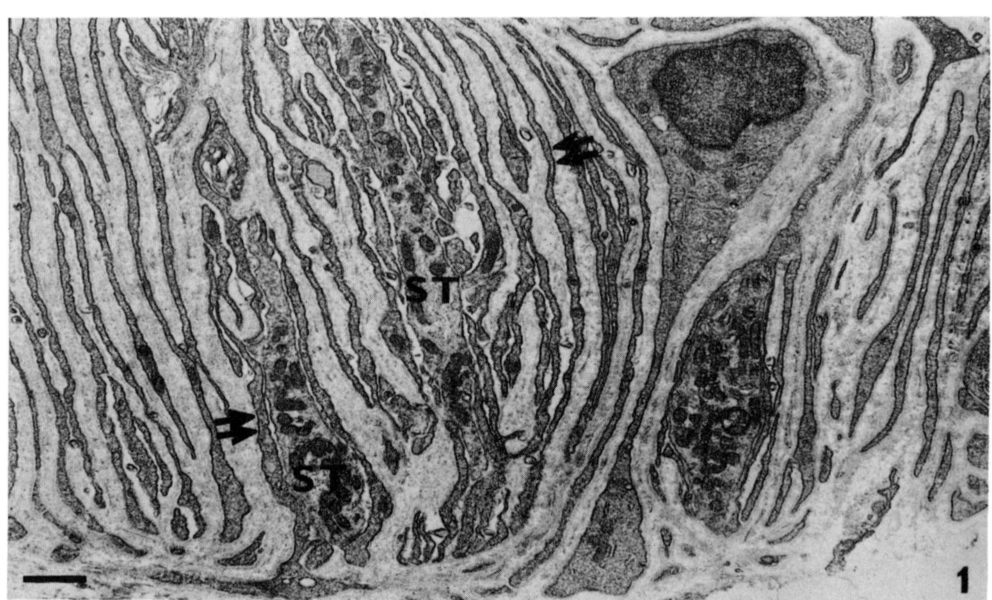

Fig. 1. An ultrathin section through a macaque Meissner corpuscle without lead citrate staining. A fine reaction product corresponding to DPP IV activity loaded the plasma membrane of Schwann cell processes (double arrows) in contrast to the axolemma of sensory terminals (ST). Bar = 1.2 μm.

REFERENCES

Dalsgaard, C.J., Jonsson, C.E., Hökfelt, T., and Cuello, A.C., 1983, Localization of substance P-immunoreactive nerve fibers in the human digital skin, Experientia, 39: 118-120.
Dubový, P., A study of the dipeptidylpeptidase IV activity in cat fungiform papillae, Acta histochem., in press.
Dubový, P., and Malinovský, L., 1984, Localization of DPP IV in sensory corpuscles by means of light and electron microscope, Histochem. J., 16:473-475.
Heymann, E., and Mentlein, R., 1978, Liver dipeptidylaminopeptidase IV hydrolyses substance P, FEBS Lett., 91: 360-364.
Lindner, G., 1984, On the effect of N- and C-terminal sequences of the substance P on nonneuronal cells in vitro, Z. mikrosk.-anat. Forsch., 98:107-118.

JUXTAORAL ORGAN: PRESENT KNOWLEDGE ON THE DEVELOPMENT AND

MORPHOLOGY OF AN ORGAN OF UNKNOWN FUNCTION

Robert Mayr

Institut für Anatomie 2
Universität Wien
Währingerstrasse 13
A - 1090 Wien, Austria

History of discovery. In the human embryo, a solid epithelial cell bud
developing earlier than the anlage of the parotid gland from the ectoderm of
the primitive oral cavity (Chievitz 1885) was considered to be a rudimentary,
prenatally degenerating anlage of a salivary gland by earlier embryologists
(Weishaupt, 1911). This anlage was then known as the "organ of Chievitz" by
embryologists (Strandberg, 1918). No correlate of this early fetal organ was
found in the older fetus, at birth, or in postnatal life, neither in man nor
in any animal, for half a century. Ramsay (1953) was the first to encounter
the epithelial cell strand in newborns, and presented an excellent study of
its development. He saw no justification to consider the cell strand as a
gland, neither an exo- nor an endocrine one. Instead he suggested it to be an
epithelial remnant, resulting from the fusion of the maxillary and mandibular
processes in a similar way and without any further function than the epithelial remnants in the raphe of the palate or in the cervical vesicles. In 1946
Massart described the persistence of the organ of Chievitz in adult chirop-
terans, in a local Italian journal. Soon thereafter, Wolfgang Zenker, then a
medical student and "Demonstrator" at the Department of Anatomy of Vienna do-
ing his first steps in scientific research, found the correlate of the fetal
epithelial cell bud known as the organ of Chievitz to persist postnatally and
to occur regularly in adult man as the central parenchymal element of a
rather small, but complex paired organ in the depth of the cheek. His discovery was first reported at the 51st meeting of the "Anatomische Gesell-
schaft" at Mainz in 1953 under the heading "Organon buccotemporale (Chievitz'
sches Organ), ein nervös-epitheliales Organ beim Menschen" (Zenker 1954). -
For more than 20 years, research on this peculiar organ of up to now unknown
function was done by Zenker and his coworkers only, presumably because of the
organ's small size, its hidden location on the medial side of the mandible
in the deep buccal region, the impossibility to see and dissect it with the
naked eye or even under the dissecting microscope, and possibly also because
of unfavourable publication of the papers of that period in German language.
In that period of intense research on this organ in Vienna, the constant
existence of the organ, its topographical relations, the abundance of its
innervation by the buccal nerve, its development in man, its comparative
anatomy, its characteristic tissue composition in all species examined, its
ultrastructure in rat and in man, have been topics of investigation; further,
a first histochemical study in man and some experimental studies in the rat,
aiming at unraveling clues to the organ's function, have been performed. It
is thence known as the "Juxtaoral Organ" or "Zenker's Organ" in literature.

Development. The early development of Chievitz' organ has been studied
by a number of embryologists between 1885 and 1950 (for references see Zenker
1982). The anlage found in embryos of man, of many mammals, of saurians and
ophidians, was believed to undergo atrophy and to disappear in the second
half of fetal life. After the discovery of the persistence of the organ
throughout adult life (Zenker 1953), Zenker and Halzl (1953) performed a
meticulous reinvestigation of the organ's ontogenesis in man, using more than
100 serially sectioned human embryos of our Viennese collections, above all
of Hochstetter's collection. They found that, in human embryos 7.5 to 12 mm
in length, the first anlage of the organ's epithelial parenchyma appears as a
thickening of the oral epithelium at the most lateral edge of the buccal
sulcus. This thickening extends for more than half the length of that groove;
from its early origin the parenchymal cord of the organ retains its elongated
form and its mesio-distal orientation in the cheek. The separation of the
parenchymal cord from the epithelial ridge covering the sulcus buccalis at
its lateral edge, is initiated by the process of innervation by the buccal
nerve. It starts aborally, proceeds orally, and is concluded in embryos 14 to
17 mm in length. After separation, the oral (= mesial) part of the epithelial
cord breaks into fragments and disappears. The distal part, as if drawn laterally by the buccal nerve, becomes dislocated into the depth of the cheek to
its definitive location and topography, and becomes invested by more or less
complex nerve terminals and connective tissue. Thus the definitive tissue
composition in the adult is being accomplished.

Morphology. The juxtaoral organ, obviously most highly developed in man,
exhibits essentially the same topography in all mammals: The elongated organ
is found (i) on the medial side of the mandibular ramus and of the insertion
of the temporalis muscle, (ii) on the lateral surface of the buccinator
muscle, (iii) in the near vicinity and almost in parallel to the buccal
vessels and the buccal nerve stem, with 2 - 4 branches of the latter entering
the organ, and (iv) in man underneath and attached to the buccotemporal
fascia, at the lower border of Bichat's fat body. - The size of the organ
varies widely between species, however, far less within a species (in man: 7 -
17 mm x 1 - 2 mm; in rodents and carnivors: 3 - 6 mm x less than 100 μm; in
the duckbill: 2 mm x 40 - 60 μm). - The characteristic tissue composition of
Zenker's organ is most evident in coronal (= cross) sections: (i) A continuous central cord of epithelial parenchyma (in man several 100 μm in diameter
with thick sprouts or thin ramifying strands, in most mammals a slender simple cord) is wrapped by (ii) an inner connective tissue capsule consisting of
fibrocytes, collagen and elastic fibres, and harbouring a high amount of
nerve terminals, (iii) a fluid space traversed by myelinated and non-myelinated nerve fibres, and (iv) an outer capsule consisting of a multi-lamellated
perineural sheath surrounded by dense connective tissue. The whole organ is
embedded in adipose tissue, except at its fixation area to the buccotemporal
fascia. - As shown in TEM (Mayr and Salzer 1967, 1968), the parenchyma is
composed of interdigitating epithelial cells connected by desmosomes and
tight junctions. The cells contain many tonofilament bundles, but do not show
any signs of keratinization. Golgi lamellae, some rough ER, but more free
ribosomes, some lipid droplets, and several groups of dense cored granules of
varying size, foremost located near the surface of the parenchyma, are found.
Only in man, this Malpighian (= keratinocyte) type cell has highly indented
nuclei, and only in man, in addition to this dominant cell type, 7% of non-
specific dendritic cells (i.e. dendritic cells lacking the characteristic
features of melanocytes or Langerhans' cells) are found.

Function. The function is unknown. Experimental data and structural indications of function are discussed in another paper (Mayr, this symposium).

Literature. All the references cited in the text are to be found in:
Zenker, W., 1982, "Juxtaoral Organ (Chievitz' Organ), Morphology & Clinical
    Aspects", Urban & Schwarzenberg, Baltimore - Munich.

310

JUXTAORAL ORGAN: ULTRASTRUCTURE AND FEATURES INDICATING

A MECHANORECEPTIVE FUNCTION

Robert Mayr

Institut für Anatomie 2
Universität Wien
Währingerstrasse 13
A - 1090 Wien, Austria

Present knowledge about the history of the discovery,
about the development and the morphology of the juxtaoral
organ is shortly summarized in a separate contribution (Mayr,
this volume). Although regularly present as a paired formation
in both cheeks in man, in all other mammals, as well as in
certain reptiles (squamates), but not found in other vertebrates
(amphibia, aves, pisces), the function of this organ is
still unknown. After an extensive study of its comparative anatomy, Salzer and Zenker (1962) introduced the descriptive designation "the juxtaoral organ", according to the most typical
common feature (i.e. the paraoral location) of the homologous
organs in mammals and reptiles. Concerning the organ's functional significance, two lines of suggestions have to be considered: (i) Indications of a secretory function of the parenchyma of the organ, and (ii) morphological features indicating a
(mechano-)receptive function of the organ (cf. Zenker,1982).
None of the two suggested functions excludes the other. Therefore, both lines of suggestions would require functional studies for verification. Experimental studies in the rat showed
that the organ undergoes a high degree of atrophy after hypophysectomy (Salzer and Zenker,1968). This result is in favour
of an endocrine function. The scarcity of capillaries within
and around the organ, however, and the modest indications of
secretory activity of the parenchymal cells in ultrastructural
studies (Salzer et al.,1964, rat; Mayr and Salzer,1967, man;
Jeanneret-Gris and Mayr,1980; Jeanneret-Gris,1980, hamster),
do not convincingly support the hypothesis of a glandular
function of the organ. - Suggestions of a receptive function
are so far exclusively based on the interpretation of morphological findings. In this paper, an electron microscopic study
of the guinea pig juxtaoral organ is presented. Its results
strongly support the idea that the juxtaoral organ is a mechanoreceptor transducing tensile forces, but this does not exclude an additional function.

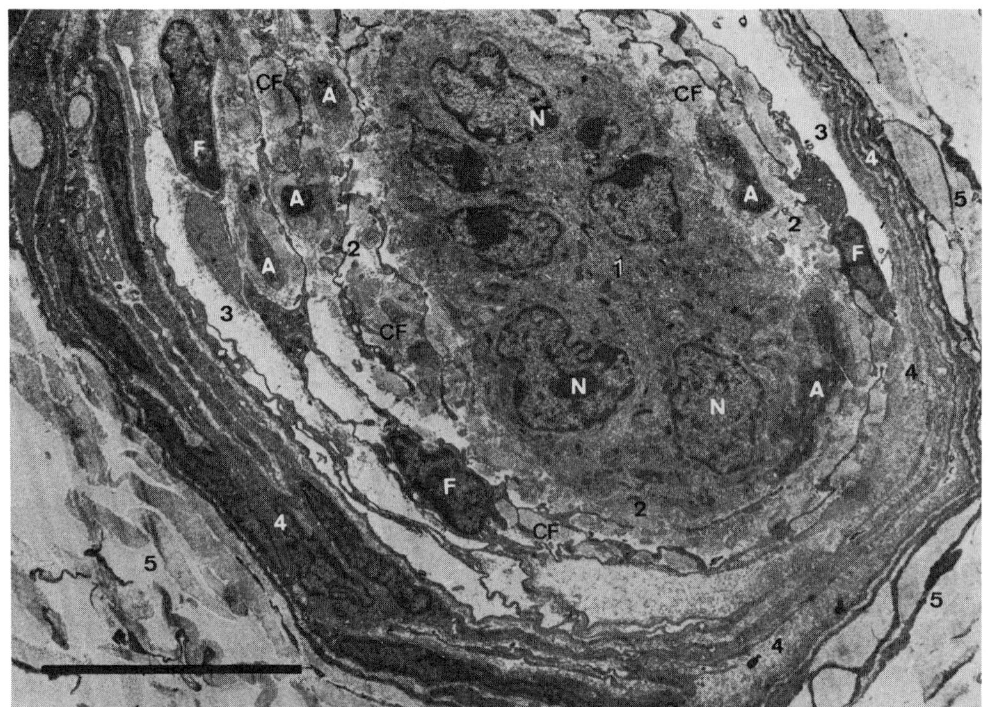

Fig. 1. Guinea pig juxtaoral organ, typical cross section. (1)
Central cord of parenchyma composed of keratinocyte like
cells with indented nuclei (N), (2) inner capsule containing terminal axons (A) amidst connective tissue fiber
compartments (CF), fibrocytes (F), (3) subperineural
space, (4) perineural capsule, and (5) outer fibrous
capsule. Bar = 10 μm.

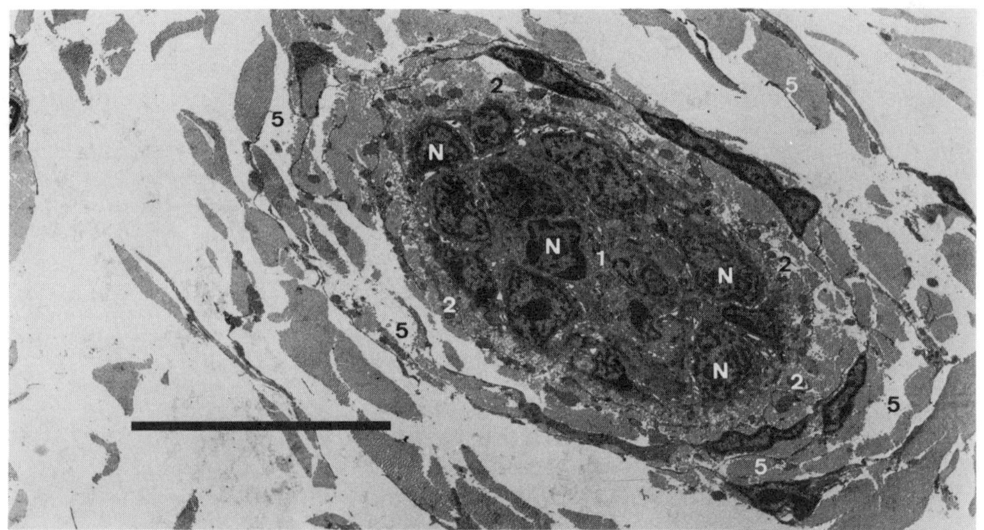

Fig. 2. Guinea pig juxtaoral organ, cross section near to its
distal end. (1) Extracapsular part of the parenchymal
cord with many nuclear profiles (N), surrounded by (2)
an inner capsule devoid of nerve fibers and (5) the
outer fibrous capsule, only. Bar = 10 μm.

Five adult guinea pigs of about 600 g body weight were
fixed by perfusion with 5 % glutaraldehyde in 0.1 M cacodylate
buffer at pH 7.4 under Nembutal anesthesia. Tissue blocks of
both cheeks treated with 2 % OsO_4 and embedded in Epon were
cut in coronary planes. Alternating series of semi- and ultrathin cross sections were cut as long as the juxtaoral organ
was found on the block. Semithin sections were stained with
toluidine blue for light microscopy, ultrathin sections were
contrasted by double staining with lead citrate and uranyl
acetate and examined in an AEI 95 electron microscope.

Results

Size, location and general structure. The juxtaoral organ
of the guinea pig is a slender structure several mm in length
and only 30 - 50 µm in diameter. It is located in loose connective tissue in the vicinity of the buccal nerve and vessels
at the lateral surface of the buccinator muscle, medial to the
mandibular ramus, with its longitudinal axis oriented in the
mesio-distal direction, about opposite to the molar teeth. Its
central element is a simple cord of epithelial parenchyma only
about 10 µm in diameter, composed of elongated keratinocyte
type cells. For most of its length, the parenchymal cord is surrounded by an inner connective tissue capsule, followed by a
subperineural fluid space of varying width, and a perineural
capsule composed of 4 - 7 cell layers. The latter is reinforced
by an outer fibrous capsule with longitudinally arranged collagen bundles. A typical cross section is shown in Fig. 1. Near
to the end of the parenchymal cord (Fig. 2), the organ has lost
its perineural sheath together with the subperineural space,
and is thus reduced in diameter. There, the parenchyma is invested with fine elastic and collagen fibers of the inner capsule and the collagen bundles of the outer capsule only. Its
diameter is approximately equal to that within the perineural
sheath; it exhibits, however, about twice the number of cells
and nuclei on cross section.

The parenchyma. The Malpighian (=keratinocyte) type cells
composing the parenchymal cord (Fig. 3) have interdigitating
surfaces and are connected by many desmosomes as well as by
some tight junctions. The elongated cells have elongated, more
or less indented nuclei. Their cytoplasm contains bundles of
tonofilaments in a predominantly longitudinal arrangement, many
free ribosomes, and moderate amounts of small mitochondrial
profiles. In the central encapsulated region the parenchyma
cells are larger on cross section, and show more stacks of Golgi
lamellae and some cisterns of rough endoplasmic reticulum.

Inner capsule and nerve terminals. A few small nerves composed of some myelinated and occasional non-myelinated nerve
fibers, enter the organ under confluence of their perineurium
with that of the organ. Coursing for various distances and undergoing some divisions, the nerve fibers traverse the subperineural space and enter the inner capsule. The latter is one to
several microns thick and consists of a rather dense layer of
longitudinally oriented collagenous fibrils, microfibrils, and
thin elastic fibers. It is immediately apposed to the basal
lamina of the parenchyma and, where it contains preterminal and
terminal axons, it is better developed and shows compartments

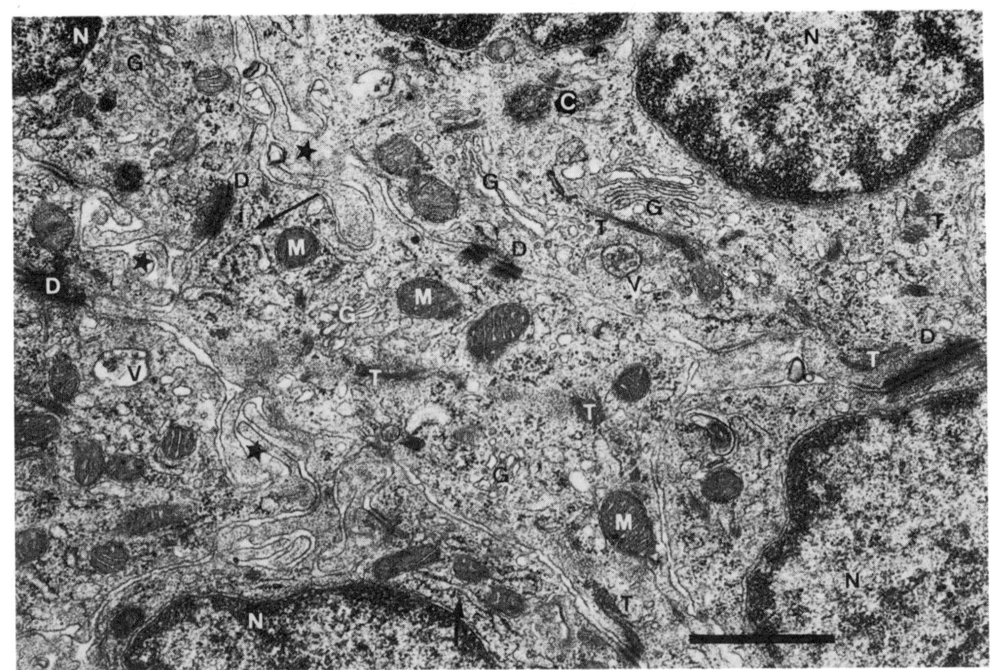

Fig. 3. Parenchyma. Epithelial cell profiles interdigitating
(*) and connected by desmosomes (D). Tonofilament
bundles (T), Golgi lamellae (G), cisterns of rough ER
(arrows), multivesicular bodies (V), mitochondria (M),
a centriol (C), and nuclei (B). Bar = 1 μm.

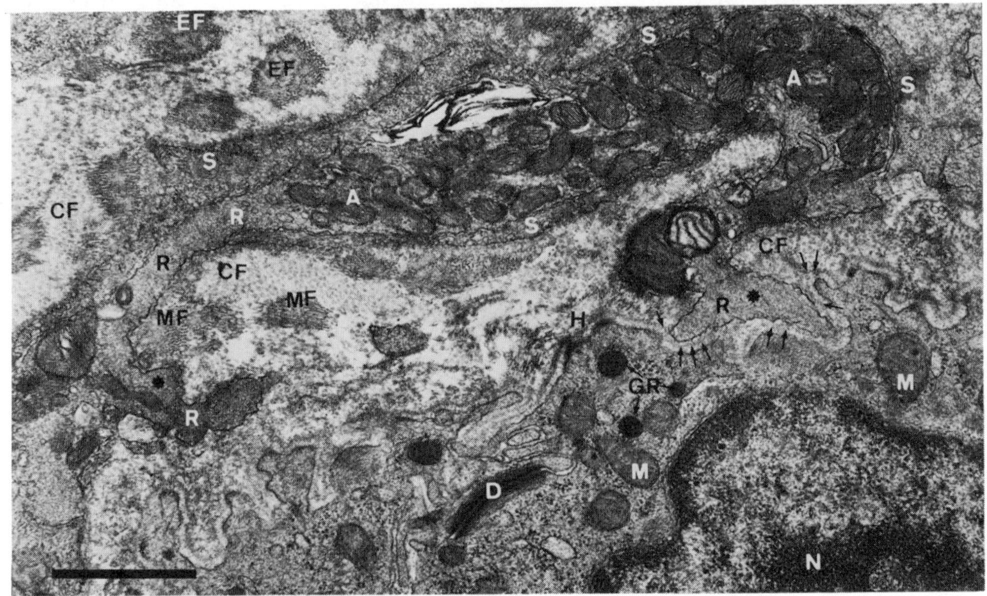

Fig. 4. Terminal axon (A) with axonal protrusions (*) containing receptorplasm (R) and contacting basal lamina of
parenchyma (arrows) or connective tissue filaments.
Schwann cell coverings (S), thin elastic fibers (EF),
collagen (CF) and microfibril bundles (MF), hemidesmosomes (H), and dense core granules (GR). Bar = 1 μm.

314

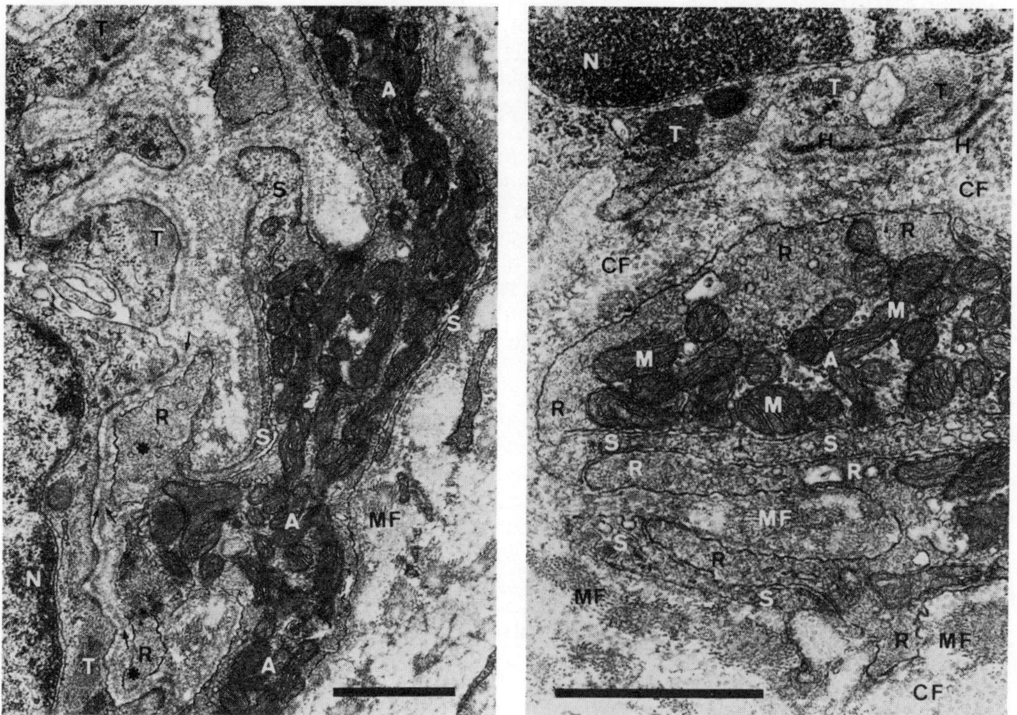

Figs. 5 and 6. Terminal axons (A). Compare caption of Fig. 4.
Bar = 1 μm.

surrounded by flat connective tissue cells. At the hemidesmosomes near the tips of basal processes of parenchyma cells,
microfilament bundles seem to be anchored to the basal lamina.
- In the inner capsule, the nerve fibers lose their myelin and
later even their neurolemmal investments. Axon terminals (Figs.
4-6) are conspicuous by accumulations of mitochondrial profiles;
subjacent to the axolemma and in axonal processes bare of
Schwann cell covering, here and there are even free of basal
lamina investment. The axoplasm has the typical appearance of
so called receptorplasm, i.e. it contains only fine filamentous
material and small irregular vesicles. These areas of axoplasm
are in immediate contact mainly with collagen and microfibril
bundles of the inner capsule, but also with the basal lamina of
the parenchyma. Often axon processes are found in indentations
of the parenchymal surface. Near to axon terminals the parenchyma may contain large dense core granules of varying size.

　　Perineural and outer capsule. The perineural capsule consists of 4 to 7 lamellae of flat perineural cells invested by
basal laminae. The interlamellar spaces contain scattered collagenous fibrils and small bundles of microfibrils. Occasionally a capillary is found between the outer lamellae, but never
inside the perineural capsule. The outer capsule is composed
of rather thick bundles of collagenous fibrils in strictly
longitudinal orientation and some fibrocytes.

<u>Discussion</u>

The juxtaoral organ of the guinea pig is, on the whole,
more similar in size and structure to that of the rat (Salzer
et al.,1964) and hamster (Jeanneret-Gris,1980) than to that of
man (Mayr and Salzer,1967). Nerve terminals are generally found
between the connective tissue fibers of the inner capsule and
may exhibit more complex structures in man. Only in the guinea
pig, direct contacts of axon processes with the basal lamina of
the parenchyma have been observed. On the other hand, no intra-
parenchymal axons were found in this species. As in the rat and
hamster, the parenchyma is a simple cord of keratinocyte type
cells. An extracapsular part of the parenchyma near the end of
the organ has been described for the first time.

Many structural features of the juxtaoral organ indicate
that it may be a complex encapsulated mechanoreceptor transducing stretch exerted in the axial direction of the organ. Its
central element, the elongated parenchymal cord, consisting of
elongated fusiform cells, is longitudinally traversed by tonofilament bundles connected by desmosomes. The inner capsule composed of strictly longitudinally arranged collagen and elastic
fibers is also apt to take up length tension and it adheres to
the parenchyma at many hemidesmosomes underlying its basal lamina. Together with the parenchymal cord it pierces the perineural capsule, which serves as a diffusion barrier for the
receptive part of the organ. On its outside, the perineurium is
reinforced by the longitudinal collagenous bundles of the outer
capsule. At the end of the perineural capsule the outer capsule
fuses with the inner one around the parenchyma. By means of the
collagen fibers of these capsules both ends of the organ are
anchored to the connective tissue of the buccotemporal space.

In its general structure, the fusiform encapsulated juxtaoral organ with its specific central element susceptive to
stretch and associated with axon terminals has many similarities
to other fusiform encapsulated mechanoreceptors such as muscle
spindles (Barker,1974), Golgi tendon organs (Schoultz and Swett,
1972), and Ruffini corpuscles (Chambers et al.,1972). In the
muscle spindle specifically differentiated parts of the terminal
axons containing receptorplasm (Andres and Düring,1973) are in
close association to the intrafusal muscle fibers only, and not
to inner capsule connective tissue fibrils. In the juxtaoral
organ axon terminals are regularly associated with inner capsule
fibril bundles comparable to the association of the branched
endings to collagen fiber bundles in Golgi tendon organs and
Ruffini corpuscles; however, processes of endaxons were also
found in association with the basal lamina of the central parenchymal cord in the guinea pig.

In the juxtaoral organ, as in Golgi tendon organs and
Ruffini corpuscles, the axon terminal itself seems to represent
the receptor or transducer (Andres and Düring,1973; Iggo,1976).
The role of the parenchymal cord known to undergo atrophy after
hypophysectomy (Salzer and Zenker,1968) remains, however, obscure. In the receptive part of the organ, parenchymal cells
show more signs of cell activity than outside the perineural
capsule. Naked intraepithelial axons have been observed in man,
rat and hamster, but not in the guinea pig. Their afferent or
efferent nature has not been established.

        The development of the juxtaoral organ starts very early
in ontogenesis (Zenker and Halzl,1953). This is in contrast to
the rather late onset of development of many other encapsulated
mechanoreceptors (Zelená,1978).

<u>References</u>

Andres, K.H., and von Düring, M., 1973, Morphology of cutaneous
        receptors, <u>in</u>:"Somatosensory Systems",A. Iggo, ed.,
        Handbook of Sensory Physiology, vol. II:2-28, Springer-
        Verlag, Berlin - Heidelberg - New York.
Barker, D., 1974, The morphology of muscle receptors, <u>in</u>:
        "Muscle Receptors", C.C. Hunt, ed., Handbook of Sensory
        Physiology, vol. III/2:1-190, Springer-Verlag, Berlin -
        Heidelberg - New York.
Chambers, M.R., Andres, K.H., von Düring, M., and Iggo, A.,
        1972, The structure and function of the slowly adapting
        type II mechanoreceptor in hairy skin, <u>Quart. J. exp.</u>
        <u>Physiol.</u>, 57:417-445.
Iggo, A., 1976, Is the physiology of cutaneous receptors de-
        termined by morphology?, <u>Progr. Brain Res.</u>, 7:15-34.
Jeanneret-Gris, B., 1980, Ultrastructure de l´organe juxta-oral,
        <u>Arch. d´Anat. microsc.</u>, 69:197-214.
Jeanneret-Gris, B., and Mayr, R., 1980, Ultrastructure de l´or-
        gane juxta-oral chez le hamster, <u>Acta anat. (Basel)</u>,
        108:264.
Mayr, R., and Salzer, G.M., 1967, Elektronenmikroskopische
        Untersuchungen am juxtaoralen Organ des Menschen, Z.
        <u>Zellforsch.</u> , 81:135-154.
Salzer, G.M., Stockinger, L., and Zenker, W., 1964, Die Ultra-
        struktur des juxtaoralen Organs der Ratte, <u>Z. Zellforsch.</u>,
        62:829-854.
Salzer, G.M., and Zenker, W., 1962, Das juxtaorale Organ, <u>Bi-</u>
        <u>bliotheca anat.</u>, Fasc. 3, S. Karger, Basel and New York.
Salzer, G.M., and Zenker, W., 1968, Veränderungen am juxta-
        oralen Organ der Ratte nach Hypophysektomie, <u>Z. Zell-</u>
        <u>forsch.</u>, 84:72-86.
Schoultz, T.W., and Swett, J.E., 1972, The fine structure of
        the Golgi tendon organ, <u>J. Neurocytol.</u>, 1:1-26.
Zelená, J., 1978, The development of Pacinian corpuscles, <u>J.</u>
        <u>Neurocytol.</u>, 7:71-91.
Zenker, W., 1982, "Juxtaoral Organ (Chievitz´ Organ), Morpho-
        logy and Clinical Aspects", Urban and Schwarzenberg,
        Baltimore - Munich.
Zenker, W., and Halzl, L., 1953, Beitrag zur Entwicklung des
        Chievitz´schen Organs beim Menschen, <u>Z. Anat. Entwickl.</u>
        <u>-Gesch.</u>, 117:215-236.

<u>Parts of this</u> study have been presented elsewhere (Mayr, R.,
1979, Verh. Anat. Ges. 73:1043-1045.

THE ULTRASTRUCTURE OF SENSORY NERVE ENDINGS

IN THE PENIS OF THE GOAT

Zdenek Halata*, Richard B.Johnson+,
Ralph L.Kitchell+ and Thomas Strasmann*

*Department of Functional Anatomy, University Hamburg, BRD and
+Department of Anatomy, Veterinary School, California State
University, Davis, USA

SUMMARY

The penis of the male goat belongs to the fibro-elastic type. Its main
components are the corpus cavernosum surrounded by a thick tunica albuginea.
At the basal aspect of the corpus cavernosum lies the urethra. The urethra
leaves the penis with its connective tissue sheath at the border between
corpus and glans penis to form the processus urethrae which is about 2,5 cm
long. Three types of nerve endings were observed: 1. free nerve endings, 2.
Ruffini corpuscles and 3. genital corpuscles. The free nerve endings occur
within all layers of the connective tissue, but mainly in the papillary layer of the dermis. The afferent axon is myelinated (diameter 1-2 μm). The
Ruffini corpuscles are mainly located at the border between the tunica
albuginea and the dermis or are arranged in groups inside the tunica albuginea. The supplying myelinated axon measures 3-5 μm in diameter. The genital
corpuscles are located in the papillary layer of the dermis as well as in
the deep dermis. They vary in size between 200 and 900 μm. The supplying
myelinated axon measures 4-6 μm in diameter. Neither Meissner corpuscles nor
typical Pacini corpuscles were observed in the glans penis.

INTRODUCTION

The penis of the male goat belongs to the fibro-elastic type which is
found in most ruminants (Nickel et al.,1979). Its main components are the
corpus cavernosum surrounded by a thick tunica albuginea. Bundles of collagenous fibres arise from the tunica albuginea and run like spokes into the
corpus cavernosum forming a tighty twisted trabecular system (Fig.1a).
Cavernous blood vessels are located between these trabeculae. On its outer
surface the tunica albuginea merges into the dermis.

In the area of the glans penis a venous plexus is found in the papillary layer of the dermis. The veins are symmetrically arranged and form
separate individual epithelial ridges (Fig.1a). At the basal aspect of the
corpus cavernosum lies the urethra. It is surrounded by fibro-elastic connective tissue and separated from the corpus cavernosum by a venous plexus.
The urethra with its connective tissue sheath leaves the penis at the border

*Supported in part by the Deutsche Forschungsgemeinschaft (Ha 477/722/86)

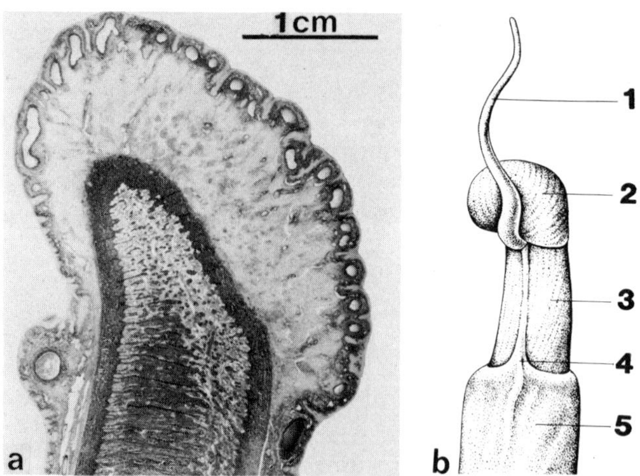

Fig. 1a. Penis in longitudinal section. The rete ridges of the epidermis
         are separated by a venous plexus of a papillary layer.
     1b. Drawing of the penis of the goat. (1) urethral protrusion,
         (2) glans, (3) corpus, (4) raphe, and (5) praeputium penis.

between corpus and glans penis to form the processus urethrae (Fig. 1b) which
is about 2,5 cm long and covered by a mucosa.

This investigation of the topography and ultrastructure of sensory
nerve endings in the corpus and glans penis in goat serves as a pilot study
for further combined morphological and physiological studies.

MATERIAL AND METHODS

The penis was taken from 3 male goats for electron microscopy and from
5 animals for light microscopy. The material for electron microscopy was
prefixed in glutaraldehyde and postfixed in OsO4 before embedding in Epon.
Semithin sections were stained according to Laczko and Levai (1975). The
ultrathin sections were stained with uranyl acetate and lead citrate. For
light microscopy, the material was fixed in Bouin´s solution and stained
with silver (Spaethe, 1984).

RESULTS

Three types of nerve endings were observed: free nerve endings, genital
corpuscles and Ruffini corpuscles.

Free nerve endings occur within all layers of the connective tissue,
but mainly in the papillary layer of the dermis. They represent the majority
of nerve endings. The afferent axon (1-2 µm in diameter) is myelinated
within the papillary layer. After losing its myelin sheath the non-myelinated preterminal fibre branches repeatedly and finally forms the typical
nerve terminals close to epithelium in the connective tissue papillae of the
papillary layer. The nerve terminals (Fig. 2a and b) show the characteristics of nerve endings: accumulation of mitochondria and vesicles, irregular-
ly arranged neurotubules and neurofilaments, and the major part of the
axolemma is covered only by a basal lamina. This type of nerve ending is
uniformly present over the whole surface of the glans penis and is only a
little less numerous in the processus urethrae.

320

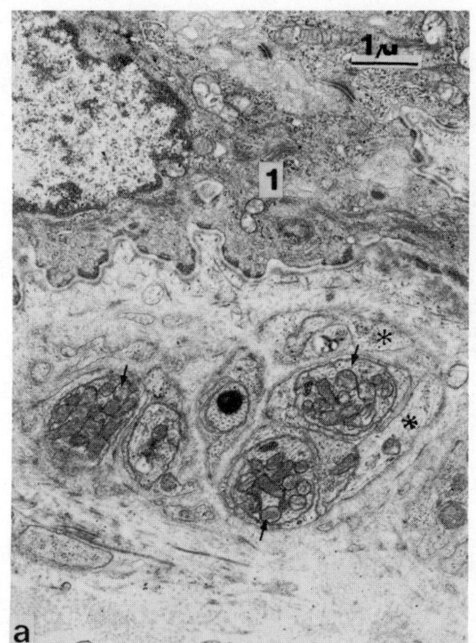
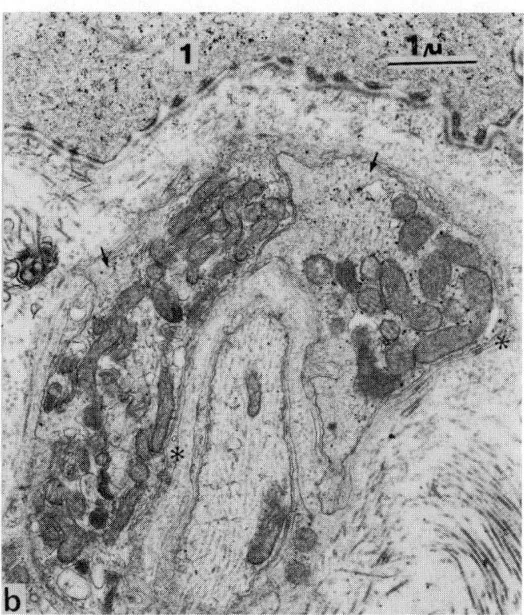

Fig. 2. Free nerve endings in the papillary layer of the glans penis
(Fig.2a) and in the urethral protrusion (Fig.2b). (1) epidermis,
(↑)nerve terminals, (*) cytoplasmic lamellae of the Schwann cell.

<u>Genital corpuscles</u> (Fig 3a-c and 4a and b) are located in the papillary
layer of the dermis near the epithelium of the glans penis as well as in the
deep dermis and near the tunica albuginea. They vary in size between 200 and
900 µm. A genital corpuscle consists of 1-4 afferent myelinated axons (4-6
µm in diameter), a perineural capsule, several inner cores and a endoneurial space. The axons lose their myelin sheaths inside the corpuscle while
the perineurium of the peripheral nerve merges into the perineurial capsule.
After losing its myelin sheath, the afferent axon may branch repeatedly.
These terminal non-myelinated portions are surrounded by several lamellae of
terminal Schwann cells with interposed collagenous fibrils. For some distance the nerve terminal sends finger-like protrusions into the endoneurial
connective tissue to form close contact with collagenous fibrils of the
endoneurial space (Fig. 4b). The cytoplasmic lamellae of the terminal
Schwann cell present numerous microvesiculations. The capsule of genital
corpuscles is complete and consists of up to 12 layers of flat perineurial
cells (Fig. 4a).

<u>Ruffini corpuscles</u> (Fig. 5a-c) are mainly located at the border between
the tunica albuginea and the dermis or inside the tunica albuginea. Within
one group of Ruffini corpuscles, up to 7 afferent axons were found, covered
by a perineurial sheath. The corpuscles are composed of an incomplete
perineurial sheath or connective tissue capsule, and an endoneurial space,
where connective tissue fibrils of the tunica albuginea and the dermis mix
with endoneurial collagen fibrils. The supplying myelinated afferent axons
measure 3-5 µm in diameter. After losing their myelin sheath inside the
corpuscle they branch repeatedly. The terminal endings show thickenings with
rather bizarre shape in cross section. Protrusions of varying length arise
from these thickenings of the terminal axons and reach into the connective
tissue of the endoneurial space and anchor between the collagenous fibrils.
The terminal Schwann cell forms flat septae which divide the endoneurial
space into compartments and may serve as additional anchorage for nerve
terminals (Fig. 5b and c). In some cases, serial sections revealed inner-

core-like lamellar systems surrounding the nerve terminals. Each Ruffini
corpuscle comprises several cylinder-like structures. The long axes of the
cylinders are parallel to the main direction of the collagenous fibrils.

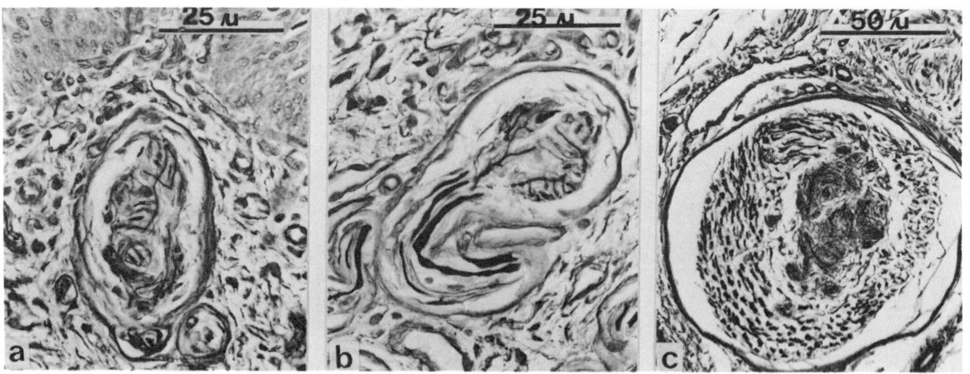

Fig. 3. Genital corpuscles from the glans penis. (Fig.3a) a corpuscle
in a papillary layer, (Fig.3b) in deep dermis and (Fig.3c)
in a nerve fascicle.

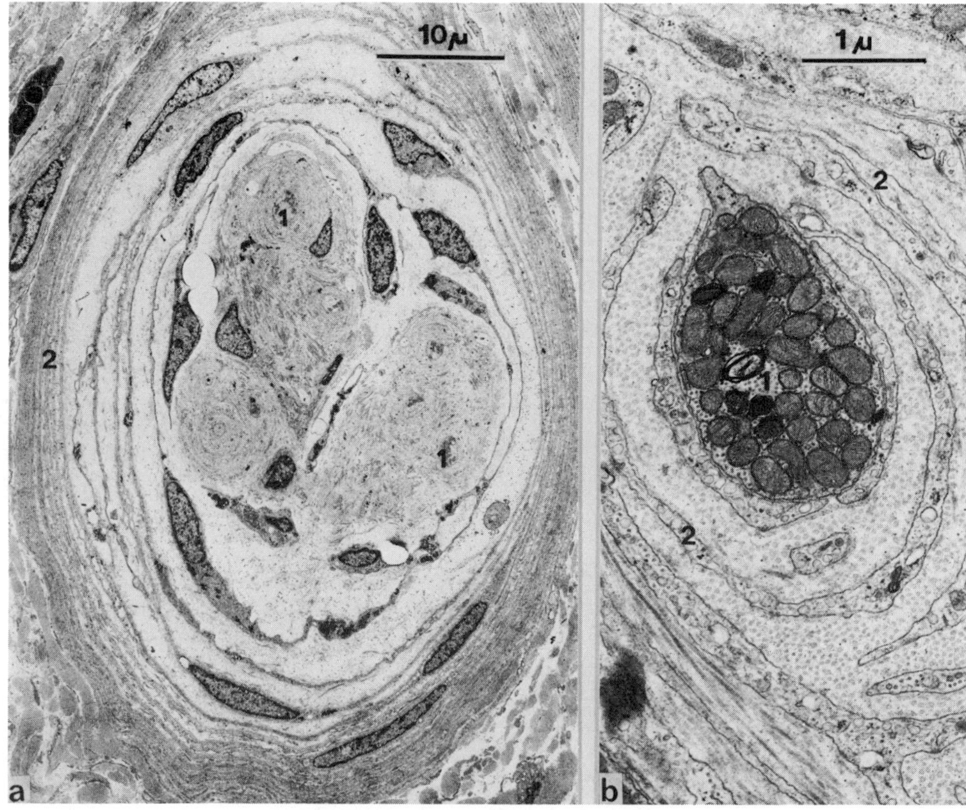

Fig. 4. Genital corpuscle from the glans penis. (Fig. 4a): (1) inner core
with cytoplasmic lamellae of the terminal Schwann cell, (2) a
perineurial capsule. (Fig. 4b): (1) nerve terminal with a finger-
like protrusion, (2) cytoplasmic lamellae of the Schwann cell.

When fibrills are orientated in the same direction, the perineurial capsule
tend to be better developed. Deep in the dermis there are also Ruffini
corpuscles with only a connective tissue capsule derived from fibroblast-
like cells.

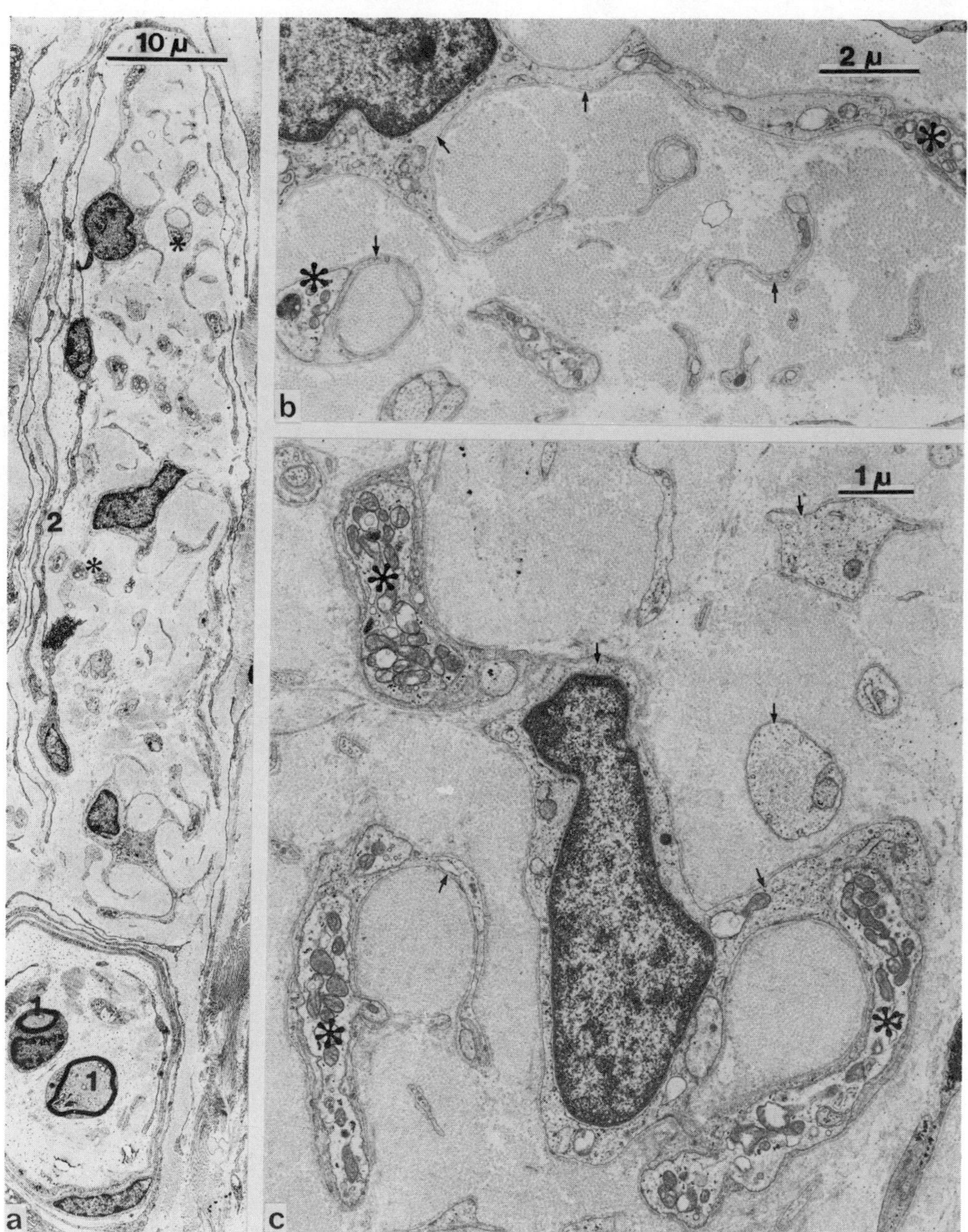

Fig. 5. Ruffini corpuscle from the deep dermis of the glans penis.
        Fig.5a - a low EM magnification. (1) myelinated fibres, (2)
        perineural capsule, (*) nerve terminals. Fig.5b and c - higher
        EM magnification. (*) nerve terminals, (↑) Schwann cells
        with thin cytoplasmic lamellae.

DISCUSSION

The ultrastructure of sensory nerve endings in the human glans and preputium penis is extensively described (McDonald and Schmitt, 1979; Halata, 1984; Halata and Munger, 1986). Compared with man, the penis of the goat displays some remarkable structural differences: a long urethral protrusion and a characteristic arrangements of fibro-elastic tissue forming the corpora cavernosa (Nickel et al., 1979) as well as of blood vessels in the dermis of the glans. Although the pattern of innervation is largely similar to man: Merkel nerve endings described in man (Halata and Munger, 1986) are absent in the epidermis as well as an abundance of free nerve endings and, somewhat fewer, genital bulbs of nearly same dimensions as in man. In contrast to the human genital bulb, a complete perineurial capsule ensheaths that of the goat and the lamellar system of terminal Schwann cells is arranged concentrically. Occasionally we found genital bulbs within a nerve fascicle.

Compared with human Ruffini corpuscles, those in the penis of the goat show remarkable differences. In the human glans (Halata and Munger, 1986) and in the preputium (Halata, 1984) Ruffini corpuscles display either an incomplete perineural capsule of 1 - 2 layers or no perineural ensheathment. In the goat, Ruffini corpuscles are situated in the fibro-elastic connective tissue and resemble Golgi tendon organs (Schoultz and Swett, 1972). Unlike the human penis, Meissner corpuscles and Vater-Pacini corpuscles are totally absent in the penis of the goat.

REFERENCES

Halata, Z., 1984, The sensory innervation of the skin of the glans penis in man. (An ultrastructural study), in: "Sensory Receptor Mechanisms". W. Hamann and A. Iggo ed., World Scientific Publ.Co., Singapore.
Halata, Z., and Munger, B. L., 1986, The neuroanatomical basis for the protopathic sensibility of the human glans penis, Brain Res., 371:205.
Laczko, J., and Levai, G., 1975, A simple differential staining method for semi-thin sections of ossifying cartilage and bone tissue embedded in epoxy resin, Mikroskopie, 31:1.
MacDonald, D. M., and Schmitt, D., 1979, Ultrastructure of the human mucocutaneous end organ, J.Invest.Dermatol., 72:181.
Nickel, R., Schummer, A., and Seiferle, E., Lehrbuch der Anatomie der Haustiere, Band II, Paul Parey Verlag, Berlin und Hamburg, 1979.
Schoultz, T. W., and Swett, J. E., 1972, The fine structure of the Golgi tendon organ , J.Neurocytol., 1:1.
Spaethe, A., 1984, Eine Modifikation der Silbermethode nach Richardson für die Axonfärbung in Paraffinschnitten, Verh.Anat.Ges., 78:101.

MECHANORECEPTORS IN HUMAN INTRA-ARTICULAR TISSUES

M. L. Zimny

Department of Anatomy
Louisiana State University Medical Center
New Orleans, LA U.S.A.

INTRODUCTION

Movements of joints in response to forces, awareness of joint position, the mechanics of joint protection and the intricate interaction of a joint with its working muscle groups, i.e., biomechanics, has been ascribed to various sensory receptors classified as mechanoreceptors. The knee is easily subjected to traumatic injuries, especially sports related, and usually involves the anterior cruciate ligament, the great stabilizer of the knee joint. Treatment of the anterior cruciate-deficient knee is still a debatable subject. In view of this, there has been renewed orthopedic interest in the ligamentous neurology of the human anterior cruciate ligament and other joint tissues associated with the knee such as the meniscus.

Most studies of neural elements associated with mammalian synovial joints were done on the knee joints of cats (Gardner, 1944; Samuel, 1952; Skoglund, 1956; Polacek, 1966; Freeman and Wyke, 1967). The nerve endings present in the tissue of the cat's knee are classified into four categories, types I, II, III, IV. Type I endings are found mainly in the fibrous capsule and resemble Ruffini endings. Type II are entirely absent from all joint ligaments but are present in the joint capsule and are considered modified Pacinian corpuscles. Type III are confined to the joint ligaments, both extrinsic and intrinsic, and represent the equivalent of Golgi tendon organs. Type IV comprises unorganized nerve terminals which function mainly as pain receptors. These four types of receptor nerve endings have also been described in the fibrous joint capsule, lateral ligament and posterior articular fat pad of the feline temporomandibular joint (Wyke, 1967).

The most extensive studies of the nerve supply of the human knee joint to date were published by Jeletsky (1931) and Gardner (1948). Both studies were able to establish the constancy of certain branches and to show definite patterns of joint innervation. For purposes of clarifying their findings and to describe new observations, a study of the nerve supply of the human knee was made from dissections of fresh amputation specimens (Kennedy, Alexander and Haynes, 1982). Related histological studies revealed that axons and receptors are rarely found in the deep fibrous substance of ligament or menisci. Pain endings predominated in adjacent connective tissue but specialized receptors were not recognized in the human anterior cruciate ligament. In a study of anterior cruciate-deficient knees (Rovere and Adair, 1983), once again, it was stated that no specialized receptors were demonstrated in the

human anterior cruciate ligament. In 1984, Schultz et al., demonstrated a
fusiform mechanoreceptor and a mechanoreceptor resembling a Golgi tendon
organ in the anterior cruciate ligament. Great controversy surrounds the
nerve supply of the meniscus. Wilson et al. (1969) suggested the penetration of neural elements into the outer third and the middle third of the body
of the meniscus. Kennedy et al. (1982) were unable to identify such fibers.
More recently (MacKenzie, 1985), it has been reported that the nerve supply
to the meniscal horns is more extensive than that seen in the body.

Materials and Methods

The present study was undertaken to identify mechanoreceptors in intra-
articular structures of human joints, such as the anterior cruciate ligament,
medial meniscus and the temporomandibular joint (TMJ) articular disk. The
structures were obtained fresh at autopsy and transported to the laboratory
in normal saline. The ligaments were cut into cross-sectional segments 1 cm
thick and kept oriented as to femoral and tibial attachments. The segments
were stained in bulk using the gold chloride technique of Gairns (1930) as
modified by Zimny et al., (1985). The menisci and articular disks were cut
into segments 1 cm thick and kept oriented anterior to posterior. The menisci and disks were pretreated with the detergent Triton X-100 to increase the
penetration of the stain and the definition of the neural elements (Zimny et
al., 1987). After staining, the segments were frozen and sectioned on a
freezing microtome at 100 microns. The sections were floated in alcoholic
gelatin, mounted on slides, dehydrated, coverslipped and studied with a light
microscope.

Results

On the basis of the classification of Freeman and Wyke (1967), all types
of receptors were found in all intra-articular structures studied. The Type
I, Ruffini endings (Fig. 1) appeared the best, morphologically, in the ligament. Type II, the modified Pacinian corpuscle, had the classic morphology
in the ligament (Fig. 2) but appeared dense and compressed in both the meniscus and the articular disk (Fig. 3). We believe that this change in configuration is due to the dense, compressed nature of the fibrocartilage composing these structures. Type III, the equivalent of a Golgi tendon organ
(Fig. 4) had the same spiralling appearance in all structures studied. Type
IV, free nerve endings (Fig. 5) were found in all the structures but the
greatest number appeared to be in the meniscus. The fact that our modified
gold chloride method was staining nerves was confirmed with transmission
electron microscopy (Fig. 6).

Our studies indicate that in the anterior cruciate ligament the greater
percentage of receptors are found in the distal (tibial) portion of the ligament. In the meniscus, nerves penetrate the outer and middle thirds of the
body and horns from the perimeniscal connective tissue with a greater concentration at the horns. Mechanoreceptors are located mainly in the outer
third and free nerve endings are mainly located in the middle third. Studies
of the TMJ articular disk show the population of mechanoreceptors to be
greatest at the periphery and progressively decreases toward the center.

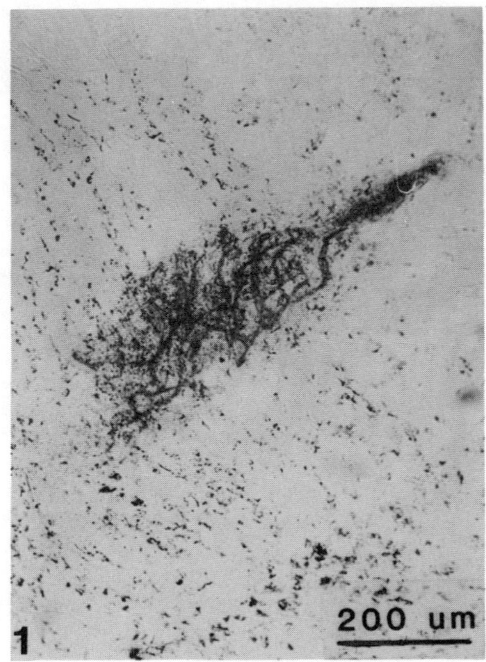

Fig. 1 A light micrograph showing a Ruffini ending in the anterior cruciate ligament.

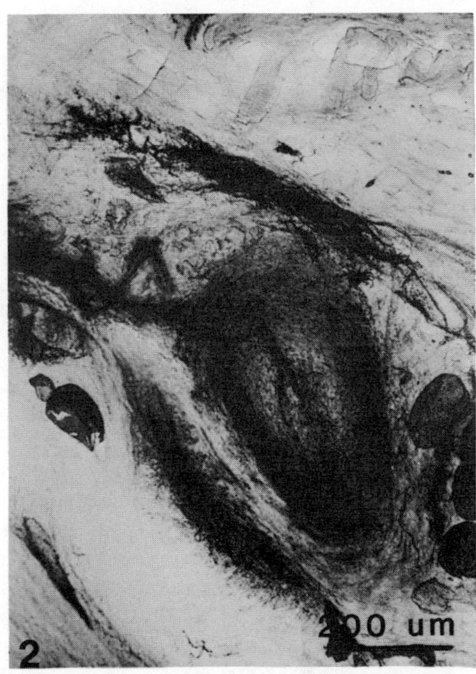

Fig. 2. A light micrograph showing a Pacinian corpuscle in the anterior cruciate ligament.

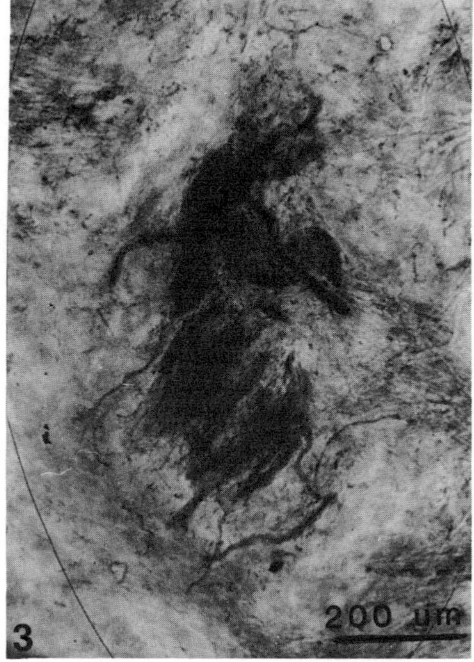

Fig. 3 A light micrograph showing Pacinian corpuscles in the medial meniscus.

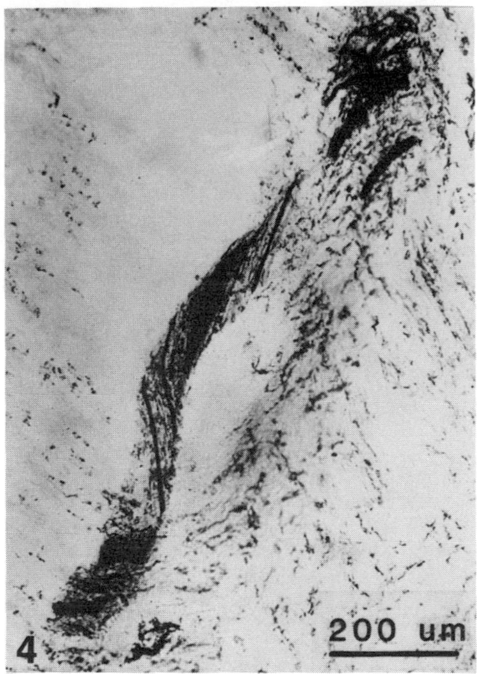

Fig. 4 A light micrograph showing a Golgi tendon organ in the anterior cruciate ligament.

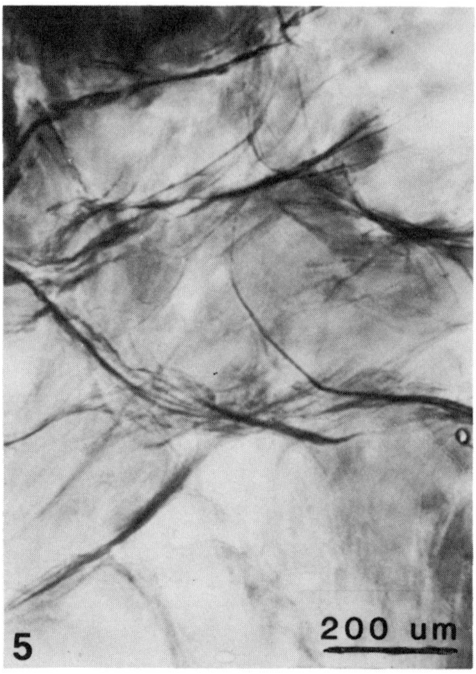

Fig. 5 A light micrograph showing
free nerve endings in the medial
meniscus.

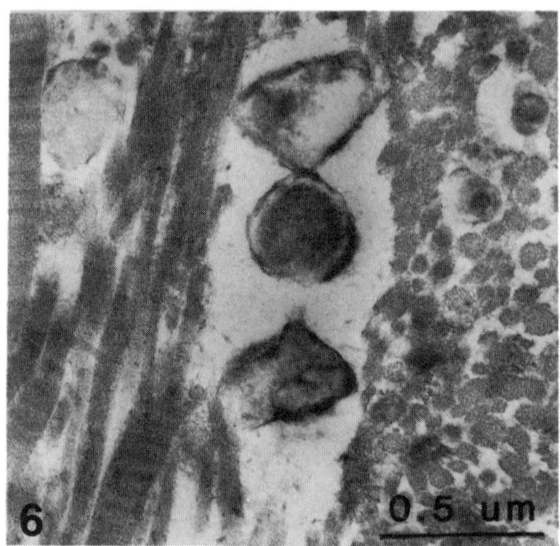

Fig. 6 An electron micrograph
showing nerve endings in the
intermediate one-third of the
medial meniscus.

Summary

Population densities of mechanoreceptors appear greater in areas related to extremes of movements. If a joint has an intra-articular structure,
then mechanoreceptors are undoubtedly present within it. These receptors
represent the first line of defense in sensing the outer limits of the range
of movement. These afferent discharges elicit support from discharging
mechanoreceptors located in the joint capsule and, subsequently, from those
in the surrounding muscles. This total afferent output alerts the central
nervous system of impending injury, which is then corrected through reflex
mechanisms. Accurate information about the relative positions of the bones
forming a joint is important to the action of applied forces and the reaction
of the joint. This information is supplied by the mechanoreceptors present
in joint related structures fibrous capsule, extrinsic and intrinsic ligaments, etc. Those joints with special intra-articular structures have an
additional, sensing mechanism for functional competency.

REFERENCES

Freeman, M.A.R., and Wyke, B., 1967, The innervation of the knee joint. An
     anatomical and histological study in the cat, J. Anat., 101:505-532.

Gairns, F.W., 1930, A modified gold chloride method for the demonstration of
     nerve endings, Q. J. Micros. Sci., 74:151-153.

Gardner, E., 1944, The distribution and termination of nerves in the knee
     joint of the cat, J. Comp. Neurol., 80:11-32.

Gardner, E., 1948, The innervation of the knee joint, Anat. Rec., 101:109-
     130.

Jeletsky, A.G., 1931, On the innervation of the capsule and epiphyses of the
     knee joint, Vestn. Khir., 22:74-112.

Kennedy, J.C., Alexander, I.J., and Hayes, K.C., 1982, Nerve supply of the
     human knee and its functional importance, Am. J. Sports Med., 10:329-
     335.

MacKenzie, W.G., Shim, S.S., Day, B., and Leung, G., 1985, The blood and
     nerve supply of the knee meniscus in man, Anat. Rec., 211:115A-116A.

Polacek, P., 1966, Receptors of the joints. Their structure, variability and
     classification, Acta Facultat. Med. Universitat Brunensis, 23:1-107.

Rovere, G.D., and Adair, D.M., 1983, Anterior cruciate-deficient knees: A
     review of the literature, Am. J. Sports Med., 11:412-419.

Samuel, E.P., 1952, The autonomic and somatic innervation of the articular
     capsule, Anat. Rec., 113:53-70.

Schultz, R.A., Miller, D.C., Kerr, C.S., and Micheli, L., 1984, Mechano-
     receptors in human cruciate ligaments, J. Bone & Joint Surg., 66A:
     1072-1076.

Skoglund, S., 1956, Anatomical and physiological studies of knee joint
     innervation in the cat, Acta Physiol. Scand., 36, Suppl., 124:1-101.

Wilson, A.S., Legg, P.G., and McNeur, J.C., 1969, Studies of the innervation
     of the medial meniscus of the knee joint, Anat. Rec., 165: 485-492.

Wyke, B., 1967, The neurology of joints, Ann. R. Coll. Surg. Engl., 41:25-50.

Zimny, M.L., St.Onge, M., and Schutte, M., 1985, A modified gold chloride
     method for the demonstration of nerve endings in frozen sections,
     Stain Technol., 60:305-306.

Zimny, M.L., St.Onge, M., and Albright, D., 1987, The use of agents which
     improve staining of fresh human fibrocartilage, Stain Technol., (In
     press).

TOPOGRAPHY OF MECHANORECEPTORS IN THE CONNECTIVE TISSUE OF THE ELBOW JOINT

REGION IN MONODELPHIS DOMESTICA, A LABORATORY MARSUPIAL

Thomas Strasmann and Zdenek Halata (1)

Institute of Anatomy, Dept. Functional Anat.
University of Hamburg, Martini Str. 52
D 2000 Hamburg, F. R. G.

SUMMARY

The elbow joint of Monodelphis domestica was studied by means of serial
semithin sections and computer-aided three dimensional reconstructions. Two
series were collected: series A consisted of the proximal part of the elbow
joint capsule over 6 mm, series B of the lateral (radial) aspect of the
joint over 2 mm. The elbow joint of Monodelphis domestica is a
ginglymo-trochoid joint with full range of pro- and supination. The
predominant part of its capsule belongs to the rigid type of joint capsules.
It is part of the surrounding connective tissue apparatus in which the
muscles are embedded. By means of light microscopy, four types of
mechanoreceptors are observed: Ruffini corpuscles (RCs), muscle spindles
(MSPs), Golgi tendon organs (GTOs) and lamellated corpuscles (LCs). In
series A (proximal part of the joint) 9 RCs, 106 MSPs, 75 GTOs and 66 LCs
are identified. Ruffini corpuscles are detected in the fibrous layer of the
capsule. MSPs and GTOs predominantely occur in the triceps, biceps and
brachialis muscles in close relation to each other. A group of 5 to 7 GTOs
without any relations to MSPs are seen in the medial intermuscular septum
proximal to the medial epicondyle. Of a total number of 66 LCs, 30% are
found in periarticular loose connective tissue, 33% in the stratum fibrosum
of the joint capsule and 24% in the muscle-tendon transition zones. The
latter two measure less than 20 µm in diameter, corpuscles of the first
group measure 50 - 70 µm in diameter. Series B displays 22 RCs, 22 MSPs, 4
GTOs and 40 LCs. The RCs predominate on the flexion side of the elbow joint
capsule, while the LCs are distributed over the whole lateral part of the
joint capsule. In contrast to the findings of series A constituting the
proximal part of the joint region, series B displays 70% of all LCs counted
in the stratum fibrosum, but only 15% in loose periarticular tissue and 15%
in muscle-tendon transition zone. Quantitative evaluations and functional
considerations are discussed.

INTRODUCTION

The spectrum of mechanoreceptor types found in joint capsules and deep
connective tissue surrounding joints is well known. In mammals, including
man, the ultrastructure and function of four types of sensory nerve endings

(1) Supported by Deutsche Forschungsgemeinschaft HA 1194/3-1.

are documented: free nerve endings as morphological equivalent for pain and temperature sensitivity, Ruffini corpuscles and Golgi tendon organs as slowly-adapting stretch organs and lamellated corpuscles as fast-adapting receptors, recording velocity. However, only minimal information is available concerning the exact localisation and quantitative relations of mechanoreceptors in the connective tissue surrounding the joint and joint capsules. This paper, therefore, documents preliminary results concerning the topographical and quantitative distribution of mechanoreceptors in joints and joint capsules of the gray short-tailed oppossum (Monodelphis domestica).

MATERIAL AND METHODS

<u>Preparation</u>. Two adult (6 and 14 months) male animals were perfused with 6 % glutaraldehyde, buffered in 0.1 M sodium-phosphate-buffer; all forelegs were dissected. After removing the skin the specimens were fixed for 24 h in the same fixative. Subsequently, the material was decalcified in 5 % EDTA (buffered in 0.1 M sodium-phosphate-buffer with NaOH up to pH 7.4) for 8 - 12 months. Slices of about 2 mm thickness were fixed in 1 % OsO4 (in 0.1 M sodium-phosphate-buffer containing 1 % sucrose at 4°C) for 2 h. The slices were dehydrated and embedded in EPON 812. Serial sections (2 μm) of the material from the younger animal provided series A. These sections were provided by a "Polycut S" with a carbide knife. Serial semithin sections (1 μm) constituted series B. Both types of sections were stained according to Laczko and Levai (1975).

<u>Reconstruction</u>. Series A provided a reconstruction over 6 mm of the proximal part of the elbow joint region, series B a reconstruction over 2 mm of the lateral aspect of the joint. At first, camera lucida drawings of different magnifications (x160 - x3200) were prepared of every 8th section and the following objects were entered: bone, connective tissue, muscles, lamellated corpuscles, muscle spindles, Ruffini corpuscles, Golgi tendon organs and nerves. Consecutive drawings of the 160 x-sections were digitalized on a 11/780 DIGITAL VAX with a COMTAL VISION ONE/20 image processing program. The specific processing programs for reading, rotation and special recording of reconstructed graphic files were developed by the IMDM, Institute for Mathematics and Dataprocessing of the University of Hamburg (for further descriptions see Boecker et al., 1985).

RESULTS

<u>Joint and joint capsule</u>. The elbow joint of Monodelphis domestica displays a composite joint of the ginglymo-trochoid type. In spite of the predominant quadrupedal existence, Monodelphis is able to perform extensive pro- and supination (about +90 and -90 degrees) in the elbow joint. Three types of joint capsule compositions were found: the rigid, the intermediate and the flaccid type. Among our reconstructions, the rigid variant predominates over the two other types. In both flexion and extension, the joint capsule exhibits a small synovial layer clearly demarcated from the dense regular connective tissue of the fibrous layer, characteristic of the rigid type of joint capsule. Meniscoid pads protrude from both lateral and medial sides into the joint cavity of the humero-radial joint. These pads display a. dense fibrous connective tissue in the fibrous layer and a small synovial layer. On both flexion and extension sides, the capsule forms folds extending towards the shoulder. Here the intermediate type of joint capsules can be observed: a wide synovial layer with adipose cells lies on a stratum fibrosum composed of bundles of collagen fibrils of varying thickness and direction. On the flexion side, the capsule shows a synovial layer at some areas, which merges into the fibrous layer composed of irregular small

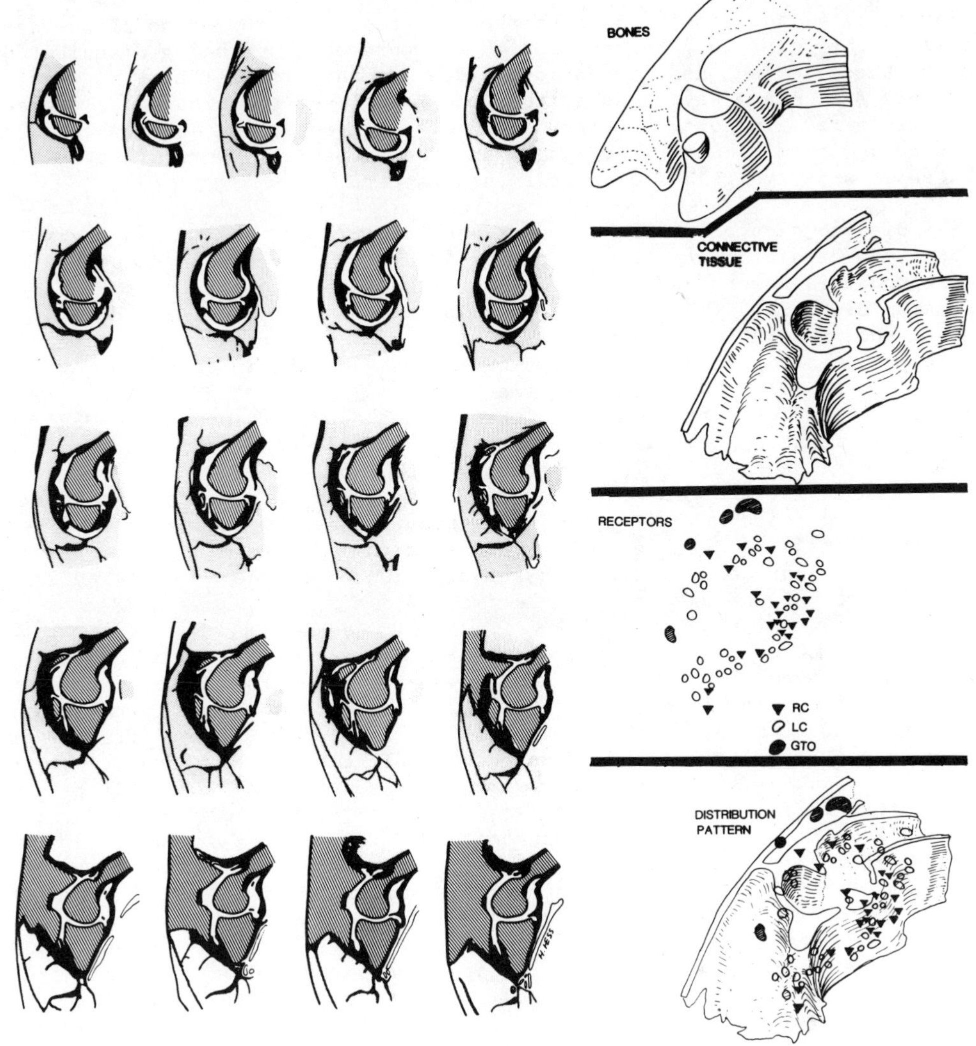

Fig. 1: <u>Left</u>: serial semithin sections of series B; black = connective tissue, hatching = bones. <u>Right</u>: drawings of computer-aided three dimensional reconstructions, mediolateral view.

bundles of collagen fibres. A clear demarcation of these two layers is lacking. Both morphological features are characteristic of the flaccid type of joint capsule. The connective tissue of the joint capsule is part of the surrounding connective tissue apparatus: series A as well as series B show that the dense connective tissue of the stratum fibrosum fuses in both epicondylar regions with dense tissue strands, serving as aponeuroses of the extensor and flexor muscle groups or as epimysial connective tissue layers. Likewise, the aponeurotic superficial septum of the supinator muscle and, on the extension side, the anconeus muscle insert in the fibrous layer (see Fig. 1).

<u>Distribution of Ruffini corpuscles</u>. By means of light microscopy 31 Ruffini corpuscles (RCs) were found in both series A and B. A receptor was distinguished as a RC when it showed each of the following characteristics: 1. a myelinated axon, 2. a capsule-like structure, 3. a smaller than 25 μm

diameter, 4. a narrow or no subcapsular space, 5. a few or no blood vessels inside the capsule. In series A, 9 RCs were observed being distributed in both the flexion and extension sides of the proximal part of the joint capsule in such a way, that neither the radial nor the ulnar side seemed to be preferred. In serie B (Fig. 1), 22 RCs were detected. 16 of these 22 RCs occurred within the flexor part of the capsule, 4 were situated in the extensor side, while 2 did not fit this pattern.

Distribution of muscle spindles and Golgi-tendon organs. In series A, a total number of 75 Golgi-tendon organs (GTOs) and 106 muscle spindles (MSPs) were counted over about 6 mm. The absolute numbers of both decrease proximo-distally. The muscles triceps, biceps and brachialis display many MSPs and GTOs in the proximal elbow joint region, whereas the epicondylarly arising muscle groups show only single GTOs and MSPs in the same region. In most cases GTOs occur in close relation to MSPs. Only a group of 4 - 7 GTOs shows no direct relations to muscles and therefore no one to MSPs: this group is situated in a ligament-like dense regular connective tissue in that area, where the medial intermuscular septum would be found. More distally, this tissue strand fuses with the ulnar epicondylar connective tissue complex, from which the flexor muscle group arises. In series B, 22 MSPs and only 4 GTOs were detected. All GTOs are situated within the triceps muscle near the ulnar bone, while MSPs are distributed in the triceps and in more distal parts of the extensor muscle group.

Distribution of lamellated corpuscles. In series A a total number of 66 lamellated corpuscles (LC) were counted over about 6 mm, in serie B (about 2 mm) 40 LCs were found. Series A shows that proximo-distally the absolute number of LCs increases. The corpuscles are located in three different areas (see Fig. 2): 1. in the periarticular loose connective tissue, near nerve fascicles and blood vessels. These measure about 50 - 70 μm in diameter and are often found as single corpuscles. 2. in the intracapsular dense fibrous connective tissue (stratum fibrosum). These corpuscles are only 10 - 20 μm in diameter and often found in clusters, situated in the outer layers of the stratum fibrosum. A smaller number of these LCs lies adjacent to the bone where the fibrous layer of the joint capsule arises from the periosteum. These periosteal LCs display the same small diameter, but are found as single corpuscles. 3. in the muscle-tendon transition zone. Adjacent to tendons and inserting muscle fibres, this third group of LCs is composed of single corpuscles of about 10 - 20 μm in diameter. These corpuscles are observed in both epicondylar regions and in the insertion zone of the triceps muscle. Series A displays 66 lamellated corpuscles, of which 30% were found in periarticular loose connective tissue, 33% in the stratum fibrosum of the joint capsule and 24% in the muscle-tendon transition zones. In contrast to the findings of series A showing the proximal part of the joint region, series B displays 70% of the 40 LCs observed in the stratum fibrosum, only 15% in the loose periarticular tissue and 15% in the muscle-tendon transition zone.

DISCUSSION

Preliminary quantitative evalutations. The above mentioned 5 criteria only allowed us to discriminate Ruffini corpuscles with a well developed capsule by means of light microscopy. However, from previous electron microscopical studies we are able to infer that in marsupials also Ruffini corpuscles (RCs) with incomplete or complete capsules or even without capsules occur (Strasmann et al., 1987). The quantitative relations of lamellated (LCs) and Ruffini corpuscles are in accordance with results of earlier studies (e.g. Polacek, 1966; Freeman and Wyke, 1967). The same is true for the close topographical relationship of muscle spindles (MSPs) and Golgi tendon organs (GTOs) (Barker, 1974). Barker also observed a close

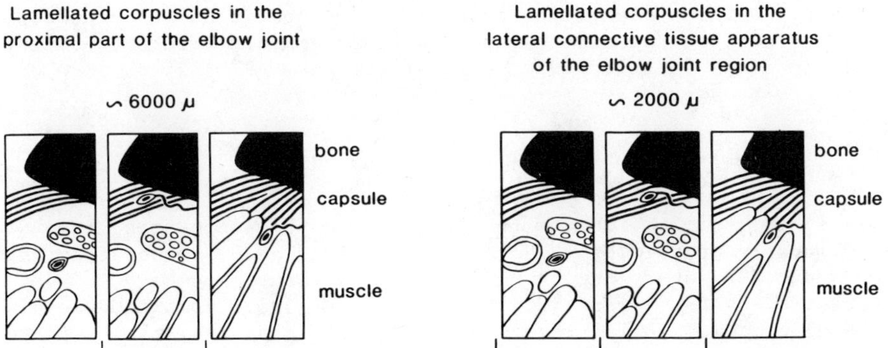

	periart.	intracaps.	muscle-tendon	
Ø	50-70 μ	≤ 20 μ	≤ 20 μ	Σ
n	20	22	24	66
%	30 %	33 %	36 %	99 %

	periart.	intracaps.	muscle-tendon	
Ø	50-70 μ	≤ 20 μ	≤ 20 μ	Σ
n	6	28	6	40
%	15 %	70 %	15 %	100 %

Fig. 2: Size and topography of lamellated corpuscles observed in series A (left) and series B (right).

relation between GTOs and LCs in the flexor hallucis longus muscle in the cat: he found 20 GTOs and 9 intramuscular LCs. In contrast to this ratio of 2.2 GTO per intramuscular LC, we found 75 GTOs and 24 lamellar corpuscles in the muscle-tendon transition zone (ratio 3.15) of different muscles in series A. The morphological pattern of the lamellated corpuscles described above resembles that of the respective corpuscles in the Kowari, observed by means of electron microscopy (Strasmann et al., 1987).

The connective tissue arrangement of the elbow joint capsule in Monodelphis domestica described here corroborates the results obtained in rat (Van der Wal et al.,1984) and in man (Van Mameren, 1983). Consequently, the mapped mechanoreceptors have to be interpreted as mechanoreceptors of the connective tissue apparatus surrounding the joint, of which the classical joint capsule seems to be only a part. Indeed, the concept of joint receptors as a basis for position sense (Polacek, 1966) was questioned on physiological grounds: e. g., Grigg et al. (1973) found no alteration in the ability to detect joint position and passive movements in sixteen patients with total hip replacement. Tracey (1978) and other physiologists maintained that the joint receptors, e.g. free nerve endings, Ruffini corpuscles and lamellated corpuscles, are only of minor importance for joint position sense. Apart from the question of whether the abundance of these classical joint receptors found in joint capsules are only provided for redundant information, the question arises, whether the mechanoreceptors described in this paper still remain after total joint replacement operations. I.e., in accordance with Van der Wal et al. (see this volume) we found numerous lamellated corpuscles in the connective tissue surrounding the capsule. 44 lamellated corpuscles (66 %) in series A are not situated in the fibrous layer of the joint capsule (see Fig. 2) but in loose periarticular (30 %) or in dense connective tissue in which muscles insert

(36 %). Within the limitations of our method it is not deduciable whether these corpuscles contribute functionally to muscle afferents recording joint position sense (Millar, 1979; McIntyre et al., 1978).

REFERENCES

Barker, D., 1974, The morphology of muscle receptors, in: "Handbook of sensory physiology," H. Autrum, R. Jung, W.R. Loewenstein, D.M. McKay, eds., Springer, Berlin Heidelberg New York.

Boecker, F.R.P., Thiede, U., Hoehne, K.-H., 1985, Combined use of different algorithms for interactive surgical planning, in: "Proceedings of computer assisted radiology," H.U. Lemke, M.L. Rhodes, C.C. Jaffe, R. Felix, eds., Springer, Berlin Heidelberg New York.

Freeman, M.A.R., Wyke, B., 1967, The innervation of the knee joint. An anatomical and histological study in the cat, J.Anat., 101:505.

Grigg, P., Finerman, G.A., Riley, L.H. ,1973, Joint position sense after total hip replacement, J.Bone Joint Surg., 55A:1016.

Laczko, J., Levai, G., 1975, A simple differential staining method for semi-thin sections of ossifying cartilage and bone tissues embedded in epoxy resin, Mikrosk., 31:1.

Mameren, H. van, 1983, Reaction forces in a model of the human elbow joint, Verh. Anat. Ges., 77:327.

McIntyre, A.K., Proske, U., Tracey, D.J., 1978, Afferent fibres from muscle receptors in the posterior nerve of the cat's knee joint, Exp.Brain Res., 33:415.

Millar, J., 1979, Convergence of joint, cutaneous and muscle afferents onto cuneate neurones in the cat, Brain Res., 175:347.

Polacek, P., 1966, Receptors of the joints. Their structure, variability and classification, Acta Fac. Med. Univ. Brunensis, 23:1.

Strasmann, T., Halata, Z., Loo, S.K., 1987, Topography and ultrastructure of sensory nerve endings in the joint capsules of the Kowari (Dasyuroides buyrnei), an Australian marsupial, Anat. Embryol., 176:1.

Tracey, D., 1978, Joint receptors - changing ideas, Trends Neurosci., 1:63.

Wal, J.C. van der, Drukker, J., Mameren, H. van, 1984, The organisation of the connective tissue in relation with muscle and nerve tissue in the lateral cubital region in the rat, Verh. Anat. Ges., 78:631.

SENSORY NERVE ENDINGS IN THE DEEP LATERAL CUBITAL REGION:

A TOPOGRAPHICAL AND ULTRASTRUCTURAL STUDY IN THE RAT

J.C. van der Wal, T. Strasmann[x], J. Drukker, and
Z. Halata[x]

Department of Anatomy and Embryology, University of
Limburg, P.O.Box 616, 6200 MD Maastricht,
The Netherlands
[x]Department of Anatomy, Division of Functional
Anatomy, University of Hamburg, F.R.G.

The connective tissue apparatus, as revealed in the human
lateral elbow region by means of an alternative method of dissection - in which capsular ligaments can not be described as
separate entities - appears to play an important role in conveying tensile stresses (Van Mameren, 1983; Van Mameren and
Drukker, 1984). In the homologous region of the rat the role of
this apparatus as to the quality of centripetal information was
questioned (Van der Wal et al., 1984). There is growing evidence
in the literature that a complete description of the morphological substrate of proprioception in a joint region should not
be restricted to the joint capsule receptors alone but has to
include the sensory nerve endings in the entire periarticular
connective tissue, the muscle tissue and even in the skin (Halata et al., 1985, for review see Abrahams, 1983). Since one of
the main conclusions of our previous investigations (Van der Wal
et al., 1984) was that in the studied region, almost all muscle
tissue functions in series with this connective tissue and thus
the usual concept - that muscles function in a parallel relation
to the capsular connective tissue - is not correct for this region, the question was raised how the organisation of the morphological substrate of proprioception is related to the found
architecture. Therefore an integral, three dimensional description of the occurrence of mechanoreceptors in relation to the
architecture of muscle and connective tissue is needed, present-
ing the functional topography of the substrate that functions
as the source of centripetal information.

This study deals with the organisation of sensory nerve
endings in the deep part of the cubital region of the rat.

MATERIALS AND METHODS

For demonstrating the presence of nerve fascicles in situ,
the investigation area in the forelegs of six 12-week-old Lewis
rats was treated with an acetylcholinesterase in toto staining
method (Baljet and Drukker, 1975). Eight other complete forelegs

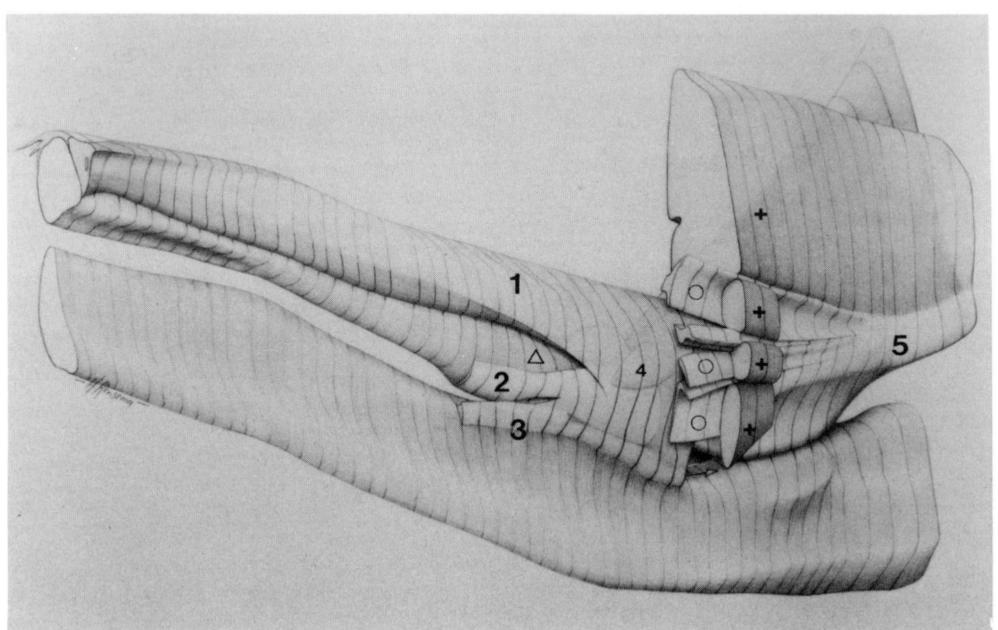

Fig. 1. 3D-reconstruction of the muscle and connective tissue
        in the deep part of the lateral cubital region of the
        rat:
        1.   Supinator septum
        2/3. Collateral ligaments
        4.   Sesamoid bone
        5.   Lateral humeral epicondyle

        + Superficial extensor muscles
        o Muscle compartment walls
        ▵ Triangular opening for the r. prof. n. radialis.

were fixed, embedded in a polyester resin, stained with a trichrome dye and cut in serial sections of 15 µm. Of two of such
series 3D-reconstructions were made by means of a computer program as to the muscle and connective tissue as well as to the
muscle spindles and Golgi tendon organs. These mechanoreceptors
were also counted in all specimens, the content per gram muscle
being calculated, based upon the mean muscle weight as measured
postmortem in six other animals. Two other supinator muscles
with neighboring connective tissue structures were dissected
after fixation in Bouin's fluid, cut in serial sections of 7 µm
and stained with the silver impregnation method modified by
Spaethe (1984). Another ten supinator muscles with neighboring
structures were fixed and prepared for semithin and ultrathin
sections. The semithin sections were stained with the method of
Laczko and Levai (1975); the ultrathin sections were contrasted
with uranyl acetate and lead citrate according to Reynolds (1963).

RESULTS

The Architecture of the Muscle and Connective Tissue

        The supinator muscle has its distal insertion to the third

quarter of the radial bone, in part via an aponeurotic tendon, in part directly to the radial periosteal layer. Proximally, all muscle fascicles insert into a broad aponeurotic connective tissue layer, which in its turn converges towards the lateral humeral epicondyle. This so called supinator septum (Fig. 1), is continuous in the distal direction with the epimysial layer on the outer edge of the muscle. Near the epicondyle the septum fuses with two "collateral" strands of collagenous tissue that run from the periosteum of the radial respectively the ulnar bones to the humeral epicondyle (numerals 2 and 3 in Fig. 1). Thus a strong connective tissue plate - adjacent to the joint cavity - is formed, which is continuous with the joint capsule connective tissue and with the humeral periosteum and in which a small sesamoid bone is incorporated. The collagenous fibers of the septum and the collateral connective tissue structures run mainly over the outside of the sesamoid bone, which in this way protrudes into the joint cavity. On the lateral (i.e. juxta-ulnar) side of the supinator muscle, the muscle tissue reaches within a distance of a few millimeters from the distal edge of the sesamoid bone. A small intermediate zone of areolarlike organized collagenous tissue is interposed between this proximal muscle border and the sesamoid bone (Fig. 1). To the medial side (i.e. on the flexor side) the septal connective tissue appears not only on the outer border of the muscle outline but also as an intramuscular septum, at the proximal level dividing the muscle into two parts.

Just proximally of the humeral epicondyle, the mentioned connective tissue structures in their turn fuse with the connective tissue layers that form the walls of the muscle compartments of the superficial extensor muscles. Nearly all the muscle fascicles of the superficial extensor muscles insert into these connective tissue layers, thus inserting indirectly onto the humeral bone (Van der Wal et al., 1984). So an "epicondylar connective (-muscle) tissue complex" can be described as a functional "entity" in which muscle tissue functions in series with the connective tissue structures - via which a.o. tensile stresses may be conveyed over the elbow joint - to which not only the main part of the four superficial extensor muscles belong, but also the complete supinator muscle and "its" septum (Fig.1).

The Topography of Nerve Fascicles in situ

A fine network of acetylcholinesterase-positive elements appeared to be present in the loose connective tissue covering the outer side of the juxtaarticular collagenous plate in which the sesamoid bone is incorporated. The nerve fascicles seem to originate from the deep branch of the radial nerve, as this branch pierces through the triangular opening that is left between the lateral edge of the supinator septum and the collateral connective tissue strand running from the radial bone to the epicondyle (Fig. 1). Some elements extend in the medial and distal direction. In the semithin sections this pattern could be confirmed, the fascicles apparently containing myelinated fibers of 2 to 4 μm in diameter. In the serial silver stained sections also nerve fascicles could be demonstrated running longitudinally between the septal connective tissue and the inserting muscle fibers.

The Occurrence of Muscle Spindles and Golgi Tendon Organs

The total content of muscle spindles of the supinator mus-

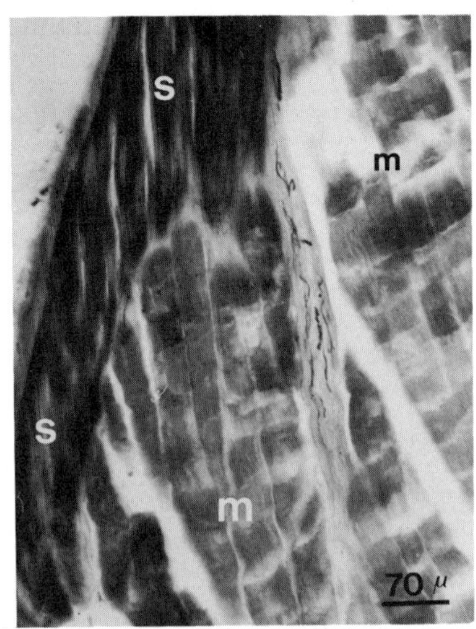

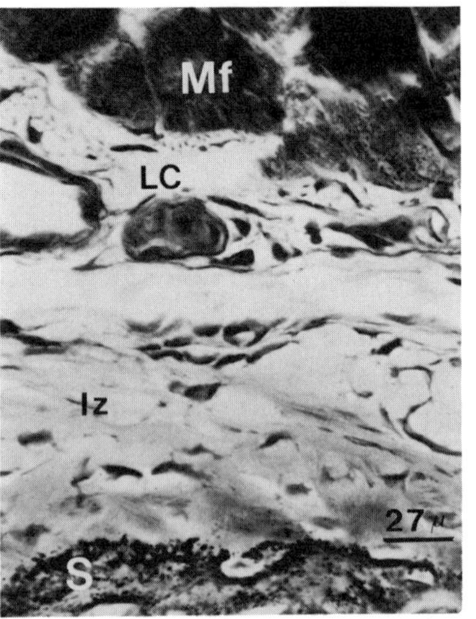

Fig. 2. GTO attached to the septal connective tissue (s) of supinator muscle fibers; silver staining.

Fig. 3. Lamellated corpuscle (LC) in intermediate zone (Iz) between supinator muscle fibers (Mf) and the sesamoid bone (S); silver staining.

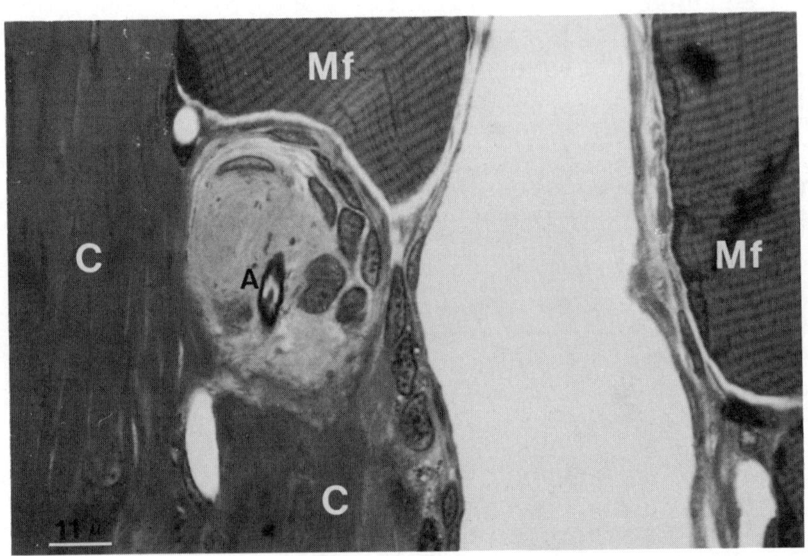

Fig. 4. Lamellated corpuscle interposed between septal collagenous tissue (C) and supinator muscle fibers (Mf). A. central axon, semithin section, stained according to Laczko and Levai.

cle is 13.38 (S.D. 1.77, range 12 - 17, n = 8) with a spindle density of 510.7 spindles per gram muscle, which concentration is high, compared to the other muscles in the area (Van der Wal et al., 1987, in press). In the 3D-reconstruction as well as in

the silver stained serial sections it could be confirmed that
most of the spindles are attached to the supinator septum with
their proximal pole.

Of the total content of Golgi tendon organs, which is 9.13
(S.D. 2.59, range 6 - 13, n = 8) - i.e. 345.8 per gram muscle
- about 80 % appear to be attached to the septal connective tissue, the rest being inserted to the distal tendon and/or the
radial periosteum. In the silver stained serial sections this
typical septum-oriented arrangement could be confirmed and in
several cases it could be demonstrated that the afferent fibers
are derived from the nerve fascicles running longitudinally between muscle tissue and septal connective tissue (Fig. 2). The
nerve endings show a spray-like arrangement within the organs.
The structures measure about 400 μm in length and 50 - 60 μm in
diameter. With respect to the spindle-like shape and the parallel arrangement of the collagenous fibers within the inner core,
we classify these "spray-like endings" (Polacek, 1966) as GTO
endings and not as Ruffini endings.

The Lamellated Corpuscles

At three characteristic spots lamellated corpuscles, mea-
suring about 20 μm in diameter and not more than 60 μm in length,
are present. Most of the corpuscles (about 15 per region) are
situated in the intermediate zone of areolarlike organized collagenous connective tissue, between the proximal border of the
supinator muscle tissue and the distal edge of the sesamoid bone
(Fig. 3). This zone is situated deeply in relation to the septum
as this runs along the outer side of the sesamoid bone and superficially to the synovial lining, bordering the joint cavity.
It also extends to the (inner) side of the collateral collagenous tissue strands that run from the humeral epicondyle to the
radial respectively the ulnar bone. The area seems to be fitted
to transduce e.g. traction forces and movements of the septal
connective tissue into compression of the corpuscles between
neighboring collagenous fibers of the mentioned intermediate
zone.

A second portion of the corpuscles (3 - 5 per area) is
situated in the loose connective tissue at the outer side of
the septum and in the triangle between the septum and the radial "collateral ligament". Finally a few corpuscles are interposed between the septal collagenous fibers and inserting muscle fibers, i.e. in the same orientation as the GTOs with this
difference that the latter are situated at the distal tapering
end of the septum and these corpuscles are only present in the
proximal 1/5 part of the muscle and its septum (Fig. 4).

At the ultrastructural level four types of nerve endings
were observed at the aforementioned intermediate zone:

1. Lamellated corpuscles (20 to 60 μm) with a single axon
in the inner core and a 2 to 4-layered perineural capsule. Besides all the usual typical features, several gap junctions
between the peripheral lamellae of the terminal Schwann cells
are noteworthy. The supplying axons measure 3 to 4 μm in diameter.

2. Small, quasi-primitive lamellated corpuscles (Fig. 5).
This type is rare. It displays several central axons covered by

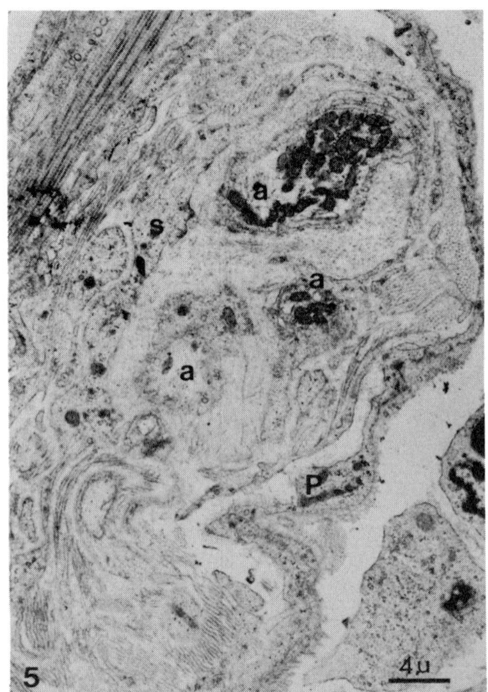
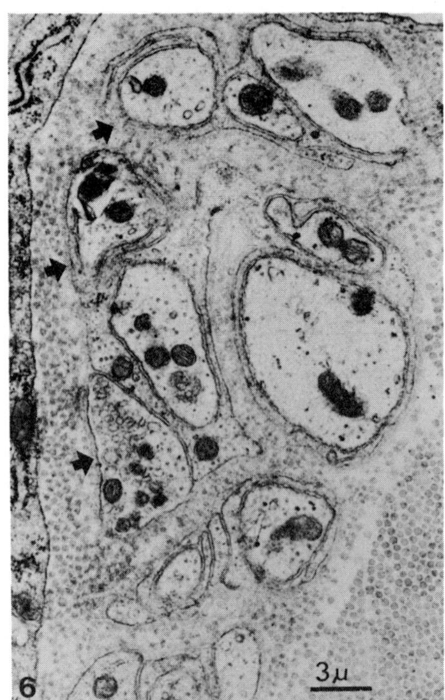

Fig. 5. Small lamellated corpuscles with three inner cores.
Only 1 to 3 lamellae of terminal Schwann cell (s) envelop the terminal axons (a). P. perineurium.

Fig. 6. Complex of nerve endings with clear vesicles and enveloped by lamellar Schwann cell processes, in dense connective tissue. Arrow: axon membrane, only covered by
basal lamina.

1 to 3 lamellar processes of terminal Schwann cells and are enveloped by a two-layered perineural capsule. This type of receptor can be classified as group-III-type-5 in the proposed
nomenclature of Vor Düring et al. (1984).

3. Free or non-encapsulated nerve endings (FNE), derived
from axons of 2 μm in diameter (Fig. 6). They are found within
the dense connective tissue and are arranged into groups of several endings, often only covered by the basal lamina. This
type of endings can be classified as group-III-type-2.

4. FNEs in the vicinity of small vessels or in close relation to mast cells. This last finding coroborates the innervation of these cells.

DISCUSSION AND CONCLUSIONS

In view of the fact that more evidence is becoming available that perception of joint movement is not restricted to information, originating from joint capsule structures, it is
noteworthy that in the way it is described here, a strict dis-
crimination between capsular connective tissue structures and
other connective tissue is irrational in this region. Follow-

ing this description - which evidently shows a more in series
organisation of muscle and connective tissue that does not coincide with the usually described muscles and ligaments as separate morphological entities - the supinator septum can be
classified as a tendon, a ligament, a part of the capsule, as
well as an epimysial layer, with the types of mechanoreceptors
usually contributed to such entities all being present in the
complex.

Describing the connective tissue as a continuum gives a
better understanding of the way in which tensile stresses, for
example, are conveyed as well as the way in which the proprioceptive substrate is organized in functional relationship with
the architecture of connective and muscle tissue. The whole
complex in this deep part of the lateral cubital region may be
considered as a kind of dynamic "muscle-ligament-complex". This
"muscle-ligament" in its turn is a part of an "epicondylar con-
nective(-muscle) tissue complex" in the area - with a similar
organisation of muscle and connective tissue - of which different tracts of connective tissue can be kept under tension according to the mechanical needs of a given situation.

It seems evident that, in the deep part of the lateral cubital region of the rat, the muscle and connective tissue is
organized and (well) equipped in such a way as to enable the
perception of transmission of forces acting on the elbow joint.
As to the type and organisation of the sensory nerve endings
present in the area, it can be stated that they are strictly
related to the functional architecture of the muscle and connective tissue and that they exhibit features, usually contributed to ligaments and tendons.

(With support by "Netherlands Organization for the Advancement
of Pure Research Z.W.O" and "Deutsche Forschungsgemeinschaft"
HA 1194/3-1).

REFERENCES

Abrahams, V.C., 1983, Neck Muscle Proprioception and Motor Con-
        trol, in: "Proprioception, Posture and Emotion", D. Gar-
        lick, ed., pp. 103-120.
Baljet, B., and Drukker, J., 1975, An acetylcholinesterase
        method for in toto staining of peripheral nerves, Stain
        Techn., 50:31-36.
Düring, M. von, Andres, K.H., and Schmidt, R.F., 1984, Ultra-
        structure of fine afferent fibre terminations in muscle
        and tendon of the cat, in: "Proceedings of the Int. Symp.
        on Sensory Receptor Mechanisms, W. Hamann and A. Iggo,
        eds., World Sci. Publ. Corp. Singapore, pp. 15 - 23.
Halata, Z., Rettig, T., and Schultze, W., 1985, The ultrastruc-
        ture of sensory nerve endings in the human knee joint
        capsule, Anatomy and Embryology, 172:265-275.
Laczko, J., and Levai, G., 1975, A simple differential staining
        method for semithin sections of ossifying cartilage and
        bone tissue embedded in epoxy resin, Mikroskopie, 31:
        1-4.
Mameren, H. van, 1983, Reaction forces in a model of the human
        elbow joint, Verh. Anat. Ges., 77:327-328.
Mameren, H. van, Drukker, J., 1984, A functional basis of in-
        juries to the ligaments and other soft tissue around

the elbow, <u>Int. J. Sports Med.</u>, 5 (Suppl.): 88-92.
Polacek, P., 1966, Receptors of the joints. Their structure,
        variability and classification, <u>Acta Fac. Med. Univ.
        Brunenesis</u>, 23:1-107.
Reynolds, E.S., 1963, The use of lead citrate at high pH as an
        electron opaque stain in electron microscopy, <u>J. Cell
        Biology</u>, 17:208-212.
Spaethe, A., 1984, Eine Modifikation der Silbermethode nach
        Richardson für die Axonfärbung in Paraffinschnitten,
        <u>Verh. Anat. Ges.</u>, 78:101-102.
Wal, J.C. van der, Drukker, and J. Mameren, H. van, 1984, The
        organization of the connective tissue in the lateral
        cubital region of the rat, <u>Verh. Anat. Ges.</u>, 78:631-632.
Wal, J.C. van der, and Drukker, J., 1987, The occurrence of
        muscle spindles in relation to the architecture of the
        connective tissue in the lateral cubital region of the
        rat,(this volume).

THE OCCURRENCE OF MUSCLE SPINDLES IN RELATION TO THE ARCHITECTURE OF THE

CONNECTIVE TISSUE IN THE LATERAL CUBITAL REGION OF THE RAT

J.C. van der Wal and J. Drukker

Department of Anatomy and Embryology
University of Limburg, P.O.Box 616, 6200 MD Maastricht
The Netherlands

In earlier investigations into the functional significance of the architecture of muscle and connective tissue in the lateral cubital region of the rat, one of the main architectural features found in the area, is the presence of an "epicondylar connective(-muscle) tissue complex" (van der Wal et al., 1984). This complex consists of several connective tissue layers, converging towards the lateral humeral epicondyle, of which the deepest are part of the (peri)articular connective tissue. The main part of the muscle fascicles of the superficial Extensor muscles and the Supinator muscle are inserting to this structure. The complex may be considered as a functional "entity" via which a.o. tensile stresses are conveyed towards the lateral humeral epicondyle and in which the muscle and connective tissue function in an in series situation. This view challenges the usual ideas e.g. about guiding of forces, since the usual concept – that the muscle tissue is organized in a parallel relation to the capsular connective tissue structures – obviously seems not to be correct.

The question was raised, how the spatial organisation of the mechano-receptors in the region is related to this architecture and whether it is such as to enable the perception of transmission of forces over the elbow joint. Therefore an exact **3D-reconstruction** of the distribution of the mechanoreceptors in relation to the architecture of the muscle and connective tissue was necessary. This presentation deals with the occurrence of muscle spindles in the investigated area.

MATERIALS AND METHODS

Eight complete forelegs of 12 week old Lewis rats were fixed, embedded in polyester resin, stained with a trichrome dye and cut in serial sections of 15 μm. Of **two** series, each tenth section was redrawn as to muscle and connective tissue as well as to mechanoreceptors and digitized by means of an XY-tablet. Line drawings were produced with a 3D-reconstruction program. Of **four** other series, projection figures of the spindle distribution along the Z-axis on standardized sections were made. In all specimens the spindles were counted. The wet weight of the muscles was measured in six other rats.

RESULTS

The different muscles in the region show relatively high densities of muscle spindles per gram muscle (102.2 – 510.7), with the various muscles showing quite different numbers, compared to each other (See Table 1.).

Table 1. Number of muscle spindles in the lateral cubital region

Name of the muscle / (with symbolic abbrevation)	Number of m.spindles			M.spindles per gram muscle
	Mean	S.D.	Range	
m. ext. carpi ulnaris (EU)	14.25	2.03	11-17	261.6
m. ext. dig. lateralis (EL)	9.00	1.31	7-10	281.3
m. ext. dig. communis (EC)	21.63	2.88	16-25	343.3
m. ext. carpi radialis (ER)	22.38	3.46	18-29	102.2
m. ext. dig. primi (EP)	12.38	1.69	10-15	339.0
m. ext. indicis (EI)	5.00	0.93	4- 6	510.2
m. supinator (ES)	13.38	1.77	12-17	510.7

In all the superficial muscles (EU, EL, EC and ER muscle) the muscle spindles are concentrated in relatively small zones, situated excentrically in the muscles. A relatively large area - on the cross section of the foreleg situated superficially as a lunar shaped zone from the ulnar to the radial side - is practically devoid of spindles: **The spindle rich zones extend between the distal tendons and the proximal connective tissue layers** of the afore mentioned "epicondylar connective - muscle tissue complex". To this complex also belongs the Supinator muscle (ES) with a high content of spindles, which are also mainly oriented (attached) to the proximal connective tissue, here represented by the so called Supinator septum - a broad aponeurotic layer of connective tissue converging to the humeral epicondyle, to which all the muscle fascicles of the ES muscle are inserting.

The 3D-reconstruction of the distribution of the spindles also threw new light on the significance of the "density" of spindles in the muscles. E.g two muscles with the **same spindle density** per gram muscle (i.e EI and ES muscle) show a completely **different spatial distribution** of the spindles: in the ES muscle they run more or less parallel to each other in a broad zone of the muscle from deep (distal tendon) to superficial (proximal Supinator septum), in the EI muscle they are organized in series, in a small longitudinal zone. On the other hand two muscles with quite **different spindle densities** (i.e. EU en ER muscle) show a **similar distribution**, with spindle rich zones oriented excentrically, between the proximal connective tissue layers, that they share with neighbouring muscles, and their distal tendons.

CONCLUSIONS

In the studied region the muscle spindle rich zones are oriented to the connective tissue architecture rather than to muscle morphology. **Differences in parameters** of the occurrence of spindles can better be understood in terms of differences in functional architecture of the connective tissue.

The spindle rich zones may be considered as "kinesiological monitor zones", as ment by Peck et al. (1984) with this extension to their concept that it is plausible that the zones are also involved in monitoring stresses and that such zones **do not coincide** with muscles as morphological entities.

The parameter "**Number of muscle spindles per gram muscle**" can only be interpreted in functional respect if the exact distribution of the spindles in relation to the architecture of muscle and connective tissue is known.

REFERENCES

Peck, D., Buxton, D.F., and Nitz, A., 1984, A Comparison of Spindle
    Concentrations in Large and Small Muscles Acting in Parallel
    Combinations, Journal of Morphology, 180: 243 -252.
Wal, J.C. van der, Drukker, J., and Mameren, H. van, 1984, The organization
    of the connective tissue in relation to muscle and nerve tissue in the
    lateral cubital region of the rat, Verh. Anat. Ges., 78: 631 -632.

DISTRIBUTION AND STRUCTURE

OF MECHANORECEPTORS IN THE MANDIBULAR JOINT OF STR/1N-MICE

Dagmar Dreessen, Zdenek Halata and Thomas Strasmann

Department of Functional Anatomy
University Hamburg
BRDeutschland

SUMMARY

The structure and innervation of the mandibular joint were investigated
in the STR/1N-mice by light and electron microscopy. In the mouse the mandibular joint consists of the fossa mandibularis, the condylus mandibulae, and
an articular disk which completely separates both articular surfaces. The
joint capsule is flaccid and comprises a synovial layer and a membrana fibrosa that is usually very thin. Two types of nerve endings were observed:
1. **free nerve endings** inside the joint capsule ventromedial and dorsolateral
in the membrana fibrosa (myelinated afferent axons of 1-2 μm in diameter)
and 2. **Ruffini corpuscles** at the insertion of the lateral pterygoid muscle
at the articular disk and the collum mandibulae (axon diameter 3-5 μm). A
Pacinian corpuscle was not observed in eight joint capsules studied.

MATERIAL AND METHODS

A total of 8 temporomandibular joints obtained from 4 STR/1N mice were
investigated. The animals were perfused with 6% glutaraldehyde, buffered in
0.1M sodium-phosphate-buffer. The skin was removed and the whole head of the
mice was fixed for 24 hours in the same solution. For 6 months the material
was decalcified with 5% EDTA-solution buffered in 0.1M sodium-phosphate-
solution. The heads of the mice were cut into small slices. The specimens
were postfixed in a 1% OsO4 solution in 0.1M phosphate buffer with 1% sucrose, dehydrated and embedded in EPON 812. Semithin sections were stained
after a modified method of Laczko and Levai (1975). Based on these semithin
section series, a computer-aided reconstruction of one mandibular joint was
obtained.

RESULTS AND DISCUSSION

In contrast to the joint capsule of man which is reinforced by ligaments, the joint capsule in mice is loose. The nerve fibres, which were
mainly found in the dorso-lateral part of the capsule, originate from muscle
branches of the mandibular nerve and also from the nervus auriculotempora-
lis. Most of the nerve fibres were observed in the stratum fibrosum and are

____

Supported in part by the DFG (Ha 1194/3-1) and the "Verein für die Erfor-
schung und Bekämpfung rheumatischer Krankheiten e.V." in Bad·Bramstedt.

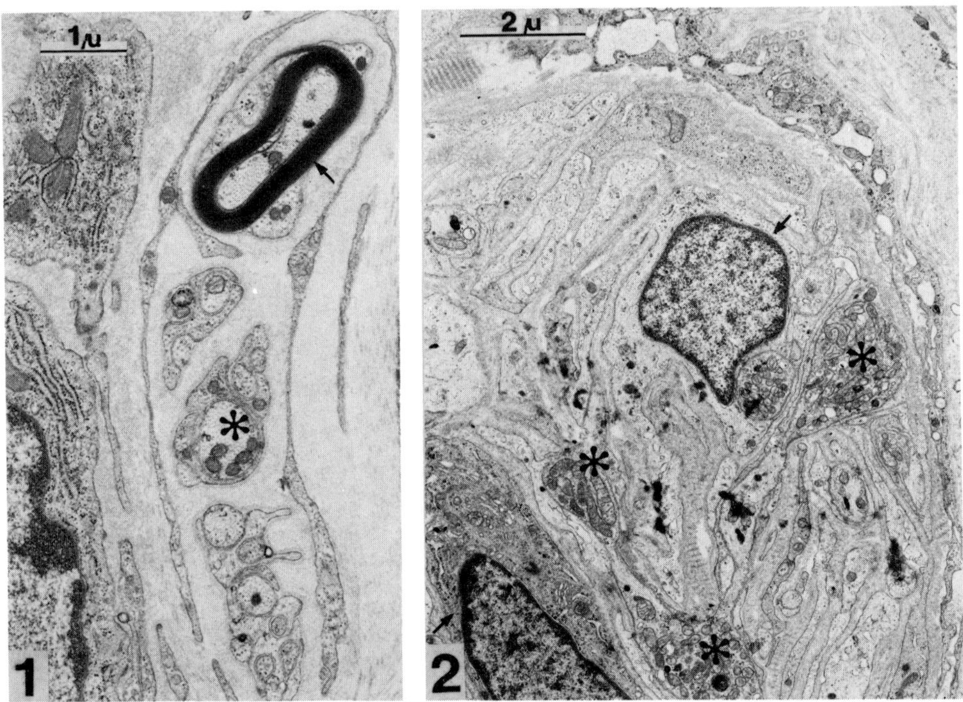

Fig. 1. Subsynovial layer of the joint capsule. (*) free nerve ending,
(↑) myelinated axon.
Fig. 2. A Ruffini-like corpuscle from the insertion of pterygoideus lat.
muscle. (*) nerve terminals, (↑) Schwann cells.

myelinated (axon-diameter about 2 μm), a few axons are unmyelinated (Fig.1).
Free nerve endings were found in the fibrous layer tightly underneath the
synovia. Most of them are situated in the dorsolateral part of the capsule
between the mandibular head and the insertion of the capsule at the discus
articularis. A few Ruffini corpuscles could be observed (Fig. 2) within the
connective tissue of the insertion of the lateral pterygoid muscle. The
diameter of their myelinated afferent axons measures about 4 μm. Pacinian
corpuscles were found neither in the capsule nor in the periarticular loose
connective tissue. Frommer and Monroe (1966) investigated the innervation of
the temporomandibular joint in mice. By means of light microscopy they also
found only free nerve endings within the joint capsule. Thus it is quite
possible that for kinaesthesia the innervation of the joint capsule is less
important than the innervation of ligaments and muscles including the fascia
(McCloskey, 1978). This thesis becomes tenable if one regards the joint
capsule as a continuation of connective tissue of muscles which moves the
temporomandibular joint.

LITERATURE

Frommer, J., Monroe, C.W., 1966, The morphology and distribution of nerve
    fibres and endings associated with the mandibular joint of the mouse,
    J.Dent.Res., 45:1762.
Laczko, J., Levai, G., 1975, A simple differential staining method for
    semi-thin sections of ossifying cartilage and bone tissues embedded
    in epoxy resin, Mikroskopie, 31:1.
McCloskey, D.I., 1978, Kinesthetic sensibility, Physiol. Rev., 58:763.

348

FUNCTIONAL SIGNIFICANCE OF BLOOD SINUSES AROUND SENSE ORGANS

IN SOME MAMMALS

Takeshi Yohro

Department of Anatomy, Faculty of Medicine
University of Tokyo
Hongo, Tokyo, Japan 113

INTRODUCTION

Anatomically the presence of venous sinuses around hair
follicles characterizes vibrissae or "sinus" hairs, which are
mechano-receptors widely distributed in mammals, except man.
The functional role of the sinus, however, is not yet clear,
and there have been several suggestions on its significance in
previous literatures on the anatomy of vibrissae. The purpose
of the present paper is to review briefly the probable functional roles of the venous sinus of vibrissae and to describe
in this context examples of venous sinuses around sense organs
in some mammals, especially in insectivores.

About four possibilities have been suggested in the literature on the anatomy of vibrissae as functional interpretations of the venous sinus. Those are:
1) An erectle body, modifying pressure and thereby acting on
   receptors in some way or other (Bonnet 1878; Messenger,
   1890; Melaragno and Montagna, 1953; Andres, 1966).
2) A protective structure keeping receptors from mechanical
   stimuli from the surroundings, i.e. other than the hair
   itself (Botezat, 1897).
3) A damper in a vibratory system (Yohro, 1977).
4) A supporting structure transmitting mechanical stimuli to
   nerve endings, e.g. via ringwulst (Stephens et al., 1973).

The first assumption is rather popular, but probably can
be denied by the absence of any specialized arteries or veins
usually found in erectile tissues such as corpus cavernosum
penis and the venous plexus in the nasal mucosa. The capsule
of the sinus is also quite hard, often as hard as a cartilage.
The third and the fourth assumptions will not be dealt with in
the present paper, since they can be proved or refuted by pro-
perly designed experiments. The second one, however, is rather
vague, and probably worth considering from morphological aspects
more in detail at least to know whether the assumption is possible or not. For that purpose I have tried to find some other
circumstances which suggest similar functions in venous sinuses,
e.g. those around sense organs other than vibrissae.

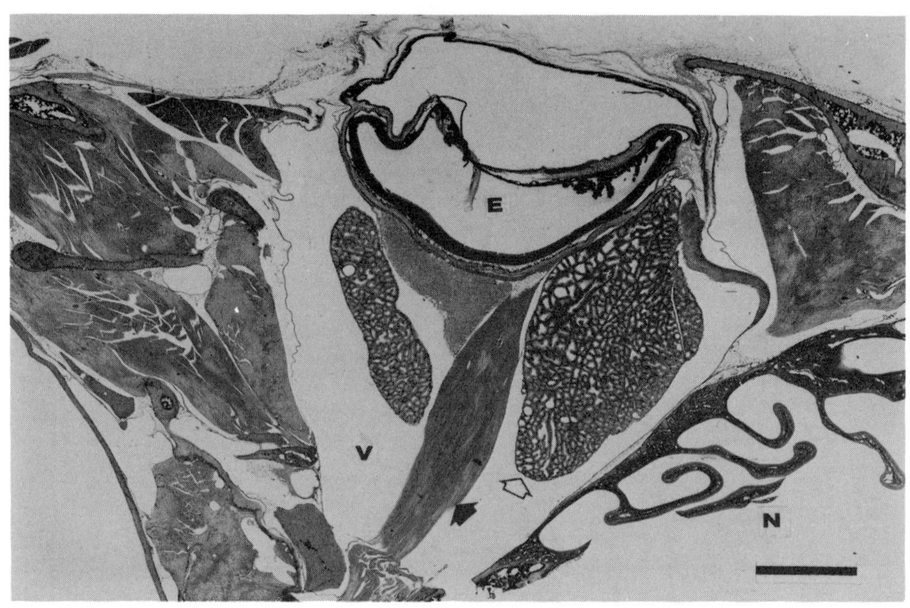

Fig. 1. The orbital region of Mongolian gerbil. Photomicrograph
of a section stained with hematoxylin and eosin. Note the
presence of a large venous sinus (V) behind the eye (E).
Black arrow indicates the extraocular muscle (M. rectus
superior), and the white arrow the Harderian gland which is
divided into two parts by the muscle on this micrograph.
N: nasal cavity. Scale 1 mm.

OBSERVATIONS AND DISCUSSIONS

The first example pertaining to the present subject is
the orbital venous sinus observed in the Mongolian gerbil (Sakai
and Yohro, 1981). The sinus is widely distributed in rodents
and lagomorphs. It usually surrounds the Harderian gland, thus
making the glandular surface quite smooth on macroscopical
observations in these mammals. In the Mongolian gerbil it
surrounds the Harderian gland, extraocular muscles and the
optic nerve to a large extent (Fig. 1). The functional role of
this venous sinus has not yet been much discussed, but it can
be suggested from the comparative anatomy of structures surrounding the eyes.

In primates, as is well-known, the lateral wall of the
orbit is bony, and it separates the orbit from the temporal
fossa. This is one of the typical features of primates, while
in some primitive primates the orbit and the temporal fossa
communicate directly with each other since the lateral bony
wall is not present as in other mammals. This structural separation of the orbit from the temporal fossa immediately suggests
the functional separation of masticatory movements from eye
movements. The completion of the lateral bony wall of the orbit
in the ascending scale of primates is associated with the position of eyes looking straight ahead, and appears to be correlated with elaboration of binocular vision in these forms.
The role of the orbital venous sinus in rodents and lagomorphs

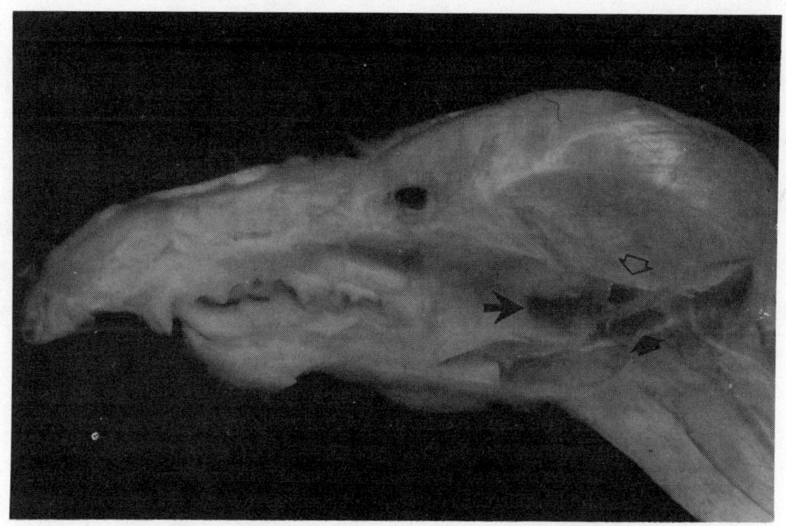

Fig. 2. Macroscopic preparation of the head of the musk-shrew,
   showing the venous sinus (long arrow), the tympanic cavity
   (short arrow) and the external auditory meatus (white arrow).
   Note the large temporalis muscle behind the eye.

may be compared to that of the bony lateral wall of the orbit
of primates, in the sense that eye movements can be separated
from masticatory movements by the presence of the orbital venous
sinus which easily collapses and refills according to the situation.

     The second example is the presence of venous sinus between
the tympanic cavity and the temporalis muscle in the musk-shrew,
Suncus murinus, an insectivore (Fig. 2). In this species the
tympanic cavity is not surrounded by the bulla, a bony structure
found in most mammals. The preparation of the skull of this
mammal clearly shows that the only bone surrounding the middle
ear cavity is the tympanic bone that supports the tympanic membrane. The author's view is that the venous sinus is apparently
intercalated between the tympanic cavity and the masticatory
apparatus including the temporalis muscle (Fig. 3), thereby
blocking the mechanical stimuli resulting from masticatory movements. The temporalis muscle plays a major role in this situation, the fact which can be supported by the following anatomical features of the head region in this mammal.

     In the musk-shrew there are several macroscopical peculia-
rities in the anatomy of head region. The head seen from above
is triangular in shape in adults, while it is long and rectan-
gular in the post-natal period. Eyes are very small as in most
of the species in Soricidae. Large salivary glands including the
parotid, the submandibular and the sublingual, lacrimal glands
and the extra-orbital part of Harderian glands, are located
mainly behind the external ears. In the skull, the following
features are immediately noticed:
1) Deficient zygomatic arch.
2) Well-developed saggital and nuchal crests (in such a small
   mammal!).
3) Presence of a specialized articular process for the jaw joint

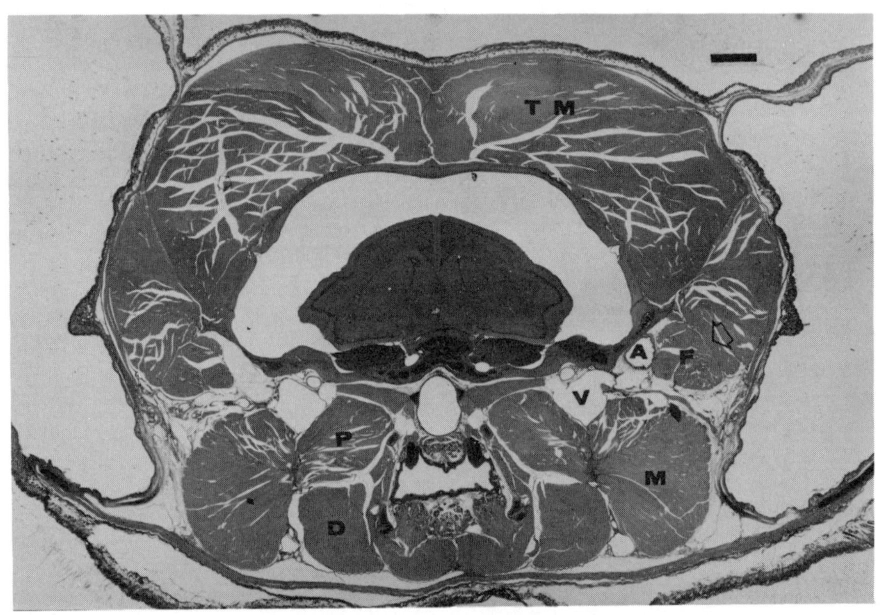

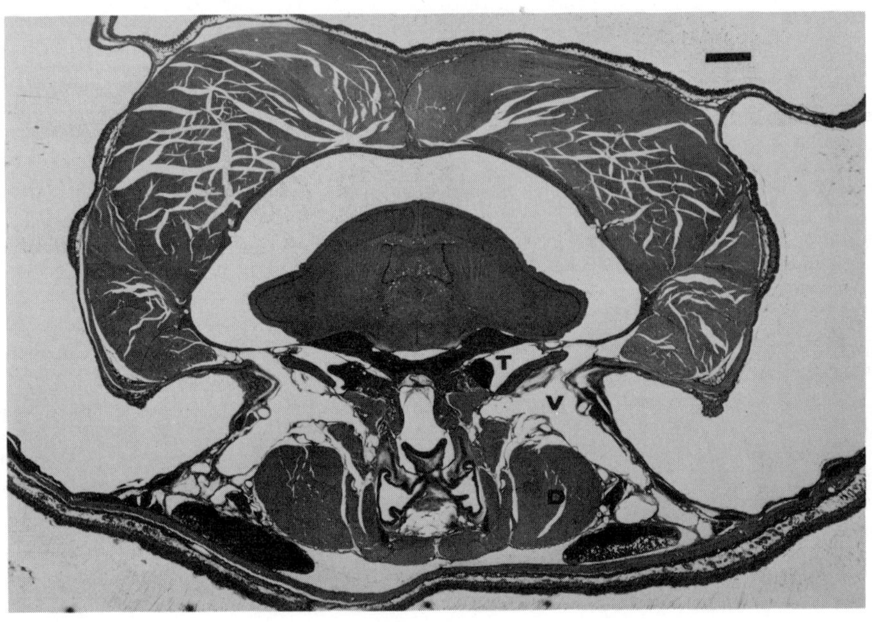

Fig. 3. Two cross sections from serial sections of the head of
the musk-shrew, showing the venous sinus intercalated between
the tympanic cavity and masticatory muscles. The top figure
is more anterior to the lower one. A: Articular cavity of the
jaw joint; D: M.digastricus; F: Muscles innervated by N.fa-
cialis; M: M.masseter; P: M.pterygoideus medialis; T: Tympanic
cavity; TM: M.temporalis; V: Venous sinus. Scale 1 mm.

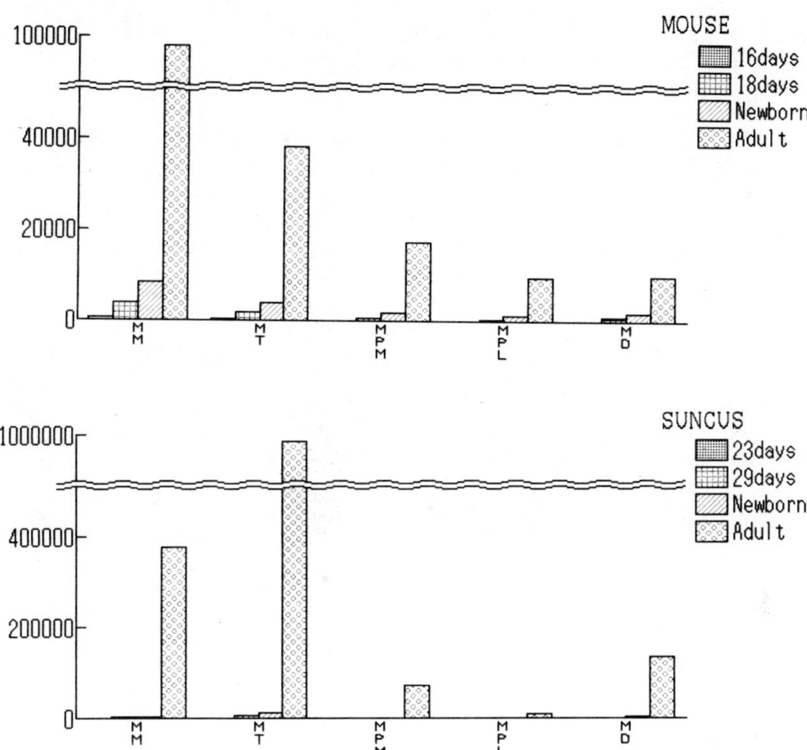

Fig. 4. Pre- and post-natal development of masticatory muscles
in the Suncus murinus and the mouse. The ordinate shows the
volume of muscles, expressed by the number of voxcels which
are obtained from an image processor. The contour of each
muscle is delineated on the photomicrograph of each section,
then projected on CRT of the image processor, thereby the
number of voxcels in the area of contour is obtained. Those
numbers are summed up for all serial sections of each specimen and the volume of the muscle is obtained as the final
value. Note the ordinate is one order higher in Suncus than
in the mouse, showing the extraordinary development of masticatory muscles in the former, especially when one compares
the volume of muscles in the adult and in the newborn. The
adult/newborn ratio is about 6 times greater in the temporalis muscle of Suncus than that in the mouse. However, whether
the post-natal development of masticatory muscles in Suncus
is retarded at birth as compared to that in the mouse cannot
be decided on the basis of the present results. Gestation
period in Suncus is 30 days as compared to 20 days in the
mouse. The relative volume of the temporalis muscle as compared to that of masseter is greater in Suncus than in the
mouse.
MM: M.masseter; MT: M.temporalis; MPM: M.pterygoideus medialis;
MPL: M.pterygoideus lateralis; MD: M.digastricus.

with an octagonally-shaped articular surface.
4) Presence of a triangular fossa on the medial side of the
   mandibula.

All of these features can be explained on the basis of the
extreme development of the temporalis muscle. The development
of this muscle begins after birth, and the fact accounts for
the difference of the head shape in the adult and the new-born.
The eyes lie at the anterior front margin of the temporalis
muscle, and eye movements must be greatly influenced by masticatory movements. The posterior part of the head is mostly
occupied by the temporalis and the masseter muscle, which may
cause the more posterior location of bodies of glands usually
lying more anteriorly in other mammals. The zygomatic arch would
confine the muscle in the temporal fossa and restricts development of the muscle by making the outer bony wall of the temporal
fossa. Thus, it is rather natural that it is missing if the
muscle develops extraordinarily. The saggital and the nuchal
crests serve as attachment sites for the muscle. The triangular
fossa of the mandibula also serves as the attachment site for
a particular bundle of muscle fibers apparently belonging to
the temporalis. The peculiar shape of the articular surface of
the jaw joint can also be accounted for by the strong pull
exerted by the temporalis and the masseter muscles.

The post-natal development of masticatory muscles of the
musk-shrew was examined to know whether they are actually well-
developed as compared to those of the laboratory mouse and to
obtain quantitative data if possible (Fig. 4). The results are
rather definite. The temporalis and the masseter muscles in
the musk-shrew grow to 6 times greater size than those of the
laboratory mouse on the basis of the adult/newborn ratio of
the volume of these muscles. The other masticatory muscles including the pterygoideus lateralis, the pterygoideus medialis
and the digastiricus grow similarly but to a lesser extent.

CONCLUSION

These two examples show that, in some cases, venous sinuses actually have a role in blocking mechanical stimuli from
surrounding tissues. The point is, however, what kind of stimuli are actually present in each case. In the case of vibrissae, we probably cannot specify them. The snout moves quite
well with muscles innervated by the facial nerve, and especially well in insectivores. In addition to this, vibrissae are
equipped with striated muscles as well as some smooth muscles,
both of which are attached to the capsule of the sinus (Yohro,
1977). One can suppose that movements caused by those muscles
may affect reception of mechanical stimuli by vibrissae and
some devices are necessary to block their effects. However,
the presence of the hard capsule around the sinus would suffice
for that purpose. Thus, as the functional interpretations of
the sinus of vibrissae, the other two possibilities appear to
remain tenable for the moment from available anatomical data.

REFERENCES

Andres, K.H., 1966, Über die Feinstruktur der Rezeptoren an
     Sinushaaren, Z. Zellforsch., 75:339-365.

Botezat, E., 1897, Die Nervenendigungen an der Tasthaaren von
        Säugetieren, Arch. f. mikr. Anat., 50:142-169.
Bonnet, J., 1878, Studien über die Innervation der Haarbälge
        der Haustiere, Morph. Jb., 4:329-398.
Melaragno, H.P., and Montagna, W., 1953, The tactile hair fol-
        licle in the mouse, Anat. Rec., 115:129-142.
Messenger, J.H., 1890, The vibrissae of certain mammals, J.Comp.
        Neur., 10:399-402.
Sakai, T., and Yohro, T., 1981, A histological study of the
        Harderian gland of Mongolian gerbils, Meriones meridia-
        nus, Anat. Rec., 200:259-270.
Stephens, R.J., Beebe, I.J., and Poulter, T.C., 1973, Innerva-
        tion of the vibrissae of the California sea lion, Zalo-
        phus californianus, Anat. Rec., 176:421-442.
Yohro, T., 1972, Structure of the sinus hair follicle in the
        big-clawed shrew, Sorex unguiculatus, J. Morph., 153:
        333-354.
Yohro, T., 1977, Arrangement and structure of sinus hair mus-
        cles in the big-clawed shrew, Sorex unguiculatus, J.
        Morph., 153:317-332.

THE STRUCTURE AND DEVELOPMENT OF MECHANORECEPTOR COMPLEXES IN

ANSERIFORM BIRDS AS SHOWN BY SEM

K. V. Avilova

Biological Department of the Moscow University
Lenin Hills
119899 Moscow, USSR

The structure of the bill tip organ (BTO) - a term introduced by Gottschaldt and Lausmann, 1974 - in the waterfowl
(order Anseriformes) was examined at the SEM level. The epidermal structures of BTO are described in species with a different mode of life and at a different phylogenetic level:
suborder Anchimae (Chauna torquata), suborder Lamellirostres
or Anseres (family Anatidae, subfamilies Anseranatinae, Anse-
rinae and Anatinae). The directions of BTO evolution are revealed and ontogenic development of BTO in the mallard (Anas
platyrynchos) is described.

Pieces of the bill tips of ducks and geese after formalin
fixation were carefully cleaned under water, defatted in alco-
hol and acetone and dried up to the critical point in a HCP-2
drier. Samples were goldcoated in the ion sputter IB-3. Photo-
graphs were made using "JEOL-JSM-50A" scanning electron microscope (SEM).

BTO consists of two parts situated in tips of the upper
and lower jaws. Anatomically they are similar and contain series of dermal papillae, which include two types of encapsulated mechanoreceptors: Herbst corpuscles proximally and
Grandry corpuscles distally. Each papilla or mechanoreceptor
complex protrudes into a keratin cap or papilla with an alve-
olus around it. The number, density of distribution per square
mm and shape variability of the epidermal structures were
compared.

The phylogenetic development of BTO from the most ancient
forms to the youngest were examined. Phylogenetic relationships
in the waterfowl were taken according to Delacour (1954-1959)
and Johnsgard (1978).

Transformations of the upper and lower BTO move in opposite directions. In the upper part of BTO the number, density
and shape variability of structures decreases in order of
screamers - geese - ducks. The absence of horny papillae is
typical of the subfamilies Anseranatinae and of most species
of the subfamily Anatinae. The number of epidermal structures

decreases from 450 in the screamer to 35-40 in dabbling ducks.
The lower BTO number and density of epidermal structures increase in the order of screamers - geese - ducks and the variability of their shapes in the order of screamers - geese -
shelducks - dabbling ducks. The increase of density from 2 to
20/mm^2 was observed. The presence of horny papillae in the
lower BTO is typical for the whole suborder Lamellirostres.
This morphological and perhaps functional asymmetry of BTO increases in waterfowl evolution.

The ecological specialization and environmental conditions
of the waterfowl also affects BTO structure. From the nonspe-
cialized ancestors closely related to screamers, waterfowl developed in three main directions.

Most species of geese are plant-eating forms, collecting
food on the ground. The complexes in the lower bill are twice
as large as in the upper bill. Their density is rather high
especially in small species. Short papillae are situated in
the upper BTO.

Sea ducks and eiders collect animal food on the bottom or
in deep water. They have a low number and density of BTO structures, the papillae are situated only in the lower BTO, they
are big, rough and all of the same shape.

Dabbling or river ducks are adapted to filter-feeding or
extraction of food particles from the water or mud for the
process of quick sucking in and pushing out little portions of
water, sometimes in complete darkness. For such a mode of feeding, effective tactile control is required. In dabbling ducks,
BTO is quite asymmetrical, papillae density is high, central
papillae are thin and long. Upper BTO alveoli are small, not
numerous and all of the same shape. Adaptations in other filter-feeding species, such as shelducks, some species of sea
and perching ducks, are similar.

In the suborder Lamellirostres as a whole two types of
BTO structure are clearly revealed. There is a "goose type"
with well developed upper and lower BTO and with morphological
variability of epidermal structures in both parts. Further
there is a "duck type" with well developed lower BTO and weakly
developed upper one. Series of transition forms are connected
both with the phylogenetic position and with the ecological
pecularieties of the species.

Epidermal structures of BTO develop in the mallard (Anas
platyrhynchos, Anatini) in the second part of incubation. Up
to the 13th day, a skin bolster forms on the inner surface of
the bill tip. It then becomes loose and knobs appear on it.
From the 21st day the surface of the bill tip becomes squamous.
During the last four days of incubation squamae form epidermal
alveoli of the future BTO. Their number is equal to that in
the adult bird. On hatching, the asymmetry of BTO is expressed
only qualitatively. Epidermal papillae of the lower BTO have
not grown yet.

Characteristic filter-feeding in mallard ducklings develops simultaneously with the lower BTO development. Two- or
three-day-old ducklings begin to filter water through the bill,
but do it rather seldom and for short periods of time. On the

9th day of age one filtering action lasts 40 s and the
full time of filter-feeding increases up to 40 % of the total
time of observation. With the increase of filter-feeding
periods the length of epidermal papillae increases. During the
second week they reach approximately a half of their full
length and by the end of the first month they become as long
as in the adult bird. By the end of the first month, besides
pecking and dabbling, such modes of feeding as straining and
up-ending appear. By the second month, ducklings begin to
forage in complete darkness.

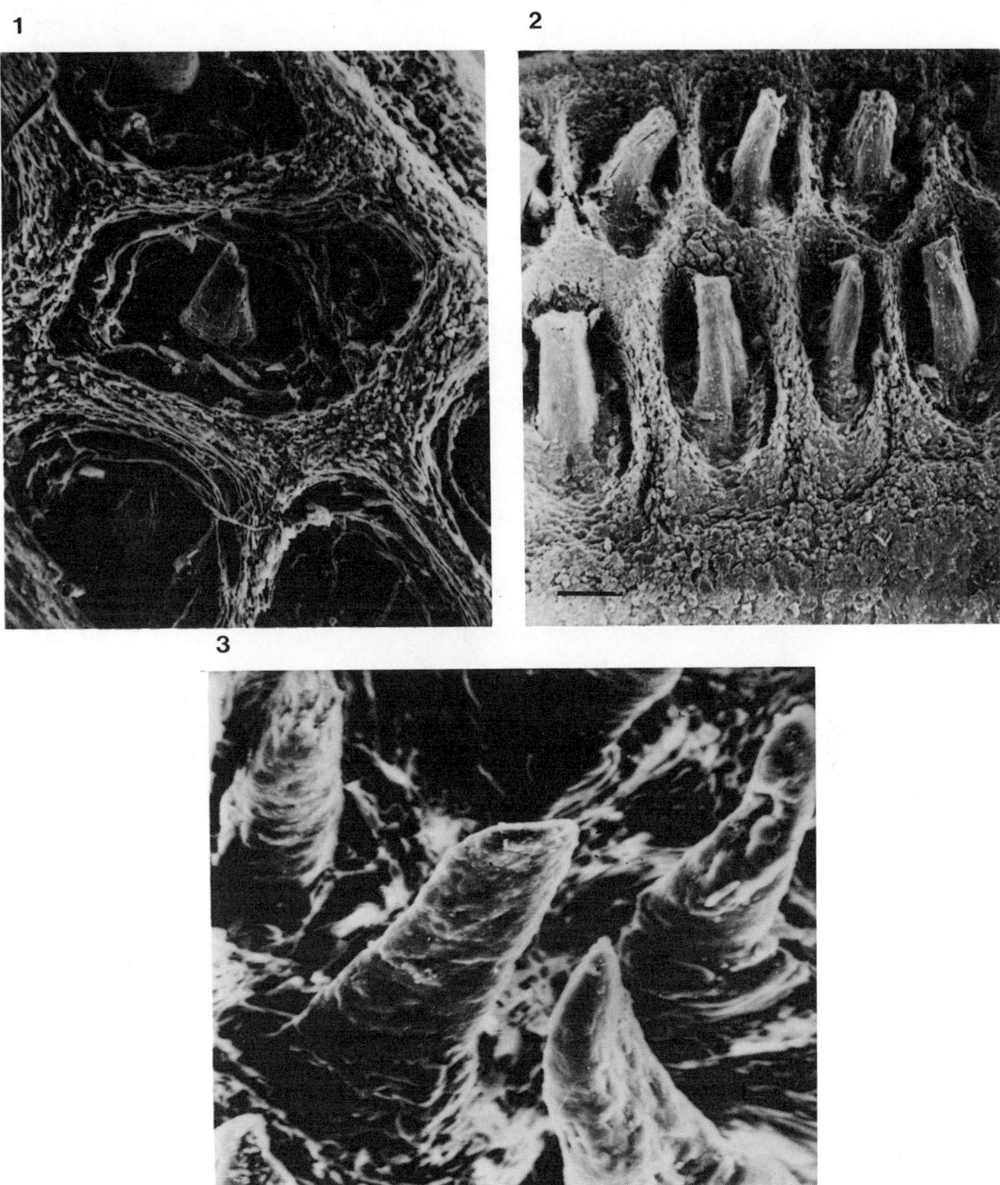

Figs. 1-3. A part of lower bill tip organ in three species of
waterfowl: king eider, Somateria spectabilis (1),
black scoter, Melanitta nigra (2) and mute swan,
Cygnus olor (3). Bar = 100 µm.

REFERENCES

Delacour, J., 1954-1959, "The Waterfowl of the World", London.
Johnsgard, P., 1978, "Ducks, Geese and Swans of the World",
      Nebraska-London.
Gottschaldt, K.M., and Lausmann, S., 1974, The peripheral mor-
      phological basis of tactile sensibility in the beak of
      geese, Cell Tiss. Res., 153:477-496.

MECHANORECEPTOR CELLS OF THE LAMPREY SPINAL CORD: DIRECT
CONNECTIONS WITH IDENTIFIED SEGMENTAL NEURONS

I.V. Batueva

Sechenov Institute
USSR Academy of Sciences
SU-194223 Leningrad, USSR

The organization of primary segmental afferents which act
on motoneurons in various species of vertebrates is not quite
clear. As is known, direct connections with motoneurons in
mammals are formed by afferents from intrafusal muscle fibers.
As to the lower (aquatic) vertebrates, they lack muscle spindles (Bone, 1964). Therefore the suggestion was made that they
can have no direct connection between the primary afferents
and motoneurons. Thus, only polysynaptic responses of motoneurons were described in bony fishes under dorsal root stimulation (Bando, 1975). However, monosynaptic responses were
obtained in rays (Leonard et al., 1978) and lampreys (Birnberger and Rovainen, 1971).

The aim of the present research was to study the excitatory postsynaptic potentials (EPSP) in motoneurons and in giant
interneurons of the spinal cord of the river lamprey, <u>Lampetra
fluviatilis</u>, elicited during stimulation of individual dorsal
sensory cells (DSC), and also the morphological structure of
labeled DSC and peripheral parts of their branches.

Two preparations, one including isolated spinal cord
(12-15 mm), and the other (4-5 cm) shell-coated cord, notochord
and adjacent myotomal muscles, were used in the experiments.
In both cases the preparation was perfused with oxygenated solution at 10-12 ^{o}C. Intracellular recordings were made with
glass microelectrodes filled with 3 M KCl. Single and rhythmic
action potentials (AP) were evoked in DSC by current injection
and the unitary EPSPs elicited in motoneurons were recorded
simultaneously and monitored on a dual beam oscilloscope. The
cells were identified by antidromic stimulation of the ventral
and dorsal roots and by their morphological structure after
labeling with HRP.

The DSC membrane potential was from 70 to 85 mV. Under
stimulation of the rostral and caudal parts of the cord they
generated APs of 75-111 mV with a stable latency (Fig. 1A1)
and reproduced rhythmic stimulation of the dorsal root (Fig.
1A2). In contrast to Müller axons, as the polarizing current
through the membrane was increased, they generated AP dis-

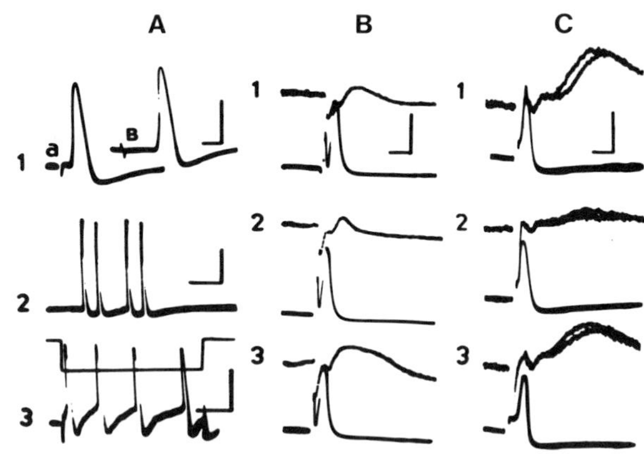

Fig. 1. Monosynaptic EPSPs in the neurons of the lateral
        column elicited by action potentials in dorsal cells.
        A - action potentials in DSC under stimulation of
        rostral (a) and caudal (B) parts of the isolated
        cord (1); rhythmic APs in DSC under dorsal root
        stimulation (2) and current (5 nA) injection (3).
        C, B - EPSPs in motoneuron and interneuron, respect-
        ively, (upper traces) and action potentials in
        DSC (lower traces) in the normal (1) and Ca^{2+}-free
        (2) solution. Restoration of initial responses
        15 min after perfusion with the normal solution (3).
        Scale bar: A - 40 mV, 2 ms (1), 20 mV, 10 ms (2, 3);
        B, C - 2 mV (upper traces), 30 mV (lower traces),
        2 ms.

charges (Fig. 1A3). After identification of DSC another micro-
electrode was inserted into the lateral column and when the
second neuron was impaled and identified, it was possible to
ascertain whether both neurons are directly connected.

Ninety-six DSC-motoneuron pairs were studied. During intracellular stimulation of DSC, EPSPs were obtained in 32 motoneurons. In half of them a synaptic delay was, in fact, absent.
This and the absence of EPSP "dropout", as well as weak poten-
tiation of EPSPs are indication of a direct link between the
DSC and the motoneuron. In most cases unitary EPSPs had a simple configuration and in the normal solution were not seen as
consisting of different components (Fig. 1B1). During perfusion
with Ca^{2+}-free solution containing 2 mM Mn^{2+}, the unitary EPSPs
were reduced but not completely blocked (Fig. 1B2). In 20 motoneurons the EPSP component resistant to a Ca^{2+}-free medium had
a constant amplitude for 2 h and amounted to 30-66 % of the
initial value. These data allow the conclusion that DSC-motoneuron EPSPs have a mixed mechanism of transmission (electrical
and chemical). The statistical analysis of fluctuation of the
chemical component amplitude, as well as the determination of
the morphological structure of DSC-motoneuron pairs revealed

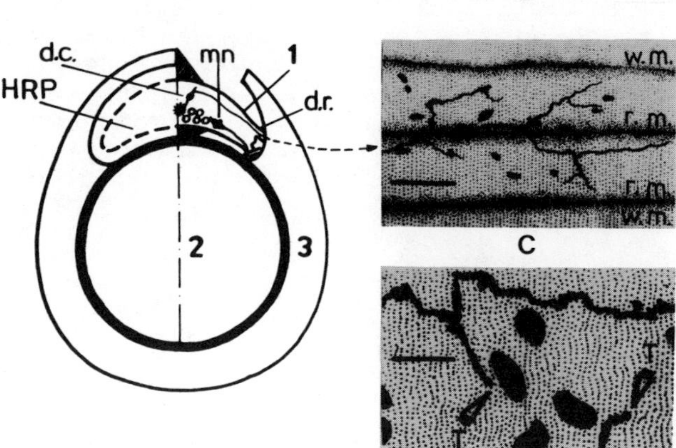

Fig. 2. Anterograde transport of HRP in dorsal cells of the
        spinal cord. A - scheme of the experiments. Frontal
        section through the spinal cord (1), notochord (2)
        and cartilage canal (3). mn - motoneuron, d.c. -
        dorsal cell, d.r. - dorsal root. Ventral root is
        dissected. Left part of the cord is removed and the
        cavity is filled with HRP. Hatching - vaseline iso-
        lation. B - dichotomic branching of the sensory
        nerve along two red muscle fibers (r.m.). White
        muscle fibers (w.m.) have no sensory branches.
        C - end terminals (T) of the sensory nerve on the
        red muscle fibers. Scale bar: 100 µm in B; 40 µm
        in C.

that each pair has 4 to 11 contact zones (Shapovalov and
Batueva, 1984).

     Five labeled DSC-interneuron pairs were investigated. On
the branches of marked DSC 18 to 30 contacts were observed
with the dendrites of the marked interneurons. This was in
good agreement with electrophysiological parameters of the
unitary DSC-interneuronal EPSPs. The quantity of points of
mediator realization was 10 to 28. The electrical component
of the unitary EPSPs in the interneurons had the amplitude
0.7-1.0 mV and was clearly seen in the normal solution (Fig.
1C1). Judging from the numerous contacts of marked interneurons
with branches of DSC on both ipsi- and contralateral sides of
the cord, they can be regarded as sensory neurons of the second order.

     It was important to determine whether there are any termi-
nals of DSC branches on the muscle fibers. To answer this
question, the method of anterograde transport of HRP was used.
Experimental conditions are given in Fig. 2A. DSC peripheral
branches were revealed in the dorsal roots, myoseptum and as
passing along the red muscle fibers to the skin. They branch
dichotomically (Fig. 2B), and their thin collaterals supply
the red muscle fibers.

The most important findings of the present study are as
follows. Firstly, demonstration of DSC direct links not only
with interneurons, but also with motoneurons. Secondly, the
discovery of DSC terminals on red muscle fibers, which allows
DSC to be regarded as primary afferents of both cutaneous and
muscle sensitivity. The obtained results testify to the role
of short latency reflex connections in the function of segmental neurons in cyclostomata.

REFERENCES

Bando, T., 1975, Synaptic organization in teleost spinal moto-
        neurons, Japan J. Physiology, 25:317-332.
Birnberger, K.L., and Rovainen, C.M., 1971, Behavioral and
        intracellular studies of a habituating fin reflex in
        the sea lamprey, J. Neurophysiol., 34:983-989.
Bone, Q., 1964, Patterns of muscular innervation in lower chor-
        dates, Int. Rev. Neurobiol., 6:99-147.
Leonard, R.B., Rudomin, P., and Willis, W.D., 1978, Central
        effects of volleys in sensory and motor components of
        peripheral nerve in the Stingray, Dasyatis sabina, J.
        Neurophysiol., 41:108-125.
Shapovalov, A.I., and Batueva, I.V., 1984, Interaction between
        segmental primary afferents and motoneurons in the spi-
        nal cord of the lamprey, Sechenov Physiol. J. USSR,
        70:1178-1188.

COXAL SETAL ORGANS IN ARCHAEOGNATHA AND ZYGENTOMA (INSECTA)

František Weyda

Institute of Entomology
České Budějovice
Czechoslovakia

MORPHOLOGY AND ULTRASTRUCTURE OF COXAL SETAL ORGANS

Coxal setal organs (CSO) have been described by Weyda and
Štys (1974). They occur on the apices of coxae of the first to
the seventh abdominal segment. Topographically and functionally the CSO are associated with eversible vesicles (organs ab-
sorbing water from a wet substrate; this type of water uptake
is essential for water economy of Archaeognatha and some Zygentoma). CSO differ slightly in structure in various species
of Archaeognatha. There are significant differences in the
number of mechanosensitive sensillae between some species of
individual genera. But this taxonomic character is of low
practical value because of the difficulties in counting the
total number of sensillae (small size, several hundred of sensillae per one insect). CSO are missing only in Kuschelochilis,
which also lacks eversible vesicles (adaptation to the specific very humid environment of the island Juan Fernandez).
Eversible vesicles as well as CSO are present only in primitive
groups of Zygentoma. For example, Tricholepidion gertschi
("living fossil" from California) has CSO fully developed.
Nicoletia phytophila (Nicoletiidae) has CSO in a reduced form.
In other more advanced species of Zygentoma, CSO are either
greatly reduced or mostly completely absent (Figs. 1 - 3).

Each CSO is formed by two sets of main components: a row
of mechanoreceptive sensillae and a row of microtrichia (Fig.
1). The structure of the sensillae corresponds to that of
thick-walled mechanoreceptive hairs of arthropods. They are
inserted in characteristic, bilaterally symmetrical, poste-
riorly elongated and elevated rims. Each sensilla possesses a
trichogen, a tormogen and a thecogen cell which envelop a bipolar sensory cell. The dendrite of the sensory cell is composed of two parts: an inner and an outer segment (the latter
is surrounded by a dendritic sheath secreted by the thecogen
cell - Fig. 1). A specific tubular body is situated on the tip
of the dendrite. This tubular body forms a very complex unit
together with the dendritic sheath, the joint membrane and the
socket septum (last two cuticular structures are located in
the joint region of the sensilla). The axons of individual sen-

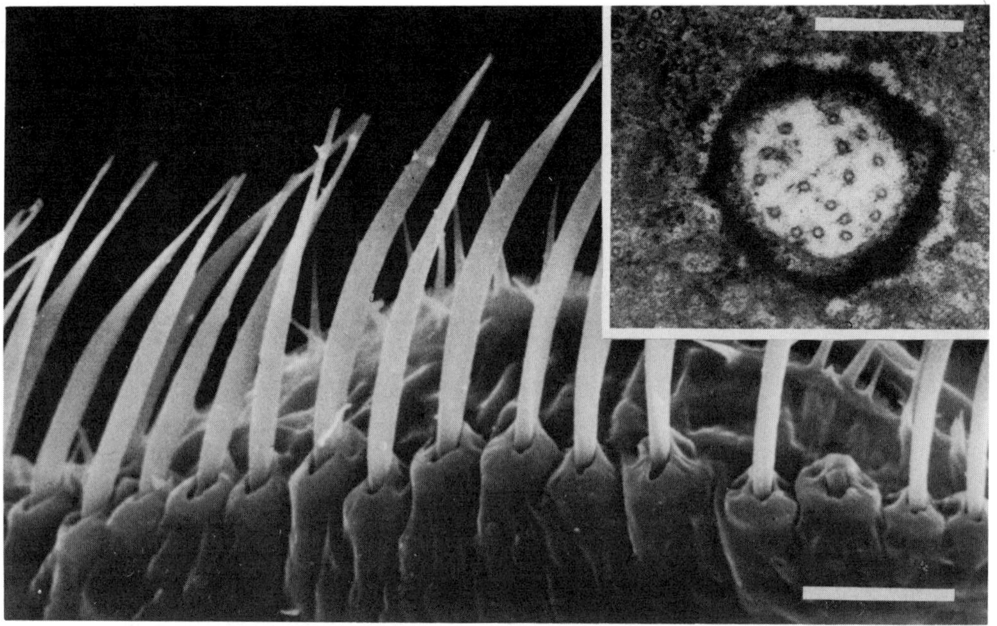

Fig. 1. Coxal setal organ in Machilis helleri (Archaeognatha).
Bar: 2 μm. Inset: Outer dendritic segment of sensory
cell innervating the mechanosensitive sensilla. Bar:
30 nm.

sory cells become associated in a nerve connecting each CSO
to the nerve cord ganglion of the corresponding abdominal
segment.

FUNCTION OF CSO

Archaeognatha and some Zygentoma are able to evert their
eversible vesicles by means of hemolymph pressure based on the
contraction of abdominal muscles. Everted vesicles are used
for absorption of water into the body cavity (passive movement of water associated with active transport of ions by a
highly specialized transporting epithelium). All eversible
vesicles located out of the water or a wet substrate are retracted within several seconds (prevention of water loss by
transpiration). CSO control the state of the eversible vesicles in a simple way (everted, retracted) and inform the
central nervous system. The mechanical construction of the
rim (socket) of the sensilla allows deflection of the tubular
body only in one direction. The microtrichia of CSO function
as accessory structures restricting the excessive swing of
mechanosensitive sensillae laterally, which ensure an optimal
transfer of mechanical stimuli.

DEVELOPMENT OF CSO

CSO are present in the functional state in the first in-
star of Archaeognatha. The number of sensillae as well as

their length exhibit negative allometric growth during ontogenetic development of Machilis helleri (Dyar´s law could be
applied for growth of CSO). Information on the development of
CSO in phylogeny of Archaeognatha and Zygentoma is completely
lacking. Probably they arose from numerous mechanosensitive
sensillae and nonsensory microtrichia occurring all over the
body.

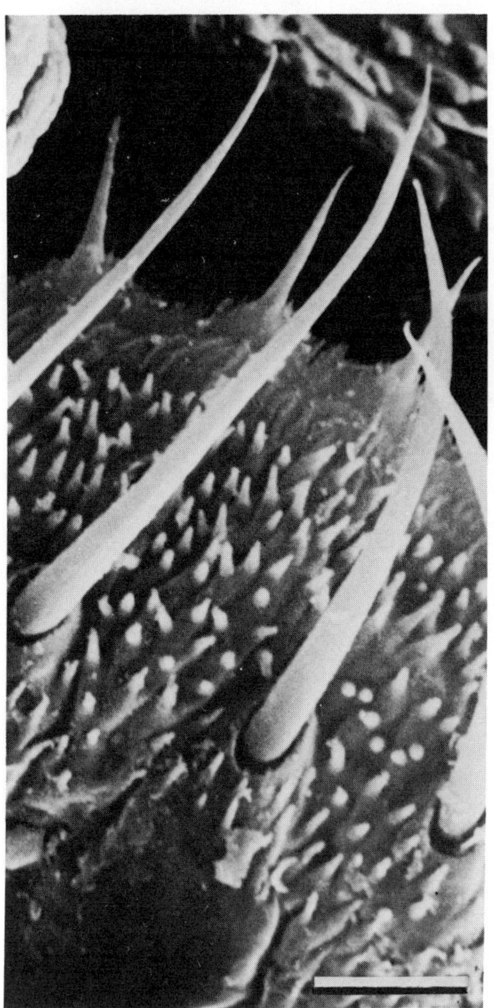

Fig. 2. CSO in Tricholepidion
gertschi (Zygentoma).
Bar: 15 um.

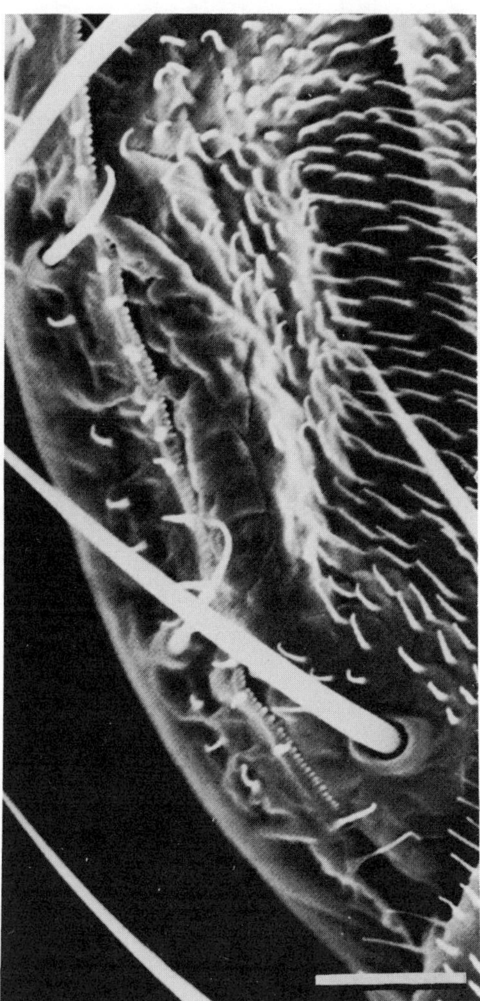

Fig. 3. Reduced CSO in Nicoletia
phytophila (Zygentoma).
Bar: 10 um.

REFERENCE

Weyda, F., and Štys, P., 1974, Coxal setal organs in Machilidae
and their homologues on the genitalia, Acta ent. bohemoslov., 71:51.

PART VIII

FUNCTIONAL SIGNIFICANCE
OF PROPRIOCEPTIVE FEED-BACK

DISCHARGES OF TENDON ORGANS DURING UNFUSED MUSCLE CONTRACTIONS

G. Horcholle-Bossavit, L. Jami, J. Petit, R. Vejsada and
D. Zytnicki

Laboratoire de Neurophysiologie
Collège de France
75231 Paris cédex 05, France

The mechanoreceptor complement of mammalian skeletal muscle
includes Golgi tendon organs, which are innervated by Ib afferent fibres
and considered as contraction sensors because their sensitivity for this
stimulus is very high (Houk and Henneman,1967). The typical location of
a tendon organ is at the myotendinous junction. Within the capsule of a
tendon organ, sensory endings contact collagen bundles connected at one
end with a small fascicle of muscle fibres (in-series muscle fibres),
while the other end is in continuity with either tendon or aponeurosis.
Several motor units contribute fibres to the fascicle and the tendon
organ can signal the contraction of each unit because the twitch of a
single in-series muscle fibre, pulling on the receptor, is enough to
elicit a discharge (Fukami,1981). Each tendon organ thus monitors the
activity of a small set of 10-15 motor units, belonging to different
physiological types (Reinking et al.,1975). Since, in addition, each of
these units usually activates several tendon organs, a comparison of the
numbers of motor units and of tendon organs in cat leg muscles suggests
that the contraction of every single motor unit in these muscles is
monitored by at least one tendon organ (Jami and Petit,1976a).

It is not certain, however, whether the force of the contraction is
the parameter actually monitored by individual tendon organs, because
their discharge frequencies rarely display a clear relation with the
tensions developed by contracting motor units as recorded at the muscle
tendon (Houk and Henneman,1967; Reinking et al.,1975; Stauffer and
Stephens,1975; Jami and Petit,1976b; Gregory and Proske,1979). The
discharge frequency of a particular tendon organ can be higher in
response to a weak slow unit than in response to a fast unit producing
strong forces. A possible clue to the difficulty of relating tendon
organ discharges and contractile forces was provided by the hypothesis
of Houk and Henneman (1967; see also Reinking et al.,1975), that
contractile forces were encoded by the ensemble discharge of all the
tendon organs in a muscle.

In experiments on the cat peroneus tertius muscle, in which there
are only 10 tendon organs on average (Scott and Young,1987), it was
possible to examine this hypothesis because all the Ib afferents from
this muscle could be functionally isolated for recording in dorsal root
filaments. The individual discharges of several (3-6) tendon organs

activated by a single motor unit were recorded simultaneously with the
tension developed by the motor unit. Data were collected on an IBM-PC XT
microcomputer and subsequent processing was used to pool the individual
responses of tendon organs and to display their ensemble discharge.

The stimulation frequencies employed were similar to the usual
firing rates of alpha-motoneurones under natural conditions (Hoffer et
al. 1987). Such frequencies elicit unfused tetanic contractions, at which
the developed tension displays oscillations, at the frequency of
stimulation, superimposed on a steady level of static tension.
Modulations in the mean level of static tension were produced by varying
the stimulation frequency. Fig.1 shows the tension profile elicited by
stimulating a fast-contracting motor unit with a "sawtooth" pattern of
stimulation in which the frequency varied linearly between 20 and 40/s.
The mean level of tension increased from 6.5g to 26g, when the
stimulation frequency increased from 20 to 40/s, while the amplitude of
superimposed oscillations decreased from 8g to 1.6g. In the response of
the tendon organ, there was one impulse for each oscillation of the
tension, so that the discharge frequency exactly reproduced the
stimulation pattern. This type of response, locked on the stimulation
frequency, is termed "1 : 1 driving", as the response of some spindles
to fusimotor activation initially observed by Kuffler et al. (1951).
Unfused tetani of fast-contracting motor units very often elicit driving
of tendon organ discharges, because oscillations of tension represent a
powerful stimulus for the dynamic sensitivity of the receptor (Jami et
al.,1985). In this type of response, the tendon organ discharge carries
information about the frequency of tension oscillations but not about
their amplitude or the mean level of tension upon which they are
superimposed.

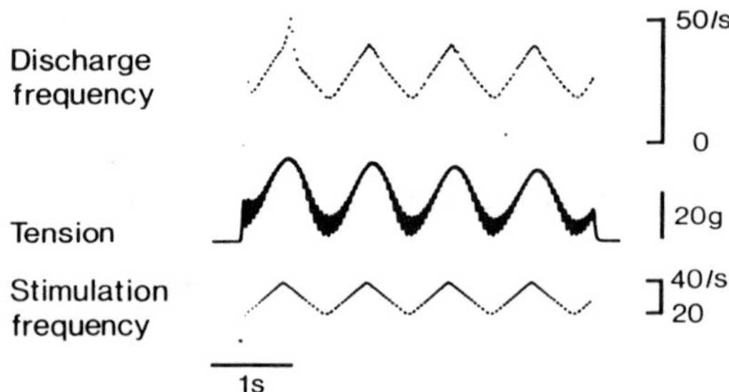

Fig. 1. Response of a peroneus tertius tendon organ
to a fast-contracting motor unit. Discharge
of the tendon organ and stimulation are
represented in instantaneous frequency.
Stimulation pattern described in text.
Conduction velocities of the afferent fibre,
90m/s and of the motor axon, 92m/s.

The same motor unit activated in total four tendon organs in the
muscle and their responses to the "sawtooth" tension profile are shown
in Fig. 2. It is clear that activation of different tendon organs by a
single motor unit is not uniform. In the example shown, the motor unit
was more efficient with GTO1 than with the other three tendon organs. At

the peak of tension, i.e., when the frequency of stimulation of the
motor unit reached 40/s, GTO1 was quite strongly activated, but the
correlation between the receptor discharge and the tension profile was
not obvious. Note, for instance, that maximal variations in discharge
frequency occurred when the oscillations superimposed on the mean level

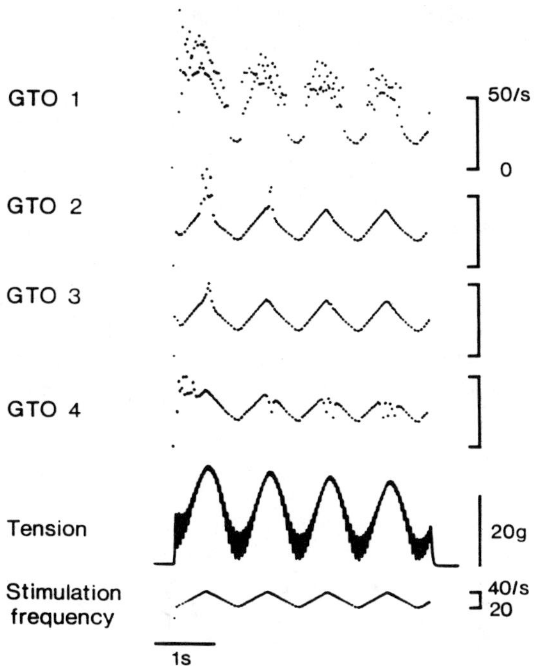

Fig. 2. Responses of four tendon organs
(GTO1-4) activated by the same
motor unit. GTO3 is the same as
in Fig.1. Conduction velocities
of afferent fibres ranged between
88 and 96 m/s.

of tension were minimal. Several factors may account for the non-uniform
activation of different tendon organs by a single motor unit. A first
possibility is that the motor unit contributed two muscle fibres to the
fascicle in-series with GTO1 and a single fibre to each of the fascicles
in-series with GTO2 and GTO3 respectively (see e.g. Spielmann and
Stauffer, 1986). The inequal efficiency of a motor unit with different
tendon organs could also be related to the non-homogeneous distribution
of sensory terminals within a tendon organ (Zelena and Soukup, 1983) : a
muscle fibre connected to a richly innervated collagenous bundle is
likely to elicit a stronger activation of the receptor than a fibre
connected to a poorly innervated bundle. Finally, it is known that a
number of muscle fibres situated in-parallel with the tendon organ and
inserting around its tendinous end can exert unloading effects on the
receptor (Houk and Henneman, 1967). Conceivably, the efficiency of a
motor unit in activating a tendon organ could be reduced (see for
instance, GTO4 in Fig. 2) if this unit had some fibres in-parallel with
the receptor, in addition to in-series fibres (Fukami, 1981).

The contractile force produced in infused tetanus present the
tendon organ with a composite stimulus, including static and dynamic
components, i.e., respectively, the mean level of tension and the
superimposed oscillations. Fig. 2 shows that individual tendon organs
responded essentially to the dynamic component of the stimulus, although
they did not provide much information about the amplitude of tension
variations. However, the ensemble discharge of the four receptors did
contain some information on this point, and also on the static component
of the tension. The discharges of all the tendon organs activated by a
motor unit were pooled and processed in different ways. In a first step,
the response of each tendon organ was divided in bins of equal
durations. Different bin widths were tested, ranging between 10 and
50ms. Then, within each bin, either the number of impulses in the pooled
responses could be counted or the instantaneous frequency of the pooled
discharges could be calculated. A third mode of processing was used to
obtain the results presented in Fig. 3 : the "average instantaneous
frequency" was calculated within each bin for each tendon organ, which
gave one figure per tendon organ (four figures in the present instance);
the average of these four figures was then calculated for each bin and
this gave the average instantaneous frequency of the pooled responses of
the four tendon organs. Fig. 3 shows that the average instantaneous
frequency of the four tendon organs displayed a profile comparable to
the tension profile, with variations reflecting both the rapid and slow

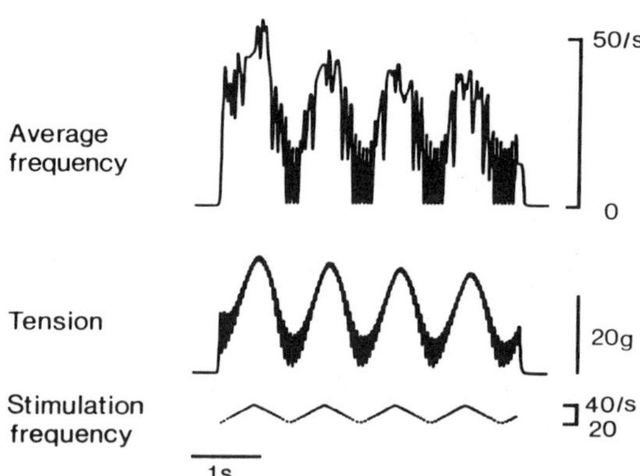

Fig. 3. Instantaneous average frequency of the
discharges of the four tendon organs
shown in Fig. 2. The method used to
calculate the average frequency is
explained in text. Bin width, 25ms.

variations of tension. This was possible because activation of the four
tendon organs by the motor unit was not uniform. The average
instantaneous frequency of the collective discharge of the tendon organs
activated by a single motor unit provided informations about the unit
tension profile which were not detectable in the instantaneous discharge
frequency of individual tendon organs.

However, when two motor units were stimulated in combination, although there was approximately linear summation of tensions, the ensemble discharges of tendon organs did not display linear summation, whatever mode of processing was employed. The properties of individual tendon organs can account for this non-linear summation, starting with the characteristics of receptor potentials (Fukami and Wilkinson,1977, Wilkinson and Fukami,1983). Moreover, there are possibilities of mechanical interactions between tendinous components within the tendon organ (Gregory and Proske,1979; Zelena and Soukup,1983) and interactions may also occur within the receptor between the multiple myelinated branches of the sensory Ib fibre (Gregory et al.,1985). In the present experiments, non linear summation of the effects of two motor units on a group of tendon organs was often due to unloading of receptors by in-parallel components of contractile force (Stuart et al.,1972). Unloading influences were exerted by one of the motor units upon some of the tendon organs activated by the other unit, and consequently, the ensemble discharge of tendon organs in response to both units contracting together was less than the sum of the ensemble discharges elicited by each unit activated on its own.

When considering the location and the structure of tendon organs, as well as their transducing characteristics, several explanations are available to account for the non-linearity of the relation between the frequencies of tendon organ discharges and the tensions developed by motor units. Nevertheless, the ensemble discharge of Ib afferents from a muscle contains information about the variations of tension and this is likely to represent, for the Central Nervous System, the most pertinent information on contractile force.

REFERENCES

Fukami, Y., 1981, Responses of isolated Golgi tendon organs of the cat muscle to contraction and electrical stimulation, J. Physiol. Lond., 318:723-748.
Fukami, Y.,and Wilkinson, R.S., 1977, Responses of isolated Golgi tendon organs of the cat, J. Physiol. Lond., 265:673-689.
Gregory, J.E.,and Proske, U., 1979, The responses of Golgi tendon organs to stimulation of different combinations of motor units, J. Physiol. Lond., 295:251-262.
Gregory, J.E., Morgan, D.L.,and Proske, U., 1985, Site of impulse initiation in tendon organs of cat soleus muscle, J. Neurophysiol., 54:1383-1395.
Hoffer, J.A., Sugano, N., Loeb, G.E., Marks, W.B., O'Donovan, M.J.,and Pratt, C.A., 1987, Cat hindlimb motoneurones during locomotion; II. Normal activity patterns, J. Neurophysiol., 57:530-553.
Houk, J.C.,and Henneman, E., 1967, Responses of Golgi tendon organs to active contraction of the soleus muscle of the cat, J. Neurophysiol., 30:466-481.
Jami, L.,and Petit, J., 1976a, Heterogeneity of motor units activating single Golgi tendon organs in cat leg muscles, Exp. Brain Res., 24:485-493.
Jami, L.,and Petit, J., 1976b, Frequency of tendon organ discharges elicited by the contraction of motor units in cat leg muscles, J. Physiol. Lond., 261:633-645.
Jami, L., Petit, J., Proske, U.,and Zytnicki, D., 1985, Responses of tendon organs to unfused contractions of single motor units, J. Neurophysiol., 53:32-42.

Kuffler, S.W., Hunt, C.C., and Quilliam, J.P., 1951, Function of
    medullated small nerve fibers in mammalian ventral roots : efferent
    muscle spindle innervation, J. Neurophysiol., 14:29-54.
Reinking, R.M., Stephens, J.A., and Stuart, D.G., 1975, The tendon organ
    of cat medial gastrocnumius muscle : significance of motor unit
    type and size for the activation of Ib afferents, J. Physiol.
    Lond., 250:491-512.
Scott, J.J.A., and Young, H., 1987, The number and distribution of muscle
    spindles and tendon organs in the peroneal muscles of the cat,
    J. Anat., 151:143-155.
Spielmann, J.M., and Stauffer, E.K., 1986, Morphological observations of
    motor units connected in-series to tendon organs, J. Neurophysiol.,
    55:147-162.
Stauffer, E.K., and Stephens, J.A., 1975, The tendon organs of cat
    soleus : static sensitivity to active force, Exp. Brain Res.,
    23:279-291.
Stuart, D.G., Mosher, C.C., Gerlach, R.L., and Reinking, R.M., 1972,
    Mechanical arrangement and transducing properties of Golgi tendon
    organs, Exp. Brain Res., 14:274-292.
Wilkinson, R.S., and Fukami, Y., 1983, Responses of isolated tendon
    organs of cat to sinusoidal stretch, J. Neurophysiol., 49:976-988.
Zelena, J., and Soukup, T., 1983, The in-series and in-parallel
    components in rat hindlimb tendon organs, Neuroscience, 9:899-910.

ACKNOWLEDGEMENTS

This work was supported by C.N.R.S. (ATP n° 960140 and a fellowship to
R. Vejsada)

A PROPOSED MECHANORECEPTOR ROLE FOR THE SMALL REDUNDANT MUSCLES WHICH ACT IN PARALLEL WITH LARGE PRIME-MOVERS

David Peck, Donald F. Buxton, and Arthur J. Nitz

Dept. of Anatomy, U.K.M.C., Lexington, KY, USA; Dept. of Anatomy and Histology, Auburn University, Auburn, AL, USA; Dept. of Physical Therapy, U.K.M.C., Lexington, KY

INTRODUCTION

Small, apparently redundant, muscles frequently act in parallel with vastly larger prime-movers and we call this morphological entity the "parallel muscle combination" (PMC). The small muscles of PMCs have much higher muscle spindle densities than their large counterparts, and we call the ratio of mean spindle density of the small muscle/s to that of the large muscle/s in a PMC, the PMC SPINDLE RATIO (PMC-SR). In view of this characteristic of PMCs we proposed that the small muscles thereof play a primarily proprioceptive role (Peck et al.,1984).

Since PMCs occur in different species we decided to compare homologous PMCs in mammalian pectoral limbs specialized for manipulation and plantigrade locomotion by using existing data for the muscles involved, namely those in man (Voss, 1971) and in the guinea pig (Martini and Palmieri, 1971). In addition, we examined spindle densities of the forelimb muscles of the dog in order to add a pectoral limb specialized for digitigrade locomotion to this comparison.

Furthermore, because there is a correlation between the number of degrees of freedom (kinematic complexity) of a joint and the number of spinal levels for its motor control in man (Peck and Brower, 1987) we decided to determine: 1.) whether the absolute number of spindles for a human joint shows a similar correlation, and 2.) if the mean spindle densities or PMC-SRs associated with joints in these digitigrade, plantigrade, and manipulative species are correlated with their kinematic complexity.

MATERIALS AND METHODS

Using standard procedures we collected the intrinsic muscles from the pectoral limbs of six dogs and took five mm thick cross-sectional samples from the middle of each muscle belly. Following routine histological preparation, muscle samples were sectioned perpendicular to the fiber direction at 7 μm, stained

with 0.1% toluidine blue and muscle spindles were counted at
2.5X, 10X and 25X magnifications. The grain-counting program of
the Bioquant system (R & M Biometrics, Nashville, TN) was used
to measure the area of each section and the number of
spindles/mm^2 was then calculated.

Eight PMCs of the dog forelimb were selected to compare
with homologous PMCs in the guinea pig (using data reported by
Martini and Palmieri, 1971) and man (using data from Voss, 1971)
as shown in table 1. Each PMC contained one or more large muscles controlling a joint in parallel with one or more small
muscles. Student's t-test was used to determine the significance of differences between the following means:
1) mean spindle density of the large muscles and mean spindle
density of the small muscles in each PMC;
2) mean spindle density of the pooled large muscles and mean
spindle density of the pooled small muscles for all PMCs in each
species;
3) mean spindle density of the pooled PMCs controlling digital
joints and mean spindle density of the pooled PMCs controlling
the more proximal joints in each species.

Correlation coefficients were calculated to test for
relations between the following pairs of parameters:
1) kinematic complexity (degrees of freedom) of joints and the
absolute number of spindles in muscles controlling them in man;
2) absolute number of spindles in muscles controlling joints and
the mean spindle density for those muscles in man;
3) absolute number of spindles in muscles controlling joints and
the mean PMC-SR for PMCs controlling them in man;
4) kinematic complexity of joints and the mean spindle density
for muscles controlling them in all three species (absolute
spindle numbers were not available for the dog and guinea pig);
5) mean spindle density for muscles controlling joints and the
mean PMC-SR for PMCs controlling those joints in all three
species.

Alpha levels for the correlation coefficients were
determined from a table in Snedecor and Cochran (1980).

RESULTS

Comparisons of spindle densities for pectoral limb PMCs of
the dog, guinea pig and man are presented in table 1. The mean
PMC-SR (N = 20) was 3.64 with a standard deviation of 3.04, and
ranged from 0.55 for the forearm pronation PMC in the dog, to
13.40 for the metacarpo-phalangeal flexion PMC in the dog. The
differences between the means of spindle densities for the
pooled large muscles and similar means for the pooled small
muscles of PMCs in pectoral limbs were significant for all three
species. The difference between the mean spindle density for
the pooled muscles of digital PMCs and that for the pooled
muscles of PMCs controlling joints proximal to the digits was
also statistically significant in the forelimbs of all three
species. In addition, the mean PMC-SRs for digits I and digits
II-V in the dog were 2.00 and 13.40 respectively, while in man
they were 5.22 and 2.11.

These findings will be considered in the discussion.

378

TABLE 1.   COMPARISON OF SPINDLE DENSITIES FOR PMCs IN THE
PECTORAL LIMBS OF THE DOG, GUINEA PIG AND MAN.

PMC	DOG (Buxton, 1985) $\#spindles/mm^2$			GUINEA PIG (Martini & Palmieri, 1971) $\#spindles/g$			MAN (Voss, 1971) $\#spindles/g$		
	N	MEAN	S.D.	N	MEAN	S.D.	N	MEAN	S.D.
PECTORAL GIRDLE									
PROTRACTION									
large mm. (X_L)	----			----			1	1.52	
Pectoralis major							1	1.52	
small mm. (X_S)	----			----			1	5.19	
Subclavius							1	5.19	
"T" VALUE	----			----			----		
ALPHA LEVEL	----			----			N.S.		
PMC-SR (X_S/X_L)	----			----			3.41		
SHOULDER.									
EXTENSION									
large mm. (X_L)	24	0.097	0.061	2	85.8	63.8	2	1.00	1.00
Supraspinatus	12	0.058		1	40.7		1	0.54	
Biceps brachii	12	0.135		1	130.9		1	1.95	
small mm. (X_S)	12	0.300	0.161	1	166.6		1	3.09	
Coracobrachialis	12	0.300		1	166.6		1	3.09	
"T" VALUE	5.523			1.034			1.511		
ALPHA LEVEL	<0.01			>0.10			>0.10		
PMC-SR (X_S/X_L)	3.09			1.94			3.09		
SHOULDER									
FLEXION									
large mm. (X_L)	42	0.056	0.033	4	74.1	42.3	4	0.72	0.59
Deltoid	6	0.083		1	46.8		1	0.51	
Triceps (longum)	12	0.022		1	32.9		1	1.60	
Teres major	12	0.086		1	91.2		1	0.36	
Infraspinatus	12	0.033		1	125.5		1	0.39	
small mm. (X_S)	12	0.181	0.090	1	171.4		1	1.26	
Teres minor	12	0.181	0.090	1	171.4		1	1.26	
"T" VALUE	5.701			2.056			0.821		
ALPHA LEVEL	<0.01			<0.10			>0.10		
PMC-SR (X_S/X_L)	3.23			2.31			1.75		
ELBOW									
EXTENSION									
large mm. (X_L)	48	0.042	0.028	3	45.4	17.6	3	1.59	0.53
Triceps longum	12	0.022		1	32.9		1	1.60	
Tric. laterale	12	0.017		1	57.8 (mean		1	2.12	
Tric. mediale	12	0.054			of lat. &		1	1.06	
Tr. accessorium	12	0.076			med. heads)				
small mm. (X_S)	12	0.173	0.147	1	150.0		no data		
Anconeus	12	0.173	0.147	1	150.0		no data		
"T" VALUE	3.110			4.853			----		
ALPHA LEVEL	<0.01			<0.10			----		
PMC-SR (X_S/X_L)	4.12			3.30			----		
FOREARM									
SUPINATION									
large mm. (X_L)	6	0.135		1	130.9		1	1.95	
Biceps brachii	6	0.135		1	130.9		1	1.95	
small mm. (X_S)	4	0.148		no data			1	8.0	
Supinator	4	0.148		no data			1	8.0	
"T" VALUE	----			----			----		
ALPHA LEVEL	N.S.			----			N.S.		
PMC-SR (X_S/X_L)	1.20			----			4.10		

(continued)

TABLE 1.(Continued)

PMC	DOG #spindles/mm^2			GUINEA PIG #spindles/g			MAN #spindles/g		
	N	MEAN	S.D.	N	MEAN	S.D.	N	MEAN	S.D.
FOREARM									
PRONATION									
large mm. (X_L)	6	0.087		no data			4	4.8	0.08
Pronator teres	6	0.087		no data			4	4.8	
small mm. (X_S)	4	0.048		no data			1	10.0	
Pronator quad.	4	0.048		no data			1	10.0	
"T" VALUE	----			----			----		
ALPHA LEVEL	N.S.			----			N.S.		
PMC-SR (X_S/X_L)	0.55			----			2.08		
METACARPO-PHALANGEAL									
FLEXION (DIGITS II-V)									
large mm. (X_L)	24	0.045	0.032	2	127.9	120.0	2	3.3	0.57
flex. dig. sup.	12	0.048		1	42.8		1	3.7	
flex. dig. prof.	12	0.042		1	213.0		1	2.9	
small mm. (X_S)	24	0.603	0.543	2	1,119.0	185.0	39	6.97	4.84
interossei	12	0.219		1	1,250.0		35	5.88	3.67
lumbricales	12	0.988		1	988.9		4	16.48	3.15
"T" VALUE	1.453			6.363			1.483		
ALPHA LEVEL	>0.10			<0.025			>0.10		
PMC-SR (X_S/X_L)	13.40			8.75			2.11		
METACARPO-PHALANGEAL									
FLEXION (DIGIT I)									
large mm. (X_L)	12	0.042	0.031	-----			1	5.5	
fl. poll. long.	12	0.042	0.031	-----			1	5.5	
small mm. (X_S)	12	0.179	0.183	-----			1	15.1	
fl. poll. brev.	12	0.179	0.183	-----			1	15.1	
"T" VALUE	0.183			-----			----		
ALPHA LEVEL	N.S.			-----			N.S.		
PMC-SR (X_S/X_L)	1.24			-----			2.74		
CARPO-METACARPAL									
ABDUCTION (DIGIT I)									
large mm. (X_L)	12	0.135	0.078	-----			1	3.8	
abd. poll. long.	12	0.135	0.078	-----			1	3.8	
small mm. (X_S)	10	0.372	0.586	-----			1	29.3	
abd. poll. brev.	10	0.372	0.586	-----			1	29.3	
"T" VALUE	0.183			-----			----		
ALPHA LEVEL	N.S.			-----			N.S.		
PMC-SR (X_S/X_L)	2.76			-----			7.71		
POOLED LARGE MUSCLES OF									
ALL PMCs (X_L)	13	0.067	0.038	9	86.8	60.0	12	2.30	1.81
POOLED SMALL MUSCLES OF									
ALL PMCs (X_S)	9	0.290	0.277	5	545.4	532.2	9	10.48	8.71
"T" VALUE	2.883			2.642			3.189		
ALPHA LEVEL	<0.005			<0.025			<0.005		
X PMC-SR (X_S/X_L)	4.33			6.28			4.56		
POOLED MUSCLES OF PROXIMAL									
PMCs (X_{PX})	16	0.095	0.076	10	101.4	54.1	13	3.02	3.11
POOLED MUSCLES OF METACARPO-PHALANGEAL & CARPO-METACARPAL									
PMCs (X_{MP})	8	0.253	0.318	4	623.7	586.4	8	10.33	9.30
"T" VALUE	1.928			2.973			2.642		
ALPHA LEVEL	<0.05			<0.01			<0.01		
X_{MP}/X_{PX}	2.66			6.15			3.43		

TABLE 2. CORRELATION OF KINEMATIC COMPLEXITY (DEGREES OF
FREEDOM) OF HUMAN JOINTS WITH ABSOLUTE SPINDLE NUMBERS
AND WITH SPINDLE DENSITIES (SP.%)

JOINT/S	upper limb degrees of freedom	abs. sp.#	sp.%	lower limb degrees of freedom	abs. sp.#	sp.%
Glenohumeral + girdle/Hip	6	2,844	1.21	3	3,517	1.20
Elbow/Knee	2	2,325	2.28	2	2,636	0.90
Wrist/Ankle	2	1,895	4.06	2	1,967	1.43
MCP/MTP joints Digits II-V	1.5	1,362	4.54	1.5	987	4.90
DIP joints Digits II-V	1	666	4.47	1	604	4.73
CM/TM joints Digit I	2	567	7.01	1.5	447	2.50
IP joint Digit I	1	199	6.22	1	244	2.03
Corr. coeff.		0.753	−0.688		0.942	−0.608
Alpha Level		=0.05	<0.10		<0.01	>0.10

In addition, there is a strong correlation (0.925, **alpha
level <0.01**) between absolute spindle numbers and spindle
densities associated with joints of the upper extremity. For
the lower extremity it is **0.614** and is not statistically
significant.

TABLE 3. CORRELATION OF KINEMATIC COMPLEXITY OF JOINTS WITH
SPINDLE DENSITIES AND WITH MEAN PMC SPINDLE RATIOS

JOINT/S	DOG dg. fr.	X sp/ mm2	X PMC-SR	GUINEA PIG dg. fr.	X sp%	X PMC-SR	MAN dg. fr.	X sp%	X PMC-SR
Gleno-humeral	3	.11	3.16	3	92.1	2.13	3	1.08	2.42
Elbow	2	.11	2.51	2	162.1	3.30	2	2.28	3.09
Digits II - V	1.5	.53	13.40	1.5	562.0	8.75	1.5	4.54	2.11
Digit I	1.5	.35	2.0	1.5	-----	----	2	7.01	5.22
CORREL. COEFF.		−0.772	−0.393		−0.839	−0.854		−0.618	−0.091
ALPHA LEVEL		N.S.	N.C.		N.S.	N.S.		N.S.	N.C.

In addition, there are good correlations between spindle
densities and **PMC-SRs** associated with joints was as follows:
dog = 0.867, guinea pig = 0.9996, and **man = 0.725.**

DISCUSSION

The significant differences between the mean spindle
densities for the pooled large muscles and for the pooled small
muscles of pectoral limb PMCs in all three species, lends
credence to a proprioceptive role for the small muscles of PMCs.

A correlation of digital PMC specialization with the primitive traction role of digits in locomotion seems consistent with the significantly higher mean spindle densities found in the pooled muscles of digital PMCs compared to those of proximal PMCs in all three species. PMC specializations of the digits apparently accompany their kinematic adaptation to new roles in the following cases: The highest mean PMC-SR (13.40), of digits II-V of the dog, suggests a "hair trigger" function for these PMCs to rapidly stabilize the metacarpals and flex the phalanges for weight bearing and traction in swift digitigrade locomotion. Comparing the difference between mean PMC-SRs for digits I and digits II-V in the dog (2.00 and 13.40 respectively) with this difference between means in man (5.22 and 2.11), also suggests PMC-SR correlations with the specialized kinematics of these digitigrade and manipulative digits. In addition, the lowest PMC-SRs (0.55 and 1.20) for pronation and supination in the dog compared to these in man (2.08 and 4.10), seem to correlate with the specializations of these extremities for digitigrade locomotion and manipulation. Finally, the intermediate values seen in the plantigrade guinea pig forelimb appear to corroborate a relation between kinematic specializations of forelimbs and proprioceptive adaptations of PMCs therein.

CONCLUSIONS

1- The more kinematically complex a joint, the more proximally it is located and the more spinal cord levels there are involved in its motor innervation.
2- The more kinematically complex a joint, the more spindles there are associated with it.
3- The higher the absolute spindle number associated with a joint, the lower the spindle densities in the muscles controlling it.
4- The higher the mean spindle density for muscles associated with a joint, the higher the mean PMC-SR for PMCs controlling it (with exceptions, probably based on specializations of the extremity).

REFERENCES

Buxton, D. F., 1985, Neuromuscular spindle density in forelimb muscles of the dog, Anat. Rec., 211:30A.
Martini, E., and Palmieri, G., 1971, Innervazione propriocettiva dei muscoli dell'arto anteriore di Cavia cobaya, Riv. Biol., 64:373-396.
Peck, D., and Brower, T. D., 1987, Algorithms for the segmental motor innervation of the extremities, Am. Surg., 53(5):270-273.
Peck, D., Buxton, D. F., and Nitz, A. J., 1984, A comparison of spindle concentrations in large and small muscles acting in parallel combinations, J. Morph., 180:243-252.
Snedecor, G. W. and Cochran, W. G., 1980, Testing the null hypothesis for correlation coefficients and table A 11(i), in "Statistical Methods, 7th ed." G. W. Snedecor and W. G. Cochran, eds., Iowa State University Press, Ames, Iowa.
Voss, H., 1971, Tabell der absoluten und relativen Muskelspindlezahlen der menschlichen Skelettmuskulatur, Anat. Anz., 129:562-572.

DISCHARGE CHARACTERISTICS OF JOINT RECEPTORS IN RELATION TO
THEIR PROPRIOCEPTIVE ROLE

William R. Ferrell

Institute of Physiology
University of Glasgow
Glasgow, G12 8QQ, Scotland, U.K.

INTRODUCTION

Since the observation that muscle vibration produces
illusions of movement and results in significant impairment in
kinaesthesis (Goodwin et al.,1972) increasing emphasis has been placed on
their contribution to this sensation whilst the contribution of joint
receptors has concurrently been diminished. It has even been suggested
that the latter make no significant contribution to the conscious awareness
of limb movement and position (Matthews,1982). The object of this work is
to demonstrate that certain experimental manipulations can result in
disturbances of kinaesthesis (awareness of limb movement) and stataesthesis
(awareness of limb position) and that these deficits can most readily be
explained by the response of joint receptors to these procedures.

Alteration of kinaesthesis at the distal interphalangeal joint

Gandevia & McCloskey (1976) showed that by suitable positioning of
the fingers, the distal phalanx of the middle finger can effectively be
disengaged from its muscles, thus allowing assessment of the kinaesthetic
contribution of other afferent sources. It was found that subjects could
correctly detect the direction of imposed movements with muscles disengaged,
and that digital nerve block impaired kinaesthetic sensibility. More
recently, Ferrell, Gandevia & McCloskey (1987) extended this to examine
the specific contribution of joint receptors to kinaesthetic sensation at
this joint. The object of these experiments was to attempt to enhance the
discharge of these receptors by fluid distension of the joint, and
subsequently silence these by injection of local anaesthetic. The results
from one subject are shown in fig 1A. With muscles disengaged, detection
of the direction if imposed movements improves with increasing velocity.
When 2% lignocaine is injected into the synovial cavity of the joint there
is a significant deterioration in performance compared to control,
suggesting that at this joint, articular receptors contribute to kinaesthetic
sensibility.

The other interesting feature is that prior injection of dextran to
distend the joint and thus enhance articular receptor discharge resulted in
improved kinaesthetic acuity which significantly exceeds control performance
This phenomenon is explicable when the behaviour of joint receptors in a
distended joint is examined (fig. 1B). The discharge of a single slowly
adapting mechanoreceptor in the cat knee joint is shown in response to a
five degree displacement of the joint (fig. 1B, upper trace). After
introduction of 0.5ml of saline into the synovial cavity of the knee, apart

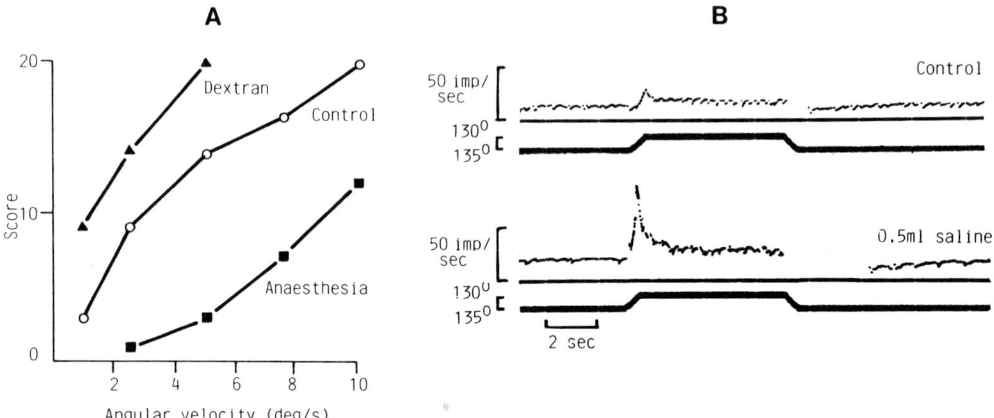

Fig. 1.   A: Kinaesthesis at the interphalangeal joint of the middle
           finger. Five degree displacements (ten flexion and ten extension)
           were imposed on the terminal phalanx with its muscles disengaged.
           To score, the subject had to correctly detect the direction of
           displacement. This was repeated at a number of different angular
           velocities. Raw scores plotted. Intra-articular injection of
           dextran improved kinaesthetic sensibility whilst subsequent
           injection of local anaesthetic produced impairment.

           B: Response of a slowly adapting receptor in the cat knee joint
           to a five degree displacement at constant velocity (upper trace).
           This same receptor produced a more vigorous response during
           movement after 0.5ml 0.9% saline was injected into the synovial
           cavity (lower trace). Acute joint distension increases its
           velocity sensitivity. Receptor located in dorsal aspect of
           capsule (cond. vel. 42m/s).

from the overall elevation in discharge, there is a marked enhancement of
the dynamic index (the difference between the peak discharge attained
during displacement and that prevailing 0.5s later) which indicates that
the velocity sensitivity of the receptor is increased (fig. 1B, lower
trace). Assuming that receptors in the human interphalangeal joints have
similar properties to those in the cat knee, and there is some evidence to
support this (Macefield, Gandevia & Burke,1987), then improvement of
kinaesthetic performance with acute joint distension observed by Ferrell
et al.(1987) is readily understood.

<u>Disturbance of stataesthesis at the knee</u>

      The fact that tonic joint receptor discharge remains elevated whilst fluid is distending the joint (see fig. 1B) provides a means of assessing whether this produces any systematic distortion of stataesthesis (position sense). This has recently been investigated by Baxendale & Ferrell (1987), who tested the ability of subjects to match knee joint position. This consisted of the experimenter moving and then holding one leg at a given position (target knee) and then requiring the subject to match this position accurately with the contralateral limb (matching knee). As shown in fig. 2A, under control conditions (joint cannulated but prior to fluid

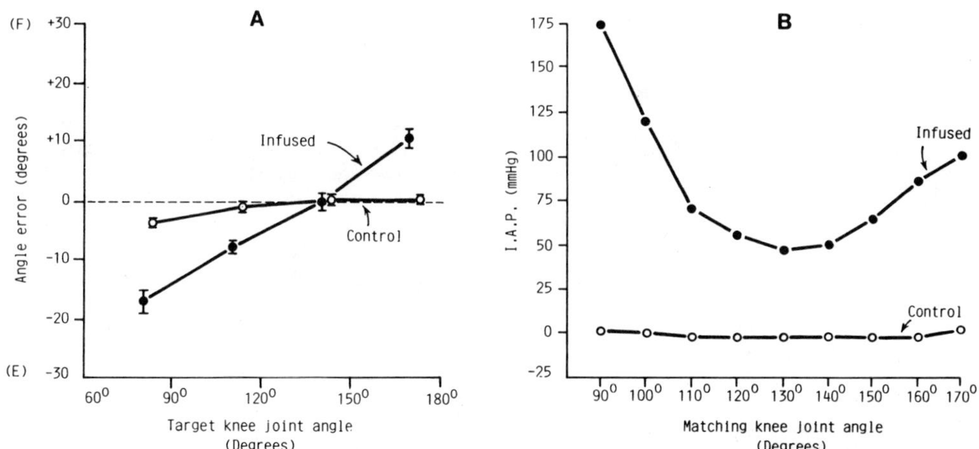

Fig. 2.   A: Human knee position matching. Angle error is the difference between the target and matching angles. Following intra-articular 0.9% saline injection, position matching is obviously affected at the extremes. A negative error indicates that matching knee is extended with respect to target knee. Means $\pm$ S.E.M. (n = 5).

B: Intra-articular pressure recording of the matching knee from the same subject, showing the large increase in pressure near the extremes when 15ml saline is present in the synovial cavity.

injection) position matching at the knee is accurate with only small errors occurring. However, after injection of 15ml of saline into the matching knee joint cavity, systematic errors occur, particularly noticeable at the extremes. This greater effect at the extremes may be related to the effect of fluid injection on intra-articular pressure (IAP) as shown in fig. 2B. Under control conditions IAP remains relatively constant – just subatmospheric at most angles except for a slight rise at the extremes. With 15ml saline present in the joint cavity there is marked elevation in IAP, particularly at the extremes. Figs. 2A and 2B are from the same subject. In animal experiments it has been demonstrated that the discharge

of slowly adapting (group II) knee joint receptors increases with rising
IAP (Wood & Ferrell,1985), and that during movement of the distended joint
these receptors are most affected at the extremes (Ferrell, Nade & Newbold,
1986). Thus it is likely that the systematic distortion of position
matching produced by acute joint distension can be attributed to altered
pattern of joint afferent discharge. This suggests that these receptors
provide positional information which is used by the central nervous system
in judgements of limb position.

<u>Contribution of joint receptors to stataesthesis at finger joints</u>

        Whether joint receptors contribute to the awareness of finger position
has proved controversial. The ability of subjects to accurately match
finger position has been investigated by Rymer and D'Almeida (1980) who
found that over a small range of displacements, ring block of the index
finger at its base did not produce obvious impairment in position matching
at the proximal interphalangeal joint. However, in more recent experiments
where testing was performed over a larger range of positions, digital nerve
block produced obvious matching deficits (Ferrell and Smith,1987). In
these experiments a servomotor positioned one index finger (target finger)
such that the proximal interphalangeal joint was held at one of four positions
(100, 125, 150 and 175 degrees). The subject's task was to match this by
active positioning of the proximal interphalangeal joint of the contralateral (matching) index finger. The results for one subject are shown in
fig. 3. Cumulative error is the sum of all angle errors occurring at each
of the four target positions. Under control conditions, matching is
relatively uniform across the whole range of positions, with the cumulative
error not exceeding 50 degrees at any position. When the matching finger
is completely anaesthetised by ring block, the subject's ability to match
finger position is equally impaired at all positions tested, so that again
errors are uniformly distributed. However, this anaesthetic procedure
removed both cutaneous and joint afferent input. In order to separate
out the specific contribution of joint receptors, in another experiment on
the same subject, the local anaesthetic was injected into the synovial
cavity of the matching finger. As shown in fig. 3, during joint anaesthesia
large errors again occur at the extremes, but position matching at intermediate angles approximates control performance. This suggests that joint
receptors provide positional information principally at the extremes of
the normal range of movement, perhaps acting as "limit detectors". This
is consistent with the discharge characteristics of receptors in animal
joints which tend to fire maximally towards the extremes (Burgess and
Clark,1969; Grigg and Greenspan,1977; Ferrell,1980; Ferrell et al.,1986).
Very recent experiments suggest that such discharge patterns also occur in
human finger joint receptors (Macefield et al.,1987).

<u>Discussion</u>

        The results of these various experiments indicate that joint
receptors make a specific contribution to the detection of both movement
and position of joints. The findings that acute knee joint distension
produces largest errors at the extremes and that finger joint anaesthesia
also produces matching deficits at the extremes suggest that these
receptors may serve to signal that the joint is approaching the end of its
normal working range. This could be an important protective function of
these receptors, not only from viewpoint of sensation, but also in the
reflex control of musculature. The current view which suggests that
proprioceptive sensation is entirely mediated by muscle receptors appears
to be too narrow. Joint receptors can certainly contribute, and there
is even some evidence for a role for cutaneous receptors in stataesthesis

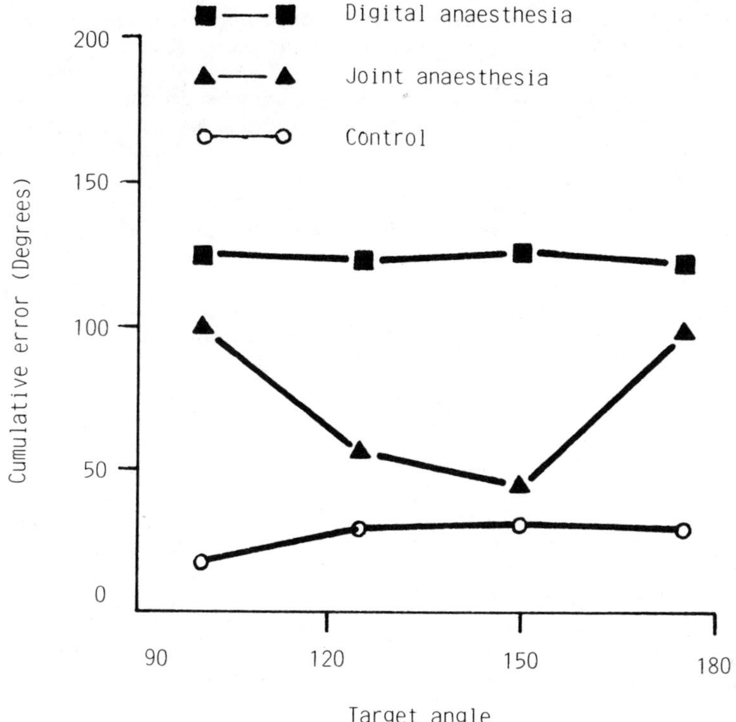

Fig. 3.   Comparison of cumulative errors at each of the four target
positions under different conditions. Errors are uniform
across all tested positions under control conditions.
With digital anaesthesia larger errors occur but remain
uniformly distributed. Upon joint anaesthesia, errors at
intermediate positions approach control values, but remain
large at extremes.

$$\text{Cumulative error} = \sum \sqrt{(\text{angle error})^2}$$

at the proximal interphalangeal joint as indicated in fig. 3. It is
likely that no one group of peripheral receptors can adequately signal
joint angle under all operating conditions and that optimal resolution
requires analysis of all available afferent sources by the central nervous
system.

References

Baxendale,R.H.,and Ferrell,W.R., 1987, Disturbances of proprioception
     at the human knee resulting from acute joint distension,
     J. Physiol. (in Press).

Burgess,P.R.,and Clark,F.J., 1969, Characteristics of knee joint
    receptors in the cat, J. Physiol.,203: 317-335.

Ferrell,W.R., 1980, The adequacy of stretch receptors in the cat knee
    joint for signalling joint angle throughout a full range of movement,
    J. Physiol.,299: 85-99.

Ferrell,W.R., Gandevia S.C., and McCloskey D.I., 1987, Role of joint
    receptors in human kinaesthesis when intramuscular receptors cannot
    contribute, J. Physiol.,386: 63-71.

Ferrell W.R., Nade S., and Newbold P.J., 1986, The inter-relation of
    neural discharge, intra-articular pressure and joint angle in the
    knee of the dog, J. Physiol. 373: 353-365.

Gandevia S.C., and McCloskey D.I., 1975, Joint sense, muscle sense and
    their combination as position sense, measured at the distal inter-
    phalangeal joint of the middle finger, J. Physiol. 260: 387-407.

Goodwin G.M., McCloskey D.I., and Matthews P.B.C., 1972, The contribution
    of muscle afferents to kinaesthesia shown by vibration induced
    illusions of movement and by the effects of paralysing joint
    afferents, Brain 95: 705-748.

Grigg P., and Greenspan B.J., 1977, Response of primate joint afferent
    neurons to mechanical stimulation of the knee joint, J. Neurophysiol.
    40: 1-8.

Macefield G., Gandevia S.C., and Burke D., 1987, Electrophysiological
    recordings from human joint afferents, Proc. Austr. Phys.
    Pharm. Soc.,16: 34P.

Matthews P.B.C., 1982, Where does Sherrington's "muscular sense"
    originate? Muscles, joints, corollary discharges? Ann. Rev.
    Neurosci.,5: 189-218.

Rymer W.Z., and D'Almeida A., 1980, Joint position sense. The effects
    of muscle contraction, Brain 103: 1-22.

Wood L., and Ferrell W.R., 1985, Fluid compartmentation and articular
    mechanoreceptor discharge in the cat knee joint, Quart. J. Expt.
    Physiol.,70: 329-335.

EFFECTS OF MECHANICAL STIMULATION OF KNEE JOINT MECHANORECEPTORS ON

FIRING OF QUADRICEPS MOTOR UNITS

Leslie Wood, Ronald H. Baxendale, William R. Ferrell, Jay R.
Rosenberg and David Halliday

Institute of Physiology, The University of Glasgow, G12 8QQ

In the past, experiments performed to observe the effects of joint
afferents on motor control have relied on monosynaptic reflex testing
(Skoglund, 1956; Grigg, Harrigan and Fogarty, 1978) or electrical stimulation of joint nerves (Lundberg, Malmgren and Schomburg, 1978). Although these experiments demonstrated that joint receptors make reflex
connections with motoneurones supplying lower limb muscles, the observed
effects tended to be weak and somewhat variable. This could have arisen
because these approaches might not be suitable for determining the
effects of a natural pattern of reflex actions arising from joints on
motor control. The present experiments use a new technique involving
periodic mechanical stimulation of only a few joint afferents whilst recording from tonically discharging motor units. Sensitive statistical
tests are then employed to obtain frequency- and time-domain measures of
the association between the separate joint afferent and motor unit spike
trains. These experiments do not rely on the methodological assumptions
inherent in previous experiments and the statistical tests allow more
powerful measures of association to be obtained than from simple cross-
correlations.

The methods used to activate joint receptors in these experiments
have been described in detail by Baxendale, Ferrell and Wood (1987) and
the statistical tests employed have been described by Rosenberg, Murray-
Smith and Rigas (1982). The basic experimental approach used activation
of small numbers of joint afferents in the posterior articular nerve (PAN)
by periodic mechanical indentation of the posterior aspect of the joint
capsule via a 1mm diameter perspex probe, whilst quadriceps motor unit
discharge and joint afferent discharge were recorded. The stimulus was
applied to the capsule at a fixed frequency of 6.7 Hz. The statistical
correlations between the joint afferent spike trains and the motor unit
spike trains were calculated in the time- and frequency-domain.

The strength of coupling between the probe input (activating the PAN
afferents) and the motor unit discharge is described by the cross-intens-
ity function and coherence for interactions in the time- and frequency-
domains respectively. The latter has values between 0 and 1 with 0
occurring in the case of independence of the two spike trains. The cross-
intensity function for the activation of five PAN afferents shows a large
peak in motor unit response occurring at 12 msec indicating a strong
coupling between the stimulus and the motor unit response with this delay.

The linear phase relation shows that there is a pure delay of 11.6 msec.
The coherence between the probe input and the motor unit response has
significant peaks at 6.7 Hz and its higher harmonics indicating a strong
coupling between the PAN units and the motor unit discharge at this
frequency.

If the analysis is further extended to examine the discharge of 2
motor units, the coherence between the probe input and each motor unit
and the synchronisation of motor unit firing demonstrates that not only
are the motor units coupled to the PAN input, but the motor units are
coupled to each other at the frequency (6.7 Hz) of the applied stimulus.

After elimination of joint afferent input by the application of 2%
lignocaine directly to PAN, motor unit discharge was no longer synchronised to the applied stimulus, even though the amplitude of indentation
remained constant. The synchrony previously seen between the motor units
was also abolished during joint nerve anaesthesia. These results indicate
that the source of the coupling must have arisen from articular structures
and was not due to spread of the mechanical stimulus to extra-capsular
receptors (e.g. muscle spindles).

These results provide strong evidence for a class of joint receptors
which can influence the behaviour of tonically discharging motor units.
In addition, these receptors can induce coupling between motor units independent of other sources of coupling. This demonstration that joint
receptors can entrain motor unit discharge suggests a role for joint
afferents in the reflex control of movement.

REFERENCES

Baxendale,R.H., Ferrell,W.R.,and Wood, L., 1987, The effect of mechanical
stimulation of knee joint afferents on quadriceps motor unit activity in
the decerebrate cat, Brain Research (In Press).

Grigg,P., Harrigan,E.P.,and Fogarty,K.E., 1978, Segmental reflexes
mediated by joint afferent neurone in the cat, J. Neurophysiol.,41 : 9-14.

Lundberg,A., Malmgren,K.,and Schomburg,E.D., 1978, Role of joint
afferents in motor control exemplified by effects on reflex pathways from
Ib afferents, J. Physiol.,284 : 327-344.

Rosenberg,J.R., Murray-Smith,D.J.,and Rigas,A., 1982, An introduction to
the application of system identification techniques to elements of the
neuromuscular system, Trans. Inst. M.C.,4:187-201.

Skoglund,S., 1956, Anatomical and physiological studies of knee joint
innervation in the cat, Acta Physiol. Scand.,36,Suppl. 124.

ANALYSES OF Ia - AFFERENT DISCHARGE IN HUMANS DURING

ISOTONIC POSITION HOLDING AND LOAD PERTURBATIONS

M.T. Jahnke and A. Struppler

Dept. of Neurol. and Clin. Neurophysiol. of the
Technical University of Munich
Möhlstraße 28, D-8000 München 80   (W.-Germany)

INTRODUCTION

When a subject is instructed to resist external load,
stretch-induced reflex responses can be observed in the emg
of the contracting muscles when the load is abruptly increased. As commonly known, these reflex responses are
grouped into segments denoted M1, M2 and M3 according to
the nomenclature of Tatton et al. (1975). The long latency
components have attracted particular attention; they have
been studied in normal as well as in pathologic states,
such as in parkinsonism and chorea.

However, the origin and physiological significance of
the late responses remain the subject of debate. Concepts
based on transcortical reflex pathways, spinal mechanisms
and different types of afferents have been suggested (Phillips,1969;   Marsden et al., 1971; Ghez and Shinoda, 1978;
Matthews 1983; Darton et al., 1985). The possible role of
mechanical oscillations of the muscle induced by sudden
stretch has also been discussed (Hagbarth et al., 1981).
Because of their high dynamic sensitivity, muscle spindle
primary endings might sense such oscillations. We have re-
investigated some of the findings reported by Hagbarth and
coworkers, applying abrupt load perturbations to the contracting deep finger flexor muscle while recording from
muscle spindles. Using a different experimental paradigm,
tonic Ia-afferent discharge was also analysed during load
bearing isotonic position holding similar to investigations
that have been performed in the finger extensor muscles
(Hulliger et al., 1982). An excerpt from the experimental
data is presented here.

METHODS AND SAMPLE SIZE

Teflon-coated stainless steel microelectrodes were
percutaneously inserted into the median nerve at the elbow.
64 microneurographic recordings were made in 57 healthy
male and female volunteers, 20-28 years of age. Discharge
characteristics of 11 muscle spindle primary endings were

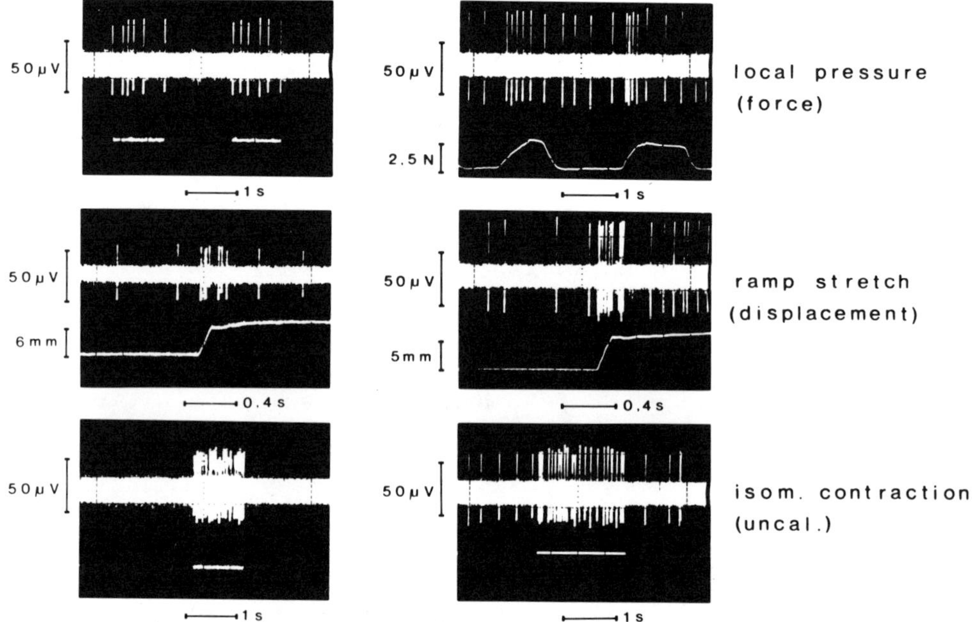

Fig.1: Demonstration of the test procedure used to identify
Ia-afferent units. The raw records show responses to
local pressure exerted on the muscle belly, ramp
stretch of the relaxed muscle and moderate isometric
contraction (uncalibrated) for two different units.

investigated quantitatively using the two experimental
paradigms. Subjects were comfortably seated in a reclining
chair and instructed to keep their fingertips in contact
with a lever attached to the axis of a torque motor capable
of producing constant and transient torques up to 1 Nm. The
device enabled the exertion of definite tractive forces on
the distal phalanges, rapid changes of load and length-
controlled extension movements of the fingers. In the position tracking task, subjects had to maintain definite positions of the lever of the torque motor, while being visual-
ly guided by a target displayed on an oscilloscope screen
together with the actual lever position. Opposing loads
varied between 0.3 and 3.4 Newtons and were adjusted according to the static sensitivity of the unit recorded from.
Whenever possible, different loads were tested with the
same unit.

Figs.2,3 (next page): PST-histograms calculated from the
afferent discharge of two different primary endings following step increase of load from 1.3 to
3.4 (fig.2) and from 1.4 to 3.8 Newtons (fig.3).
The total number of accumulated trials was 100 and
63, respectively. The lower curves show the average displacement of the fingers.

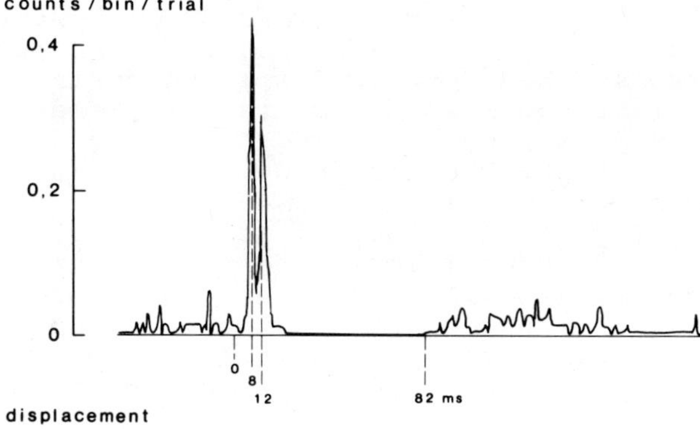
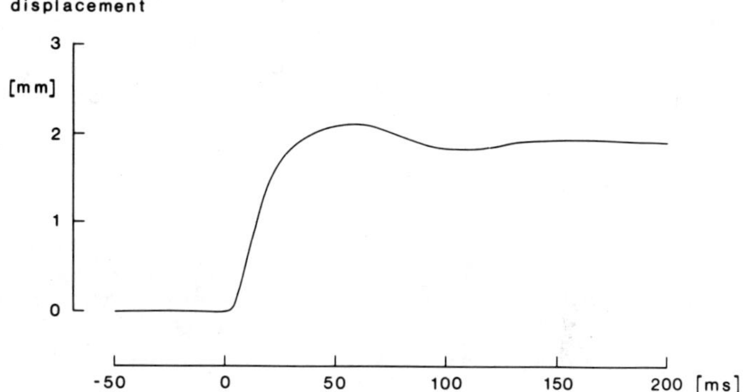

Fig.2

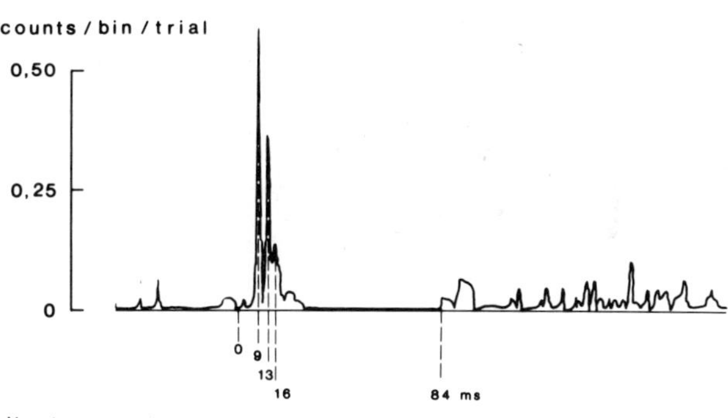
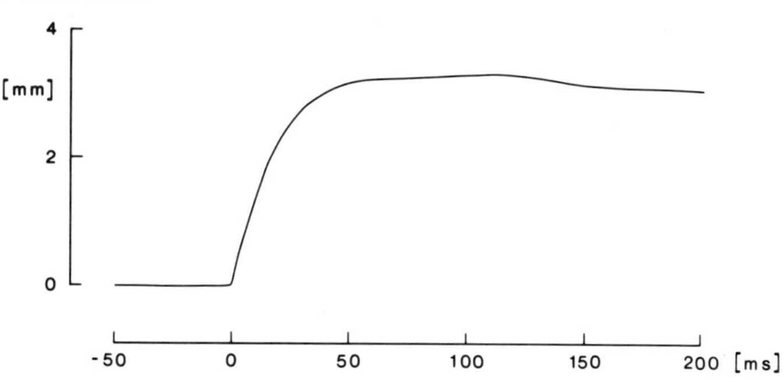

Fig.3

Units were classified as Ia-afferents provided they responded to local pressure on the muscle belly within a circumscribed area, displayed dynamic sensitivity and a deceleration response during ramp-and-hold stretches of the relaxed muscle and fired tonically during sustained moderate isometric contraction, below 10% of maximum voluntary force (cf. fig.1).

For off-line quantitative analyses original recordings were converted into trains of standard pulses using a level discriminator. Amplitude and shape of the action potentials counted by the discriminator were continuously monitored on a digital oscilloscope. The standard pulses were fed into a NICOLET series 1070 laboratory computer to evaluate post stimulus histograms (cf. figs. 2,3) or interspike interval distributions, depending on the experimental paradigm. Median values and 80%-percentiles of instantaneous frequency were then calculated from the ISI-distributions.

RESULTS AND DISCUSSION

Figs. 2 and 3 show PST-histograms calculated from the afferent discharge of two different Ia-units following step increase of load. Only the receptor bearing portion of the deep digital flexor was tested. The lower curves show the average displacement of the "optimum" finger. As can be seen in these figures, there was an initial highly synchronized burst of two and occasionally three spikes revealing instantaneous frequencies up to 250 Hz, followed by a silent period. No further spikes were counted until tonic activity resumed approximately 80 ms post stimulus.

The discharge characteristics of these two units are representative of similar findings with other muscle spindle afferents with high dynamic sensitivity as judged from the response to ramp stretches in the relaxed state. So far, we have not observed grouping in the afferent discharge of single Ia fibres comparable to that described by Hagbarth and coworkers (1981) mainly from multiunit recordings. We observed after each burst a postexcitatory subnormal state where the spindle primary ending seemed to remain unresponsive to mechanical oscillations, if they were indeed present.

These findings can probably be explained on the basis of the creep that is reported to occur in the dynamic bag1 intrafusal fibre following a ramp stretch (Boyd 1976). Of course, stretching to the extent that stable cross-bridges in the dynamic bag1 fibre are broken is a prerequisite for subsequent creep. It is not likely to occur following just small taps on the tendon or on the muscle belly.

Some typical results of the position holding task are shown in figs. 4 and 5. Black dots indicate median values, vertical bars 80% percentiles of instantaneous frequency (ordinate) as calculated from the distribution of interspike-intervals for different positions (abscissa), "0 mm" corresponding to an intermediate or "natural" position of the fingers. Note that tonic Ia-discharge rates tended to be higher in the flexed (negative length values) than in the extended position of the fingers (positive length values), particularly with small opposing loads. When the load was increased, discharge rates remained high in the extended state and finally became independent of position

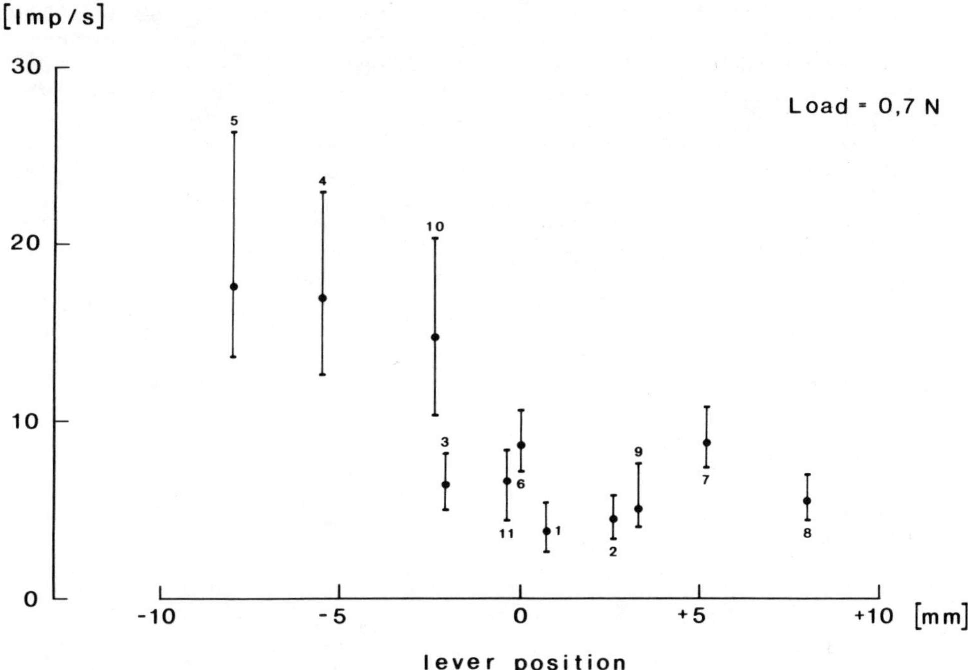

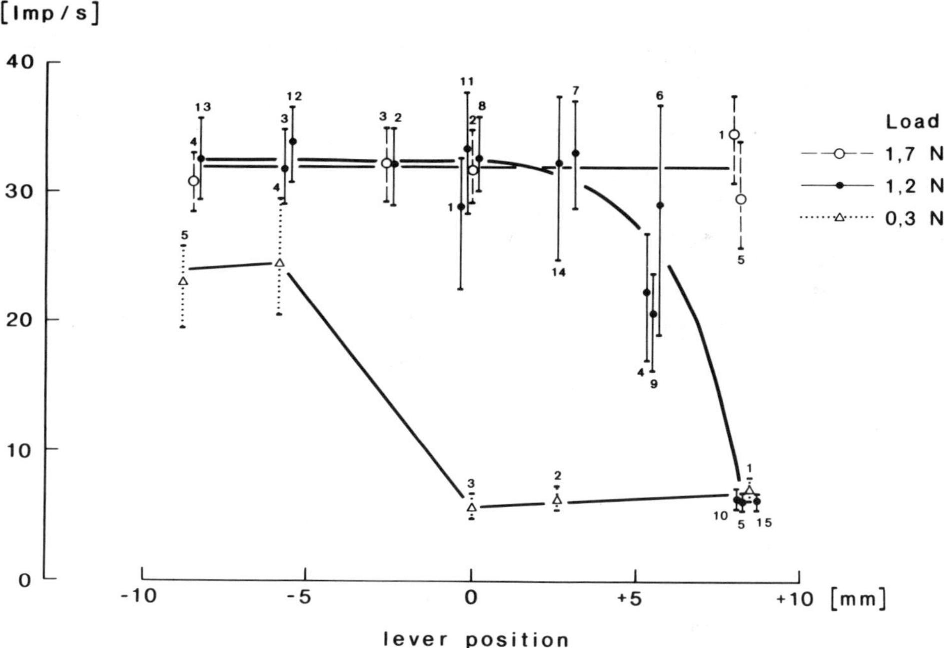

Figs.4,5: Tonic discharge of two different Ia-afferent units as related to muscle length and load. Black dots indicate median values, vertical bars 80% percentiles of instantaneous frequency. The total number of counts underlying each median value varied between 106 and 287. The small numbers represent the sequence of length changes.

(fig.5). This reversed position response (depending on the size of the load) indicates that fusimotor outflow compensated for the changes in muscle length; in some of the units displaying similar response patterns, in-series-coupling of the spindles with the extrafusal muscle fibres might also be an explanation. Mean integrated emg-activity of the digital flexor (not shown here) displayed a similar dependency on muscle length and load as the muscle spindle afferent discharge, so that under the given paradigm tonic Ia-discharge varied in approximate proportion to the internally developed force necessary to maintain position. It is interesting to speculate that during position-holding against external load muscle spindle afferents originating from the contracting muscles form part of a load-dependent servoassistance loop involving spinal reflex pathways.

REFERENCES

Boyd, I.A., 1976, The mechanical properties of dynamic nuclear bag fibres, static nuclear bag fibres and nuclear chain fibres in isolated cat muscle spindles, Prog. Brain Res., 44:33-50.

Darton, K., Lippold, O.C.J., Shahani, M., and Shahani, U., 1985, Long-latency spinal reflexes in humans, J. Neurophysiol., 53:1604-1617.

Ghez, C., and Shinoda, Y., 1978, Spinal mechanisms of the functional stretch reflex, Exp. Brain Res., 32:55-68.

Hagbarth, K.-E., Hägglund, J.V., Wallin, E.U., and Young, R.R., 1981, Grouped spindle and electromyographic responses to abrupt wrist extension movements in man, J. Physiol., 312:81-96.

Hulliger, M., Nordh, E., and Vallbo, A.B., 1982, The absence of position response in spindle afferent units from human finger muscles during accurate position holding, J. Physiol., 322:167-179.

Marsden, C.D., Merton, P.A., and Morton, H.B., 1971, Servo action and stretch reflex in human muscle and its apparent dependence on peripheral sensation, J. Physiol., 216:21P.

Matthews, P.B.C., 1983, Segmented electromyographic responses on vibration of human triceps brachii and the effect of altering the period of vibration, J. Physiol., 334:30P.

Phillips, C.G., 1969, Motor apparatus of the baboon's hand, Proc. Roy. Soc., 173:141-174.

Tatton, W.G., Forner, S.D., Gerstein, G.L., Chambers, W.W., and Liu, C.N., 1975, The effect of postcentral cortical lesions on motor responses to sudden upper limb displacements in monkeys, Brain Res., 96:108-113.

INFLUENCE OF MUSCLE AFFERENTS AND MECHANORECEPTOR CUTANEOUS INPUTS ON ALPHA MOTOR NEURONS AT REST AND DURING VOLUNTARY CONTRACTION

Manik Shahani[2], O.C.J. Lippold[1], K. Darton[1],
U. Shahani[1], R. Sangeeta[2], P.G. Patel[2], Kalpana
Sekhar[2], and D.H. Dastoor[2]

[1]University College London, Gower Street, London
WC1E 6BT, England
[2]E.C.I. Institute of Electrophysiology for
Fundamental and Applied Research
Dr. E. Borges Marg, Parel, Bombay 400 012, India

INTRODUCTION

Many studies have established a relationship between muscle afferent inputs or cutaneous inputs separately, and the excitability of alpha motor neurons. However, in normal physiological situations, more often these inputs coming from muscle spindles of agonist, antagonist or synergistic muscles operate simultaneously with mechanoreceptor inputs from the skin; besides, the behavior of the alpha motor neurons when they are being driven by supraspinal influences and otherwise, in the context of peripheral inputs, becomes of greater importance in terms of functional significance. In order to study this complex phenomenon, several experiments have been designed which may help in the better understanding of the physiology of alpha motor neurons during voluntary movements as well as in reflex activities.

METHODS AND RESULTS

1. EMG recordings from the first dorsal interosseous muscle during isometric contraction in response to prodding

The first dorsal interosseous muscle (1st d. int. m.) was held in isometric contraction by the volunteers; visual display on the meter of the integrated EMG signals helped the subjects to keep the same degree of tension. The lateral border of the index finger at the level of the distal phalanx was hit by a prod, producing a movement towards adduction of the finger, thus stretching 1st d. int. m. Surface electrodes were used for recording the EMG from the 1st d. int. m.; these signals were rectified and averaged. Details of the techniques used have been described previously (Darton et al., 1985).

These experiments were repeated but the prod was applied to the medial side of the index finger thus pushing the index

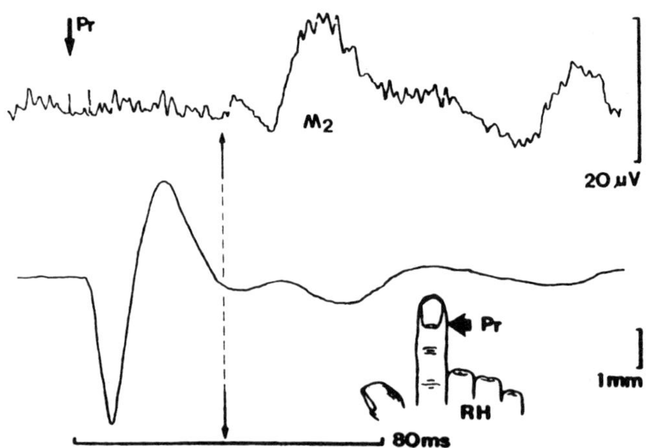

Fig. 1. Averaged rectified EMG response from the 1st int. m.
during isometric contraction on prodding the medial
side of the index finger. There is no M1 response,
but M2 is present.

finger in the direction of abduction thereby preventing stretch
of the 1st d. int. m. In another series of experiments, the
arm was cooled, with the prod being applied to the lateral
border of the index finger. Complete anesthesia of the index
finger was achieved by applying tourniquet type pressure with
a tightly tied rubber band just distal to the metacarpo-phalangeal joint; in this situation, the standard experiment was
carried out.

Instead of using the prod, the skin over the distal phalynx
of the index finger was stimulated while the 1st d. int. m. was
in isometric contraction. Later, electrical stimulation was
applied to the median nerve both at the wrist and elbow.

The abductor hallucis muscle was used in the same way as
the 1st d. int. m., i.e. when it was in isometric contraction,
the prod was applied in the direction which stretched the same
muscle.

In the standard experiment, when the first dorsal int. m.
was in isometric contraction and the prod was applied to dis-
place the index finger to adduction, the first two bursts of
EMG activity (which may be termed as M1 and M2) on averaging
were noticed at a mean latency of 32.4 (S.D. 2.4 ms) and 55.1
ms (S.D. 11.5 ms). The third burst of EMG activity (M3) occurred
at latencies longer than 100 ms.

However, when the index finger was prodded on the medial
side which shortened the muscle rather than stretched it, the
EMG burst was obtained at a latency compatible with M2; there
was no M1 (Fig. 1).

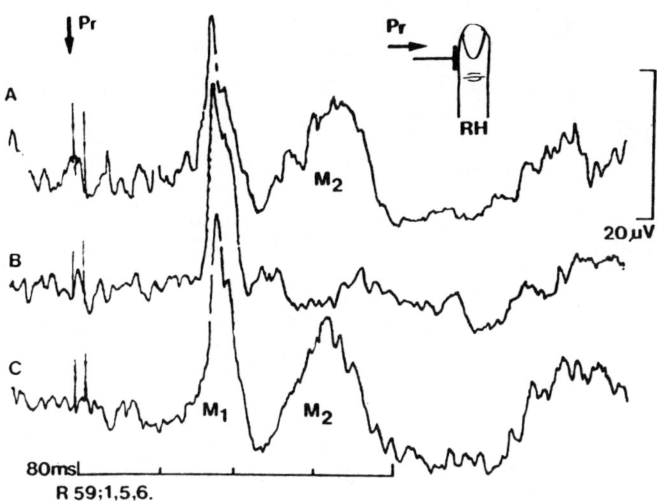

Fig. 2. Averaged rectified EMG response from the 1st int. m.
during isometric contraction on prodding the lateral
border of the index finger before and after blocking
the cutaneous nerve. M1 response is present, but M2
is absent.

Cooling of the arm delayed the latencies of M1 and M2
almost equally (about 10 ms), thereby suggesting that the afferents responsible for M1 as well as M2 are almost of equal
diameter and have similar conduction velocities. If Group II
spindle afferents were to be implicated with M2, the shift of
the latency of M2 should have been greater. Complete block was
produced for cutaneous afferents arising in the index finger
by using tourniquet pressure with rubber bands; M2 was completely abolished while M1 was present (Fig. 2). This clearly demonstrated that the cutaneous afferent input is responsible
for M2. This impression was further reinforced when a response
could be obtained at the M2 latency by electrically stimulating
the skin over the phalanx of the index finger, when the 1st d.
int. m. was in isometric contraction. By stimulating the median
nerve at the wrist and elbow and obtaining a response compatible
with M2 (1st d. int. m. in isometric contraction), it was possible to calculate the conduction velocity by the reduced latencies of the response. The conduction velocity thus calculated was approximately 50 m/s. It is important to remember
that the median nerve does not conduct spindle afferents from
the 1st d. int. m.

When the experiments were repeated in the foot in the same
fashion on the muscle abductor hallucis, M1 and M2 were obtained
at latencies 39.4 ms (S.D. 2.4 ms) and 66.3 ms (S.D. 2.9 ms).
M1 - M2 interval was found to be about 27 ms in the case of
muscle of the foot as against 23 ms in the case of the hand
muscle. This would strongly suggest that the M2 response is
mediated in the spinal cord; if it were due to a transcortical
pathway, in case of the foot muscle, the M1-M2 interval would
have had to increase by approx. 15 ms.

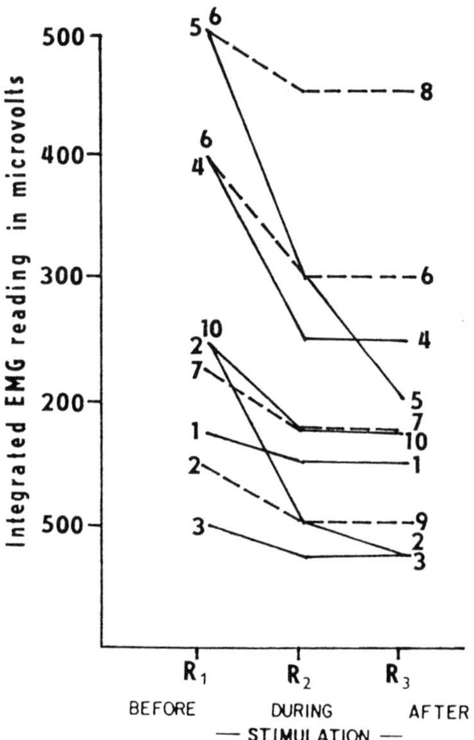

Fig. 3. IEMG values in microvolts recorded from the tibialis
anterior muscle during maximal isometric contraction
before and after stimulation of the ipsilateral medial
popliteal nerve. There is a reduction in IEMG values
after 10 min stimulation.

2. Integrated EMG activity in the tibialis anterior muscle
during maximal isometric contractions

The integrated EMG (IEMG) was recorded from 10 normal volunteers by using surface electrodes over the tibialis anterior
muscle when the subjects were asked to produce maximal isometric contraction of the same muscle for a period of 10 s.
The resistance to the dorsiflexors was offered by a spring;
by a set of pulleys, the angle of the foot was kept at 90°.
The subjects were then asked to relax; the medial popliteal
nerve was then stimulated with a pulse 0.05 ms in duration at
a frequency of 10 Hz, the intensity of the stimulus being minimal and painfree. It was presumed that large diameter fibers,
i.e. Group IA fibers were being selectively stimulated. IEMG
readings were recorded after an interval of 10 min and subsequently again after another 10 min of stimulation; while
taking IEMG readings electrical stimulation was switched off.

As a control, similar recordings were made but without
electrical stimulation (placebo effect). In some subjects, non
specific nerves such as the median nerve and lateral popliteal
nerve were stimulated. In another control group, the sural
nerve was stimulated on the same side. However, the stimulus
strength was increased and the subjects definitely felt the
sensation; IEMG readings were taken.

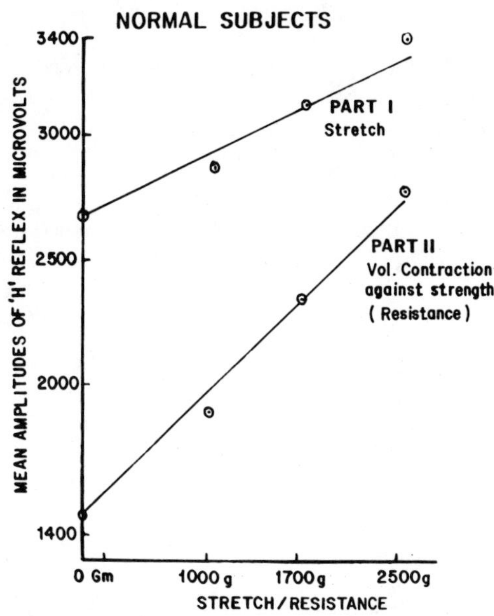

Fig. 4. Values of the "H" reflex from the soleus muscle in
normal subjects on stretching the antagonistic muscles
with progressively increasing loads with and without
contraction of the antagonists to overcome the resistance. The absolute values of the "H" reflex are smaller when the antagonists are contracting.

IEMG values recorded from the tibialis anterior during maximal
isometric contraction of dorsi flexors decreased in all the
subjects when the medial popliteal nerve was stimulated (Fig.
3). There was a mean reduction of 29.73 % at the end of 10 min
of stimulation while a further decrease was noticed only in
two subjects after the next 10 min of stimulation.

There was no decrease in IEMG values when nonspecific
nerves were stimulated. In the case of sural nerve stimulation,
there was a slight increase in the values of IEMG (mean percentage increase 23.33 %, S.D. 5.76).

3. "H" reflex responses in the soleus muscle during stretch
of antagonists

Ten normal volunteers were studied by obtaining "H" reflex
from the soleus muscle with surface electrodes; the stimuli
were of suprathreshold intensity, i.e. the "M" response was
also present along with the "H" reflex. "M" and "H" responses
were averaged for 10 sweeps, while the antagonistic muscles,
i.e. dorsi flexors of the ankle were being stretched at zero
tension and then progressively with 1000, 1700 and 2000 g. In
the second part of the experiment, the subjects were asked to
counteract the weights and hold the ankle at 90° by producing
voluntary contraction of the dorsiflexors of the ankle. Averaged "H" and "M" responses with progressively increasing
resistance from zero grams to 1000, 1700 and 2500 g respectively were recorded. The amplitudes of averaged "H" and "M"
responses were measured.

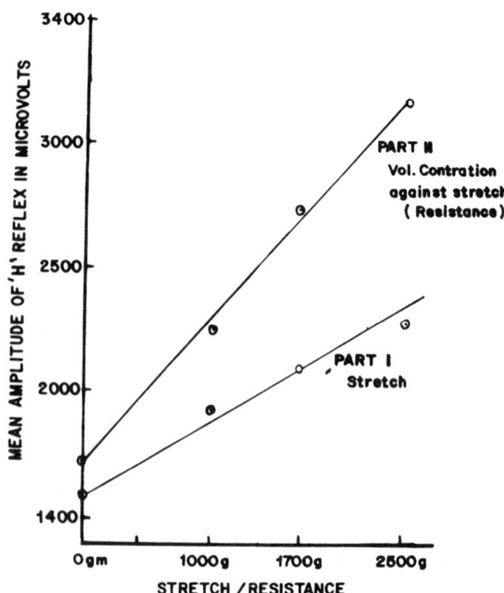

Fig. 5. Values of the "H" reflex from the soleus muscle in
hemiplegic subjects on stretching the antagonistic
muscles with progressively increasing loads with and
without the contraction of antagonists to overcome
the resistance. The absolute values of the "H" reflex
are larger when the antagonists are contracting.

These experiments were also repeated in patients (hemi-
plegics) with corticospinal symptomatology.

On progressively stretching the antagonist, there was an
increase in amplitude of the "H" reflex in all subjects. The
same behavior is also noticed when the volunteers were contracting the antagonist, i.e. ankle dorsi flexors were resisting progressively increasing weights. The amplitudes of the
"M" response, however, remained the same. It can be seen from
Fig. 4 that the absolute values of "H" reflex amplitude are
less when the subjects are producing voluntary contraction in
the antagonist to overcome the stretch applied to the antagonist compared to the stretch of antagonists alone.

However, in the case of hemiplegic patients who had some
corticospinal lesions, the behavior was just the opposite to
that in normal subjects. In other words, the mean amplitude of
the "H" reflex in hemiplegic subjects was higher when they
were counteracting the resistance by voluntary contraction of
the antagonist than when the stretch was being applied to the
antagonists (Fig. 5).

4. Influence of femoral nerve stimulation on tension in ankle
   dorsiflexors during the withdrawal reflex

The withdrawal reflex was produced in normal subjects by
painful electrical stimulation (pulse duration being 1000 ms).

Ten such stimuli were given at random to the sole of the foot
at intensities which produced the withdrawal reflex. The subjects were made to lie supine with their knees strapped; by a
specially designed arrangement of pulleys, the tension developed by dorsiflexors of the ankle could be recorded. The ipsilateral femoral nerve was stimulated for 20 min by placing two
electrodes (one square cm each) lateral to the femoral artery
in the inguinal canal. The intensity of the stimulus was minimal; just when the subjects barely started to feel the stimulation. In other words, it was assumed that the large diameter
fibers such as Group IA were preferentially stimulated. After
switching off electrical stimulation of the femoral nerve,
tension for dorsi flexors was recorded again in response to 10

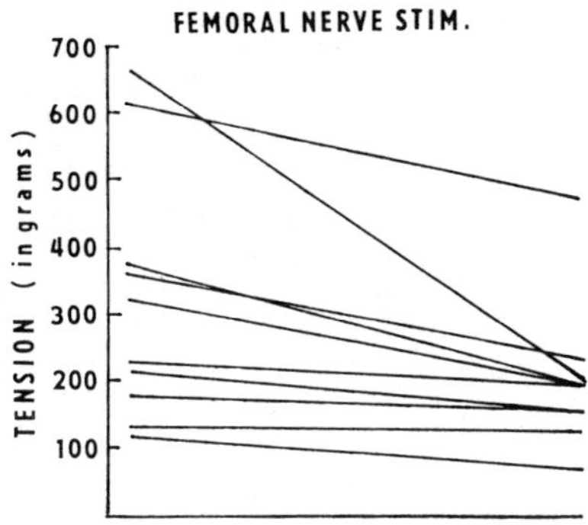

Fig. 6. Values of the tension developed by the dorsiflexors in
the withdrawal reflex before and after stimulation of
the ipsilateral femoral nerve. The tension in the dorsiflexors significantly decreased after the stimulation.

random painful stimuli producing the withdrawal reflex. In
another group of subjects, the sural nerve was stimulated,
posterior to the lateral malleolus for 20 min and the experiments were repeated in the same fashion. In the third group
of subjects, a non-specific nerve such as the median nerve was
stimulated. In the fourth group, no stimulation was given and
two sets of readings were taken (placebo effect).

The results showed that following stimulation of the femoral nerve, the tension developed by dorsiflexors decreased
in all the 10 subjects of the experimental group; the mean
tension was reduced from 323.25 g to 143.1 g a reduction equivalent to 53.51 % (S.D. = 13.26; p < 0.001).

DISCUSSION

1. Whenever a muscle performing an isometric contraction
is purturbed resulting in its stretch, three responses, M1, M2
and M3, are obtained (Lee and Tatten, 1972). M1 is due to
Group IA input, as the same is abolished when the displacement
of the finger is done in the opposite direction, i.e. producing
shortening of the muscle. M2 is due to large diameter cutaneous
afferents as the same could be abolished when complete nerve
block is achieved at the level of the proximal phalanx of the
index finger (Darton et al. 1985). The shifts in M1 and M2 in
terms of latency were also more or less equal on cooling of
the arm; this suggests that the afferents involved may have
similar diameters and conduction velocities. This rules out the
possibility that M2 is due to Group II muscle afferents. Electrical stimulation of the digit at the distal phalanx also produced a response compatible with the M2 latency. When comparing
the M1-M2 interval in the case of the finger muscle as against
the muscle of the foot, the difference obtained is nearly 4 ms
which is too short to cover the conduction time to and back
along the spinal cord from the lumbar to the cervical region;
this proves that M2 is not a transcortical response but is
mediated at the spinal cord level. There seems to be sufficient
evidence to suggest that M2 is a response due to cutaneous afferents produced at the spinal level when the central excitatory
state of the corresponding motor neuronal pool is already raised
by the corticospinal drive. The physiological significance of
M2 response may be that of retaining the ongoing activity in
the corresponding muscle. In another set of experiments when
the median nerve was stimulated at the wrist, M2 could be elicited from the triceps or biceps muscle whichever was already
held in isometric contraction; interestingly M2 could not be
obtained from either of the muscles when the biceps and triceps
were held in co-contraction (Shahani, 1987).

2. Electrical stimulation of the medial popliteal nerve
by the method described resulted in Group IA input in the spinal cord. Thus Group IA afferents from the antagonistic muscles
of the dorsi flexors of the ankle were able to reduce the
values of IEMG from tibialis anterior, one of the main dorsiflexors of the ankle. It has been described that Group IA interneurons have an inhibitory influence on the motor neuronal
pool of the antagonistic muscle (Eccles et al., 1956; Eccles,
1969). It is also known that the Group IA afferent input results
in the release of enkephalins via other interneurons. What is
of greater significance is that the IEMG was reduced even when
the stimulus was not on, thus suggesting a longer lasting influence of Group IA input which may be the result of accumulation of enkephalins and glycin which has been implicated with
Group IA interneurons (Shahani, 1987).

There was no difference in the value of IEMG when non-
specific stimulation was given, or in placebo studies. However,
when the sural nerve which has Group II and Group III fibers
from cutaneous receptors was stimulated, a slight increase in
IEMG readings was obtained. This could be due to the well-known
observation that the FRAs (flexor reflex afferents) specially
facilitate flexor motor neurons (Hugon, 1967).

3. In the experiments on the "H" reflex, the increase in
the amplitude of averaged "H" on progresively applying stretch

to the antagonist would suggest that the influence of facili-
tatory inputs coming from the Group II muscle afferents of the
antagonist and large diameter cutaneous afferents from the
dorsum of the foot are more effective (Wilson and Kato, 1965)
than the inhibitory influence exerted through Group IA interneurons of the antagonistic muscle. However, when the subjects
were voluntarily contracting, the antagonist muscle against the
same loads which were applied to stretch them, the values of
amplitudes of averaged "H" reflexes were lower. In this situation, due to alpha gamma coupling (Vallbo and Hagbarth, 1967),
Group IA input was possibly greater and therefore, IA interneuron inhibitory mechanisms were relatively more effective.
On the other hand, in hemiplegic subjects where there was impairment of the corticospinal systems, the values of the amplitude of the averaged "H" were greater when the subjects were
voluntarily counteracting resistance than when the antagonists
were just being stretched. This could be attributed to the
failure of Group IA interneuron inhibitory mechanisms or due
to the failure of the localization of the corticospinal drive
to the specific neuronal pool of the antagonistic muscle alone
(Eccles et al., 1956). These experiments further reinforce the
conclusion that the cutaneous afferents tend to facilitate that
neuronal pool which is already being driven; in this case the
programme being the "H" reflex.

4. Stimulation of the femoral nerve which involved differential Group IA fibers from the quadriceps muscle (a knee
extensor) resulted in the reduction of the withdrawal reflex
as measured by the tension developed by the dorsiflexors of
the ankle. This would strongly suggest that the stimulation of
Group IA fibers coming from the extensor muscles results in
release of such neurotransmitters through Group IA interneurons
and other interneurons, which have a longlasting inhibitory
influence on the motor neurons of flexor muscles. A slight
reduction seen in the tension developed by dorsiflexors on the
withdrawal reflex after sural nerve stimulation may be attributed to conditioning, as sural nerve comprises Group II and
Group III flexor reflex afferents.

It may thus be concluded that quite often in a physiological situation when a limb is displaced, large diameter
cutaneous afferents are also excited besides Group IA afferents
arising from muscle spindles. This results in early and late
response M1 and M2 which are both mediated at the spinal cord
level, and are due to Group IA (a muscle spindle) and large
diameter cutaneous fibers respectively. It is also seen that
when Group IA fibers are stimulated for a certain length of
time, the inhibitory effect on the antagonistic motor neurons
exerted from Group IA interneurons lasts longer than the period
of stimulation. The large diameter cutaneous afferents and
probably Group II muscle afferents play some role in helping
to keep the ongoing programmes, be they voluntary or reflex in
nature.

REFERENCES

Darton, K., Lippold, O.C.J., Shahani, M., and Shahani, U.,
        1985, Long latency spinal reflexes in humans, J. Neuro-
        physiol., 53:1604-1618.
Delwaide, P.J., 1976, Reflex Physiology in Man, in "The Motor

system - Neurophysiology and Muscle Mechanisms", Manik
    Shahani, ed., Elsevier Publishing Co., Amsterdam,
    pp. 169-180.
Eccles, J.C., Fatt, P., and Langdren, S., 1956, The central
    pathway for the direct inhibitory action of impulses
    in the largest afferent nerve fibre to muscle, J. Neuro-
    physiol., 19:75-88.
Eccles, J.C., 1969, The Inhibitory Pathways of the C.N.S.,
    Liverpool University Press, Liverpool, pp. 135.
Hugon, M., 1967, Reflexes - polysynaptiques cutanés et commandes
    volontaires. Thèse De Doctoral D´Et al Des Sciences
    Naturelles, Fac. Sci., Paris.
Lee, R.G., and Tatten, W.G., 1972, Long latency reflexes to
    imposed displacements of the human wrists; dependence
    on duration of movement, Exp. Brain Res., 45:207-216.
Lloyd, D.P.C., 1946, Integrative pattern of excitation and in-
    hibition in two-neurone reflex, J. Neurophysiol., 9:
    439-444.
Matthews, P.B.C., 1963, Does the long latency component of the
    human stretch reflex depend after all on spindle se-
    condary afferents?, J. Physiol. (London), 341:16P.
Shahani, M., 1987, M2-A long latency spinal reflex due to skin
    afferents in man, in: "Motor Control", G.N. Gantchev,
    ed., Plenum Press, New York.
Vallbo, A.B., and Hagbarth, K.E., 1967, Impulse recorded with
    microelectrodes in human muscle nerves during stimu-
    lation of mechano receptors and voluntary contractions,
    Electroenceph. Clin. Neurophysiol., 23:392.
Wilson, V.J., and Kato, M., 1965, Excitation of extensor moto-
    neurones by Group I afferent fibres in ipsilateral
    muscle nerves, J. Neurophysiol., 28:545-554.

POSTURAL RESPONSES EVOKED BY UNILATERAL VIBRATION OF LOWER
LIMB MUSCLES IN STANDING SUBJECTS

Marian Šaling and František Hlavačka

Institute of Normal and Pathological Physiology
Centre of Physiological Sciences, Slovak Academy
of Science, Bratislava, Czechoslovakia

Integration of somatosensory information from different
regions of the body is an important mechanism for maintaining
upright posture, in which proprioceptive afferentation from
the lower limb does not play a negligible role. Bilateral vibration of lower limb muscles, which activates muscle afferents
in a specific way, is able to evoke body tilts or even a falling reaction of the subject (Eklund, 1972; Gurfinkel et al.,
1977). Postural responses probably result from a change of
general somesthetic information about body position evoked by
muscle afferent activity. In the present work we have tested
the possibility of inducing postural responses by vibration of
a single muscle of the lower extremity and simultaneously, we
tried to find whether the evoked afferent activity can influence the integration of somesthetic information about body
position.

Experiments were performed on 5 healthy subjects 21-35
years old: We used the stabilographic method for recording of
body movements. Vibration stimulation (f = 90 Hz, amplitude
1 mm) was sequentially applied on the tendon or muscle belly
of m. tibialis ant. (TA), m. triceps surae (TS), m. quadriceps, (Q), and the hamstrings (H), first on the right and then
on the left lower extremity. Stabilograms in anterior-posterior (AP) and lateral (L) directions and statokinesigrams were
recorded during 50 s in a situation with the eyes closed.
Vibration (duration 20 s) was applied 10 s after the recording
had started, which was sufficient to obtain a stable effect.

Unilateral vibration of separate muscles of the lower
limb induced a proportional shift of the centre of body gravity (body tilt) up to attainment of the new equilibrium point.
After vibration was finished, a gradual return to the previous
body gravity centre was observed. In Fig. 1 A,B, the shift of
body gravity centre on vibration of right and then left
Achilles tendon is shown. A gradual shift of gravity centre
in AP direction and towards the unstimulated extremity in the
L direction was observed. Direction of the shift of body gravity centre caused by unilateral vibration of the tested
muscles is illustrated in Fig. 1 C. The effects evoked by

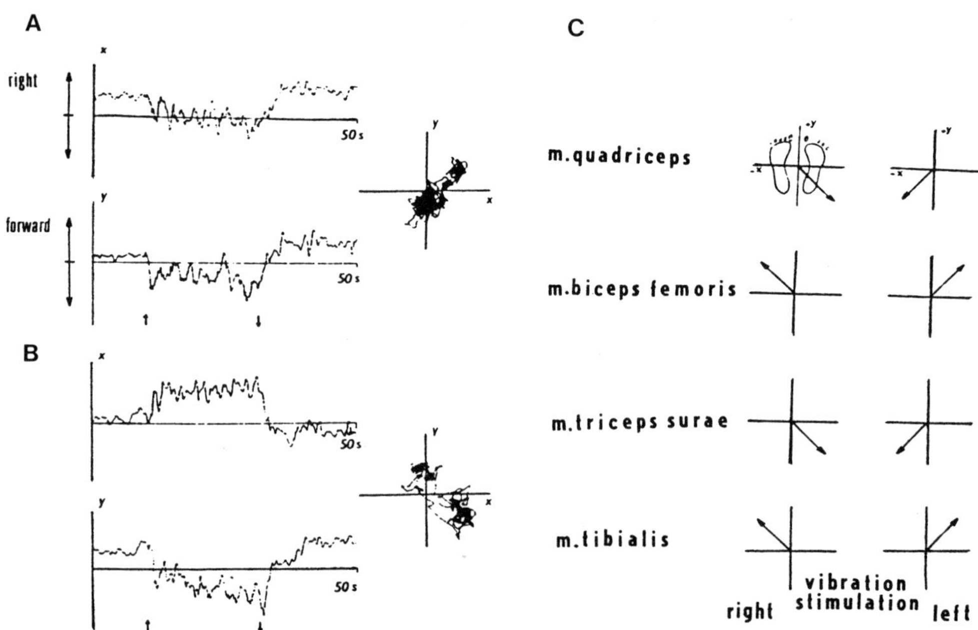

Fig. 1. Postural responses evoked by vibration of right (A) and left (B) Achilles tendon; recordings of body tilts in lateral and anterior-posterior stabilograms and in statokinesigram. Little arrows under stabilograms indicate the start and the end of vibration. (C) Direction of body tilts evoked by unilateral vibration of tested muscles.

vibration of identical muscles of right and left lower limb had the same direction along the AP axis, but opposite to that along the L axis; their amplitudes were approximately equal. Identical shifts of body gravity centre were observed on vibration of TA, H (forwards and towards the stimulated extremity) and TS, Q (backwards and towards the unstimulated extremity) (Fig. 1, C). Majority of tested subjects reported a perception of body tilt. After vibration had been discontinued, all subjects were feeling a marked sensation of return motion.

Our results documented that vibration of a single muscle of the lower extremity was able to evoke body tilting simultaneously along the AP and L axes. Identically oriented shifts of body gravity centre induced by stimulation of TS and Q or TA and H suggest that the evoked afferent impulsation is probably perceived as a body tilt. The effects observed on unilateral vibration of identical muscles of the right and left lower limb are of opposite direction along the L axes. This may be the reason why on bilateral vibration no effects in the lateral axes were observed. Our data suggest that the evoked afferent activity is processed from the view of its functional significance for providing general information about body position. This assumption is supported by perception of body tilt, as reported by the tested subjects.

REFERENCES

Eklund, G., 1972, General features of vibration-induced effects
    on balance, <u>Uppsala J. Med. Sci.</u>, 77:112-124.
Gurfinkel, V.S., Lipschic, M.I., and Popov, K.E., 1977, Issle-
    dovanie sistemy regulyatsii postvibratsionnoy stimulat-
    siei myshechnykh vereten, <u>Fiziologia tcheloveka</u>, 3:
    635-643.

A PRELIMINARY AND QUANTITATIVE REAPPRAISAL OF THE WARTENBERG TEST IN
PARKINSONIAN PATIENTS

G.C. Leslie, Alison MacArthur, W.J. Mutch and N.J. Part

Departments of Physiology and Medicine
The University of Dundee
Dundee, DD1 4HN. Scotland

INTRODUCTION

Wartenberg (1951), appreciating the need for a simple, reliable, test
to examine muscle tone in rigidity or spasticity, described such a test.
Basically he observed the duration and pattern of movement of the lower legs
as they swung freely, after simultaneous release from the horizontal.
Reintroducing this test to measure spasticity in young adults, Bajd and
Vodovnik (1984) obtained quantitative data by use of electogoniometers.
Their test has been evolved further, to include microcomputer analysis (see
acknowledgement) of the goniometer output. We have, using these advances,
reapplied the Wartenberg pendulum test to Parkinsonian patients, and present
here a preliminary account of some of our findings, together with some data
from both healthy age matched controls and elderly stroke patients, with and
without spasticity.

METHODS

Each patient, having given informed consent, was made comfortable in a
semi-supine position on an examination couch, with their legs hanging freely
over the end. Electrogoniometers were attached to measure knee joint angle.
The examiner lifted the leg to the horizontal position, assessed whether the
limb was fully relaxed and then released it, to swing freely under gravity.
A/D conversion of the goniometer's output was started about 1 second before
release and lasted for ten seconds (see fig. 1A). The test was made six
times on each leg. The computer was used to display the angle and derived
velocity data (fig. 1A,1B). Bajd and Vodovnik (1984) quantified the
intensity of spasticity with a relaxation index, the ratio of the magnitude
of the first swing (a in fig. 1A) to the magnitude of the angle between the
initial and final positions of the leg (b in fig. 1A). For Parkinsonian
patients, we have in addition considered the maximum velocity of first swing
to be relevant.

RESULTS

Twenty Parkinsonian patients, 10 females and 10 males, aged between 53
and 84 years, have been investigated. Examples of traces are shown in fig.
1A,1B. The relaxation index for this test was 1.21 and the maximum

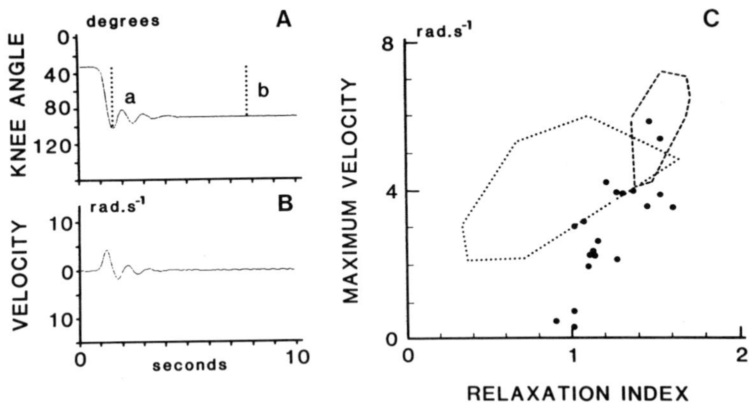

FIGURE 1. Shows (A) the angle data against time, of a
single test; (B) the derived velocities during the three
swings; (C) maximum velocity during first swing against
relaxation index for the left legs of Parkinsonian
patients (● , n = 20). Also shown (fig. 1C) are the
'hull' areas of the same relationship for normal controls
(------ , n = 32) and stroke patients (·········, n = 15).

velocity on first swing was 4.28 radians per second. Means ± standard
deviations for the distributions of relaxation index and maximum velocity on
first swing for the right leg were 1.23 ± 0.20 and 2.94 ± 1.45 radians per
second respectively.

CONCLUSION

We conclude that Wartenberg's test is simple and reliably reproducible
and that the quantitative data are in keeping with conceptions of the
neurological conditions. Furthermore the test may well be useable in
evaluating drug therapy.

REFERENCES

Bajd, T.,and Vodovnik, L., 1984, Pendulum testing of spasticity,
    J.biomed.Eng., 6:9-16.
Wartenberg, R., 1951, Pendulousness of the legs as a diagnostic test,
    Neurology, 1:18-24.

ACKNOWLEDGEMENT

We thank Dr. B.J. Andrews and his colleagues at the Bioengineering
Unit, University of Strathclyde, Glasgow, for generously making available
their software, used in the analysis of the collected data.

THE INFLUENCE OF LOAD ON THE SHORT LATENCY HIT REFLEX EVOKED

BY A MECHANICAL STIMULUS IN MAN

J.S. Ekiel, M.K. Lebiedowska[1], and M.J. Zieniewicz

Institute of Biocybernetics, KRN 55, 00-818 Warsaw
and [1]Child Hospital, Department of Rehabilitation
Warsaw, Poland

The intensity of a reflex reaction to a mechanical stimulus, which involves mechanoreceptors, may depend on the type
of stimulus: its point of application, its type and its parameters, namely its rise time, duration and its shape. A short
latency reflex reaction (Matthews, 1986) depends on the excitability conditions in motoneuron pools of activated muscles
and the mechanical stimulus parameters. Tatton (1985) concluded
that the amplitude of the reaction is proportional to the background activity and to the velocity of the applied stimulus.

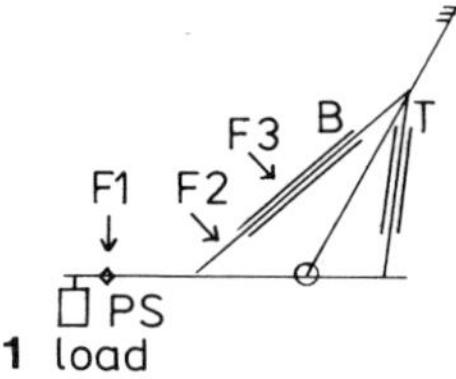

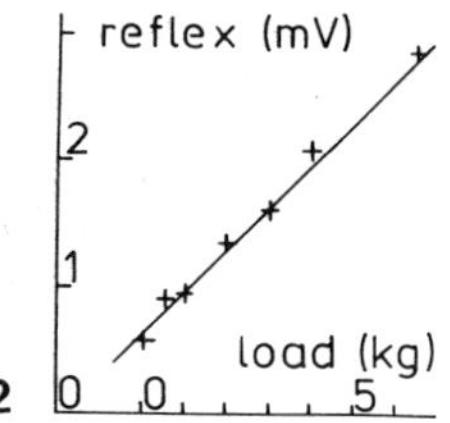

Fig. 1. Mechanical model of human arm. B represents flexor,
        T extensor muscles. F1, F2, F3 - possible stimulus,
        PS - processus styloideus radii.
Fig. 2. Dependence of reflex amplitude on external load.

It is known from clinical practice that the reflex amplitude varies from one tendon tap to another (F2 in Fig. 1).

Under such conditions a statistical method of evaluation should
be used.

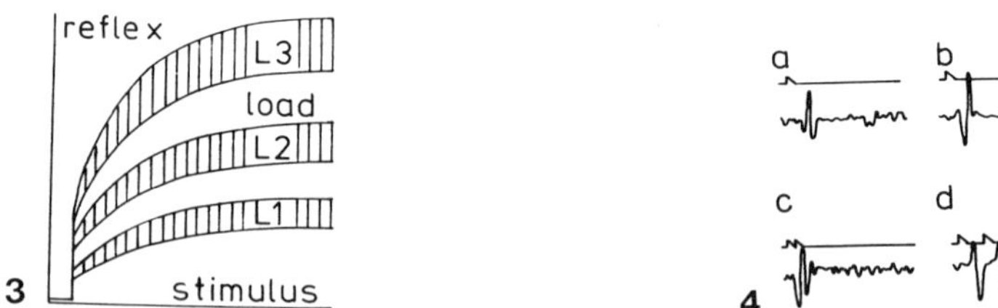

Fig. 3. Dependence of reflex amplitude on mechanical stimulus
        and external load: L3>L2>L1.
Fig. 4. Examples of experimental results. F1 is acting on PS.
        Upper trace shows mechanical stimulus (one or two),
        lower shows EMG of biceps brachii. Time of each record
        is 180 ms.

A new method of generating a noninvasive, repeatable and
controlled force stimulus of the impulse type has been developed. Fig. 1 shows a mechanical model, based on Katbab´s model
(Katbab and Hemami, 1985) of the human arm, where B represents
flexor and T extensor muscles. An electrically controled mechanical force (F1 in Fig. 1) of 5 ms duration is acting on the
subject´s processus styloideus radii (P.S. in Fig. 1). The sub-
ject´s forearm is flexed and held in the horizontal position.

The reflexes evoked by forces F1 and F2 were compared
while the forearm was externally loaded (from 0 to 7 kg). The
experimental results show that the amplitude of the bioelectric
reflex depends linearly on the external load (Fig. 2). Each
point in Fig. 2 is the mean of up to a hundred measurements.
The external load is proportional to the background bioelectric
activity of the muscles and can be measured with better ac-
curacy than this activity.

The bioelectrical reflex amplitude is an increasing
function of both the mechanical stimulus and the external load,
with a considerable nondeterministic component (Fig. 3).

Some experimental results are shown in Fig. 4. Each pair
of traces shows the mechanical stimulus (upper trace) and bioelectrical activity in the biceps brachii (lower trace) recorded with surface electrodes. The response to a single stimulus consists of: pre-existing background activity, multi-
phasic delayed response, isoelectric line (silence period) and
return to the background activity (Fig. 4A). Fig. 4B, C and D

show the results when two stimuli act on the P.S. with different interstimulus delay (15 to 50 ms).

The method presented here enables a noninvasive and repeatable investigation of short latency motor control mechanisms and, of course, of mechanoreceptors as a part of this system. It is also a very useful tool in clinical diagnostic research, but a detailed clinical report exceeds the range of this paper.

REFERENCES

Matthews, P.B.C., 1986, Observations on the automatic compensation of reflex gain on varying the pre-existing level of motor discharge in man, J. Physiol., 374:73-90.
Tatton, W.G., Bedingham, W., Verrier, M.C., Bruce, I.C. and Blair, R.D.G., 1985, Abnormalities of mechanoreceptor-evoked electromyographic activity in central motor disorders, in: "Electromyography and Evoked Potentials", A. Struppler and A. Weind, ed., Springer Verlag, Heidelberg, p. 9-18.
Katbab, A., and Hemami, H., 1985, Role of gain programming in the voluntary movement of the human forearm. Med. and Biol. Eng. and Comput., 23: 230-236.

levels in the resting animal, and winds up to moderate levels
of γ_s-dominated action in slow stepping. In faster steps,
further γ_s wind-up occurs and tonic+modulated γ_d action
becomes evident. In imposed movements, paw shakes, beam
walking (particularly when difficult), and in some accidental
slips, γ_d set increases greatly, not necessarily in parallel
with γ_s set.

We conclude that motor set is strongly expressed at the
fusimotor level, and that the fusimotor system should
therefore be added to the list of CNS structures involved in
setting or scheduling transmission parameters. Most of the
power of fusimotor action seems to be held in reserve for
novel, difficult, or strenuous movements.

<u>Retrospective evidence for $\alpha - \gamma$ independence 1950 - 1970</u>

Though linkage and co-activation of α and γ motoneurones
seemed to be an incontrovertible fact until the late 1970's,
and indeed in 1979 was referred to as a "principle in
motricity" (Granit, 1979), a substantial body of evidence
dating back to the earliest studies of fusimotor function,
suggested that at least a proportion of γ-motoneurones might
be controlled separately from α - motoneurones. Granit & Kaada
(1952) and Granit et al. (1955) found that the stimulation of
many CNS structures led to a diffuse, tonic change in
fusimotor action, often independently of muscle tension.
Stimulation within the reticular formation was found to be
particularly potent in this respect. These results led to
speculations about the role of the fusimotor system in arousal
and expectancy, notions that gradually lost favour in the face
of the $\alpha - \gamma$ linkage hypothesis (Granit, 1955).

Sears (1964) and Corda et al. (1966) found that a proportion
of respiratory γ fibres was tonically active, independently
of phasic α - activation. However, this was interpreted in
terms of a separate postural role for the muscles, rather than
a sensitivity-setting function within the context of the
respiratory movements. Vestibular stimulation (Andersson &
Gernandt, 1956; Poppele, 1967) was shown to evoke fusimotor
effects at lower thresholds than, or independently of, α-
motoneuronal activation. γ_d action often seemed less coupled
to α -activation than did γ_s action (Bergmans & Grillner,
1969; Grillner, et al., 1969), and a strong case was made for
the existence of a region of the midbrain specialised for the
activation of the γ_d system (Appelberg, 1962, 1963, 1981).

<u>Fusimotor control inferred from afferent recordings in
behaving animals</u>

The most striking feature of the early spindle afferent
recordings in freely moving cats was the deep modulation of
firing in relation to muscle displacement (Taylor & Cody,
1974; Goodwin & Luschei, 1975; Prochazka et al., 1975, 1977,
1979; Loeb & Duysens, 1979). This seemed at odds with the
expected role of α -coactivated γ activity in boosting spindle
discharge to offset unloading during shortening contractions.
From the outset a tonic activation of γ_s neurones in the
relaxed animal and during phasic α activity in some chewing
and stepping sequences was suspected (Cody et al., 1975;

418

Prochazka et al., 1975). What we would now refer to as fusimotor wind-up to stepping was described in the earliest report (Prochazka et al., 1975). It was also noted that dramatic increases in spindle sensitivity occurred when movements were imposed on an animal, often in the absence of detectable changes in EMG (Prochazka et al., 1977). However, such was the hold of the $\alpha - \gamma$ linkage concept, that it was not until such effects had been observed repeatedly that the need for an alternative hypothesis of fusimotor control became obvious (Prochazka & Wand, 1981; Prochazka, 1983; Loeb, 1984).

Inferences about fusimotor action underlying muscle spindle firing patterns in voluntary movement are difficult and prone to error. Before the development of acute verification experiments (Hulliger & Prochazka, 1983), the procedure was to compare qualitatively the afferent responses to length variations with what might be expected from the large volume of information on fusimotor action previously gained in acute experiments. Experience with the verification experiments has shown that while inferences of this sort are probably quite reliable when large changes in spindle sensitivity and bias occur, the situation is far less satisfactory for small changes. Hidden sources of error include first and foremost the inappropriate monitoring of muscle length: inaccurate alignment of length gauges, or the monitoring of joint displacement via a manipulandum not firmly attached to the extremity, and whose pivot cannot accurately follow that of the joint. Small misalignments can lead to fundamental errors in interpretation. Internal shortening caused by extrafusal contractions can modulate spindle firing, particularly in powerful contractions when length variations are small. Unrecognised, this can erroneously be ascribed to fusimotor modulation. It has recently been claimed that this effect can be a very major one (Griffiths & Hoffer, 1987), though our own control experiments with distributed α stimulation suggest that in most instances it is minor (report in preparation). Other factors modulating spindle discharge in the absence of fusimotor variation include pressure exerted by one body segment on another (e.g. during a crouch the thigh muscles press onto those of the calf); travelling pressure waves initiated at foot contact or foot-off and giving rise to bursts of firing; pressure transmitted laterally between overlying muscles (e.g. Biceps femoris posterior onto lateral gastrocnemius).

The difficulties just outlined are no doubt responsible for a certain heterogeneity in current theories of fusimotor control derived from chronic recordings. The schemes hypothesised have ranged from straightforward $\alpha - \gamma$ linkage (Hagbarth, 1981; Gandevia & Burke, 1985) through partial independence, with one type of γ linked, (Appenteng et al., 1980: γ_s linked; Taylor et al., 1985: γ_d linked; Loeb et al., 1985a,b,c: γ's linked to task groups) to virtually complete dissociation and strong task-dependence of γ action (Prochazka & Hulliger, 1983). Our own group has felt for some time that fusimotor action in routine, stereotyped movements is generally at low levels and that the details of γ control in such situations are to some extent overshadowed by the power and scope of fusimotor action in novel and difficult motor tasks. The modulations in spindle firing ascribed to fusimotor action in say stepping, pale into insignificance when compared to the massive increases in spindle sensitivity which usually occur when movements are imposed upon an animal. Similar large increases in sensitivity are often observed when cats are lifted and moved from one part of a room to another, or when they negotiate narrow walkways. After many such observations, the possibility of a common causal factor, namely novelty or vigilance, suggested itself (Prochazka & Hulliger, 1983).

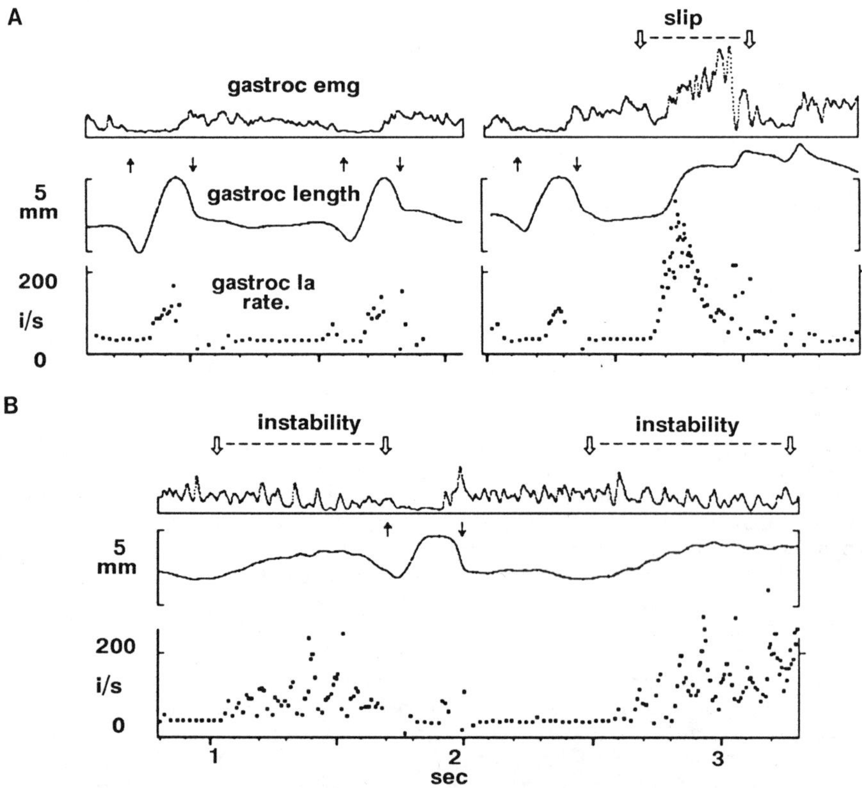

Fig. 1.   Responses of a single spindle primary ending to muscle length changes during three types of movement: normal step cycles (A, left); a step followed by a slip (A, right); two periods of instability while negotiating a narrow beam. Note the very high sensitivity to stretch during the slip and the periods of instability.

Fig. 1 shows two examples of large increases in spindle afferent sensitivity recorded in difficult and unpredictable tasks. In Fig. 1A, after some "routine" steps, the cat's foot slipped on the edge of the benchtop, which itself was about 1 m above floor level. The cat managed to recover its balance, and resumed stepping. During the recovery manoeuvre, there was a sudden and very large response to muscle stretch, indicating a surge of γ_d action.

In Fig. 1B, the cat was walking along a 3 cm wide wooden beam
spanning a 1 m gap between two pedestals, 30 cm above a flat
benchtop. The task was difficult, as evidenced in a
concomitant video recording by overt tremor, body sway, long
intervals between steps, and repeated failures in other trials
to reach the food reward. Before and after the single step in
this record, the length trace shows small oscillations,
corresponding to leg tremor and wobbles of the beam clearly
visible in the video. The gastrocnemius spindle Ia afferent
was very sensitive to the length oscillations in these
periods, especially when compared to its low sensitivity to
the stretch during the step, and its low tonic firing level
for the ensuing second or so of stability before the beam
started oscillating again.

Acute reconstruction experiments

As mentioned above, inferences about underlying fusimotor
action are subject to error unless the changes in sensitivity
are as dramatic as those in Fig. 1. We have therefore carried
out large numbers of acute verification or reconstruction
experiments to put the inferences on a firm footing. In
these, muscle length variations from the chronic recordings
were replicated by an electromagnetic length servo, and the
firing of isolated spindle endings was made to match the
original spindle firing profiles by computer-controlled
fusimotor stimulation (Hulliger et al., 1987). Starting from
arbitrary tonic levels of fusimotor stimulation, in successive
iterations the time course of fusimotor stimulation was
gradually modified so that good matches were obtained between
the acute and chronic afferent firing profiles.

The reconstruction experiments have lent strong support to the
general notion that fusimotor action is set according to task
or context. Imposed movements have generally led to γ_d-
dominated reconstructions, while steps were usually best
fitted with γ_s stimulation. Other situations in which the
reconstruction experiments indicated γ_d set were paw shake
responses, beam walking and stumbling corrective reactions (as
in Fig. 1). Wind-up at the onset of stepping, particularly of
γ_s action, was verified in a number of different chronically
recorded spindles. Step cycles were divided into groups
according to swing duration. Reconstructions based on the
averaged records indicated that as the speed of gait increased
(swing duration decreased), fusimotor action was wound up
further.

Of 7 chronically recorded triceps surae Ia afferents analysed
in detail, 3 of them, for a proportion of the time, showed
qualitative evidence (not yet verified by reconstruction
experiments) of e.m.g.-linked γ action, probably dynamic,
additional to presumed tonic γ_s action during walking. In
these 3 units the mean step duration of all steps which either
showed this effect or not was 240 ms and 270 ms respectively.
Comparing the sample means using Student's t-test revealed
that this difference was highly significant at the p<.001
level. α - linked γ_d modulation, when evident, was thus

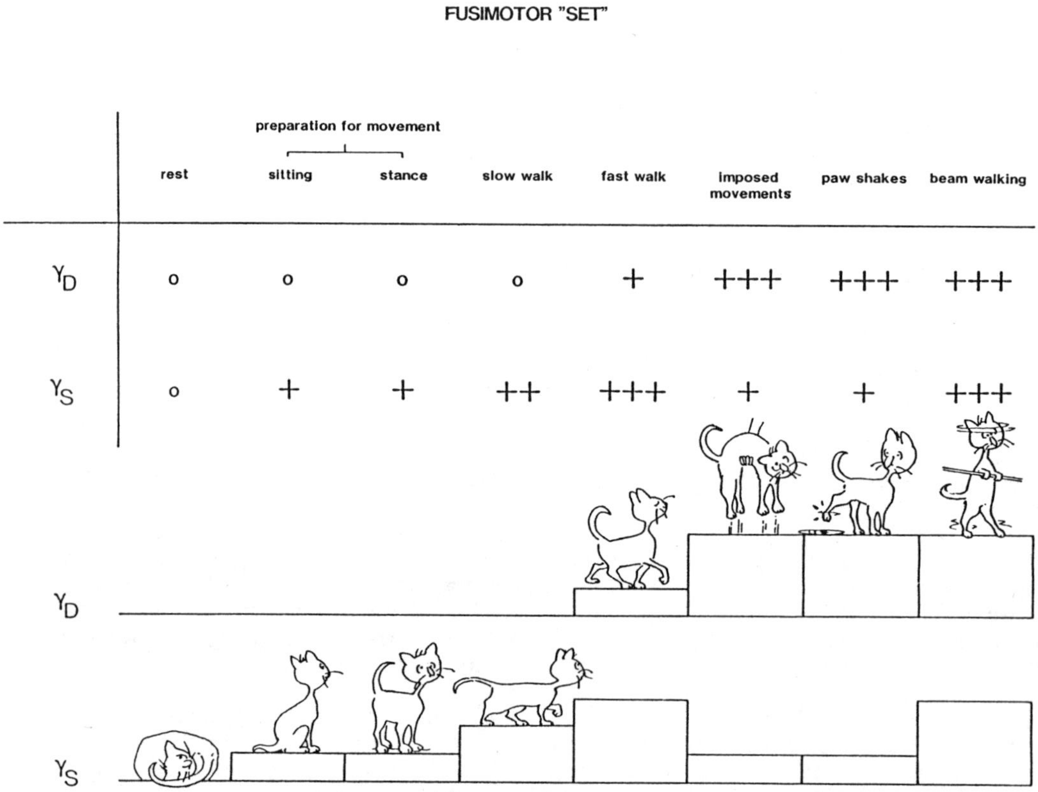

	rest	preparation for movement		slow walk	fast walk	imposed movements	paw shakes	beam walking
		sitting	stance					
γ_D	o	o	o	o	+	+++	+++	+++
γ_S	o	+	+	++	+++	+	+	+++

Fig. 2. Fusimotor set summary. The specific balance of static and dynamic fusimotor activity under different behavioural conditions is shown. At rest: no fusimotor activity. Preparation for movement: "wind-up" of fusimotor drive, predominantly static in type. Slow walking: low level static fusimotor action. Fast walking: further "wind-up" of static activity with additional α-linked dynamic fusimotor action. Imposed movements: "switch" from predominant static fusimotor activity to dynamic drive. Difficult, unpredictable or novel motor tasks: reintroduction of static fusimotor activity to a greater (beam and ladder walking) or lesser (paw shakes) extent.

associated with higher gait speeds. Only rarely did we record data at gait velocities exceeding 1.0 m/s. It seems quite likely that even higher levels of γ set occur in trotting (0.7 - 2.7 m/s) or galloping (>2.7 m/s). It should be stressed that the inferred α - linked component of γ action was always superimposed upon a tonic background of γ action, which also increased with gait speed.

In terms of parameter "set", as a first approximation one could express this as:

$$\gamma(t) = [A + Bv(t)]\,\gamma_s + [C + Dv(t)\alpha(t)]\,\gamma_d \qquad \ldots\ldots 1$$

where A,B,C,D are the "set" parameters; v(t) is gait velocity; $\alpha(t)$ and $\gamma(t)$ are the time courses of α activity and γ action (γ_s and γ_d) respectively.

On this view, in novel and difficult movements, parameter C in particular (the tonic γ_d parameter) would be reset to a much higher level. In recent experiments on intercostal fusimotor control, Sears' (1964) description of tonic and phasic γ activation has been corroborated (Greer & Stein, 1986). It was suggested that γ_s action was tonic, and γ_d action was α-linked. The strength of the γ_d action depended upon respiratory drive (altered experimentally by hypercapnia and tracheal occlusion). As increased drive leads to increased respiratory rate, and also occurs in difficult, strenuous tasks, the above equation may provide a useful description of fusimotor control in the respiratory system as well as the locomotor system in cats. If fusimotor control in jaw movements does indeed involve α- γ_s rather than α- γ_d linkage as proposed by Appenteng et al. (1980), a more general form of the above equation would be required:

$$\gamma(t) = [A + Bv(t) + C\alpha(t)]\,\gamma_s + [D + Ev(t)\alpha(t)]\,\gamma_d$$

However, it should be noted that there is contradictory evidence on this issue (Larson et al., 1981), favouring α-independent γ_s set and increased, partially α-linked γ_d set in difficult jaw movement tasks, a combination covered by equation 1.

Our present view of fusimotor set in the cat is illustrated in the cartoon of Fig. 2. Many questions remain open. If a task involves only one limb, is γ set changed only in that limb? To what extent do the parameters suggested in the above equations vary during a given task (if they are rarely constant, should they rather be viewed as variables)? What other tasks or contexts lead to predictable changes in fusimotor set?

Motor Set: participating CNS structures

In a recent monograph, Evarts et al. (1984) reviewed the evidence for set in the sensorimotor cortex, thalamus and cerebellum (see also Brooks, 1984; Martin & Ghez, 1985). Other supraspinal areas in which neuronal firing has been associated with attention and preparation for movement are: posterior

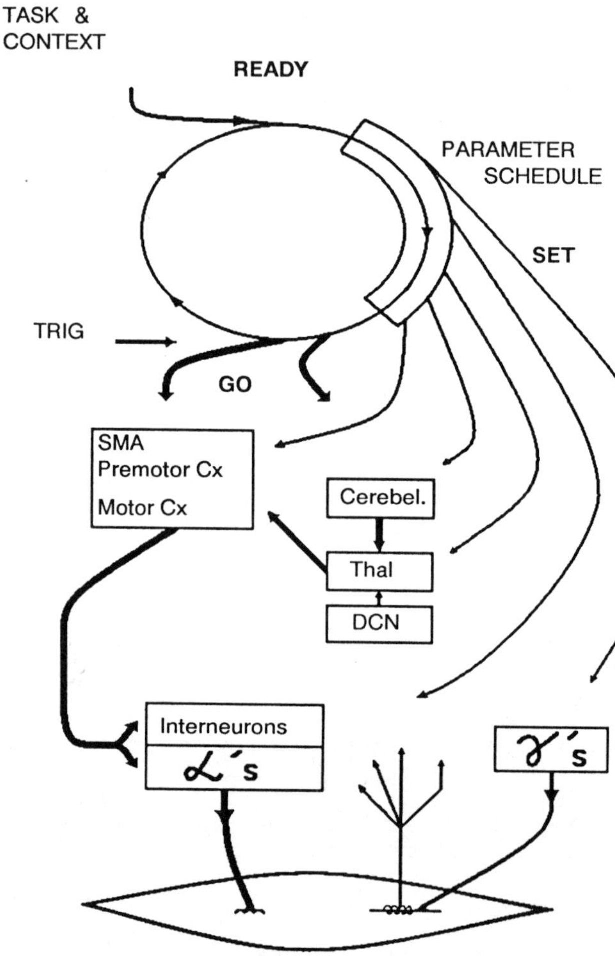

Fig. 3. Schematic showing CNS structures in which parameter setting has been implicated in motor tasks. The "hold" loop ("READY") generates a parameter schedule ("SET") appropriate to the particular task and context, and an internal or external trigger initiates the movement ("GO"). Parameters are set in the supplementary motor area (SMA), premotor cortex (Premotor Cx), motor cortex (Motor Cx), cerebellum (Cerebel.), thalamus (Thal), dorsal column nuclei (DCN), spinal cord (interneurons) and the fusimotor system ('s). In principle, gain control of afferent feedback from muscle spindle endings is set by at least three neuronal systems: fusimotor neurones, spinal interneurones and the DCN.

parietal cortex (Leinonen, 1980; Mountcastle et al., 1975; MacKay & Crammond, 1987), supplementary motor area (Brinkman & Porter, 1983; Tanji & Kurata, 1983), premotor cortex (Mauritz & Wise, 1986), thalamus and basal ganglia (Strick, 1976; Neafsey et al., 1978, Macpherson et al., 1980), medial lemniscus (Ghez & Lenzi, 1971; Coulter, 1974) and dorsal column nuclei (Ghez & Pisa, 1972).

At the spinal level, Gurfinkel, & Kots (1966), Requin & Paillard (1971), Kots (1977), Requin (1985) and Scheirs & Brunia (1985) described task-dependent changes in H-reflexes and tendon jerks (T-reflexes) some hundreds of msec. prior to a movement. Interestingly, some of these data were interpreted in terms of reflex set, but the possibility of fusimotor modulation of the tendon jerks (and therefore the existence of fusimotor set) was largely neglected. As it happens, fusimotor action does not in any case modulate spindle responses to tendon taps to the extent previously supposed (Gregory et al., 1977), and so the T/H ratio is probably a very poor indicator of fusimotor drive. The evidence for fusimotor set therefore relies largely on the chronic afferent data described above. In one instance, selective activation of the fusimotor system in the maintenance of accurate position was inferred from chronic recordings of neurons in area 3a (Tanji, 1976, see also Evarts & Fromm, 1977). So far afferent recordings in humans have generally failed to show convincing fusimotor set in the absence of α - activity (Gandevia & Burke, 1985) though in the only recordings to date from suspected fusimotor fibres in human subjects, firing increased independently of α activity, in relation to environmental disturbances, voluntary activity such as laughing, and mental computation (Ribot et al., 1986). For technical reasons, single fibre recordings in humans can only be achieved by severely restricting the range of motor tasks. It seems likely that the situations in which dramatic changes in fusimotor set occur in humans are out of the reach of current neurographic techniques.

On balance, we feel that there is now good evidence in cat and monkey that the fusimotor system is part of the neuronal circuitry which sets or schedules transmission parameters in different sensorimotor tasks (Fig. 3). On this view, the overall role of γ motoneurones should be sought in motor tasks of very widely ranging types.

<u>Acknowledgements</u>

This work was supported by the Alberta Heritage Fund for Medical Research, The British Medical Research Council, the European Science Foundation and the Swiss National Science Foundation. We thank Mr. Steven Vincent for technical help.

REFERENCES

Ach, N., 1905, "Über die Willenstätigkeit und das Denken", Göttingen.

Andersson, S., & Gernandt, B.E., 1956, Ventral root discharge in response to vestibular and proprioceptive stimulation. _J. Neurophysiol._, 19: 524.

Appelberg, B., 1962, The effect of electrical stimulation in nucleus ruber on the response to stretching primary and secondary muscle spindle afferents, _Acta Physiol. Scand._, 56: 140.

Appelberg, B., 1963, Central control of extensor muscle spindle dynamic sensitivity. _Life Sciences,_ 9: 706.

Appelberg, B., 1981, Selective central control of dynamic gamma motoneurones utilised for the functional classification of gamma cells, _in_: "Muscle Receptors and Movement", A. Taylor & A. Prochazka, eds., Macmillan, London.

Appenteng, K., Morimoto, T., & Taylor, A., 1980, Fusimotor activity in masseter nerve of the cat during reflex jaw movements, _J. Physiol._, 305: 415.

Bergmans, J., & Grillner, S., 1969, Reciprocal control of spontaneous activity and reflex effects in static and dynamic γ- motoneurones revealed by an injection of DOPA, _Acta Physiol. Scand._, 77: 106.

Boring, E.G., 1957, "A History of Experimental Psychology. 2nd Ed.," Appleton-Century-Crofts, New York.

Brinkman, C., & Porter, R., 1983, Supplementary motor area and premotor area of money cerebral cortex: functional organization and activities of single neurons during performance of a learned movement, _in_ "Motor Control Mechanisms in Health and Disease", J.E. Desmedt, ed. Raven Press, New York.

Brooks, V.B., 1984, The cerebellum and adaptive tuning of movements, _Exp. Brain Res._, Suppl. 9: 170.

Cody, F.W.J., Harrison, L.M., & Taylor, A., 1975, Analysis of the activity of muscle spindles of the jaw-closing muscles during normal movements in the cat. _J. Physiol._, 253: 565.

Corda, M., Euler, C.V., & Lennerstrand, G., 1966, Reflex and cerebellar influences on α and "rhythmic" and "tonic" activity in the intercostal muscle, _J. Physiol._ 184: 898.

Coulter, J.D, 1974, Sensory transmission through lemniscal pathway during voluntary movement in the cat, _J. Neurophysiol._, 37: 831.

Evarts E.V., & Fromm, C., 1977, Sensory responses in motor cortex neurons during precise motor control, _Neuroscience Letters,_ 5: 267.

Evarts E.V., Shinoda, Y., & Wise, S.P., 1984, "Neurophysiological Approaches to Higher Brain Functions," John Wiley, New York.

Gandevia, S., & Burke, D., 1985, Effect of training on voluntary activation of human fusimotor neurons. <u>J. Neurophysiol.</u>, 54: 1422.

Ghez, C., & Lenzi, G.L., 1971, Modulation of sensory transmission in cat lemniscal system during voluntary movement, <u>Pflügers Arch. ges. Physiol.</u>, 323: 273.

Ghez, C., & Pisa, M., 1972, Inhibition of afferent transmission in cuneate nucleus during voluntary movement in the cat, <u>Brain Res.</u>, 40: 145.

Gibson, J.J. (1941) A critical review of the concept of set in contemporary experimental psychology. <u>Psychol. Bull.</u>, 38: 781.

Goodwin, G.M., & Luschei, E.S., 1975, Discharge of spindle afferents from jaw-closing muscles during chewing in alert monkeys, <u>J. Neurophysiol.</u>, 38, 560.

Granit, R., 1955, "Receptors and Sensory Perception," Yale University Press, New Haven.

Granit, R., 1979, Interpretation of supraspinal effects on the gamma system, <u>in</u>: "Reflex Control of Posture and Movement". Progr. Brain Res., 50, Elsevier, Amsterdam.

Granit, R., & Kaada, B. R., 1952, Influence of stimulation of central nervous structures on muscle spindles in cat, Acta <u>Physiol. Scand.</u> 27: 130.

Granit, R., Holmgren, B., & Merton, P.A., 1955, The two routes for excitation of muscle and their subservience to the cerebellum, <u>J. Physiol.</u> 130: 213.

Greer, J.J., & Stein, R.B. (1986) Tonic and phasic activity of gamma motoneurons during respiration in the cat. <u>Society for Neuroscience Abstracts</u>, 12, 683.

Gregory, J.E., Prochazka, A., & Proske, U., 1977, Responses of muscle spindles to stretch after a period of fusimotor activity compared in freely moving and anaesthetised cats, <u>Neuroscience Letters</u> 4: 67.

Griffiths, R.I., & Hoffer, J.A., 1987, Muscle fibres <u>shorten</u> when the whole muscle is being stretched in the 'yield' phase of the freely walking cat. <u>Soc. Neurosci.</u> Abstr. In Press.

Grillner, S., Hongo, T., & Lund, S., 1969, Descending monosynaptic and reflex control of γ-motoneurones, <u>Acta Physiol. Scand.</u> 75: 592.

Gurfinkel, V.S., & Kots, Y.M., 1966, Dvigatelnaya prednastroyka o cheloveka. <u>in</u>: "Nervyn Mekanismy Dvigatelnoi Deyatelnosty," Akademia Nauk, SSSR.

Hagbarth, K.-E., 1981, Fusimotor and stretch reflex functions studied in recordings from muscle spindle afferents in man, in: "Muscle Receptors and Movement". A. Taylor, & A. Prochazka, eds. Macmillan, London.

Hulliger, M., Horber, F., Medved, A., & Prochazka, A., 1987, An experimental simulation method for iterative and interactive reconstruction of unknown (fusimotor) inputs contributing to known (spindle afferent) responses, J. Neurosci. Methods, In Press.

Hulliger, M., & Prochazka, A. (1983) A new simulation method to deduce fusimotor activity from afferent discharge recorded in freely moving cats. J. Neurosci. Methods, 8: 197.

Kots, Y.M., 1977, "The Organization of Voluntary Movement", Plenum, New York.

Larson, C.R., Smith, A., & Luschei, E.S., 1981, Discharge characteristics and stretch sensitivity of jaw muscle afferents in the monkey during controlled isometric bites. J. Neurophysiol., 46: 130.

Leinonen, L. (1980) Functional properties of neurones in the posterior part of area 7 in awake monkeys. Acta Physiologica Scandinavica, 108: 301
Loeb, G.E., 1984, The control and responses of mammalian muscle spindles during normally executed motor tasks, Exercise Sport Sci. Rev., 12: 158.

Loeb, G.E., & Duysens, J., 1979, Activity patterns in individual hindlimb primary and secondary muscle spindle afferents during normal movements in unrestrained cats. J. Neurophysiol., 42: 420.

Loeb, G.E., Hoffer, J.A., & Pratt, C.A., 1985a, Activity of spindle afferents from cat anterior thigh muscles. I. Identification and patterns during normal locomotion. J. Neurophysiol., 54: 549.

Loeb, G.E., & Hoffer, J.A., 1985b, Activity of spindle afferents from cat anterior thigh muscles. II. Effects of fusimotor blockade. J. Neurophysiol., 54: 565.

Loeb, G.E., Hoffer, J.A., & Marks, W.B., 1985c, Activity of spindle afferents from cat anterior thigh muscles. III. Effects of external stimuli, J. Neurophysiol., 54: 578.

MacKay, W.A., & Crammond, D.J., 1987, Neuronal correlates in posterior parietal lobe of the expectation of events, Behav. Brain Res. 24, 167.

Macpherson, J.M., Rasmusson, D.D., & Murphy, J.T., 1980, Activities of neurons in "motor" thalamus during control of limb movement in the primate. J. Neurophysiol., 44: 11.

Martin, J.H., & Ghez, C., 1985, Task-related coding of stimulus and response in cat motor cortex, Exp. Brain Res. 57: 427.

Mauritz, K.-H., & Wise, S.P., 1986, Premotor cortex of the

rhesus monkey:neuronal activity in anticipation of predictable environmental events, <u>Exp. Brain Res.</u>, 61: 229

Mountcastle, V.B., Lynch, J.C., Georgopoulos, A., Sakata, H.,& Acuna, C., 1975,   Posterior parietal association cortex of the monkey: command functions for operations within extrapersonal space, <u>J. Neurophysiol.</u>, 38: 871.

Neafsey, E.J., Hull, C.D.,& Buchwald, N.A., 1978, Preparation for movement in the cat. II. Unit activity in the basal ganglia and thalamus, <u>Electroenceph. clin. Neurophysiol.</u>, 44: 714.

Poppele, R., 1967,   Responses of gamma and alpha motor systems to phasic and tonic vestibular inputs, <u>Brain Res.</u>, 6: 535.

Prochazka, A., 1983,   The uncoupling of alpha and of static and dynamic fusimotor activity in the cat: fusimotor "set", <u>Proc. Internat. Union Physiol. Sci</u>. 15: 12.

Prochazka, A.,& Hulliger, M., 1983,   Muscle afferent function and its significance for motor control mechanisms during voluntary movements in cat, monkey and man, <u>in</u>: "Motor Control Mechanisms in Health and Disease" J.E. Desmedt, ed., Raven, New York.

Prochazka, A., Hulliger, M., Zangger, P.,& Appenteng, K., 1985,   "Fusimotor set": new evidence for α-independent control of γ-motoneurones during movement in the awake cat, <u>Brain Res.</u>, 339: 136.

Prochazka, A., Stephens, J.A.,& Wand, P., 1979,   Muscle spindle discharge in normal and obstructed movements, <u>J. Physiol.</u>,287: 57.

Prochazka, A.,& Wand, P., 1981,   Independence of fusimotor and skeletomotor systems during voluntary movement, <u>in</u>: "Muscle Receptors and Movement" A. Taylor, & A. Prochazka, eds. Macmillan, London.

Prochazka, A., Westerman, R.A.,& Ziccone, S.P., 1975,   Spindle units recorded during voluntary hindlimb movements in the cat, <u>Proc. Austral. Physiol. Pharmacol. Soc.</u>,6: 101.

Prochazka, A., Westerman, R.A.,& Ziccone, S., 1976, Discharges of single hindlimb afferents in the freely moving cat. <u>J. Neurophysiol.</u>, 39: 1090.

Prochazka, A., Westerman, R.A.,& Ziccone, S., 1977,   Ia afferent activity during a variety of voluntary movements in the cat, <u>J. Physiol.</u>, 268: 423.

Requin, J., 1985, Looking forward to moving soon: ante factum selective processes in motor control. <u>in</u>: "Attention and Motor Performance XI", M.I. Posner, & O.S.M. Marin, eds., Lawrence Erlbaum Assoc. Inc., Hillsdale, New Jersey.

Requin, J.,& Paillard, J., 1971,   Depression of monosynaptic reflexes as a specific aspect of preparatory motor set in visual reaction time, <u>in</u>: "Visual Information Processing and

Control of Motor Activity". Bulgarian Academy of Sciences, Sofia.

Ribot, E., Roll, J.-P., & Vedel, J.-P., 1986, Efferent discharges recorded from single skeletomotor and fusimotor fibres in man, <u>J. Physiol</u>., 375,: 251.

Scheirs, J.G.M., & Brunia, C.H.M., 1985, Achilles tendon reflexes and surface EMG activity during anticipation of a significant event and preparation for a voluntary movement, <u>J. Motor Behav</u>.,17: 96.

Sears, T.A., 1964, Efferent discharges in alpha and fusimotor fibres of intercostal nerves of the cat, <u>J. Physiol</u>.,174: 295.

Strick, P.L., 1976, Activity of ventrolateral thalamic neurons during arm movement, <u>J. Neurophysiol</u>., 39: 1032.

Tanji, J., 1976, Selective activation of neurons in cortical area 3a associated with accurate maintenance of limb positions, <u>Brain Res</u>., 115, 328.

Tanji, J.,& Kurata, K., 1983, Functional organization of the supplementary motor area, <u>in</u>: "Motor Control Mechanisms in Health and Disease", J.E. Desmedt, ed., Raven, New York.

Taylor, A., & Cody, F.W.J., 1974, Jaw muscle spindle activity in the cat during normal movements of eating and drinking, <u>Brain Res</u>.,71: 523.

Taylor, J., Stein, R.B.,& Murphy, P.R., 1985, Impulse rates and sensitivity to stretch of soleus muscle spindle afferent fibres during locomotion in premammillary cats. <u>J. Neurophysiol</u>. 53: 341.

Watson, R.I., 1963, "The Great Psychologists. From Aristotle to Freud", Lippincott, Philadelphia.

Watt, H.J., 1905, Experimentelle Beiträge zu einer Theorie des Denkens, <u>Arch. ges. Psychol</u>., 4: 289.

PARTICIPANTS

ALDSKOGIUS, H.
>Department of Anatomy, Karolinska Institutet, Box 60400,
S-10401, Stockholm, Sweden

ARUTYUNIAN, R.
>Sechenov Inst. of Evolutionary Physiology and Biochemistry,
Thorez pr. 44, 194223, Leningrad, USSR

ASMUSSEN, G.
>Karl-Marx-Universität, Bereich Medizin, Carl-Ludwig-Institut für Physiologie, Liebigstraße 27, 7010 Leipzig, GDR

AVILOVA, K.V.
>Biological faculty, Moscow State University, 117133 Moscow,
USSR

AZAROVA, V.S.
>Severtzov Institute of Evolutionary Animal Morphology and
Ecology, USSR Academy of Sciences, Leninsky Prospekt 33,
Moscow V-71, USSR

BANKS, R.
>Department of Zoology, University of Durham, Durham
DH1 3LE, UK

BATUEVA, I.V.
>Sechenov Inst. of Evolutionary Physiology and Biochemistry,
Thorez pr. 44, 194223, Leningrad, USSR

BOLANOWSKI, S., Jr.
>Department of Physiology, University of Rochester Medical
School, Rochester, N.Y. 14642, USA

BORECKÝ, J.
>Institute of Physiology, Czechoslovak Academy of Sciences,
Vídeňská 1083, 14220 Prague 4, CSSR

BOYD, I.A.
>Institute of Physiology, The University, Glasgow G12 8QQ,
UK

BULYAKOVA, N.V.
>Severtzov Institute of Evolutionary Animal Morphology and
Ecology, USSR Academy of Sciences, Leninsky Prospekt 33,
Moscow V-71, USSR

BUTLER, R.
>Department of Anatomy, McMaster University, 1200 Main
Street West, Hamilton, Ontario, Canada L8N 3Z5

CHOUCHKOV, C.N.
>Department of Anatomy, Histology and Embryology, M.A. Higher
Medical Inst., 6003 Stara Zagora, Bulgaria

DICKSON, M.
>Institute of Physiology, The University, Glasgow G12 8QQ,
UK

DREESSEN, D.
>Universität Hamburg, Abteilung für Funktionelle Anatomie,
UKE, Martinistrasse 52, D-2000 Hamburg 20, FRG

DUBOVÝ, P.
      Department of Anatomy, Faculty of Medicine, Purkyně Uni-
      versity, Komenského nám. 2, 66243 Brno, CSSR
ERIKSSON, P.O.
      Department of Anatomy, University of Umea, S-901 87 Umea,
      Sweden
FERRELL, W.
      Institute of Physiology, The University, Glasgow G12 8QQ,
      UK
FUKAMI, Y.
      Department of Cell Biology and Physiology, Washington
      Univ., 660 South Euclid Avenue, St. Louis, MO 63110, USA
GATAULIN, R.
      Department of Histology, Kazan State Medical Institute,
      Butlerov Street 49, 420012 Kazan 12, USSR
GLADDEN, M.
      Institute of Physiology, The University, Glasgow G12 8QQ,
      UK
HALATA, Z.
      Universität Hamburg, Abteilung für Funktionelle Anatomie,
      UKE, Martinistrasse 52, D-2000 Hamburg 20, FRG
HALLIDAY, D.
      Institute of Physiology, The University, Glasgow G12 8QQ,
      UK
HNÍK, P.
      Institute of Physiology, Czechoslovak Academy of Sciences,
      Vídeňská 1083, 14220 Prague 4, CSSR
HORÁK, V.
      Department of Genetics, Institute of Physiology and
      Genetics of Farm Animals, Czechoslovak Academy of Sciences,
      27721 Liběchov, CSSR
HUNT, C.C.
      Washington University School of Medicine, Box 8101, 660
      South Euclid Avenue, St. Louis, MO 63110, USA
IDE, CH.
      Department of Anatomy, Iwate Medical College, School of
      Medicine, Morioka 020, Japan
IGGO, A.
      Department of Preclinical Veterinary Sci., University of
      Edinburgh, Summerhall, Edinburgh EH9 1QH, UK
ILYINSKY, O.B.
      Central Institute of Medico-Biological Problems of Sports,
      Elizavetinsky proezd 10, 107005 Moscow, USSR
ITO, F.
      Department of Physiology, Nagoya University School of
      Medicine, Nagoya 466, Japan
JAHNKE, M.
      Neurologische Klinik der Technischen Universität München,
      Möhlstraße 28, D-8000 München 80, FRG
JAMI, L.
      Laboratoire de Neurophysiologie, Collège de France, 11
      Place Berthelot, 75231 Paris Cedex 05, France
JIRMANOVÁ, I.
      Institute of Physiology, Czechoslovak Academy of Sciences,
      Vídeňská 1083, 14220 Prague 4, CSSR
KUCERA, J.
      Department of Neurology, Boston University School of
      Medicine, 80 East Concord Street, Boston, MA 02118, USA
LANDON, D.
      Institute of Neurology, Queen Square, London WC1, UK

LAPORTE, Y.
	Collège de France, Laboratoire de Neurophysiologie, 11
	Place Berthelot, 75231 Paris Cedex 05, France
LESLIE, G.C.
	Department of Physiology, The University, Dundee DD1 4HN,
	UK
MAEDA, N.
	First Department of Oral Anatomy, Josai Dental University,
	Sakado, Saitama 350-02, Japan
MAIER, A.
	Department of Cell Biology and Anatomy, The University of
	Alabama, Birmingham, AL 35294, USA
MALINOVSKÝ, L.
	Department of Anatomy, Faculty of Medicine, Purkyně Uni-
	versity, Komenského nám. 2, 66243 Brno, CSSR
MASUDA, T.ˣ
	Department of Oral Diagnosis, Josai Dental University,
	Sakado, Saitama 350-02, Japan
MAYR, R.
	Institute of Anatomy 2, University of Vienna, Währinger-
	strasse 13, A-1090 Wien, Austria
MUNGER, B.L.
	Department of Anatomy, The Pennsylvania State University,
	POB 850, Hershey, PA 17033, USA
NAVA, P.B., Jr.
	Department of Anatomy, Loma Linda University, Loma Linda,
	CA 92350, USA
NURSE, C.
	Department of Biology, McMaster University, 1200 Main
	Street West, Hamilton, Ontario L8S 4K1, Canada
OAKLEY, B.
	Department of Biology, Neuroscience Laboratory Bldg.,
	University of Michigan, 1103 East Huron, Ann Arbor, MI
	48104-1687, USA
OSAWA, K.
	Department of Oral Diagnosis, Josai Dental University,
	Sakado, Saitama 350-02, Japan
OVALLE, W.
	Department of Anatomy, University of British Columbia,
	Vancouver, B.C., Canada V6T 1W5
PÁČ, L.
	Department of Anatomy, Faculty of Medicine, Purkyně Uni-
	versity, Komenského nám. 2, 66243 Brno, CSSR
PALEČEK, J.
	Institute of Physiology, Czechoslovak Academy of Sciences,
	Vídeňská 1083, 14220 Prague 4, CSSR
PART, N.
	Department of Physiology, The University, Dundee DD1 4HW,
	UK
PECK, D.
	Department of Anatomy, University of Kentucky Medical
	Center, 800 Rose Street, Lexington, KY 40536-0084, USA
POPOVA, M.F.ˣ
	Severtzov Institute of Evolutionary Animal Morphology and
	Ecology, USSR Academy of Sciences, 33 Leninsky Prospekt,
	Moscow V-71, USSR
PROCHAZKA, A.
	Department of Physiology, University of Alberta, Edmonton
	T6G 2EI, Canada
PROSKE, U.
	Department of Physiology, Monash University, Clayton,
	Victoria 3168, Australia

ROWLERSON, A.
    Institute of Physiology, The University, Glasgow G12 8QQ,
    UK
SAHGAL, V.
    Rehabilitation Institute of Chicago, 345 East Superior
    Street, Chicago, IL 60611-4496, USA
ŠALING, M.
    Institute of Normal and Pathological Physiology, Sienkie-
    wiczowa 1, 813 71 Bratislava, CSSR
SAXOD, R.
    Laboratoire de Zoologie et Biologie Animale, Université
    de Grenoble PB 68, 38402 Saint-Martin-D´Heres Cedex, France
SCHMALBRUCH, H.
    Institute of Neurophysiology, University of Copenhagen,
    Panum Institute, Blegdamsvej 3 C, 2200 Copenhagen N, Denmark
SCHRÖDER, J.M.
    Medizinische Fakultät, Klinikum der RWTH Aachen, Pauwels-
    strasse, 5100 Aachen, FRG
SCOTT, J.A.
    Department of Zoology, University of Durham, Science Labo-
    ratories, South Road, Durham DH1 3LE,UK
SAHANI, M.
    E.I.C. Institute of Electrophysiology for Fundamental and
    Applied Research, Dr. E. Borges Marg, Parel, Bombay 400012,
    India
SOUKUP, T.
    Institute of Physiology, Czechoslovak Academy of Sciences,
    Vídeňská 1083, 14220 Prague 4, CSSR
SPASSOVA, I.
    Department of Anatomy and Histology, Medical Academy,
    Sofia, Bulgaria
STEPHENS, H.
    Department of Neurology, Boston University School of Me-
    dicine, 80 East Concord Street, Boston, MA 02118, USA
STRASMANN, T.
    Universität Hamburg, Abteilung für Funktionelle Anatomie,
    UKE, Martinistrasse 52, 2000 Hamburg 20, FRG
STRUPPLER, A.
    Neurologische Klinik der Technischen Universität München,
    Möhlstrasse 28, D-8000, München 80, FRG
STUDITSKY, A.N.
    Severtzov Institute of Evolutionary Animal Morphology and
    Ecology, USSR Academy of Sciences, Leninsky Prospekt 33,
    Moscow, USSR
SUTHERLAND, F.L.
    Institute of Physiology, The University, Glasgow G12 8QQ,
    UK
SVENDSEN, K.F.
    Institute of Neurophysiology, University of Copenhagen,
    Panum Institute, Blegdamsvej 3 B, 2200 Copenhagen N,
    Denmark
THORNELL, L.-E.
    Department of Anatomy, University of Umea, 901 87 Umea,
    Sweden
TOFT, P.B.
    Institute of Neurophysiology, University of Copenhagen,
    Panum Institute, Blegdamsvej 3 C, 2200 Copenhagen N,
    Denmark
UMNOVA, M.M.
    A.N. Severtzov Institute of Evolutionary Animal Morphology
    and Ecology, 33 Leninsky Prospekt, Moscow V-71, USSR

VAN BEEK, J.A.
        Nieuweweg 39, 7241 ES Lochem, The Netherlands
VAN DER WAL, J.C.
        Rijksuniversiteit Limburg, Fac. Med., POB 616, Beeldsnij-
        dersdreef 101, 6200 MD, Maastricht, The Netherlands
VEJSADA, R.
        Institute of Physiology, Czechoslovak Academy of Sciences,
        Vídeňská 1083, 14220 Prague 4, CSSR
WALRO, J.M.
        Department of Anatomy, Northeastern Ohio Universities,
        College of Medicine, Rootstown, OH 44272, USA
WEYDA, F.
        Institute of Entomology, Czechoslovak Academy of Sciences,
        Branišovská 31, 37005 České Budějovice, CSSR
WOOD, L.
        Institute of Physiology, The University, Glasgow G12 8QQ,
        UK
YOHRO, T.
        Department of Anatomy, University of Tokyo, Faculty of
        Medicine, Hongo, Tokyo, Japan
ZELENÁ, J.
        Institute of Physiology, Czechoslovak Academy of Sciences,
        Vídeňská 1083, 14220 Prague 4, CSSR
ZIENIEWICZ, M.
        Institute of Biocybernetics, KRN 55, 00-818 Warsaw, Poland
ZIMNY, M.L.
        Louisiana State University Medical Center, 1901 Perdido
        Street, New Orleans, LA 70119, USA

---

[x]Not present personally

AUTHOR INDEX